Mathematical Research

Asymptotics for Elliptic Mixed Boundary Problems

S. Rempel
B.-W. Schulze

Volume 50

AKADEMIE-VERLAG BERLIN

In this series original contributions of mathematical research in all fields are contained, such as

— research monographs
— collections of papers to a single topic
— reports on congresses of exceptional interest for mathematical research.

This series is aimed at promoting quick information and communication between mathematicians of the various special branches.

In diese Reihe werden Originalbeiträge zu allen Gebieten der mathematischen Forschung aufgenommen wie

— Forschungsmonographien
— Sammlungen von Arbeiten zu einem speziellen Thema
— Berichte von Tagungen, die für die mathematische Forschung besonders aktuell sind.

Die Reihe soll die schnelle Information und gute Kommunikation zwischen den Mathematikern der verschiedenen Fachgebiete fördern.

Manuscripts
in English and German comprising at least 100 pages and not more than 500 pages can be admitted to this series. With respect to a quick publication the manuscripts are reproduced photomechanically. Authors who are interested in this series please turn directly to the 'Akademie-Verlag'. Here you will get more detailed information about the form of the manuscripts and the modalities of publication.

Manuskripte
in englischer und deutscher Sprache, die mindestens 100 Seiten und nicht mehr als 500 Seiten umfassen, können in diese Reihe aufgenommen werden. Im Interesse einer schnellen Publikation werden die Manuskripte auf fotomechanischem Weg reproduziert. Autoren, die an der Veröffentlichung entsprechender Arbeiten in dieser Reihe interessiert sind, wenden sich bitte direkt an den Akademie-Verlag. Sie erhalten dort genauere Informationen über die Gestaltung der Manuskripte und die Modalitäten der Veröffentlichung.

S. Rempel · B.-W. Schulze

Asymptotics for Elliptic Mixed Boundary Problems

Mathematical Research · Mathematische Forschung

Wissenschaftliche Beiträge
herausgegeben von der
Akademie der Wissenschaften der DDR
Karl-Weierstraß-Institut für Mathematik

Band 50

Asymptotics for Elliptic Mixed Boundary Problems

by

S. Rempel

and

B.-W. Schulze

Asymptotics for Elliptic Mixed Boundary Problems

Pseudo-Differential and Mellin Operators in Spaces with Conormal Singularity

by

Stephan Rempel

and

Bert-Wolfgang Schulze

Akademie-Verlag Berlin 1989

Autoren:

Dr. Stephan Rempel
Prof. Dr. Bert-Wolfgang Schulze

Karl-Weierstraß-Institut für Mathematik
der Akademie der Wissenschaften der DDR, Berlin

Die Titel dieser Schriftenreihe werden vom Originalmanuskript der Autoren reproduziert.

ISBN 3-05-500676-3
ISSN 0138-3019

Erschienen im Akademie-Verlag Berlin, Leipziger Str. 3-4, Berlin, DDR-1086

Lizenznummer: 202·100/410/89
Printed in the German Democratic Republic
Gesamtherstellung: VEB Druckerei "G. W. Leibniz", Gräfenhainichen, DDR-4450
Lektor: Dr. Reinhard Höppner
LSV 1065
Bestellnummer: 763 988 0 (2182/50)

05600

Preface

This monograph contains a pseudo-differential calculus of elliptic boundary value problems with and without the transmission property, on manifolds with conical singularities and with edges. The basic idea is to study appropriate algebras of operators with symbolic structures which contain in particular the parametrices of boundary problems for differential operators.

The analogous approach on closed compact C^∞ manifolds is standard and leads to the well-known classical pseudo-differential operators (ψDO's) and to the index theory.

The calculus of ψDO's has been developed in the past decades in many directions, e.g. with more general symbol classes, non-standard operator conventions, continuity in distribution spaces, Fourier integral operators and so on.

Another direction is the theory of pseudo-differential boundary value problems, in particular of mixed and transmission problems, and the pseudo-differential calculus on manifolds with singularities. The corresponding operator and symbolic structures are of a high complexity and far from being in a final transparent state. There are also many interesting open problems in this area.

The present exposition is an attempt of a systematic theory for the problems mentioned at the beginning in a unified way, which contains at the same time earlier results.

The intuitive idea is to establish first a calculus in direction transversal to the singularities which contains the remaining variables and covariables as parameters and then to carry out the ψDO action in the other directions. For boundary value problems this leads first to the boundary symbolic calculus along $\mathbb{R}_+$ being the normal to the boundary and then to a ψDO calculus tangent to the boundary with operator-valued amplitude functions. In the case of edges the $\mathbb{R}_+$ calculus has

already boundary value problem-valued amplitude functions but then we apply again a ψDO calculus along the edges.
This corresponds to the local model of an edge as $\mathbb{R}_+ \times \{$compact C^∞ manifold with boundary$\} \times \{$edge$\}$.

In this sense boundary value problems for ψDO's are a particular case of edge problems. The edge there is the boundary and the local model $\mathbb{R}_+ \times \{$boundary$\}$. In other words, also for edges in general we have to expect extra boundary conditions along the edge which satisfy an analogue of the Shapiro-Lopatinskij condition. In parametrices we also get potentials which motivate matrix valued operators in the sense of Boutet de Monvel's algebra or Višik/Eskin's work.

Another supporting principle is the conormal asymptotics of solutions as a part of the elliptic regularity. It also determines the structure of the smoothing operators (here called Green operators in analogy to those in Boutet de Monvel's algebra).

Our exposition just presents such an approach. This is of course a huge program which made it necessary to respect carefully the hierarchies of symbolic and operator levels. The lowest level is the $\mathbb{R}_+$ calculus with scalar symbols, based on the Mellin formulation of ψDO's on $\mathbb{R}_+$. The first two chapters are devoted to this theory. In Chapter 3 follows the theory of ψDO's (with and) without the transmission property with respect to the boundary. Chapter 4 contains first the cone theory as an operator — valued analogue of the material of the first two chapters and then the edge ψDO calculus itself. The various steps are motivated and further commented in special notes sections that also contain many references. Clearly our approach is embedded into the achievements of the ψDO calculus connected with the names of many authors such as Kohn, Nirenberg, Maslov, Hörmander, Višik, Egorov and many others. The idea of symbolic structures was already invented and widely used in the theory of singular integral operators. In particular the content of the Chapters 1,2 may be considered in this tradition. Concerning the literature,as far as it is not complete in our bibliography with respect to sources and methods of the analysis, we refer to further papers and textbooks such as Agmon, Douglis, Nirenberg [A 1], Višik, Eskin [V 1]...[V 5], [E 2], Gochberg, Krupnik [G 3], Kondrat'ev [K 5], Grubb [G 6], Hörmander [H 3], Lions, Magenes [L 4], Plamenevskij [P 7], Rempel, Schulze [R 1], Šubin [S 3], Trêves [T 2].

Acknowledgement: We are grateful to J. Leiterer (Karl-Weierstraß-Instiute of Mathematics Berlin, and M. Lorenz (Technical University Karl-Marx-Stadt) for a number of hints and technical improvements, further to the Editor, in particular Dr. Höppner from the Akademie-Verlag, for his patient cooperation.

Contents

1. Operators on the half axis

1.1. The Mellin transform

In this section we remind of some definitions and identities on the Mellin transform. Since the material is classical, we only give the statements (for proofs cf. [J 1]).

Let $\mathbb{R}_+ = \{t \in \mathbb{R} : t > 0\}$. The Mellin transform of $u \in C_0^\infty(\mathbb{R}_+)$ is defined as

$$\tilde{u}(z) = Mu(z) = \int_0^\infty u(t)t^{z-1}dt, \tag{1}$$

which is considered for $z \in \mathbb{C}$ or likewise for $\operatorname{Re} z = 1/2$. The Mellin transform (1) is an entire function in $z \in \mathbb{C}$ satisfying the estimates

$$|\tilde{u}(z)| \leq c_m(1+|z|)^{-m} \quad a^{|\operatorname{Re} z|}$$

for any $m \in \mathbb{Z}_+$ with a constant $c_m > 0$ and $\operatorname{supp} u \subseteq [a^{-1}, a]$, $a > 1$.

1. Proposition. The Mellin transform has an extension to an isomorphism

$$M : L^2(\mathbb{R}_+) \longrightarrow L^2(\operatorname{Re} z = 1/2), \tag{2}$$

where

$$\int_0^\infty |u(t)|^2dt = \frac{1}{2\pi} \int_{-\infty}^\infty |\tilde{u}(1/2+iy)|^2dy \tag{3}$$

and

$$M^{-1}f(t) = \frac{1}{2\pi} \int_{-\infty}^\infty t^{-(1/2+iy)}f(1/2+iy)dy \ . \tag{4}$$

For general $u \in L^2(\mathbb{R}_+)$ the Mellin transform $\tilde{u}(z)$ is only defined on $\operatorname{Re} z = 1/2$. If $\tilde{u}(z)$ has an extension to $\mathbb{C}$, say as a meromorphic function, we also consider this extension as the Mellin transform of u. An example is the Euler Γ function,

$$\Gamma(z) = \int_0^\infty t^{z-1}e^{-t}dt, \tag{5}$$

which is meromorphic with simple poles at $z = -j$, $j \in \mathbb{Z}_+$.

We also have an extension of M from $C_0^\infty(\mathbb{R}_+)$ to the space

$$r^+\mathcal{E}' := \{u \in \mathcal{D}'(\mathbb{R}_+) : u = r^+v \text{ for some } v \in \mathcal{E}'(\mathbb{R})\},$$

r^+ being the restriction to $\mathbb{R}_+$. Then for $u = r^+v$

$$Mu(z) := \langle v, r^+t^{z-1} \rangle \tag{6}$$

is defined and analytic for Re $z > r$ for some $r \in \mathbb{R}$ and independent of the choice of v with $u = r^+v$. (6) will be denoted as the Mellin transform of u. Let $\mathcal{F}_+$ be the space of all functions $f(z)$ that are defined for Re $z > r$, for some $r \in \mathbb{R}$, and are analytic there, satisfying an estimate of the form

$$|f(z)| \leq c(1+|z|)^m a^{\mathrm{Re}\, z}$$

with constants $c, a > 0$, $m \in \mathbb{Z}$. If f_1, f_2 are equal in some half plane Re $z >$ const, f_1 will be identified with f_2. Then $\mathcal{F}_+$ is a commutative algebra with respect to the addition and multiplication of functions in Re $z >$ const. On the other hand, $r^+\mathcal{E}'$ is a commutative algebra with the convolution $*$ as product

$$\langle u * v, \varphi \rangle := \langle u(s) \otimes v(t), \varphi(st) \rangle,$$

$\varphi \in C_0^\infty(\mathbb{R}_+)$, $u, v \in r^+\mathcal{E}'$. The Dirac measure δ_1 at $t = 1$ is the unit.

2. <u>Proposition</u>. $r^+\mathcal{E}' \ni u \longrightarrow Mu$ defines an isomorphism of algebras

$$M : r^+\mathcal{E}' \longrightarrow \mathcal{F}_+, \tag{7}$$

in particular, $M\delta_1 = 1$.

Clearly (7) coincides with (1) for $u \in L^2(\mathbb{R}_+)$ and supp u bounded. We now establish some identities for the Mellin transform in the sense of (7). They are valid in a similar sense for $L^2(\mathbb{R}_+)$, too, providad the involved functions are in L^2.

We introduce the notation

$$\vartheta := t\frac{d}{dt} \quad . \tag{8}$$

Then $\vartheta : r^+\mathcal{E}' \longrightarrow r^+\mathcal{E}'$ is an isomorphism. Furthermore, $u \longrightarrow t^p u$, $p \in \mathbb{C}$, defines an isomorphism $t^p : r^+\mathcal{E}' \longrightarrow r^+\mathcal{E}'$. Moreover, $u \in r^+\mathcal{E}'$ implies $(\log t)u \in r^+\mathcal{E}'$.

3. <u>Lemma</u>. We have

$$M \vartheta u(z) = -zMu(z), \tag{9}$$

$$M\, t^{p}u(z) = Mu(z+p),\ p \in \mathbb{C}, \tag{10}$$

$$(M(\log t)u)(z) = \frac{d}{dz}\ Mu(z). \tag{11}$$

We are mainly concerned here with the Mellin transform on $L^2(\mathbb{R}_+)$. Let us mention for completeness that (1) (first given on $C_0^\infty(\mathbb{R}_+)$) has a continuous extension to an isomorphism

$$M_\gamma\ : L^2(\mathbb{R}_+,\ t^{2\gamma}\,dt) \longrightarrow L^2(\mathrm{Re}\ z = \tfrac{1}{2} + \gamma) \tag{12}$$

for every $\gamma \in \mathbb{R}$. Here $u \in L^2(\mathbb{R}_+,\ t^{2\gamma}\,dt) \Longleftrightarrow \int |u(t)|^2 t^{2\gamma}\,dt < \infty$. All statements in connection with the Mellin transform have a natural generalization to the L^2 spaces with weight. For simplicity in the basic calculus we restrict ourselves to $\gamma = 0$. The general case is obtained by applying the isomorphism

$$t^{-\gamma} : L^2(\mathbb{R}_+, dt) \longrightarrow L^2(\mathbb{R}_+,\ t^{2\gamma}\,dt) =: L^{2,-\gamma}(\mathbb{R}_+)$$

and a translation in the z plane, cf. (10).

The Mellin transform can be expressed in terms of the Fourier transform and some substitution. For $u \in C_0^\infty(\mathbb{R}_+)$ and $t = e^s$, $g(s) = e^{s/2}u(e^s)$ we have

$$(Mu)(1/2+iy) = 2\pi\,(F^{-1}g)(y). \tag{13}$$

Here F^{-1} denotes the inverse Fourier transform

$$F^{-1}g(y) = \frac{1}{2\pi}\int_{-\infty}^{\infty} e^{iys}g(s)\,ds. \tag{14}$$

4. <u>Theorem</u>. A function $f \in \mathcal{F}_+$ is the Mellin transform of a distribution $u \in r^+\mathcal{E}'$ with $\mathrm{supp}\ u \subseteq [a^{-1},a]$, $a \geq 1$, iff f is an entire function and there exists an $m \in \mathbb{Z}_+$ and a constant $c > 0$ such that

$$|f(z)| \leq c(1+|z|)^m a^{|\mathrm{Re}\ z|},\ z \in \mathbb{C}. \tag{15}$$

$f \in \mathcal{F}_+$ is the Mellin transform of some C^∞ function u with $\mathrm{supp}\ u \subseteq [a^{-1},a]$, $a \geq 1$, iff f is an entire function and for any $m \in \mathbb{Z}_+$ there is a constant $c_m > 0$ such that

$$|f(z)| \leq c_m(1+|z|)^{-m} a^{|\mathrm{Re}\ z|},\ z \in \mathbb{C}. \tag{16}$$

5. <u>Theorem</u>. The following conditions are equivalent

(i) $f(z)$ is holomorphic in $\operatorname{Re} z > 1/2$, the function $f_x(y)$: $y \longrightarrow f(x+1/2+iy)$ belongs to $L^2(\mathbb{R}_y)$ for all $x > 0$, and there are constants $a, c > 0$ such that

$$\| f_x \|_{L^2(\mathbb{R})} \leq c a^x \quad \text{for all } x \in \mathbb{R}_+, \tag{17}$$

(ii) $f(z)$ is the Mellin transform of some function $u \in L^2(\mathbb{R}_+)$, supp $u \subseteq [0,a]$.

If (i) or (ii) are satisfied, f_x tends to some limit f_0 in $L^2(\mathbb{R})$ as $x \longrightarrow +0$ and then

$$\| f_x \|_{L^2(\mathbb{R})} = \sqrt{2\pi}\, \| t^x u(t) \|_{L^2(\mathbb{R}_+)}, \quad x \geq 0. \tag{18}$$

Let

$$L^2_{(0)} = \{ u \in L^2(\mathbb{R}_+) : \text{supp } u \text{ bounded} \},$$
$$L^2_{(\infty)} = \{ u \in L^2(\mathbb{R}_+) : \text{dist } (0, \text{supp } u) > 0 \}.$$

Then $I : u(t) \longrightarrow t^{-1} u(t^{-1})$ defines a bijection

$$I : L^2_{(0)} \longrightarrow L^2_{(\infty)}$$

and

$$Mu(z) = MIu(w) \qquad \text{for } w = -z + 1,$$

where obviously $\operatorname{Re} z \geq 1/2 \Longleftrightarrow \operatorname{Re} w \leq 1/2$. This gives us an analogue of the latter theorem for $L^2_{(\infty)}$, where the half plane $\operatorname{Re} z > 1/2$ is to be replaced by $\operatorname{Re} z < 1/2$. Throughout this exposition a cut-off function means some $\omega \in C_0^\infty(\overline{\mathbb{R}}_+)$ which is identically 1 near $t = 0$. For $u \in L^2(\mathbb{R}_+)$ and any cut-off function ω we may write $u = \omega u + (1-\omega) u$ and

$$Mu(z) = M\omega u(z) + M(1-\omega)u(z), \quad \operatorname{Re} z = 1/2.$$

Since $\omega u \in L^2_{(0)}$, $(1-\omega)u \in L^2_{(\infty)}$, $Mu(z)$ is decomposed into a sum $f_+ + f_-$, where $f_\pm$ has an analytic extension into $\operatorname{Re} z \gtrless 1/2$ with the corresponding properties mentioned in Theorem 5 and its analogue for f_-.

6. <u>Corollary</u>. The following conditions are equivalent:

(i) $f(z)$ is an entire function and the function $f_x : y \longrightarrow f(x+1/2+iy)$

belongs to $L^2(\mathbb{R}_y)$ for all $x \in \mathbb{R}$, and there are constants $a, c > 0$ such that

$$\|f_x\|_{L^2(\mathbb{R})} \leq c a^{|x|} \quad \text{for all } x \in \mathbb{R},$$

(ii) $f(z)$ is the Mellin transform of some function $u \in L^2_{(0)} \cap L^2_{(\infty)}$.

7. <u>Corollary</u>. A function v on $\mathbb{R}_+$ is of the form

$$v = t^{\delta} u, \qquad \delta \in \mathbb{R}, \qquad u \in L^2_{(0)}$$

iff $\tilde{v}(z)$ is holomorphic in $\operatorname{Re} z > 1/2 - \delta$ and $v_x : y \longrightarrow v(x+1/2-\delta+iy)$ belongs to $L^2(\mathbb{R}_y)$ for all $x \in \mathbb{R}_+$ and $\|v_x\|_{L^2(\mathbb{R})} \leq c a^x$, $x \in \mathbb{R}_+$, with constants $a, c > 0$.

Finally let us mention the following simple

8. <u>Lemma</u>. Let $f(z)$ be a holomorphic function in the strip $\rho_1 < \operatorname{Re} z < \rho_2$ and

$$\int |f(x+iy)|^2 \, dy \leq a^2 \quad \text{for } \rho_1 < x < \rho_2.$$

Then for any $\varepsilon > 0$

$$|f(z)| \leq \frac{a}{\sqrt{\pi\varepsilon}} \quad \text{for } \rho_1 + \varepsilon < \operatorname{Re} z < \rho_2 - \varepsilon.$$

For a proof cf. [J1]. Note that a similar lemma is valid for uniform norms, cf. [B2]. If $f(z)$ is holomorphic in the strip $0 < \operatorname{Re} z < 1$ and bounded and continuous in $0 \leq \operatorname{Re} z \leq 1$, then

$$|f(iy)| \leq a_0, \qquad |f(1+iy)| \leq a_1, \qquad -\infty < y < \infty$$

implies

$$|f(x+iy)| \leq a_0^{1-x} \, a_1^x, \quad -\infty < y < \infty .$$

An L^2 version of the above statement can easily be proved. If $f(z)$ is holomorphic in $\rho_1 \leq \operatorname{Re} z \leq \rho_2$ we have for all x, $\rho_1 < x < \rho_2$,

$$\int |f(x+iy)|^2 \, dy \leq \int |f(\rho_1+iy)|^2 \, dy + \int |f(\rho_2+iy)|^2 \, dy.$$

In fact, there exists a function $u \in L^2(\mathbb{R}_+, t^{2x-1}dt)$ such that $f(x+iy) = M(t^{x-1/2} u(t))$ and

$$\int |f(x+iy)|^2 \, dy = 2\pi \int_0^\infty t^{2x-1} |u(t)|^2 \, dt.$$

The estimate follows from the inequality

$$t^{2x-1} \leq t^{2\varrho_1-1} + t^{2\varrho_2-1}, \qquad t \in \mathbb{R}_+ .$$

1.2. Spaces of functions with discrete conormal singularity

1.2.1. The spaces $\mathcal{S}_{\mathcal{P}}(\mathbb{R}_+)$ and $C^{\infty}_{\mathcal{P},\mathcal{P}'}(\mathbb{R}_+)$

The following sections develop the functional analysis in spaces with asymptotics in the simplest case. However they contain basic elements of the more general theory that is developed in 2.1. Moreover, throughout the calculus we have frequently to return to the discrete asymptotics.

Let $\mathbb{Z}_+ = \{j \in \mathbb{Z} : j \geq 0\}$ and

$$\mathcal{P} = \{(p_j, m_j) : j \in \mathbb{Z}_+\} \qquad (1)$$

be a sequence of pairs $(p_j, m_j) \in \mathbb{C} \times \mathbb{Z}_+$ with

$$\operatorname{Re} p_j \longrightarrow -\infty \quad \text{for } j \longrightarrow \infty, \ \operatorname{Re} p_j \geq \operatorname{Re} p_{j+1}. \qquad (2)$$

We introduce a space of functions $\mathcal{S}_{\mathcal{P}}(\mathbb{R}_+) \subseteq C^{\infty}(\mathbb{R}_+)$ which have an asymptotic expansion (a so-called 'conormal singularity')

$$u(t) \sim \sum_{j=0}^{\infty} \sum_{k=0}^{m_j} \zeta_{jk} t^{-p_j} \log^k t \quad , \qquad (3)$$

as $t \longrightarrow 0$ (to be defined in a more precise form) and such that $e^+(1-\omega)u$ belongs to the Schwartz space $\mathcal{S}(\mathbb{R})$. Here

$$e^+ f(t) := \begin{cases} f(t) & \text{for } t > 0, \\ 0 & \text{for } t \leq 0 \end{cases} \qquad (4)$$

if f is given on $\mathbb{R}_+$ and $\omega(t)$ is a cut-off function, i.e. $\omega \in C_0^{\infty}(\overline{\mathbb{R}_+})$ and $\omega(t) = 1$ for t in a neighbourhood of the origin. $\mathcal{S}_{\mathcal{P}}$ will be equipped with a Fréchet topology. It is convenient for us to use semi-norms derived from the L^2-scalar product. For this reason we suppose here

$$\operatorname{Re} p_j < 1/2, \quad j \in \mathbb{Z}_+ \ . \qquad (5)$$

The general case easily follows by applying the one-to-one map $u(t) \longrightarrow$

$t^{\alpha}u(t)$ with appropriate α.

If

$$T^{\alpha}_{\varphi} := \{(p_j + \alpha, m_j) : j \in \mathbb{Z}_+\}, \quad \alpha \in \mathbb{C}, \tag{6}$$

we set

$$\mathcal{S}_{T^{\alpha}_{\varphi}} := \{t^{\alpha}u : u \in \mathcal{S}_{\varphi}\} .$$

Similar asymptotics will be considered with respect to $t \to \infty$. This leads us to the spaces $C^{\infty}_{\varphi,\varphi'}(\mathbb{R}_+)$.

Now we pass to the precise definitions. Set

$$\vartheta = t \frac{d}{dt} . \tag{7}$$

Let $\delta \geq 0$ and $N = N(\delta,\varphi) \in \mathbb{Z}_+$ be the smallest number for which

$$\operatorname{Re} p_j < 1/2 - \delta \quad \text{for } j > N(\delta,\varphi).$$

1. __Definition__. By $\mathcal{S}_{\varphi}(\mathbb{R}_+)$ we denote the subspace of all $u \in C^{\infty}(\mathbb{R}_+)$ with the following properties. There exists a sequence

$$\{\zeta_{jk} = \zeta_{jk}(u) : 0 \leq k \leq m_j,\ j \in \mathbb{Z}_+\} \tag{8}$$

such that for

$$v^{\varphi}_{\delta}u(t) := u(t) - \sum_{j=0}^{N} \sum_{k=0}^{m_j} \zeta_{jk} t^{-p_j} \log^k t\, \omega(t) \tag{9}$$

with arbitrary $\delta \geq 0$ and $N = N(\delta,\varphi)$

$$t^{\eta-\delta}\vartheta^{l} v^{\varphi}_{\delta} u \in L^2(\mathbb{R}_+) \tag{10}$$

for arbitrary $\eta \geq 0$, $l \in \mathbb{Z}_+$. If φ is fixed, for abbreviation we set $v_{\delta} := v^{\varphi}_{\delta}$. The uniqueness of the ζ_{jk} is established in Proposition 6. It is clear that we have $N(\delta,\varphi) = N(\delta_0,\varphi)$ for all δ in a small neighbourhood of some δ_0. Further

$$\|t^{\eta-\delta}\vartheta^{l} v_{\delta_0}u\|_{L^2} \leq c_1\|t^{\eta_1-\delta_1}\vartheta^{l} v_{\delta_0}u\|_{L^2} \leq c_2\|t^{\eta_0-\delta_0}\vartheta^{l} v_{\delta_0}u\|_{L^2} \tag{11}$$

for η, η_1 close to η_0 and δ, δ_1 close to δ_0, $\eta_1 - \delta_1 \geq \eta - \delta \geq \eta_0 - \delta_0$, with constants c_1, c_2. Thus, if (10) is satisfied for all rational δ, η, it follows for δ, $\eta \in \mathbb{R}_+$, too. Thus we may restrict ourselves to a countable set $\underline{\alpha}$ of indices $\alpha = (\eta,\delta,l) \in \mathbb{R}_+ \times \mathbb{R}_+ \times \mathbb{Z}_+$.

The definition of the space $\mathcal{S}_\varphi$ shows that the numeration of the points in φ is not essential. More precisely, we should speak about countable sets of points with given multiplicities. For notational convenience we fix numerations and keep in mind that each other numeration could be taken as well.

As a simple consequence of (11) we get

2. Proposition. $\mathcal{S}_\varphi$ equipped with the system of norms

$$\pi_\alpha(u) := (\|u\|^2_{L^2} + \| t^{\gamma-\delta}\vartheta^l\, r_\delta u\|^2_{L^2})^{1/2} \quad , \alpha \in \underline{\alpha} \, , \tag{12}$$

is a Fréchet space.

If some $u \in C^\infty(\mathbb{R}_+)$ has the property that $u\omega \in \mathcal{S}_\varphi$, ω a cut-off function, we speak of the conormal singularity near $t = 0$ or also of the conormal asymptotics for $t \to +0$ of the type φ.

Now we consider another sequence

$$\varphi' = \{(p_j', m_j') : \quad j \in \mathbb{Z}_+\} \tag{13}$$

with

$$\operatorname{Re} p_j' > 1/2, \quad \operatorname{Re} p_j' \to \infty \quad \text{as } j \to \infty . \tag{14}$$

Let $\delta' \geq 0$ and $N' = N'(\delta', \varphi') \in \mathbb{Z}_+$ be the smallest number for which

$$\operatorname{Re} p_j' > 1/2 + \delta' \quad \text{for } j > N'(\delta', \varphi').$$

3. Definition. By $C^\infty_{\varphi,\varphi'}(\mathbb{R}_+)$ we denote the subspace of all $u \in C^\infty(\mathbb{R}_+)$ with the following properties. There exist sequences (8) and

$$\{ \zeta_{jk}' = \zeta_{jk}'(u) : 0 \leq k \leq m_j', \; j \in \mathbb{Z}_+\} \tag{15}$$

such that for

$$r^{\varphi,\varphi'}_{\delta,\delta'} u(t) := u(t) - \sum_{j=0}^{N} \sum_{k=0}^{m_j} \zeta_{jk} t^{-p_j} \log^k t \;\omega(t)$$

$$- \sum_{j=0}^{N'} \sum_{k=0}^{m_j'} \zeta_{jk}' t^{-p_j'} \log^k t \;(1 - \omega(t))$$

with arbitrary $\delta, \delta' \geq 0$ and $N = N(\delta, \varphi)$, $N' = N'(\delta', \varphi')$

$$t^\gamma \vartheta^l r^{\varphi,\varphi'}_{\delta,\delta'} u \in L^2(\mathbb{R}_+) \tag{16}$$

for all $\gamma \in [-\delta, \delta']$ and $l \in \mathbb{Z}_+$. If φ, φ' are fixed, we also write

$$\upsilon_{\delta,\delta'} := \upsilon_{\delta,\delta'}^{\varphi,\varphi'} .$$

Similarly as above we have the

4. Proposition. $C^\infty_{\varphi,\varphi'}$ is a Fréchet space with the system of norms

$$\rho_\beta(u) := (\|u\|^2_{L^2} + \|t^\gamma \vartheta^l \upsilon_{\delta,\delta'} u\|^2_{L^2})^{1/2} \tag{17}$$

$\beta = (\delta, \delta', \gamma, l) \in \overline{\mathbb{R}}_+ \times \overline{\mathbb{R}}_+ \times \mathbb{R} \times \mathbb{Z}_+$, $\gamma \in [-\delta, \delta']$.

The assumption Re $p_j < 1/2$, Re $p_j' > 1/2$ was made for convenience. In this case we have $C^\infty_{\varphi,\varphi'}(\mathbb{R}_+) \subset L^2(\mathbb{R}_+)$. A more general definition then follows by posing

$$C^\infty_{T^\alpha \varphi, T^{\alpha'} \varphi'} = \{ g^{\alpha,\alpha'} u \; : u \in C^\infty_{\varphi,\varphi'} \} , \quad \alpha, \alpha' \in \mathbb{C},$$

with a non-vanishing function $g^{\alpha,\alpha'}$ satisfying

$$g^{\alpha,\alpha'}(t) = \begin{cases} t^\alpha & \text{for } t < 1, \\ t^{\alpha'} & \text{for } t > 2. \end{cases} \tag{18}$$

It is our next goal to characterize the functions in $\mathcal{S}_\varphi$ and $C^\infty_{\varphi,\varphi'}$ in terms of the Mellin transform and to introduce equivalent systems of norms. To this end we first evaluate the Mellin transforms of certain particular functions.

5. Lemma. Set

$$\omega^{p,k}(t) := t^{-p} \log^k t \;\; \omega(t),$$

$p \in \mathbb{C}$, Re $p < 1/2$, $k \in \mathbb{Z}_+$. Then the Mellin transform $\tilde{\omega}^{p,k}(z)$ is meromorphic with exactly one pole at $z = p$ of order $k+1$. If $\chi \in C^\infty(\mathbb{C})$ is an arbitrary function with

$$\chi = \begin{cases} 1 & \text{for } |z-p| > c, \\ 0 & \text{for } |z-p| < \frac{c}{2} \end{cases}$$

and $c > 0$ some constant, we have for all $N \in \mathbb{Z}_+$

$$\int | z^N \chi(z) \;\; \tilde{\omega}^{p,k}(z) |^2 \, dy < \infty , \quad z = x+iy .$$

<u>Proof:</u> We have

$$(-\vartheta-p)^{k+1}\; t^{-p}\log^k t\;\;\omega(t) = v(t) \in C_0^\infty(\mathbb{R}_+) \quad .$$

Then the assertion follows immediately from the fact that

$$\tilde{\omega}^{p,k}(z) = (z-p)^{-(k+1)}\,\tilde{v}(z)$$

and $\tilde{v}(z)$ is an entire function with

$$\int |z^N\,\tilde{v}(z)|^2\,dy \;<\; \infty$$

for arbitrary $N \in \mathbb{Z}_+$. $\square$

Let $\varepsilon(t)$ be the characteristic function of the interval $[0,1]$ and

$$\varepsilon^{p,k}(t) := t^{-p}\log^k t\;\;\varepsilon(t) \quad . \tag{19}$$

Then a simple calculation shows

$$\tilde{\varepsilon}^{p,k}(z) = k!\;(z-p)^{-(k+1)}. \tag{20}$$

A $\chi \in C^\infty(\mathbb{C})$ is called a φ ($\varphi+\varphi'$) excision function if

(i) $|\chi(z)| \leq$ const for all $z \in \mathbb{C}$,

(ii) there are open neighbourhoods U, V of $\bigcup_{j=0}^{\infty}\{p_j\}$ ($\bigcup_{j=0}^{\infty}\{p_j\} \cup \bigcup_{j=0}^{\infty}\{p_j'\}$) with $\overline{U} \subset V$ and

$$\chi(z) = \begin{cases} 1 & \text{for } z \in \mathbb{C}\setminus V, \\ 0 & \text{for } z \in U . \end{cases}$$

By $\mathcal{L}^2(\mathbb{R}_+)$ we denote the subspace of all $u \in L^2(\mathbb{R}_+)$ for which $\tilde{u}(z)$ (first given for Re $z = 1/2$) has a meromorphic extension in $\mathbb{C}$ with poles at $z = p_j$ and multiplicities m_j+1, and that further

$$\int |u(t)|^2 t^N dt < \infty \text{ for all } N \in \mathbb{Z}_+, \quad \int_{\mathrm{Re}\, z=\gamma} |\chi(z)u(z)|^2 dy < \infty$$

for all $\gamma \in \mathbb{R}$ and every φ-excision function χ. Set

$$\mathcal{H}^r(\mathbb{R}_+) := \left\{ u \in \mathcal{D}'(\mathbb{R}_+) : \vartheta^l u \in L^2(\mathbb{R}_+),\;\; 0 \leq l \leq r \right\},$$

$$\mathcal{H}^r_\varphi(\mathbb{R}_+) := \left\{ u \in \mathcal{H}^r(\mathbb{R}_+) : \vartheta^l u \in \mathcal{L}^2_\varphi(\mathbb{R}_+),\;\; 0 \leq l \leq r \right\},$$

$r \in \mathbb{Z}_+$.

Similar notations can be introduced with respect to φ, φ'. Denote by $\mathcal{L}^2_{\varphi,\varphi'}(\mathbb{R}_+)$ the subspace of those $u \in L^2(\mathbb{R}_+)$ for which $\tilde{u}(z)$ has a meromorphic extension in $\mathbb{C}$ with poles at $z = p_j$ and $z = p_j'$, with multiplicities m_j+1 and $m_j'+1$, respectively, and $\int_{\mathrm{Re}\, z=\gamma} |\chi(z)u(z)|^2 dy < \infty$

for all $\gamma \in \mathbb{R}$ and every $\varphi + \varphi'$ - excision function χ. Then define

$$\mathcal{H}^r_{\varphi,\varphi'}(\mathbb{R}_+) = \{u \in \mathcal{H}^r(\mathbb{R}_+) : \vartheta^l u \in \mathcal{L}^2_{\varphi,\varphi'}(\mathbb{R}_+), \quad 0 \leq l \leq r\}.$$

6. Proposition. Let $u \in L^2(\mathbb{R}_+)$, $r \in \mathbb{Z}_+$. Then the following conditions are equivalent

(i) $u \in \mathcal{H}^r_\varphi(\mathbb{R}_+)$

(ii) $\tilde{u}(z)$ is meromorphic with poles at $z = p_j$ of multiplicities m_j+1 and for any φ excision function χ

$$\int_{\mathrm{Re}\, z = x} |z^l \tilde{u}(z) \ \chi(z)|^2 dy < \infty \quad \text{for all } x \in \mathbb{R},\ 0 \leq l \leq r, \tag{21}$$

(iii) there exist unique coefficients $\zeta_{jk} \in \mathbb{C}$, $j \in \mathbb{Z}_+$, $0 \leq k \leq m_j$, such that for arbitrary $\delta \geq 0$ there is an $N = N(\delta, \varphi) \in \mathbb{Z}_+$ such that for all $\eta \geq 0$

$$t^{\eta - \delta}\, \vartheta^l\, \mathscr{r}_\delta\, u \in L^2(\mathbb{R}_+), \qquad 0 \leq l \leq r. \tag{22}$$

Proof. The definition of $\mathcal{H}^r_\varphi$ implies that $M\, \vartheta^l u(z)$ is meromorphic with the described poles, $0 \leq l \leq r$. Using Lemma 5 we find coefficients ζ_{jk} and an $N = N(\delta, \varphi)$ such that

$$f^l(z) := M\vartheta^l\, \mathscr{r}_\delta\, u(z)$$

is holomorphic for $\mathrm{Re}\, z > 1/2 - \delta$, $0 \leq l \leq r$. Since both $t^\eta\, \vartheta^l u$ and $t^\eta\, \omega^{p,k}$ belong to $L^2(\mathbb{R}_+)$ for any $\eta \geq 0$, we conclude (22). The ζ_{jk} are precisely the Laurent coefficients of the meromorphic function $\tilde{u}(z)$ and thus uniquely determined. This proves (i) $\Longrightarrow$ (iii).

Conversely (iii) implies that $\tilde{u}(z)$ is meromorphic with the described poles, cf. Lemma 5 and 1.1, Corollary 7. For $\delta = 0$ the relation (22) is equivalent to $\int t^\eta\, |\vartheta^l u|^2 dt < \infty$, $0 \leq l \leq r$, i.e. $u \in \mathcal{H}^r_\varphi$, which proves (iii) $\Longrightarrow$ (i).

Next we show (iii) $\Longrightarrow$ (ii), where the meromorphy is already clear. From (22) it follows that $f^l(z + \eta - \delta) \in L^2(\mathrm{Re}\, z = 1/2)$, i.e.

$$\int_{\mathrm{Re}\, z = 1/2} |M\vartheta^l\, \mathscr{r}_\delta\, u(z + \eta - \delta)|^2\, dy < \infty .$$

Thus $\int_{\operatorname{Re} z = x} |M\vartheta^1 \mathcal{N}_\delta u(z)|^2 \, dy < \infty$ for all $x \in \mathbb{R}$, with $x + \delta - 1/2 \geqq 0$.

Using the definition of $\mathcal{N}_\delta$ and Lemma 5 we obtain

$$\int_{\operatorname{Re} z = x} |M\vartheta^1 \tilde{u}(z) \quad \chi(z)|^2 \, dy < \infty \tag{23}$$

for $x + \delta - 1/2 \geqq 0$. Since δ is arbitrary, we obtain (21).

The converse (ii) $\Longrightarrow$(iii) follows from (23) and Lemma 5, since the above conclusions are valid in the other directions, too. $\square$

In a similar way we can prove

7. Proposition. Let $u \in L^2(\mathbb{R}_+)$, $r \in \mathbb{Z}_+$. Then the following conditions are equivalent

(i) $u \in \mathcal{H}^r_{\varphi,\varphi'}(\mathbb{R}_+)$,

(ii) $\tilde{u}(z)$ is meromorphic with poles at $z = p_j$ and $z = p_j'$ of multiplicities m_j+1 and $m_j'+1$, respectively, and for any $\varphi + \varphi'$ excision function χ

$$\int_{\operatorname{Re} z = x} |z^l \tilde{u}(z) \quad \chi(z)|^2 dy < \infty \quad \text{for all } x \in \mathbb{R},\ 0 \leqq l \leqq r , \tag{24}$$

(iii) there exist unique coefficients $\zeta_{jk}, \zeta'_{jk'} \in \mathbb{C}$, $j \in \mathbb{Z}_+$, $0 \leqq k \leqq m_j$, $0 \leqq k' \leqq m_j'$, such that for arbitrary $\delta \geqq 0$, $\delta' \geqq 0$ there are $N = N(\delta, \varphi)$, $N' = N'(\delta', \varphi') \in \mathbb{Z}_+$ such that $t^\gamma \vartheta^l \mathcal{N}_{\delta,\delta'} u \in L^2(\mathbb{R}_+)$ for all l, $0 \leqq l \leqq r$, and $\gamma \in [-\delta, \delta']$.

It is obvious that it suffices to verify the condition (ii) in the Propositions 6, 7 for some fixed excision function χ .

8. Proposition. $\mathcal{H}^r_\varphi$ is a Fréchet space with the system of norms (12), $\alpha = (\eta, \delta, l) \in \overline{\mathbb{R}}_+ \times \overline{\mathbb{R}}_+ \times \mathbb{Z}_+$, $0 \leqq l \leqq r$. The coefficients $\zeta_{jk} = \zeta_{jk}(u)$ continuously depend on u in the sense that every ζ_{jk} converges if u converges. A similar proposition holds for $\mathcal{H}^r_{\varphi,\varphi'}$.

Proof. As in the Propositions 2 and 4 we may restrict ourselves to countable sets of norms. The completeness is a simple consequence of that of $L^2(\mathbb{R}_+)$ and Proposition 6 . Now let $(u_n)_{n \in \mathbb{Z}_+}$ be a convergent sequence in $\mathcal{H}^r_\varphi$, $u = \lim u_n$. Then we have the convergence of $\mathcal{N}_{\delta_1} u_n$ and $\mathcal{N}_{\delta_2} u_n$

in $L^2(\mathbb{R}_+)$ for any δ_1, δ_2. Thus $(\mathscr{V}_{\delta_1} - \mathscr{V}_{\delta_2})u_n$ converges, too. i.e. any finite sum $\sum_{j=0}^{N} \sum_{k=0}^{m_j} \zeta_{jk}(u_n)\, \omega^{p_j,k}$ converges in $L^2(\mathbb{R}_+)$. The linear span of the functions $\omega^{p_j,k}$, $0 \leq j \leq N$, $0 \leq k \leq m_j$, is an $\sum_{j=0}^{N}(m_j+1)$ dimensional subspace E of $L^2(\mathbb{R}_+)$, since the $\omega^{p_j,k}$ are linearly independent. The space $L^2(\mathbb{R}_+)$ induces in E the Euclidean topology and hence $\zeta_{jk}(u_n) \to \zeta_{jk}(u)$ for all j,k. A similar conclusion holds for $\mathcal{H}^r_{\mathscr{P},\mathscr{P}'}$.
It is clear that

$$\mathcal{S}_{\mathscr{P}} = \bigcap_{r \in \mathbb{Z}_+} \mathcal{H}^r_{\mathscr{P}}, \tag{25}$$

$$C^{\infty}_{\mathscr{P},\mathscr{P}'} = \bigcap_{r \in \mathbb{Z}_+} \mathcal{H}^r_{\mathscr{P},\mathscr{P}'} \tag{26}$$

with the projective limit topologies. From now on we mainly consider $\mathcal{S}_{\mathscr{P}}$. The simple generalizations to $C^{\infty}_{\mathscr{P},\mathscr{P}'}$ are left to the reader.
Set

$$\mathcal{S}_0 = \left\{ r^+u \; : \; u \in \mathcal{S}(\mathbb{R}),\ \operatorname{supp} u \subseteq \overline{\mathbb{R}}_+ \right\} \tag{27}$$

equipped with the induced Schwartz space topology and let $\mathcal{Z}$ be the space of all sequences $(\zeta_{jk})_{0 \leq k \leq m_j}$, $j \in \mathbb{Z}_+$, with arbitrary $\zeta_{jk} \in \mathbb{C}$ considered in the topology of componentwise convergence. Then we have a continuous inclusion $\iota : \mathcal{S}_0 \to \mathcal{S}_{\mathscr{P}}$ and a continuous map $\zeta : \mathcal{S}_{\mathscr{P}} \to \mathcal{Z}$, cf. Proposition 8, that assigns to each $u \in \mathcal{S}_{\mathscr{P}}$ the sequence of coefficients $\zeta_{jk}(u)$. Below (cf. Theorem 14) we shall prove

9. <u>Theorem</u>. The sequence $0 \longrightarrow \mathcal{S}_0 \xrightarrow{\iota} \mathcal{S}_{\mathscr{P}} \xrightarrow{\zeta} \mathcal{Z} \longrightarrow 0$ is exact.

The only point in Theorem 9 which remains to be proved is that ζ is surjective.
Let $\delta \geq 0$ and

$$\mathscr{P}(\delta) = \left\{ (p_j, m_j) \in \mathscr{P} \; : \; \operatorname{Re} p_j < 1/2 - \delta \right\}, \tag{28}$$

and $N(\delta, \mathscr{P})$, $\mathscr{V}_\delta$ as above. It is obvious that $\mathcal{S}_{\mathscr{P}(\delta)} \subseteq \mathcal{S}_{\mathscr{P}}$ is a subspace with finite codimension and that $u \in \mathcal{S}_{\mathscr{P}}$ belongs to $\mathcal{S}_{\mathscr{P}(\delta)}$ if $\zeta_{jk}(u) = 0$ for $0 \leq j \leq N(\delta, \mathscr{P})$, $0 \leq k \leq m_j$.

10. <u>Lemma</u>. The operator

$$\pi_\delta : \mathcal{S}_\varphi \longrightarrow \mathcal{S}_{\varphi(\delta)} \tag{29}$$

is a continuous projection. Further

$$\vartheta^l : \mathcal{S}_\varphi \longrightarrow \mathcal{S}_\varphi \tag{30}$$

is continuous for all $l \in \mathbb{Z}_+$, and the operator of multiplication by t^γ

$$t^\gamma : \mathcal{S}_\varphi \longrightarrow \mathcal{S}_{T^{-\gamma}\varphi} \tag{31}$$

is a topological isomorphism for all $\gamma \in \mathbb{C}$, $\operatorname{Re} p_j + \operatorname{Re}\gamma < 1/2$, $j \in \mathbb{Z}_+$.

<u>Proof</u>. We have $\omega^{p_j,k} \in \mathcal{S}_\varphi$ for all p_j and $0 \leq k \leq m_j$. Thus the map

$$\kappa_\delta : u \longrightarrow \sum_{j=0}^{N} \sum_{k=0}^{m_j} \zeta_{jk}(u)\, \omega^{p_j,k}, \quad N = N(\delta,\varphi),$$

is a continuous projection $\kappa_\delta : \mathcal{S}_\varphi \longrightarrow \mathcal{S}_\varphi$ with finite-dimensional range, where $\pi_\delta = 1 - \kappa_\delta$ is the complementary projection. Since $\operatorname{im} \pi_\delta \subseteq \mathcal{S}_{\varphi(\delta)}$, (29) is proved.

For proving (30) it is sufficient to consider the case $l = 1$. We first prove that $\vartheta u \in \mathcal{S}_\varphi$ for $u \in \mathcal{S}_\varphi$. Per definitionem we know that

$$\| t^{\eta-\delta} \vartheta^{l+1} \pi_\delta u \|_{L^2} = \| t^{\eta-\delta} \vartheta^l (\vartheta u - \vartheta \kappa_\delta u) \|_{L^2}$$

for all $l \in \mathbb{Z}_+$, $\eta \geq 0$, $\delta \geq 0$. Moreover,

$$\vartheta \omega^{p_j,k} = -p_j \omega^{p_j,k} + k\, \omega^{p_j,k-1} + \vartheta(\omega) t^{-p_j} \log^k t .$$

Thus we have

$$\kappa_\delta \vartheta u = \sum_{j=0}^{N} \sum_{k=0}^{m_j} (-p_j \zeta_{jk}(u) + (k+1)\, \zeta_{j,k+1}(u))\, \omega^{p_j,k},$$

where $\zeta_{j,m_j+1} := 0$. Then $\vartheta \kappa_\delta u = \kappa_\delta \vartheta u + \sum_{j=0}^{N} \sum_{k=0}^{m_j} \zeta_{jk}(u) t^{-p_j} \log^k t\, \vartheta(\omega)(t)$.

Since $\mathcal{S}_\varphi \ni u \longrightarrow \zeta_{jk}(u) \in \mathbb{C}$ is continuous, the mapping $\mathcal{S}_\varphi \ni u \longrightarrow \vartheta \pi_\delta u - \pi_\delta \vartheta u = \vartheta \kappa_\delta u - \kappa_\delta \vartheta u \in \mathcal{S}_\varphi$ is continuous, too. Then the continuity of ϑ follows from $t^{\eta-\delta} \vartheta^l \pi_\delta(\vartheta u) = t^{\eta-\delta} \vartheta^{l+1} \pi_\delta u + t^{\eta-\delta} \vartheta^l (\pi_\delta \vartheta u - \vartheta \pi_\delta u)$.

It remains to consider (31), which is a simple consequence of

$$t^{\eta-(\delta+\gamma)} \pi_{\delta+\gamma}^{T^{-\gamma}\varphi} (t^\gamma u) = t^{\eta-\delta} \pi_\delta^\varphi u. \quad \square$$

Later on it will turn out that $\mathcal{K}_\delta$ is a Green operator, cf. 1.2.2. Remember that the bijection $u \longrightarrow t^\gamma u$ gives a definition of $\mathcal{S}_{T^\gamma \mathcal{P}}$ for all $\gamma \in \mathbb{C}$ with a topology following from the requirement that $t^\gamma : \mathcal{S}_{\mathcal{P}} \longrightarrow \mathcal{S}_{T^\gamma \mathcal{P}}$ is an isomorphism. The third statement of Lemma 10 shows that there is no contradiction if $\mathrm{Re}\, p_j < 1/2$, $\mathrm{Re}(p_j+\gamma) < 1/2$.

11. **Remark.** Assume

$$T^{j}\mathcal{P} \subseteq \mathcal{P}, \quad j \in \mathbb{Z}_+ \quad , \tag{32}$$

i.e. $(p_i,m_i) \in \mathcal{P} \Longrightarrow (p_i - j, m_i) \in \mathcal{P}$ for all $i,j \in \mathbb{Z}_+$. Then, in the definition of the conormal asymptotics we may replace $\omega^{p_j,k}$ by

$$\eta^{p_j,k}(t) := t^{-p_j} \log^k t \; e^{-t} \quad . \tag{33}$$

12. **Remark.** Set

$$\mathcal{S}(\overline{\mathbb{R}}_+) = \left\{ r^+ u \; ; \; u \in \mathcal{S}(\mathbb{R}) \right\} \tag{34}$$

with the Schwartz space $\mathcal{S}(\mathbb{R})$,

$$\mathcal{P}_0 = \left\{ (p_j, m_j) \; : \; p_j = -j, \; m_j = 0, \quad j \in \mathbb{Z}_+ \right\} . \tag{35}$$

Then

$$\mathcal{S}(\overline{\mathbb{R}}_+) = \mathcal{S}_{\mathcal{P}_0}(\mathbb{R}_+) \quad . \tag{36}$$

The conormal 'singularity' of the type $\mathcal{P}_0$ corresponds to the Taylor expansion of a function which is C^∞ up to $t = 0$. It is then convenient to talk about a Taylor asymptotics (of some order j_1 if we have a Taylor expansion up to $j = j_1$).

Remember that the topology of $\mathcal{S}_{\mathcal{P}}$ was derived from a countable system of scalar products

$$\pi_\alpha(u,v) := (u,v)_{L^2(\mathbb{R}_+)} + (t^{\eta-\delta}\vartheta^l \mathcal{V}_\delta u, \; t^{\eta-\delta}\vartheta^l \mathcal{V}_\delta v)_{L^2(\mathbb{R}_+)}, \tag{37}$$

$\alpha = (\eta, \delta, l) \in \underline{\alpha}$, cf. Proposition 7. In this way $\mathcal{S}_{\mathcal{P}}$ is a countable normed space with Hilbert norms.

If $\mathcal{S}_{\mathcal{P}}^{[\alpha]}$ is defined as the Hilbert space being the completion of $\mathcal{S}_{\mathcal{P}}$ with respect to the scalar product π_α, we have

$$\mathcal{S}_{\mathcal{P}} = \varprojlim \mathcal{S}_{\mathcal{P}}^{[\alpha]} \quad , \tag{38}$$

where $\varprojlim$ denotes the projective limit over $\alpha \in \underline{\alpha}$.

Now we introduce another countable system of Hilbert norms in $\mathcal{S}_{\mathfrak{p}}$. Let ρ_j, $j \in \mathbb{Z}$, be a sequence of real numbers with the property

$$\rho_{-j} \neq \operatorname{Re} p_k \quad \text{for } j,k \in \mathbb{Z}_+, \tag{39}$$

$$\rho_0 = 1/2, \quad \rho_{-j} \longrightarrow -\infty \quad \text{as } j \longrightarrow \infty,$$

$$\operatorname{Re} p_{j+1} < \rho_{-j'} < \operatorname{Re} p_j \tag{40}$$

for at least one $j' \in \mathbb{Z}_+$ and

$$\rho_k = k \quad \text{for } k = 1,2,\ldots . \tag{41}$$

Set $\underline{\beta} = \mathbb{Z}_+ \times \mathbb{Z}_+ \times \mathbb{Z}_+$,

$$\mu_\beta(u,v) := \int_{-\infty}^{\infty} \tilde{u}(1/2+iy)\, \overline{\tilde{v}(1/2+iy)}\,dy + \int_{-\infty}^{\infty} (\rho_{h-j}+iy)^l \tilde{u}(\rho_{h-j}+iy) \overline{(\rho_{h-j}+iy)^l \tilde{v}(\rho_{h-j}+iy)}\,dy \tag{42}$$

and

$$\mu_\beta(u) := \mu_\beta(u,u)^{1/2}, \quad \beta = (h,j,l) \in \underline{\beta}. \tag{43}$$

Clearly, μ_β only depends on h-j, but our notation reminds of $\eta - \delta$ in π_α.

13. <u>Lemma</u>. The countable sets of Hilbert norms

$$(\pi_\alpha)_{\alpha \in \underline{\alpha}}, \quad (\mu_\beta)_{\beta \in \underline{\beta}} \tag{44}$$

define the same Fréchet topology on $\mathcal{S}_{\mathfrak{p}}$ which is thus independent of the particular choice of the ρ_i, $i \in \mathbb{Z}$.

<u>Proof</u>. In order to prove the equivalence, we have to show that

$$(\mathcal{S}_{\mathfrak{p}},(\pi_\alpha)) \longrightarrow (\mathcal{S}_{\mathfrak{p}},(\mu_\beta))$$

is continuous in both directions, which means that

(i) to any β there are finitely many α_k, $k = 1,\ldots,k_\beta$, with

$$\mu_\beta(u) \leq c \sup_k \pi_{\alpha_k}(u), \quad u \in \mathcal{S}_{\mathfrak{p}}, \tag{45}$$

with some constant $c = c(\beta,\alpha_1,\ldots,\alpha_{k_\beta})$.

(ii) to any α there are finitely many β_l, $l = 1,\ldots,l_\alpha$, with

$$\pi_\alpha(u) \leq c \sup_l \mu_{\beta_l}(u), \quad u \in \mathcal{S}_{\mathfrak{p}}, \tag{46}$$

with some constant $c = c(\alpha, \beta_1, \ldots, \beta_{l_\alpha})$.

(i) follows if for any sequence (u_n) which tends in $(\mathcal{S}_\wp, (\pi_\alpha))$ to zero, we get $\mu_\beta(u_n) \longrightarrow 0$ if we only use $\pi_{\alpha_k}(u_n) \longrightarrow 0$, $k = 1, \ldots, k_\beta$. (ii) follows in a similar way.

First observe that the map

$$\zeta : \mathcal{S}_\wp \longrightarrow \mathcal{Z} \tag{47}$$

is continuous with respect to (π_α) as well as (μ_β). The continuity of ζ with respect to (π_α) immediately follows from Proposition 8. The continuity with respect to (μ_β) is also simple. Choose $\rho' = \rho_{i'}$, $\rho'' = \rho_{i''}$ in such a way that

$$\rho_{i''} > \operatorname{Re} p_{j_0} > \rho_{i'} \tag{48}$$

for some given j_0 and a polynomial $g(z)$ for which $f(z) = g(z)\tilde{u}(z)$ has precisely one simple pole at $z = p_{j_0}$. Then the integrals $\int_{\operatorname{Re} z = \rho} |z^r f(z)|^2 dy$ are finite for $\rho = \rho'$, $\rho = \rho''$, and $r \in \mathbb{Z}_+$. Thus

$$\frac{1}{2\pi i}\left\{ \int_{\operatorname{Re} z = \rho'} f(z)dz - \int_{\operatorname{Re} z = \rho''} f(z)dz \right\} \tag{49}$$

is finite, too, and equals the residue of f at the point $z = p_{j_0}$, which is up to a constant factor the top coefficient in the Laurent series of $\tilde{u}(z)$ at $z = p_{j_0}$. It is clear that (49) converges if finitely many norms μ_β converge, i.e. the considered coefficient converges. For the lower order coefficients in the Laurent series at $z = p_{j_0}$ we conclude in a similar manner by replacing u by $u - \sum_{k=k'}^{m_{j_0}} \zeta_{j_0,k}\, \omega^{p_{j_0},k}$, $k' = m_{j_0}, m_{j_0}-1, \ldots, 0$, $j_0 = 0, 1, 2, \ldots$. Let $u_n \in \mathcal{S}_\wp$, $u_n \longrightarrow 0$, with respect to (π_α). Since ζ is continuous, we get for $\rho = 1/2 - \delta + \eta$, $\rho \neq \operatorname{Re} p_j$,

$$\int_{\operatorname{Re} z = \rho} |z^l (M \mathcal{H}_\delta u_n)(z)|^2 dy \longrightarrow 0 \tag{50}$$

as $n \longrightarrow \infty$, cf. Lemma 5, and $t^{\eta-\delta} \vartheta^l \mathcal{H}_\delta u_n \longrightarrow 0$ in $L^2(\mathbb{R}_+)$, i.e.

$$\int_{\operatorname{Re} z = \rho} |z^l (M \mathcal{H}_\delta u_n)(z)|^2 dy \longrightarrow 0,$$

which implies

$$\int_{\mathrm{Re}\, z = \varrho} |z^l \tilde{u}_n(z)|^2 dy \longrightarrow 0 \quad \text{as } n \longrightarrow \infty .$$

This gives us (i) for arbitrary fixed ß.

Conversely let $u_n \longrightarrow 0$ in $(\mathcal{S}_\varphi, (\mu_\beta))$. Then we may conclude (50), since ζ is continuous with respect to (μ_β) and thus,

$$2\pi \int_0^\infty |t^{\eta-\delta} \mathcal{N}_\delta u(t)|^2 dt$$

$$= \int_{\mathrm{Re}\, z = \varrho} |z^l M \mathcal{N}_\delta u_n(z)|^2 dy$$

$$\leq \int_{\mathrm{Re}\, z = \varrho} |z^l \tilde{u}_n(z)|^2 dy + \int_{\mathrm{Re}\, z = \varrho} |z^l \mathcal{K}_\delta \tilde{u}_n(z)|^2 dy$$

$$\longrightarrow 0$$

for $\varrho = 1/2 - \delta + \eta$. This yields (ii). □

The following assertion is an analogue of the Borel theorem for C^∞ functions.

14. <u>Theorem</u>. Let φ be a sequence (1) with (2), (5). Then, for arbitrary coefficients $\zeta_{jk} \in \mathbb{C}$, $0 \leq k \leq m_j$, $j \in \mathbb{Z}_+$, there exists a $u \in \mathcal{S}_\varphi$ with $\zeta_{jk} = \zeta_{jk}(u)$, in other words, the mapping $\zeta : \mathcal{S}_\varphi \longrightarrow \mathcal{Z}$ is surjective.

<u>Proof</u>. It suffices to show that there exists a sequence of numbers $c_j > 0$ such that

$$u(t) := \sum_{j=0}^{\infty} \sum_{k=0}^{m_j} \zeta_{jk} t^{-p_j} \log^k t \; \omega(c_j t) \tag{51}$$

converges in $\mathcal{S}_\varphi$, where, of course, the c_j depend on the ζ_{jk}. The convergence will follow from

$$\mu_\beta(u) = \sum_{j=0}^{\infty} \sum_{k=0}^{m_j} |\zeta_{jk}| \mu_\beta(t^{-p_j} \log^k t \; \omega(c_j t)) < \infty . \tag{52}$$

Let $p \in \mathbb{C}$, $k \in \mathbb{Z}_+$, $c > 0$ and

$$f_c^{p,k}(z) := M(t^{-p} \log^k t\, \omega(ct))(z) .$$

Then

$$\begin{aligned}
f_c^{p,k}(z) &= \int_0^\infty t^{z-p-1} \log^k t\, \omega(ct)dt \\
&= \int_0^\infty (s/c)^{z-p-1} \log^k(s/c)\, \omega(s)c^{-1}ds \\
&= c^{-(z-p)} \int_0^\infty s^{z-p-1}(\log s - \log c)^k\, \omega(s)ds \\
&= c^{-(z-p)} \sum_{n=0}^k \beta_n \log^{k-n}c \int_0^\infty s^{z-p-1}(\log s)^n\, \omega(s)ds \\
&= c^{-(z-p)} \sum_{n=0}^k \beta_n \log^{k-n}c\, f_1^{p,n}(z)
\end{aligned}$$

with fixed constants β_n. For $\varrho \neq \operatorname{Re} p$ we have for all $l \in \mathbb{Z}_+$

$$\int |(\varrho + iy)^l f_1^{p,n}(\varrho + iy)|^2 dy < C(l,p)$$

with constants $C(l,p)$, $0 \leq n \leq k(p)$. With $c^{-(z-p)} = c^{-(\varrho - \operatorname{Re} p)} c^{-i(y - \operatorname{Im} p)}$ $c \geq 1$, we have

$$\begin{aligned}
&\left(\int |(\varrho + iy)^l f_c^{p,k}(\varrho + iy)|^2 dy\right)^{1/2} \\
&= \left(\int |(\varrho + iy)^l c^{-(z-p)} \sum \beta_n \log^{k-n}c\, f_1^{p,n}(\varrho + iy)|^2 dy\right)^{1/2} \\
&\leq c^{-(\varrho - \operatorname{Re} p)} \sum |\beta_n| \log^{k-n}c \left(\int |(\varrho+iy)^l f_1^{p,n}(\varrho+iy)|^2 dy\right)^{1/2} \\
&\leq c^{-(\varrho - \operatorname{Re} p)} \sum |\beta_n| \log^{k-n}c\, C(l,p) .
\end{aligned}$$

From this we obtain for $\varrho > \operatorname{Re} p$

$$\int |(\varrho + iy)^l f_c^{p,k}(\varrho + iy)|^2 dy \longrightarrow 0 \quad \text{as } c \longrightarrow \infty .$$

Let $p = p_j$, $0 \leq k \leq m_j$. Since $\operatorname{Re} p_j \longrightarrow -\infty$ for $j \longrightarrow \infty$, for any $r \in \mathbb{Z}$ there is some j such that $\varrho_r > \operatorname{Re} p_j$. Thus, for any $r = h-i$ and l we can choose the sequence c_j in such a way that (52) holds, $\beta = (h,i,l)$. Denote this sequence by $c_j^{(r,l)}$. Let $\nu = \nu(r,l)$ be a bijective map $\mathbb{Z} \times \mathbb{Z}_+ \longrightarrow \mathbb{Z}_+$ and denote the corresponding sequence by c_j^ν. Then the diagonal sequence c_j^j obviously has the property that (52) holds for arbitrary β, which proves (51). □

With the sequence of $\varrho_j \in \mathbb{R}$, $j \in \mathbb{Z}_+$, which was involved in the definition of $(\mu_B)_{B \in \underline{B}}$, we can associate another system of semi-norms, namely

$$\bar{\mu}_B(u) := \sup_{y \in \mathbb{R}} |(\varrho_{h-j}+iy)^l \tilde{u}(\varrho_{h-j}+iy)|, \quad u \in \mathcal{S}_{\varphi},$$

$B \in \underline{B}$. On the other hand, we also have a system of semi-norms

$$\bar{\pi}_\alpha(u) := \sup_{t \in \mathbb{R}_+} |t^{1/2+\eta-\delta} \partial_t^l \, v_\delta u(t)|, \quad u \in \mathcal{S}_{\varphi},$$

$\alpha = (\eta, \delta, l) \in \underline{\alpha}$.

15. __Lemma.__ The countable sets of semi-norms

$$(\bar{\pi}_\alpha)_{\alpha \in \underline{\alpha}}, \quad (\bar{\mu}_B)_{B \in \underline{B}} \tag{53}$$

define the same Fréchet topology on $\mathcal{S}_{\varphi}$ as $(\pi_\alpha)_{\alpha \in \underline{\alpha}}$.

__Proof.__ It is easy to see that $\mathcal{S}_{\varphi}$ is complete with respect to $(\bar{\mu}_B)$ and the corresponding topology is stronger or equal to that of (μ_B). Since comparable Fréchet topologies on a space are equivalent, we obtain the equivalence of $(\bar{\mu}_B)$ and (μ_B). In a similar way we argue for $\bar{\pi}_\alpha$. If $v \in L^2(\mathbb{R}_+)$, $\partial v \in L^2(\mathbb{R}_+)$,

$$\sup_{\substack{t \in \mathbb{R}_+ \\ t \geq t_0 > 0}} |v(t)| \leq \sup_{\substack{t \in \mathbb{R}_+ \\ t \geq t_0 > 0}} \left|\int_t^\infty v'(\tau)d\tau\right| \leq \sup_{\substack{t \in \mathbb{R}_+ \\ t > t_0 > 0}} t^{-1/2} \|\partial_t v\|_{L^2(\mathbb{R}_+)}$$

Thus, the topology defined by (π_α) is stronger or equal to that of $(\bar{\pi}_\alpha)$. The completeness of $(\mathcal{S}_{\varphi}, (\bar{\pi}_\alpha))$ is clear, too, i.e. (π_α) and $(\bar{\pi}_\alpha)$ are equivalent. Using Lemma 13 we obtain that $(\bar{\pi}_\alpha)$ and $(\bar{\mu}_B)$ are equivalent. □

Let us mention another way of introducing a Fréchet topology in $\mathcal{S}_{\varphi}$, namely by the system of semi-norms

$$\lambda_\gamma(u) = \|t^{\eta-\delta-1/2} \partial^l v_\delta u\|_{L^1}, \quad \gamma \in \underline{\gamma}, \tag{54}$$

where $\underline{\gamma}$ consists of the indices $\gamma = (\eta, \delta, l) \in \mathbb{R}_+ \times \mathbb{R}_+ \times \mathbb{Z}_+$.

16. __Lemma.__ The systems $(\pi_\alpha)_{\alpha \in \underline{\alpha}}$ and $(\lambda_\gamma)_{\gamma \in \underline{\gamma}}$ define the same Fréchet topologies on $\mathcal{S}_{\varphi}$.

Since we do not use the latter statement explicitly, the proof is left to the reader.

The definition of the spaces $\mathcal{S}_{\mathfrak{p}}$ and $C^{\infty}_{\mathfrak{p},\mathfrak{p}'}$ suggests generalizations for systems of poles of infinite order in the Mellin image. In Section 2.1.1. we systematically study spaces of that type.

Let us finish this section with a description of the topological dual $\mathcal{S}'_{\mathfrak{p}}$ defined as the space of continuous linear forms on $\mathcal{S}_{\mathfrak{p}}$. From Theorem 9, an exact sequence

$$0 \longrightarrow \mathcal{Z}' \xrightarrow{\zeta'} \mathcal{S}'_{\mathfrak{p}} \xrightarrow{\iota'} \mathcal{S}'_0 \longrightarrow 0 \qquad (55)$$

follows by duality. Here ι' simply restricts functionals to $\mathcal{S}_0$, and ζ' is defined by

$$\langle \zeta'(\varrho), u \rangle = \langle \varrho, \zeta(u) \rangle, \quad \varrho \in \mathcal{Z}', \ u \in \mathcal{S}_{\mathfrak{p}} .$$

$\mathcal{Z}'$ may be identified with the space of all finite sequences

$$\{ \nu_{jk} : 0 \leq j \leq N, \ 0 \leq k \leq m_j \} \cong C^{m(N)}$$

$m(N) = \sum_{j=0}^{N} (m_j+1)$, equipped with the inductive limit topology for $N \longrightarrow \infty$. For $\mathcal{S}'_0$ we have a natural identification

$$\mathcal{S}'_0 \cong \{ f|_{\mathbb{R}_+} : f \in \mathcal{S}'(\mathbb{R}) \}.$$

Let ω be a cut-off function. Then, for $u \in \mathcal{S}_{\mathfrak{p}}$, $f \in \mathcal{S}'_{\mathfrak{p}}$

$$\langle f,u \rangle = \langle f,\omega u \rangle + \langle f,(1-\omega)u \rangle,$$

and it is natural to set

$$\langle f,\omega u \rangle = \langle \omega f,u \rangle, \quad \langle f,(1-\omega)u \rangle = \langle (1-\omega)f,u \rangle.$$

In view of $(1-\omega)u \in \operatorname{im} \iota$ the functional $(1-\omega)f$ may be identified with some element in $\mathcal{S}'_0$. Thus, the specific properties of elements in $\mathcal{S}'_{\mathfrak{p}}$ are concentrated on a neighbourhood of the origin.

17. <u>Definition</u>. Let $\mathcal{S}'^{,l}_{\mathfrak{p}}$, $l \in \mathbb{Z}_+$, be the space of all $f \in \mathcal{S}'_{\mathfrak{p}}$ for which there is a cut-off function $\omega = \omega_f$ such that $\iota'(\omega f) \in \mathcal{S}'_0$ is of order $\leq l$. Similarly, let $\mathcal{S}'^{,l}_0$ be the space of all $f \in \mathcal{S}'_0$ for which ωf is of order $\leq l$ for some $\omega = \omega_f$. Then

$$\mathcal{S}'_{\mathcal{P}} = \bigcup_{l \in \mathbb{Z}_+} \mathcal{S}'^{,l}_{\mathcal{P}} \quad .$$

For $\delta > 0$ we set

$$\mathcal{P}(\delta) = \{(p_j, m_j) \in \mathcal{P} : \operatorname{Re} p_j < -\delta + 1/2\}.$$

18. <u>Lemma.</u> For any $l \in \mathbb{Z}_+$ there exists a $\delta > 0$ such that

$$\iota_\delta : \mathcal{S}_0 \longrightarrow \mathcal{S}_{\mathcal{P}(\delta)}$$

induces a bijection

$$\iota'_\delta : \mathcal{S}'^{,l}_{\mathcal{P}(\delta)} \longrightarrow \mathcal{S}'^{,l}_0 \quad . \tag{56}$$

<u>Proof.</u> Let $f \in \mathcal{S}'^{,l}_0$. Then ωf for suitable ω can be represented in the form

$$\langle \omega f, u \rangle = \sum_{i=0}^{l} \int g_i(t) \frac{\partial^i}{\partial t^i} u(t) dt \tag{57}$$

with continuous functions $g_i(t)$. Since for $\delta > 0$ large enough $\frac{\partial^i}{\partial t^i}$: $\mathcal{S}_{\mathcal{P}(\delta)} \longrightarrow L^2(\mathbb{R}_+)$ is continuous, $0 \leq i \leq l$, (57) can be extended to $\mathcal{S}_{\mathcal{P}(\delta)}$ as a continuous linear functional. Denote that extension by $\varkappa_\delta(\omega f)$. Furthermore, $(1-\omega)f$ is automatically given on $\mathcal{S}_{\mathcal{P}(\delta)}$. Thus

$$\varkappa_\delta(f) := \varkappa_\delta(\omega f) + (1-\omega)f$$

defines an extension of f to $\mathcal{S}_{\mathcal{P}(\delta)}$ and $\varkappa_\delta \circ \iota'_\delta$, $\iota'_\delta \circ \varkappa_\delta$ are obviously the identities on the corresponding spaces. This follows immediately by completion of $\mathcal{S}_0$ and $\mathcal{S}_{\mathcal{P}(\delta)}$ with respect to Sobolev norms of order l and extension of the functionals to the completions which coincide. □

We have

$$\mathcal{S}_{\mathcal{P}} = \mathcal{S}_{\mathcal{P}(\delta)} \oplus \mathcal{K}_\delta \mathcal{S}_{\mathcal{P}}$$

(cf. the proof of Lemma 10) and mappings

$$\mathcal{S}'_{\mathcal{P}(\delta)} \longrightarrow \mathcal{S}'_{\mathcal{P}} \;, \quad (\mathcal{K}_\delta \mathcal{S}_{\mathcal{P}})' \longrightarrow \mathcal{S}'_{\mathcal{P}}$$

which are canonically induced by π_δ and $\mathcal{K}_\delta$, respectively. In view of Lemma 18 we have a bijection (56). Thus the elements in $\mathcal{S}'^{,l}_0$ and $\mathcal{S}'^{,l}_{\mathcal{P}(\delta)}$ may be denoted by the same letters.

From

$$\langle f,u\rangle = \langle f, \mathcal{r}_\delta u\rangle + \langle f, \mathcal{k}_\delta u\rangle,$$

$f \in \mathcal{S}'_{\varphi}$, $u \in \mathcal{S}_{\varphi}$, we thus obtain the

19. <u>Theorem.</u> For any $l \in \mathbb{Z}_+$ there exists a $\delta > 0$ such that for any $f \in \mathcal{S}'^{,l}_{\varphi}$ we have a unique decomposition

$$f = g + h$$

with $g \in \mathcal{S}'^{,l}_{0}$, $h \in (\mathcal{k}_\delta \mathcal{S}_\varphi)' \cong \mathbb{C}^m$, $m = \sum_{j=0}^{N} (m_j+1)$, $N = N(\delta,\varphi)$.

1.2.2. Tensor products and Green operators

Let $\underline{\varphi}$ be the set of all sequences $\{(p_j,m_j) \in \mathbb{C} \times \mathbb{Z}_+ : j \in \mathbb{Z}_+\}$ with

$$\operatorname{Re} p_j \longrightarrow -\infty \quad \text{as } j \longrightarrow \infty, \quad \operatorname{Re} p_j < 1/2,\ j \in \mathbb{Z}_+ .$$

Similarly $\underline{\varphi}'$ denotes the set of all sequences $\{(p'_j,m'_j) \in \mathbb{C} \times \mathbb{Z}_+ : j \in \mathbb{Z}_+\}$ with

$$\operatorname{Re} p'_j \longrightarrow \infty \quad \text{as } j \longrightarrow \infty, \quad \operatorname{Re} p_j > 1/2,\ j \in \mathbb{Z}_+ .$$

We will study a class of operators, called Green operators, the kernels of which belong to $\mathcal{S}_{\varphi_1} \otimes_\pi \mathcal{S}_{\varphi_2}$ for certain $\varphi_1, \varphi_2 \in \underline{\varphi}$, or more generally to $\mathcal{S}_\varphi \otimes_\pi L^2(\mathbb{R}_+)$ for some $\varphi \in \underline{\varphi}$ ($\otimes_\pi$ always stands for the projective tensor product). Analogous constructions are possible for tensor products $C^\infty_{\varphi_1,\varphi'_1} \otimes_\pi C^\infty_{\varphi_2,\varphi'_2}$ or $C^\infty_{\varphi,\varphi'} \otimes_\pi L^2(\mathbb{R}_+)$.

We first prove that $\mathcal{S}_\varphi$ is a nuclear space for each $\varphi \in \underline{\varphi}$.

Let $(\delta_k)_{k\in\mathbb{Z}_+}$, $(\gamma_k)_{k\in\mathbb{Z}_+}$ be sequences of real numbers with $\operatorname{Re} p_j > -\delta_j > \operatorname{Re} p_{j+1}$, $j \in \mathbb{Z}_+$, $\gamma_k > 1/2$, $\gamma_k < \gamma_{k+1}$, $\gamma_k \longrightarrow +\infty$ as $k \longrightarrow \infty$. Then

$$p_0, p_1, \dots, p_{n(k)} \in B_k := \{z \in \mathbb{C} : -\delta_k < \operatorname{Re} z < \gamma_k\},$$

$n(k) = N(\delta_k,\varphi)$. In Section 1.2.1 we discussed an isomorphism

$$\mathcal{S}_\varphi \cong \mathcal{r}_\delta \mathcal{S}_\varphi \oplus \mathcal{k}_\delta \mathcal{S}_\varphi, \quad \delta = \delta_k + 1/2.$$

$\mathcal{k}_\delta \mathcal{S}_\varphi$ is of finite dimension and isomorphic to the space of Laurent coefficients belonging to $z_j = p_j$, $j = 0,\dots,k-1$, which may be identified with $\mathbb{C}^{m(k)}$, $m(k) = \sum_{j=0}^{n(k)} (m_j+1)$. Denote the mentioned isomorphism by

$$\zeta_\delta : \mathcal{K}_\delta \mathcal{S}_\varphi \longrightarrow \mathbb{C}^{m(k)} .$$

The Mellin image of $\mathcal{V}_\delta \mathcal{S}_\varphi$ consists of analytic functions in Re z > $1/2 - \delta = -\delta_k$ and

$$\varkappa'_{k,l}(v) = \{\int |(-\delta_k+iy)^l \tilde{v}(-\delta_k+iy)|^2 dy + \int |(\gamma_k+iy)^l \tilde{v}(\gamma_k+iy)|^2 dy\}^{1/2} < \infty$$

for $v \in \mathcal{V}_\delta \mathcal{S}_\varphi$, $k,l \in \mathbb{Z}_+$. Set $\zeta_\delta u := \zeta_\delta \mathcal{K}_\delta u$. Then

$$\varkappa_{k,l}(u) := (\varkappa'_{k,l}(\mathcal{V}_\delta u)^2 + |\zeta_\delta u|^2 + \|u\|^2_{L^2(\mathbb{R}_+)})^{1/2}$$

is a countable set of norms on $\mathcal{S}_\varphi$ for $k,l \in \mathbb{Z}_+$.

Set

$$\varkappa_{(k)}(u) = (\sum_{i=0}^{k} \varkappa_{k,i}(u)^2)^{1/2}.$$

Then $\varkappa_{(k)}$ is again a Hilbert norm on $\mathcal{S}_\varphi$. Let $\mathcal{S}_\varphi^{(k)}$ be the completion of $\mathcal{S}_\varphi$ with respect to $\varkappa_{(k)}$. Let $\mathcal{A}^{(k)}$ be the Hilbert space of all holomorphic functions f(z) in the strip B_k for which

$$\sup_{-\delta_k < \rho < \gamma_k} \int |(\rho+iy)^j f(\rho+iy)|^2 dy < \infty$$

for $j = 0,1,\dots,l(k)$.

From the results of the previous section then easily follows the

1. <u>Proposition</u>. The map $u \longrightarrow M\mathcal{V}_\delta u|_{B_k} + \zeta_\delta u$, $u \in \mathcal{S}_\varphi$; $\delta = \delta_k + 1/2$, has a unique extension to a topological isomorphism

$$\mathcal{S}_\varphi^{(k)} \xrightarrow[\cong]{} \mathcal{A}^{(k)} \oplus \mathbb{C}^{m(k)} \qquad (1)$$

and

$$\mathcal{S}_\varphi = \bigcap_{k \in \mathbb{Z}_+} \mathcal{S}_\varphi^{(k)} \qquad (2)$$

where the projective limit topology is equivalent to the Fréchet topology generated by $(\pi_\alpha)_{\alpha \in \underline{\alpha}}$.

2. Proposition. The embeddings

$$\mathcal{S}_{\varphi}^{(k+1)} \longrightarrow \mathcal{S}_{\varphi}^{(k)} \tag{3}$$

are Hilbert-Schmidt and consequently, $\mathcal{S}_{\varphi}^{(k+2)} \longrightarrow \mathcal{S}_{\varphi}^{(k)}$ are nuclear, $k \in \mathbb{Z}_+$.

Proof. We have

$$\mathcal{S}_{\varphi}^{(k)} \cong \mathcal{A}^{(k)} \oplus \mathbb{C}^{m(k)}, \quad \mathcal{S}_{\varphi}^{(k+1)} \cong \mathcal{A}^{(k+1)} \oplus \mathbb{C}^{m(k+1)},$$

where $m(k+1) = m(k) + m_k + 1$. Set

$$\mathcal{L}^{(k)} = \left\{ u = \sum_{l=0}^{m_k} \zeta_l \omega^{p_k, l} \ : \ \zeta_l \in \mathbb{C} \right\}.$$

Then $\mathcal{L}^{(k)} \cong \mathbb{C}^{m_k+1}$ and we may write

$$\mathcal{S}_{\varphi}^{(k+1)} \cong \mathcal{A}^{(k+1)} \oplus \mathcal{L}^{(k)} \oplus \mathbb{C}^{m(k)}.$$

Let $v = v_1 + v_2 + w$, $v_1 \in \mathcal{A}^{(k+1)}$, $v_2 \in \mathcal{L}^{(k)}$, $w \in \mathbb{C}^{m(k)}$. Then the embedding (3) is induced by

$$v \longrightarrow (v_1 + v_2)\big|_{B_k} + w.$$

Since $V \longrightarrow v_2\big|_{B_k} + w$ is a finite-dimensional operator, it suffices to show that $\mathcal{A}^{(k+1)} \longrightarrow \mathcal{A}^{(k)}$ induced by the restriction to B_k, is Hilbert-Schmidt. We have

$$\partial B_k = \{ \operatorname{Re} z = -\delta_k \} \cup \{ \operatorname{Re} z = \gamma_k \}.$$

Consider ∂B_k with the clockwise orientation. Then, for any $z \in \partial B_k$, $f \in \mathcal{A}^{(k+1)}$,

$$(1 + z^{l(k+1)}) f(z) = \frac{1}{2\pi i} \int_{\partial B_{k+1}} \frac{(1+w^{l(k+1)}) f(w)}{z-w} \, dw,$$

or equivalently,

$$(1 + z^{l(k)}) f(z) = \frac{1}{2\pi i} \, \frac{1+z^{l(k)}}{1+z^{l(k+1)}} \int_{\partial B_{k+1}} \frac{(1+w^{l(k+1)}) f(w)}{z-w} \, dw. \tag{4}$$

The space $\mathcal{A}^{(k+1)}$ may be identified with a closed subspace of functions f in

$$L^2(\partial B_{k+1}) \cong L^2(\operatorname{Re} z = -\delta_{k+1}) \oplus L^2(\operatorname{Re} z = \gamma_{k+1})$$

for which

$$\int_{\partial B_{k+1}} |(1 + z^{l(k+1)})f(z)|^2 dz < \infty .$$

Then $\mathcal{A}^{(k+1)} \longrightarrow \mathcal{A}^{(k)}$, $f \longrightarrow f|_{B_k}$, can equivalently be expressed by the formula (4), which may be regarded as an integral operator $L^2(\partial B_{k+1}) \longrightarrow L^2(\partial B_k)$ with the kernel

$$K(z,w) = \frac{1}{2\pi i} \frac{1+z^{l(k)}}{1+z^{l(k+1)}} \frac{1}{z-w} .$$

$K(z,w)$ is obviously square integrable. Thus $\mathcal{A}^{(k+1)} \longrightarrow \mathcal{A}^{(k)}$ is Hilbert-Schmidt and Proposition 2 is proved. □

As a corollary we thus obtain

3. Theorem. The space $\mathcal{S}_{\varphi}$ is nuclear.

Proof. $\mathcal{S}_{\varphi}$ is the projective limit of Hilbert spaces $\mathcal{S}_{\varphi}^{(k)}$. The nuclearity holds per definitionem if for any k there is some k' such that $\mathcal{S}_{\varphi}^{(k')} \longrightarrow \mathcal{S}_{\varphi}^{(k)}$ is Hilbert-Schmidt. This was just established in Proposition 2. □

Theorem 3 reminds of results on the nuclearity of spaces of analytic functions.

Let $\varphi, \psi \in \underline{\varphi}$ and consider the spaces $\mathcal{S}_{\varphi}$ and $\mathcal{S}_{\psi}$ with the systems of norms $(\pi_{\alpha})_{\alpha \in \underline{\alpha}}$ and $(\rho_{\beta})_{\beta \in \underline{\beta}}$, respectively. Here

$$\pi_{\alpha}(u) = \left\{ \| t^{\eta-\delta} \vartheta^{l} \varkappa_{\delta}^{\varphi} u \|^2_{L^2(\mathbb{R}_+)} + \|u\|^2_{L^2(\mathbb{R}_+)} \right\}^{1/2}, \quad \alpha = (\eta, \delta, l)$$

and ρ_{β} is similarly defined with respect to ψ.

Set for abbreviation

$$P_{\alpha} u := t^{\eta-\delta} \vartheta^{l} \varkappa_{\delta}^{\varphi} u, \quad \alpha = (\eta, \delta, l),$$

$$Q_{\beta} v := t^{\eta'-\delta'} \vartheta^{l'} \varkappa_{\delta'}^{\psi} v, \quad \beta = (\eta', \delta', l').$$

Then

$$\mathcal{S}_{\varphi} = \left\{ u \in L^2(\mathbb{R}_+) : \| P_{\alpha} u \|_{L^2(\mathbb{R}_+)} < \infty , \ \alpha \in \underline{\alpha} \right\} ,$$

$$\mathcal{S}_{\psi} = \left\{ v \in L^2(\mathbb{R}_+) : \| Q_{\beta} v \|_{L^2(\mathbb{R}_+)} < \infty , \ \beta \in \underline{\beta} \right\}.$$

If P_{α} (Q_{β}) acts with respect to the variable t (s), we write $P_{\alpha} = P_{\alpha}(t)$

$(Q_\beta = Q_\beta(s))$.

Let us briefly remind of the definition of the tensor product between locally convex spaces, cf. [S 1], here for $\mathcal{S}_\varphi, \mathcal{S}_\eta$.

Denote by $\mathcal{S}_\varphi \otimes \mathcal{S}_\eta$ the algebraic tensor product of $\mathcal{S}_\varphi, \mathcal{S}_\eta$, i.e. the vector space of all finite linear combinations

$$w = \sum u_j \otimes v_j, \quad u_j \in \mathcal{S}_\varphi, \ v_j \in \mathcal{S}_\eta \quad . \tag{5}$$

Introduce in $\mathcal{S}_\varphi \otimes \mathcal{S}_\eta$ the system of norms

$$\tau_{\alpha\beta}(w) := \inf \sum \pi_\alpha(u_j)\, \rho_\beta(v_j),$$

where the infimum is taken over all representations of w in the form (5). The (projective) tensor product $\mathcal{S}_\varphi \otimes_\pi \mathcal{S}_\eta$ is defined as the completion of $\mathcal{S}_\varphi \otimes \mathcal{S}_\eta$ with respect to the system $(\tau_{\alpha\beta})_{\alpha \in \underline{\alpha}, \beta \in \underline{\beta}}$.

Since $\mathcal{S}_\varphi, \mathcal{S}_\eta$ are Fréchet spaces, we have the following property of the tensor product, which is proved, for instance, in [S 1].

4. <u>Proposition</u>. Any $w \in \mathcal{S}_\varphi \otimes_\pi \mathcal{S}_\eta$ can be written in the form

$$w = \sum_{j=0}^{\infty} \lambda_j f_j \otimes g_j \tag{6}$$

with $\sum |\lambda_j| < \infty$ and sequences $(f_j)_{j \in \mathbb{Z}_+}$, $(g_j)_{j \in \mathbb{Z}_+}$ tending to zero in $\mathcal{S}_\varphi$ and $\mathcal{S}_\eta$, respectively. Then

$$\tau_{\alpha\beta}(w) = \inf \sum |\lambda_j|\, \pi_\alpha(f_j) \rho_\beta(g_j),$$

where the infimum is to be taken over all representations of w in the form (6).

Note that $\mathcal{S}_\varphi \otimes_\pi \mathcal{S}_\eta$ is a dense subspace of $L^2(\mathbb{R}_+ \times \mathbb{R}_+)$.

5. <u>Proposition</u>. $w(t,s) \in \mathcal{S}_\varphi \otimes_\pi \mathcal{S}_\eta$ is equivalent to

$$P_\alpha(t) Q_\beta(s) w(t,s) \in L^2(\mathbb{R}_+ \times \mathbb{R}_+) \tag{7}$$

for all $\alpha \in \underline{\alpha}$, $\beta \in \underline{\beta}$, and $w(t,s) \in \mathcal{S}_\varphi \otimes_\pi L^2(\mathbb{R}_+)$ is equivalent to

$$P_\alpha(t) w(t,s) \in L^2(\mathbb{R}_+ \times \mathbb{R}_+), \quad \alpha \in \underline{\alpha} \quad . \tag{8}$$

In order to prove Proposition 5, we shall give another description of $\mathcal{S}_\varphi \otimes_\pi \mathcal{S}_\eta$ and $\mathcal{S}_\varphi \otimes_\pi L^2(\mathbb{R}_+)$, respectively. By $\mathcal{S}_\varphi^{[\alpha]}$ we denote the Hilbert space obtained as completion of $\mathcal{S}_\varphi$ with respect to the norm π_α generated by the scalar product

$$\pi_\alpha(u,v) = (u,v)_{L^2(\mathbb{R}_+)} + (P_\alpha u, P_\alpha v)_{L^2(\mathbb{R}_+)}, \tag{9}$$

$\alpha \in \underline{\alpha}$. Similarly we consider $\mathcal{S}_\eta^{[\beta]}$, $\beta \in \underline{\beta}$.

A simple reformulation of Proposition 2 says that for any $\alpha \in \underline{\alpha}$ there is an $\alpha' \in \underline{\alpha}$ for which

$$\mathcal{S}_\varphi^{[\alpha']} \longrightarrow \mathcal{S}_\varphi^{[\alpha]} \tag{10}$$

is Hilbert-Schmidt (nuclear).

Besides the projective tensor product $\mathcal{S}_\varphi^{[\alpha]} \otimes_\pi \mathcal{S}_\eta^{[\beta]}$ we have another tensor product between the Hilbert spaces induced by the scalar product

$$\eta_{\alpha\beta}(u \otimes f, v \otimes g) = \pi_\alpha(u,v)\rho_\beta(f,g), \tag{11}$$

$u,v \in \mathcal{S}_\varphi^{[\alpha]}$, $f,g \in \mathcal{S}_\eta^{[\beta]}$. The completion of $\mathcal{S}_\varphi^{[\alpha]} \otimes \mathcal{S}_\eta^{[\beta]}$ with respect to the norm $\eta_{\alpha\beta}(h,h)^{1/2}$ associated to (11), will be denoted by $\mathcal{S}_\varphi^{[\alpha]} \otimes_H \mathcal{S}_\eta^{[\beta]}$. Then we have a continuous embedding

$$\mathcal{S}_\varphi^{[\alpha]} \otimes_\pi \mathcal{S}_\eta^{[\beta]} \longrightarrow \mathcal{S}_\varphi^{[\alpha]} \otimes_H \mathcal{S}_\eta^{[\beta]}, \tag{12}$$

where the topology in $\mathcal{S}_\varphi^{[\alpha]} \otimes_\pi \mathcal{S}_\eta^{[\beta]}$ is stronger than the Hilbert-Schmidt topology induced by $\mathcal{S}_\varphi^{[\alpha]} \otimes_H \mathcal{S}_\eta^{[\beta]}$. But the projective limits over all $\alpha \in \underline{\alpha}$, $\beta \in \underline{\beta}$ are equal, i.e.

$$\mathcal{S}_\varphi \otimes_\pi \mathcal{S}_\eta = \underleftarrow{\lim}\ \mathcal{S}_\varphi^{[\alpha]} \otimes_\pi \mathcal{S}_\eta^{[\beta]} = \underleftarrow{\lim}\ \mathcal{S}_\varphi^{[\alpha]} \otimes_H \mathcal{S}_\eta^{[\beta]}. \tag{13}$$

The first equality in (13) is just the definition of $\mathcal{S}_\varphi \otimes_\pi \mathcal{S}_\eta$. The second one follows from the fact that for any $\alpha \in \underline{\alpha}$ we find an $\alpha' \in \underline{\alpha}$ for which (10) is nuclear, which means that

$$\mathcal{S}_\varphi^{[\alpha']} \otimes_H \mathcal{S}_\eta^{[\beta]} \longrightarrow \mathcal{S}_\varphi^{[\alpha]} \otimes_\pi \mathcal{S}_\eta^{[\beta]}$$

is continuous and consequently also

$$\underleftarrow{\lim}\ \mathcal{S}_\varphi^{[\alpha]} \otimes_H \mathcal{S}_\eta^{[\beta]} \longrightarrow \underleftarrow{\lim}\ \mathcal{S}_\varphi^{[\alpha]} \otimes_\pi \mathcal{S}_\eta^{[\beta]}.$$

The converse inclusion follows from (12). In a similar way we obtain

$$\mathcal{S}_\varphi \otimes_\pi L^2(\mathbb{R}_+) = \underleftarrow{\lim}\ \mathcal{S}_\varphi^{[\alpha]} \otimes_\pi L^2(\mathbb{R}_+) = \underleftarrow{\lim}\ \mathcal{S}_\varphi^{[\alpha]} \otimes_H L^2(\mathbb{R}_+). \tag{14}$$

By interchanging factors an analogous statement follows for $L^2(\mathbb{R}_+) \otimes_\pi \mathcal{S}_\eta$.

Then Proposition 5 is an immediate consequence.

Note that we did not use any specific feature of the spaces $\mathcal{S}_\varphi$, but only the nuclearity. In particular, instead of $\mathcal{S}_\varphi^{[\alpha]}$ we could as well have taken the space $\mathcal{S}_\varphi^{(k)}$ introduced in the beginning.

By $\mathcal{L}(\mathcal{S}_\varphi, \mathcal{S}_\eta)$ we denote the space of all linear continuous operators $A : \mathcal{S}_\varphi \longrightarrow \mathcal{S}_\eta$. Let $\underline{\mathcal{B}}$ be the set of all bounded subsets $\mathcal{B} \subset \mathcal{S}_\varphi$,

$$\gamma_{\mathcal{B},B}(A) = \sup\{ \rho_B(Au) : u \in \mathcal{B}\}, \mathcal{B} \in \underline{\mathcal{B}}, B \in \underline{B}. \tag{15}$$

Then we consider $\mathcal{L}(\mathcal{S}_\varphi, \mathcal{S}_\eta)$ in the topology generated by the system of semi-norms (15).

We similarly consider $\mathcal{L}(L^2(\mathbb{R}_+), \mathcal{S}_\eta)$ with the topology connected with the system $\underline{\mathcal{C}}$ of all bounded sets $\mathcal{C}$ in $L^2(\mathbb{R}_+)$ and $B \in \underline{B}$. The restriction of operators to $\mathcal{S}_\varphi$ then yields an embedding

$$\mathcal{L}(L^2(\mathbb{R}_+), \mathcal{S}_\eta) \longrightarrow \mathcal{L}(\mathcal{S}_\varphi, \mathcal{S}_\eta) \tag{16}$$

which is continuous, since $\underline{\mathcal{B}} \subseteq \underline{\mathcal{C}}$. Of course, in $\underline{\mathcal{C}}$ we only need one set, namely $\mathcal{C} = \{ u : \|u\|_{L^2(\mathbb{R}_+)} \leq 1\}$.

Note that we have a natural map

$$\mathcal{L}(\mathcal{S}_{\varphi_1}, \mathcal{S}_{\eta_1}) \times \mathcal{L}(\mathcal{S}_{\varphi_2}, \mathcal{S}_{\eta_2}) \longrightarrow \mathcal{L}(\mathcal{S}_{\varphi_1} \hat{\otimes}_\pi \mathcal{S}_{\varphi_2}, \mathcal{S}_{\eta_1} \hat{\otimes}_\pi \mathcal{S}_{\eta_2}),$$

the tensor product of operators, obtained as the closure of $A_1 \otimes A_2 : \mathcal{S}_{\varphi_1} \otimes \mathcal{S}_{\varphi_2} \longrightarrow \mathcal{S}_{\eta_1} \otimes \mathcal{S}_{\eta_2}$ for $A_i \in \mathcal{L}(\mathcal{S}_{\varphi_i}, \mathcal{S}_{\eta_i})$, $i = 1,2$. In particular, for $A \in \mathcal{L}(\mathcal{S}_\varphi, \mathcal{S}_\eta)$ and the identity in $\mathcal{S}_\psi$ we have

$$A \otimes 1 \in \mathcal{L}(\mathcal{S}_\varphi \hat{\otimes}_\pi \mathcal{S}_\psi, \mathcal{S}_\eta \hat{\otimes}_\pi \mathcal{S}_\psi). \tag{17}$$

Here we may substitute other spaces as well.

6. <u>Definition.</u> Let $w \in \mathcal{S}_\eta \hat{\otimes}_\pi \mathcal{S}_\varphi$ and

$$op_G(w) : L^2(\mathbb{R}_+) \longrightarrow L^2(\mathbb{R}_+) \tag{18}$$

be defined by

$$op_G(w)u(t) = \int_0^\infty w(t,s)u(s)ds.$$

Then (18) is called a <u>Green operator</u> with the kernel w. Similar notations are used for $w \in \mathcal{S}_\eta \hat{\otimes}_\pi L^2(\mathbb{R}_+)$ and $w \in L^2(\mathbb{R}_+) \hat{\otimes}_\pi \mathcal{S}_\varphi$.

Of course, the Green operators (on the half axis) are nothing else than particular Hilbert-Schmidt operators in $L^2(\mathbb{R}_+)$.

The class of Green operators is closed under taking adjoints, where the corresponding $\varphi, \sigma \in \underline{\varphi}$ have to be indicated. In particular, $op_G(w)^* = op_G(w^*)$ with $w^*(t,s) = \overline{w(s,t)}$ and $w^* \in \mathcal{S}_{\bar{\varphi}} \otimes_\pi \mathcal{S}_{\bar{\sigma}}$ if $w \in \mathcal{S}_\sigma \otimes_\pi \mathcal{S}_\varphi$. Here $\bar{\varphi} \in \underline{\varphi}$ is defined as $\{(\bar{p}_j, m_j): j \in \mathbb{Z}_+\}$ with the complex conjugate $\bar{p}_j$ of p_j. Note that

$$w_1 \in \mathcal{S}_\sigma \otimes_\pi L^2(\mathbb{R}_+), \quad w_2 \in L^2(\mathbb{R}_+) \otimes_\pi \mathcal{S}_\varphi$$

implies

$$op_G(w_1) op_G(w_2) = op_G(w)$$

with

$$w(t,s) = \int w_1(t,s') w_2(s',s) ds' \in \mathcal{S}_\sigma \otimes_\pi \mathcal{S}_\varphi .$$

For purposes below let us introduce the notation

$$op_G(F \otimes_\pi E) \tag{19}$$

for the space of all Green operators on the half axis with kernel in $F \otimes_\pi E$, where E, F are $\mathcal{S}_\varphi$, $\varphi \in \underline{\varphi}$ or $L^2(\mathbb{R}_+)$ (or some more general space, cf. Section 1.3.1).

7. <u>Proposition</u>. op_G induces a topological isomorphism

$$op_G : \mathcal{S}_\sigma \otimes_\pi L^2(\mathbb{R}_+) \longrightarrow \mathcal{L}(L^2(\mathbb{R}_+), \mathcal{S}_\sigma).$$

<u>Proof</u>. Let $w \in \mathcal{S}_\sigma \otimes_\pi L^2(\mathbb{R}_+)$ and Q_β as above. Then

$$Q_\beta [op_G(w)u] = op_G(Q_\beta(t)w)u$$

implies

$$\| Q_\beta [op_G(w)u] \|_{L^2(\mathbb{R}_+)} \leq \| Q_\beta(t)w \|_{L^2(\mathbb{R}_+ \times \mathbb{R}_+)} \| u \|_{L^2(\mathbb{R}_+)} .$$

This is the continuity of op_G.

For any given $k \in \mathbb{Z}_+$ the operator

$$op_G(w) : L^2(\mathbb{R}_+) \longrightarrow \mathcal{S}_\sigma^{(k+1)}$$

is continuous by assumption. Then

$$op_G(w) : L^2(\mathbb{R}_+) \longrightarrow \mathcal{S}_\sigma^{(k)}$$

is Hilbert-Schmidt, because the embedding $\mathcal{S}_\sigma^{(k+1)} \hookrightarrow \mathcal{S}_\sigma^{(k)}$ is so (cf.

Proposition 2). Hence $op_G(w)$ has its kernel in $\mathcal{S}_{\mathfrak{q}}^{(k)} \otimes_H L^2(\mathbb{R}_+)$. Since k was arbitrary, the kernel is even in $\mathcal{S}_{\mathfrak{q}} \otimes_\pi L^2(\mathbb{R}_+) = \varprojlim \mathcal{S}_{\mathfrak{q}}^{(k)} \otimes_H L^2(\mathbb{R}_+)$. This proves the surjectivity of op_G. The injectivity is clear, since a Hilbert-Schmidt operator in $L^2(\mathbb{R}_+)$ has a uniquely determined kernel function. □

Proposition 7 basically is a kernel theorem.

Denote by $\mathcal{S}_{\underline{\mathfrak{p}}}$ the inductive limit of all $\mathcal{S}_{\mathfrak{p}}$, $\mathfrak{p} \in \underline{\mathfrak{p}}$, which is again a subspace of $C^\infty(\mathbb{R}_+) \cap L^2(\mathbb{R}_+)$.

In the context of the algebra $\mathcal{L}$ to be studied below we set

$$\mathcal{L}_G = \varinjlim_{\mathfrak{q},\mathfrak{p} \in \underline{\mathfrak{p}}} op_G(\mathcal{S}_{\mathfrak{q}} \otimes_\pi \mathcal{S}_{\mathfrak{p}}). \tag{20}$$

Let us introduce an addition between the sequences $\mathfrak{p}, \mathfrak{q} \in \underline{\mathfrak{p}}$,

$$\mathfrak{p} = (p_j, m_j)_{j \in \mathbb{Z}_+}, \quad \mathfrak{q} = (q_k, n_k)_{k \in \mathbb{Z}_+}$$

by

$$\mathfrak{p} + \mathfrak{q} := \mathfrak{r} = (r_l, h_l)_{l \in \mathbb{Z}_+} , \tag{21}$$

where $(r_l) = (p_j) \cup (q_k)$ is defined as the set-theoretic union of systems of points, whereas h_l belonging to $r_l \in \mathbb{C}$ is defined by

$h_l = m_j + n_k + 1$ for $r_l = p_j = q_k$ for certain j, k,

$h_l = m_j$ for $r_l = p_j$ with some j and $r_l \neq q_k$ for all k,

$h_l = n_k$ for $r_l \neq p_j$ for all j and $r_l = q_k$ with some k.

This notation corresponds to the behaviour of the systems of poles of meromorphic functions under multiplication.

Another natural operation between $\mathfrak{p}, \mathfrak{q}$ is the product that corresponds to the product of functions in $\mathcal{S}_{\mathfrak{p}}$ and $\mathcal{S}_{\mathfrak{q}}$ itself, cf. Section 1.3.1, namely

$$\mathfrak{p} \cdot \mathfrak{q} := \mathfrak{s} = (s_l, g_l)_{l \in \mathbb{Z}_+} , \tag{22}$$

where $\mathfrak{s}$ is defined as the set of all pairs $(p_j + q_k, m_j + n_k)$, $j, k \in \mathbb{Z}_+$.

In several situations we shall use the following simple observation. Let H be a Hilbert space, $\mathcal{L}(H)$ the algebra of bounded operators in H, $\mathcal{C}(H)$ the norm-closed two-sided ideal of compact operators. Consider a left ideal $\mathcal{A} \subset \mathcal{C}(H)$, not necessarily norm-closed. Then $1 + \mathcal{A} \subset \mathcal{L}(H)$ is a subalgebra consisting of Fredholm operators with index zero.

8. Proposition. Let $A \in \mathcal{A}$. Then the finite-dimensional space $\ker(1+A)$ is a subspace of $\operatorname{im} \mathcal{A} \subset H$ and $\operatorname{coker}(1+A) = \ker(1+A^*)^{\perp}$, $\ker(1+A^*) \subset \operatorname{im} \mathcal{A}^* \subset H$. If $1+A$ is surjective, then each right inverse $(1+A)_r^{(-1)}$ belongs to $1+\mathcal{A}$.

Proof. Each $u \in \ker(1+A)$ satisfies $u = -Au$, hence $u \in \operatorname{im} A$, and $v \in \ker(1+A^*)$ belongs to $\operatorname{im} A^*$. Set $A_1 = (1+A)_r^{(-1)} - 1$. Then $(1+A)(1+A_1) = 1$ gives $A_1 = -A(1+A_1) \subset \mathcal{A}\mathcal{L}(H) \subset \mathcal{A}$. □

We apply now Proposition 8 to $H = L^2(\mathbb{R}_+)$ and ideals of Green operators.

9. Theorem. Let $w \in \mathcal{S}_q \hat{\otimes}_\pi L^2(\mathbb{R}_+)$. Then

$$1 + \mathrm{op}_G(w) : L^2(\mathbb{R}_+) \longrightarrow L^2(\mathbb{R}_+)$$

is Fredholm with index zero and $\ker(1+\mathrm{op}_G(w)) \subset \mathcal{S}_q$. If $1+\mathrm{op}_G(w)$ is surjective on $L^2(\mathbb{R}_+)$ then each right inverse belongs to $1+\mathrm{op}_G(\mathcal{S}_q \hat{\otimes}_\pi L^2(\mathbb{R}_+))$.

If $w \in \mathcal{S}_q \hat{\otimes}_\pi L^2(\mathbb{R}_+) \cap L^2(\mathbb{R}_+) \hat{\otimes}_\pi \mathcal{S}_p$ and $1+\mathrm{op}_G(w)$ bijective in $L^2(\mathbb{R}_+)$, we get that $(1+\mathrm{op}_G(w))^{-1} = 1+\mathrm{op}_G(w_1)$, where $w_1 \in \mathcal{S}_q \hat{\otimes}_\pi (L^2(\mathbb{R}_+) \cap L^2(\mathbb{R}_+) \hat{\otimes} \mathcal{S}_p$.

Proof. For the first statement we apply Proposition 8 to the left ideal $\mathrm{op}_G(\mathcal{S}_q \hat{\otimes}_\pi L^2(\mathbb{R}_+))$. In the second case we do the same in addition for the adjoints which have kernels in $\overline{\mathcal{S}}_p \hat{\otimes}_\pi L^2(\mathbb{R}_+)$. □

It is convenient to have a criterion for the Green property of a Hilbert-Schmidt operator $op_G(b)$, $b \in L^2(\mathbb{R}_+) \otimes_H L^2(\mathbb{R}_+)$, in terms of the mapping properties

$$op_G(b) \in \mathcal{L}(L^2(\mathbb{R}_+), \mathcal{S}_{\mathfrak{q}}), \quad op_G(b)^* \in \mathcal{L}(L^2(\mathbb{R}_+), \mathcal{S}_{\bar{\mathfrak{p}}}) \tag{23}$$

for certain $\mathfrak{p}, \mathfrak{q} \in \underline{\mathfrak{P}}$. From (23) and Proposition 7 it follows

$$b \in \mathcal{S}_{\mathfrak{q}} \otimes_\pi L^2(\mathbb{R}_+) \cap L^2(\mathbb{R}_+) \otimes_\pi \mathcal{S}_{\mathfrak{p}}$$

and then $op_G(b)$ is of Green type in virtue of the following

<u>Proposition</u> 10. We have

$$\mathcal{S}_{\mathfrak{q}} \otimes_\pi \mathcal{S}_{\mathfrak{p}} = \{\mathcal{S}_{\mathfrak{q}} \otimes_\pi L^2(\mathbb{R}_+)\} \cap \{L^2(\mathbb{R}_+) \otimes_\pi \mathcal{S}_{\mathfrak{p}}\}$$

for arbitrary $\mathfrak{p}, \mathfrak{q} \in \underline{\mathfrak{P}}$, and the topologies are equivalent.

<u>Proof</u>: First we obviously have a continuous embedding

$$\mathcal{S}_{\mathfrak{q}} \otimes_\pi \mathcal{S}_{\mathfrak{p}} \hookrightarrow \{\mathcal{S}_{\mathfrak{q}} \otimes_\pi L^2(\mathbb{R}_+)\} \cap \{L^2(\mathbb{R}_+) \otimes_\pi \mathcal{S}_{\mathfrak{p}}\}.$$

For $\mathfrak{p} = \{p_j, m_j\}_{j \in \mathbb{Z}_+}$, $\mathfrak{q} = \{q_j, n_j\}_{j \in \mathbb{Z}_+}$ and arbitrary fixed $\delta > 0$ we define

$$P_\delta(z) = \prod \frac{(z - p_j)^k}{(z - \gamma)^k} \qquad Q_\delta(z) = \prod \frac{(z - q_i)^k}{(z - \rho)^k}$$

$\gamma < \frac{1}{2} - \delta$ and $\rho < \frac{1}{2} - \delta$ are fixed, the products are taken over all j, i for which $\operatorname{Re} p_j \geq \frac{1}{2} - \delta$, $\operatorname{Re} q_i \geq \frac{1}{2} - \delta$, and $k \in \mathbb{Z}_+$, $k = \max \{m_j + 1, n_i + 1\}$, where the maximum is taken over all indices i, j occurring in the products. Define $P_\delta(\vartheta)$ by $P_\delta(\vartheta) = M^{-1} P_\delta(-z) M$ and in the same way $P_\delta^{-1}(\vartheta)$, $Q_\delta(\vartheta)$, $Q_\delta^{-1}(\vartheta)$. Let $u \in \mathcal{S}_{\mathfrak{q}} \otimes_\pi L^2(\mathbb{R}_+) \cap L^2(\mathbb{R}_+) \otimes_\pi \mathcal{S}_{\mathfrak{p}}$ and set $v(s,t) = Q_\delta(\vartheta_s) P_\delta(\vartheta_t) u(s,t)$. Then

$$v(s,t) \in \mathcal{S}_0^{[\alpha]} \otimes_\pi L^2(\mathbb{R}_+) \cap L^2(\mathbb{R}_+) \otimes_\pi \mathcal{S}_0^{[\beta]} \tag{24}$$

for k large enough and all

$$\alpha = (\eta, \delta, 1), \; \beta = (\eta', \delta, 1'), \quad \eta, \eta' \in \mathbb{R}_+ ,$$

$1, 1' \in \mathbb{Z}_+$. We shall prove that for any given $\tilde{\alpha} = (\tilde{\eta}, \tilde{\delta}, \tilde{1})$, $\tilde{\beta} = (\tilde{\eta}', \tilde{\delta}, \tilde{1}')$ there exist finitely many multi-indices α, β as above so that (24) for these α, β implies

$$v(s,t) \in \mathcal{S}_0^{[\tilde{\alpha}]} \otimes_\pi \mathcal{S}_0^{[\tilde{\beta}]}$$

Then we get

$$u(s,t) = Q_\delta^{-1}(\vartheta_s)\ P_\delta^{-1}(\vartheta_t)\quad v(s,t) \in \mathcal{S}_\eta^{[\tilde{\alpha}]} \otimes_\pi \mathcal{S}_\varphi^{[\beta]}$$

and hence, from (13), the assertion. For proving the desired property of v we use the elementary inequality $x^a y^b \leq c(x^{2a} + y^{2b})$ for all $x, y \in \mathbb{R}_+$ and $a, b \in \mathbb{R}$ with some $c > 0$. Then

$$\left\| s^{\tilde{\eta}-\tilde{\delta}}\, t^{\tilde{\eta}'-\tilde{\delta}}\quad v(s,t) \right\|_{L^2(\mathbb{R}_+ \times \mathbb{R}_+)}$$

$$\leq c \left\| s^{2(\tilde{\eta}-\tilde{\delta})}\quad v(s,t) \right\|_{L^2(\mathbb{R}_+ \times \mathbb{R}_+)} + \left\| t^{2(\tilde{\eta}'-\tilde{\delta})}\quad v(s,t) \right\|_{L^2(\mathbb{R}_+ \times \mathbb{R}_+)}$$

and in the same way

$$\left\| \vartheta_s^{1} \vartheta_t^{1'}\quad v(s,t) \right\|_{L^2(\mathbb{R}_+ \times \mathbb{R}_+)}$$

$$= (2\pi)^{-2} \left\| w^{\tilde{1}}\, z^{\tilde{1}'}\quad v(w,z) \right\|_{L^2(\Gamma_{\frac{1}{2}} \times \Gamma_{\frac{1}{2}})}$$

$$\leq c\left\{ \left\| w^{2\tilde{1}}\quad \tilde{v}(w,z) \right\|_{L^2(\Gamma_{\frac{1}{2}} \times \Gamma_{\frac{1}{2}})} + \left\| z^{21'}\quad v(w,z) \right\|_{L^2(\Gamma_{\frac{1}{2}} \times \Gamma_{\frac{1}{2}})} \right\}$$

$$= c\left\{ \left\| \vartheta_s^{2\tilde{1}}\quad v(s,t) \right\|_{L^2(\mathbb{R}_+ \times \mathbb{R}_+)} + \left\| \vartheta_t^{2\tilde{1}'}\quad v(s,t) \right\|_{L^2(\mathbb{R}_+ \times \mathbb{R}_+)} \right\}$$

and

$$\left\| \vartheta_s^{\tilde{1}}\, t^{\tilde{\eta}'-\tilde{\delta}}\quad v(s,t) \right\|_{L^2(\mathbb{R}_+ \times \mathbb{R}_+)}$$

$$= (2\pi)^{-1} \left\| w^{\tilde{1}}\, t^{\tilde{\eta}'-\tilde{\delta}}\quad v(\tilde{w},t) \right\|_{L^2(\mathbb{R}_+ \times \Gamma_{\frac{1}{2}})}$$

$$\leq c\left\{ \left\| \vartheta_s^{21}\quad v(s,t) \right\|_{L^2(\mathbb{R}_+ \times \mathbb{R}_+)} + \left\| t^{2(\tilde{\eta}'-\tilde{\delta})}\quad v(s,t) \right\|_{L^2(\mathbb{R}_+ \times \mathbb{R}_+)} \right\}.$$

An analogous estimate for

$$\left\| s^{\tilde{\eta}-\tilde{\delta}}\, \vartheta_t^{\tilde{1}'} v(s,t) \right\|_{L^2(\mathbb{R}_+ \times \mathbb{R}_+)}$$

combined with the other estimates finally yields

$$\left\| s^{\tilde{\eta}-\tilde{\delta}} t^{\tilde{\eta}'-\tilde{\delta}} \vartheta_s^{\tilde{1}} \vartheta_t^{\tilde{1}'} v(s,t) \right\|_{L^2(\mathbb{R}_+ \times \mathbb{R}_+)}$$

$$\leq c\left\{ \left\| s^{2(\tilde{\eta}-\tilde{\delta})} \vartheta_s^{2\tilde{1}} v(s,t) \right\|_{L^2(\mathbb{R}_+ \times \mathbb{R}_+)} + \left\| t^{2(\tilde{\eta}'-\tilde{\delta})} \vartheta_t^{2\tilde{1}'} v(s,t) \right\|_{L^2(\mathbb{R}_+ \times \mathbb{R}_+)} \right\}. \tag{25}$$

The estimates in the proof of Proposition 10 may be modified by using the inequality

$$x^a y^b \leq c\left\{ x^{\operatorname{sgn} a(|a|+\varepsilon)} + y^{\varepsilon^{-1} b(|a|+\varepsilon)} \right\}$$

for all $x,y \in \mathbb{R}_+$, $a,b \in \mathbb{R}$ and arbitrary $\varepsilon > 0$ with a constant $c = c(a,b,\varepsilon) > 0$. The proof easily follows from the obvious case when $a,b \in \overline{\mathbb{R}}_+$ by replacing $x(y)$ by $x^{-1}(y^{-1})$.

Then (25) can be replaced by estimates where on the right, for instance, we have an exponent of s close to $\tilde{\eta}-\tilde{\delta}$ instead of $2(\tilde{\eta}-\tilde{\delta})$. This has to be compensated by enlarging the other exponents. In other words, the s (or t strip) needed for estimating the mixed terms in (25) may be close to the original one.

The assertion of Proposition 13 is well-known in the special case $\varphi = \eta = \varphi_0$. Then it follows immediately that Proposition 13 is valid for $\varphi, \eta \in \underline{\varphi}$ containing only a finite number of points. Using the notion of infinite sums of topological vector spaces (cf. 2.1.5.) one could derive another proof of Proposition 10 from this.

1.3. Mellin operators

Now we introduce operators of Mellin type which give a first expansion of the algebra of Green operators discussed in 1.2.2. At a first glance the Mellin operators are similar to ψDO's with Fourier transform replaced by the Mellin transform. Note, however, that they are smoothing outside the origin and are localized modulo Green operators in an arbitrary small neighbourhood of 0. Moreover, no Mellin symbols with explicit t-dependence will occur.

1.3.1. Definition and basic properties

In this section we shall introduce a class of operators in $\mathcal{L}(\mathcal{S}_{\mathfrak{p}}, \mathcal{S}_{\mathfrak{q}})$ for $\mathfrak{p}, \mathfrak{q} \in \underline{\mathfrak{P}}$, which are closely connected with the Mellin convolution. The simplest so-called Mellin operator has the form

$$\omega \, op_M[f] \omega : \mathcal{S}_{\mathfrak{p}} \longrightarrow \mathcal{S}_{\mathfrak{p}+\mathfrak{q}} \quad , \tag{1}$$

where ω means the multiplication with a cut-off function on $\mathbb{R}_+$, $f \in C^\infty_{\mathfrak{q},\mathfrak{q}'}$ and $op_M[f]$ is the Mellin convolution

$$op_M[f]u(t) = \int_0^\infty f(t/t')u(t')dt'/t', \; u \in L^2(\mathbb{R}_+) \quad . \tag{2}$$

Below we shall prove that the choice of ω does not affect the operator modulo Green operators. More generally, we shall deal with operators of the form $t^{j-\gamma}\omega op_M[f]\omega t^{\gamma}$, $j \in \mathbb{Z}_+$, $\gamma > 0$, and converging sums

$\sum_{j=0}^{\infty} t^{j-\gamma_j} \omega_j \, op_M[f_j] \, \omega_j t^{\gamma_j}$ with Mellin symbols f_j that are not necessarily in $C^\infty_{\mathfrak{q},\mathfrak{q}'}$.

First we concentrate on the multiplication by the function ω occurring in (1). This is a simple special case of the multiplication by a function with conormal singularity.

1. __Lemma.__ Let $\mathfrak{q} \in \underline{\mathfrak{P}}$ and $f \in \mathcal{S}_{\mathfrak{q}}$, then the operator M_f of multiplication by f defines continuous operators

$$M_f : \mathcal{S}_{\mathfrak{p}} \longrightarrow \mathcal{S}_{\mathfrak{p}\mathfrak{q}} \; , \; M_f : C^\infty_{\mathfrak{p},\mathfrak{p}'} \longrightarrow \mathcal{S}_{\mathfrak{p}\mathfrak{q}}$$

for any $\mathfrak{p} \in \underline{\mathfrak{P}}$, with $\max(\mathrm{Re}\,p_j + \mathrm{Re}\,q_k) < 1/2$, and $f \to M_f$ induces continuous mappings

$$\mathcal{S}_{\mathfrak{q}} \longrightarrow \mathcal{L}(\mathcal{S}_{\mathfrak{p}}, \mathcal{S}_{\mathfrak{p}\mathfrak{q}}), \; \mathcal{S}_{\mathfrak{q}} \longrightarrow \mathcal{L}(C^\infty_{\mathfrak{p},\mathfrak{p}'}, \mathcal{S}_{\mathfrak{p}\mathfrak{q}}).$$

For $f \in C^\infty_{\mathfrak{q},\mathfrak{q}'}$ we get continuous operators

$$M_f : \mathcal{S}_{\mathfrak{p}} \longrightarrow \mathcal{S}_{\mathfrak{p}\mathfrak{q}} \quad , \quad M_f : C^\infty_{\mathfrak{p},\mathfrak{p}'} \longrightarrow C^\infty_{\mathfrak{p}\mathfrak{q},\mathfrak{p}'\mathfrak{q}'} \quad ,$$

and $f \longrightarrow M_f$ induces continuous mappings

$$C^\infty_{\mathfrak{q},\mathfrak{q}'} \longrightarrow \mathcal{L}(\mathcal{S}_{\mathfrak{p}}, \mathcal{S}_{\mathfrak{p}\mathfrak{q}}), \; C^\infty_{\mathfrak{q},\mathfrak{q}'} \longrightarrow \mathcal{L}(C^\infty_{\mathfrak{p},\mathfrak{p}'}, C^\infty_{\mathfrak{p}\mathfrak{q},\mathfrak{p}'\mathfrak{q}'}).$$

Proof: Let $\varphi = \{(p_j, m_j)\}_{j \in \mathbb{Z}_+}$, $\eta = \{(q_k, n_k)\}_{k \in \mathbb{Z}_+}$, $\varphi\eta = \{(r_l, d_l)\}_{l \in \mathbb{Z}_+}$, $\tilde{\omega}^{z,j}(t) = t^{-z} \log^j t \; \omega^2(t)$. Then (with obvious meaning of subscripts) for all $\delta \geq 0$ we obtain

$$r_\delta^{\eta}(f) = f - \sum_{j=0}^{N(\delta,\eta)} \sum_{k=0}^{n_j} \xi_{jk}^{\eta}(f)\, \omega^{q_j,k}, \quad r_\delta^{\varphi}(u) = u - \sum_{j=0}^{N(\delta,\varphi)} \sum_{k=0}^{m_j} \xi_{jk}^{\varphi}(u) \times \omega^{p_j,k}$$

$$\tilde{r}_\delta^{\varphi\eta}(fu) = fu - \sum_{j=0}^{N} \sum_{k=0}^{d_j} \xi_{jk}^{\varphi\eta}(fu)\, \tilde{\omega}^{r_j,k}$$

$$= r_\delta^{\eta}(f)u + \Big\{ \sum_{j=0}^{N(\delta,\eta)} \sum_{k=0}^{n_j} \xi_{jk}^{\eta}(f)\, \omega^{q_j,k} \Big\}\, r_\delta^{\varphi}(u),$$

$N = N(\delta, \varphi\eta)$, $\tilde{r}_\delta^{\varphi\eta}$ defined by means of $\tilde{\omega}^{r_j,k}$ Applying $t^{\eta-\delta}\vartheta^l$, $\eta \geq 0$, $l \in \mathbb{Z}_+$, to both sides we easily obtain estimates of $\| t^{\eta-\delta}\vartheta^l \tilde{r}_\delta^{\varphi\eta}(fu)\|_{L^2(\mathbb{R}_+)}$ in terms of finite sums of products of analogous semi-norms in $\mathcal{S}_\varphi$ and $\mathcal{S}_\eta$. Here we use that the coefficients in the asymptotics continuously depend on the functions. □

2. Corollary. Let ω be an arbitrary cut-off function. Then $M_\omega : \mathcal{S}_\varphi \longrightarrow \mathcal{S}_\varphi$ is continuous for each $\varphi \in \underline{\varphi}$ and has a continuous extension M_ω : $C^\infty_{\varphi,\varphi'} \longrightarrow \mathcal{S}_\varphi$, $\varphi' \in \underline{\varphi}'$ arbitrary.

Below we shall again write ω instead of M_ω.

Now we investigate Mellin convolution operators defined by (2). From

$$M(op_M[f]u)(z) = \tilde{f}(z)\tilde{u}(z), \quad u \in L^2(\mathbb{R}_+),$$

by the Mellin inversion formula we obtain

$$op_M[f]u(t) = (2\pi)^{-1} \int_{-\infty}^{\infty} t^{-1/2-iy}\, \tilde{f}(1/2+iy)\tilde{u}(1/2+iy)dy .$$

3. Proposition. Let $f \in C^\infty_{\eta,\eta'}$, and define $op_M[f]$ by

$$op_M[f]u(t) = \int_0^\infty f(t/t')u(t')dt'/t'.$$

Then $op_M[f]$ defines continuous mappings

$$\mathrm{op}_M[f] : \mathcal{S}_{\varphi} \longrightarrow C^{\infty}_{\varphi+q, q'}, \quad \mathrm{op}_M[f] : C^{\infty}_{\varphi,\varphi'} \longrightarrow C^{\infty}_{\varphi+q, \varphi'+q'} .$$

Moreover, $f \longrightarrow \mathrm{op}_M[f]$ induces continuous mappings

$$C^{\infty}_{q,q'} \longrightarrow \mathcal{L}(\mathcal{S}_{\varphi}, C^{\infty}_{\varphi+q, q'}), \quad C^{\infty}_{q,q'} \longrightarrow \mathcal{L}(C^{\infty}_{\varphi,\varphi'}, C^{\infty}_{\varphi+q, \varphi'+q'}) .$$

Proof. We have

$$M(\mathrm{op}_M[f]u) = \tilde{f}(z)\tilde{u}(z).$$

It follows that

$$\mu_{\beta}(\mathrm{op}_M[f]u) \leq c \sum_{|\beta'|, |\beta''| \leq |\beta|} \bar{\mu}_{\beta''}(f)\bar{\mu}_{\beta'}(u) ,$$

where β'' has the form $\beta'' = (h, j, 0)$ and c is a constant independent of h and u. Taking the norms $\mu_{\beta}, \bar{\mu}_{\beta}$ on $\mathcal{S}_{\varphi}, C^{\infty}_{\varphi+q, q'}, \ldots$, we get the assertion. □

4. Definition. A meromorphic function h on the complex plane is called a Mellin symbol if there exists a $\gamma > 0$ such that

$$T^{\gamma}h(z) := h(z+\gamma) \in M(C^{\infty}_{q,q'})$$

for some q, q' depending on γ. Denote by $\mathfrak{M}$ the set of all Mellin symbols.
The characterization of $M(C^{\infty}_{q,q'})$ in Section 1.2.1 shows that $h \in \mathfrak{M}$ has the following properties:

(i) each strip $c \leq \mathrm{Re}\, z \leq c'$ only contains a finite number of poles of h,

(ii) for each $k \in \mathbb{N}$ and any $c, c' \in \mathbb{R}$ there is a constant c_k such that

$$|h(z)| \leq c_k(1+|z|)^{-k}, \quad c \leq \mathrm{Re}\, z \leq c',$$

$|\mathrm{Im}\, z|$ sufficiently large.

The results of Section 1.2.1 immediately show that (i), (ii) characterize Mellin symbols among all meromorphic functions. $\mathfrak{M}$ is larger than $M(C^{\infty}_{q,q'})$, since there may be poles at $\mathrm{Re}\, z = 1/2$. The best constants c_k yield a system of semi-norms on $\mathfrak{M}$, defining a Fréchet space topology. It is convenient to set

$$\mathfrak{z} = (T^{\gamma}q + T^{\gamma}q') ,$$

where the sum is understood in a sense analogous to 1.2.2. (21) and

$$\mathfrak{M}_{\mathfrak{z}} = \{ h : T^{\gamma}h \in \mathfrak{M}_{\mathfrak{z}^1} \}, \quad \mathfrak{M}_{\mathfrak{z}^1} = M(C^{\infty}_{q,q'}),$$

where $\mathfrak{z}^1 := q + q' = T^{-\gamma}\mathfrak{z}$. Then the translation

$$T^{\gamma} : \mathcal{M}_{\mathfrak{z}} \longrightarrow \mathcal{M}_{T^{-\gamma}\mathfrak{z}}$$

is an isomorphism and may be used to define the topology of $\mathcal{M}_{\mathfrak{z}}$ in terms of the topology of $\mathcal{M}_{\mathfrak{z}^1}$, which was defined above. $\mathfrak{z}$ is nothing else than a system of pairs

$$(q_j, n_j) \in \mathbb{C} \times \mathbb{Z}_+, \ j \in \mathbb{Z}$$

with $|\mathrm{Re}\ q_j| \longrightarrow \infty$ as $j \longrightarrow \pm\infty$. Denote by $\underline{\mathfrak{z}}$ the set of all such systems. Then, instead of $\mathcal{M}$ we also write $\mathcal{M}_{\underline{\mathfrak{z}}}$ being the inductive limit of $\mathcal{M}_{\mathfrak{z}}$ over all $\mathfrak{z} \in \underline{\mathfrak{z}}$. Clearly for every $\mathfrak{z} \in \underline{\mathfrak{z}}$ we find some $\gamma > 0$ such that $T^{-\gamma}\mathfrak{z}$ is of the form $\mathfrak{q} + \mathfrak{q}'$ with certain $\mathfrak{q} \in \underline{\mathfrak{p}}$, $\mathfrak{q}' \in \underline{\mathfrak{p}}'$.

5. Definition. Let $h \in \mathcal{M}$, $j \in \mathbb{Z}_+$, and $j \geq \gamma \geq 0$ such that $T^{\gamma}h \in M(C^{\infty}_{\underline{\mathfrak{p}},\underline{\mathfrak{p}}'}) = \tilde{C}^{\infty}_{\underline{\mathfrak{p}},\underline{\mathfrak{p}}'}$. We call an operator of the form $t^{j-\gamma}\omega\, op_M(T^{\gamma}h)\,\omega t^{\gamma}$ a Mellin operator of conormal order $-j$ and h its conormal symbol. The class of all these operators is denoted by $t^{j-\gamma}\,\omega\, op_M(\tilde{C}^{\infty}_{\underline{\mathfrak{p}},\underline{\mathfrak{p}}'})\omega t^{\gamma}$.

For every $\mathfrak{p} \in \underline{\mathfrak{p}}$ we have a continuous operator

$$t^{j-\gamma}\omega\, op_M(T^{\gamma}h)\omega t^{\gamma} : \mathcal{S}_{\mathfrak{p}} \longrightarrow \mathcal{S}_{T^{-j}\mathfrak{p} + T^{-j+\gamma}\mathfrak{q}} \tag{1}$$

The assumption $0 \leq \gamma \leq j$ ensures bounded operators in $L^2(\mathbb{R}_+)$.

The proof of Proposition 3 shows that we might admit more general meromorphic functions as symbols of Mellin operators satisfying (i), and

(ii)' for given c, c' there exist $\eta = \eta(c,c') \in \mathbb{R}$ and a constant $C(c,c')$ such that

$$|h(z)| \leq C(1+|z|)^{\eta} \quad \text{for } c \leq \mathrm{Re}\ z \leq c',\ |\mathrm{Im}\ z| \text{ sufficiently large.}$$

It is not necessary for our purposes to use such Mellin symbols in full generality. Below, in Section 1.4.1., we extend the class $\mathcal{M}_{\underline{\mathfrak{z}}}$ by more special non-strongly decreasing Mellin symbols. It is obvious that the continuity (1) remains valid for the non-strongly decreasing Mellin symbols. That will be used in Section 1.4.2.

6. Lemma. For the L^2-adjoint $op_M[f]^*$ of $op_M[f]$, $f \in C^{\infty}_{\mathfrak{q},\mathfrak{q}'}$, we have

$$op_M[f]^* = op_M[f^*] \,,$$

where $f^*(t) = \overline{f(1/t)}\,\frac{1}{t} \in C^{\infty}_{-T^{-1}\bar{\mathfrak{q}}',\,-T^{-1}\bar{\mathfrak{q}}}$, $\bar{\mathfrak{q}}$ means the complex conjugate of $\mathfrak{q}$.

Proof. For $u,v \in L^2(\mathbb{R}_+)$ we have

$$(op_M[f]u,v) = \int_0^\infty \int_0^\infty f(t/t')\,\tfrac{1}{t'}\,u(t')\overline{v(t)}\,dt'dt$$

$$= \int_0^\infty \int_0^\infty u(t')\overline{\overline{f(t/t')}\,\tfrac{t}{t't}\,v(t)}\,dtdt'$$

$$= (u,\ op_M[f^*]v). \quad \square$$

From

$$op_M[f] = op_M(m) \text{ with } m(z) = (Mf)(z)\ .$$

it follows that

$$op_M(m)^* = op_M(m^*) \text{ with } m^*(z) = (M\bar{f})(1-\bar{z}).$$

The following Lemma shows that the choice of the cut-off function ω in the definition of Mellin operators is not essential modulo Green operators.

7. Lemma. Let $h \in \tilde{C}^\infty_{\mathfrak{q},\mathfrak{q}'}$, and ω_i cut-off functions, $i = 1,2,3,4$. Then we have

(i) $\omega_1\, op_M(h)(1-\omega_2) \in op_G(\mathcal{S}_{\mathfrak{q}} \otimes_\pi C^\infty_{0,-T^{-1}\mathfrak{q}})$,

(ii) $(1-\omega_1)\, op_M(h)\omega_2 \in op_G(C^\infty_{0,\mathfrak{q}'} \otimes_\pi \mathcal{S}_{-T^{-1}\mathfrak{q}'})$,

(iii) $\omega_1\, op_M(h)(\omega_2-\omega_3) \in op_G(\mathcal{S}_{\mathfrak{q}} \otimes_\pi \mathcal{S}_0)$,

(iv) $(\omega_1-\omega_2)\, op_M(h)\omega_3 \in op_G(\mathcal{S}_0 \otimes_\pi \mathcal{S}_{-T^{-1}\mathfrak{q}'})$,

(v) $\omega_1\, op_M(h)\omega_2 - \omega_3\, op_M(h)\omega_4 \in op_G(\mathcal{S}_{\mathfrak{q}} \otimes_\pi \mathcal{S}_{-T^{-1}\mathfrak{q}'})$

Proof. (i) $\Rightarrow$ (ii), (iii) $\Rightarrow$ (iv) follows by taking adjoints. (iii), (iv) $\Rightarrow$ (v), because

$$\omega_1\, op_M(h)\omega_2 - \omega_3\, op_M(h)\omega_4 = \omega_1\, op_M(h)(\omega_2-\omega_4) + (\omega_1-\omega_3)\, op_M(h)\omega_4\ .$$

Now we shall prove (iii); (i) is treated in the same manner.

According to 1.2.2. Proposition 7 the quality of the kernel can be derived from the extension of the operator to $L^2(\mathbb{R}_+)$. Obviously the multiplication by $\omega_2 - \omega_3$ is continuous in $L^2(\mathbb{R}_+)$ and $M((\omega_2-\omega_3)u)(z)$ $\Leftarrow u \in L^2(\mathbb{R}_+)$, is an entire function with finite $L^2(\Gamma_\rho)$ norms for all $\rho \in \mathbb{R}$. Then $h(z)\, M((\omega_2-\omega_3)u)(z)$ is meromorphic and rapidly de - creasing on each Γ_ρ for $|z| \to \infty$. Therefore $\omega_1 \, op_M(h)(\omega_2-\omega_3) \in \mathcal{L}(L^2(\mathbb{R}_+), \mathcal{S}_q)$. In the same way $(\omega_2-\omega_3)\, op_M(h)^* \, \omega_1 \in \mathcal{L}(L^2(\mathbb{R}_+), \mathcal{S}_o)$ may be proved. In view of 1.2.2. Proposition 13 the kernel then belongs to $\mathcal{S}_q \otimes_\pi \mathcal{S}_o = \mathcal{S}_q \otimes_\pi L^2(\mathbb{R}_+) \cap L^2(\mathbb{R}_+) \otimes_\pi \mathcal{S}_o$. □

8. <u>Proposition</u>. Let $h \in \mathcal{M}_{\mathfrak{z}}$, $\gamma, \lambda \geq 0$, $j \in \mathbb{Z}_+$, $0 \leq \gamma \leq \lambda \leq j$, $T^{-\gamma}\mathfrak{z} = T^{-\gamma}q + T^{-\gamma}q'$ with certain $T^{-\gamma}q \in \underline{\mathfrak{P}}$, $T^{-\gamma}q' \in \underline{\mathfrak{P}}'$, and let $T^\lambda h, T^\gamma h$ be holomorphic near $\mathrm{Re}\, z = 1/2$. Then

$$t^{j-\gamma}\omega op_M(T^\gamma h)\omega t^\gamma - t^{j-\lambda}\omega op_M(T^\lambda h)\omega t^\lambda \in op_G(\mathcal{S}_{T^{-j}\nu_{\gamma,\lambda}} \otimes_\pi \mathcal{S}_{p_{\gamma,\lambda}})$$

with

$$\nu_{\gamma\lambda} = \{(q_k', m_k') \in q' : \tfrac{1}{2} + \gamma < \mathrm{Re}\, q_k' < \tfrac{1}{2} + \lambda\},$$

$$p_{\gamma\lambda} = \{(q_k^*, m_k^*) \in T^{-1-2\gamma} q' : \tfrac{1}{2} - \gamma < -\mathrm{Re}\, q_k' < \tfrac{1}{2} - \lambda\}.$$

<u>Proof</u>. Per definitionem we have

$$(t^{j-\gamma}\omega op_M(T^\gamma h)\,\omega t^\gamma u)(t)$$

$$= \omega(t)\frac{1}{2\pi}\int_{-\infty}^{\infty} t^{j-\gamma-1/2-iy} h(1/2+iy+\gamma)\widetilde{\omega u}(1/2+iy+\gamma)\,dy\,,$$

and by the Residue Theorem the difference

$$t^{j-\gamma}\omega op_M(T^\gamma h)\omega t^\gamma u - t^{j-\lambda}\omega op_M(T^\lambda h)\omega t^\lambda u$$

has a finite-dimensional image which is spanned by $t^j \omega^{q_i', k}$, $1/2+\gamma < \mathrm{Re}\, q_i' < 1/2 + \lambda$. The same conclusion yields the assertion for the adjoint, cf. the remarks after 1.2.2, Definition 6. □

Let $h_j \in \mathfrak{M}_{\mathfrak{z}^j}$, $\mathfrak{z}^j \in \underline{\mathfrak{z}}$, $j \in \mathbb{Z}_+$, and choose constants γ_j, $0 \le \gamma_j \le j$, with $\gamma_j \longrightarrow \infty$, $j - \gamma_j \longrightarrow \infty$ as $j \longrightarrow \infty$ and $T^{\gamma_j} h_j \in MC^\infty_{\mathfrak{q}^j, \mathfrak{q}'^j}$ for certain $\mathfrak{q}^j \in \underline{\mathfrak{p}}$, $\mathfrak{q}'^j \in \underline{\mathfrak{p}}'$. Then there is another sequence $m_j \in \mathfrak{M}_{\mathfrak{w}^j}$, $\mathfrak{w}^j \in \underline{\mathfrak{z}}$, $j \in \mathbb{Z}_+$, $\delta_j := j - \gamma_j$, such that $T^{\delta_j} m_j \in MC^\infty_{\mathfrak{r}^j, \mathfrak{r}'^j}$ for certain $\mathfrak{r}^j \in \underline{\mathfrak{p}}$, $\mathfrak{r}'^j \in \underline{\mathfrak{p}}'$, namely $m_j(z) := T^{\delta_j}(T^{\gamma_j} h_j)^*(z)$. Thus

$$(t^{j-\gamma_j} \omega_j \, op_M(T^{\gamma_j} h_j) \, \omega_j t^{\gamma_j})^* = t^{j-\delta_j} \omega_j \, op_M(T^{\delta_j} m_j) \omega_j \, t^{\delta_j},$$

$\omega_j(t) := \omega(c_j t)$, $c_j > 0$ some constant. Let $\mathfrak{p} \in \underline{\mathfrak{p}}$ be arbitrary with $T^{-j}\mathfrak{p} \subseteq \mathfrak{p}$ and $\mathfrak{q}, \mathfrak{r} \in \underline{\mathfrak{p}}$ be such that $T^{-j+\gamma_j} \mathfrak{q}^j \subseteq \mathfrak{q}$, $T^{-j+\delta_j} \mathfrak{r}^j \subseteq \mathfrak{r}$ for all $j \in \mathbb{Z}_+$.

9. Definition. Denote by $\mathcal{L}_M$ the set of all operators

$$\mathcal{M} = \sum_{j=0}^{\infty} t^{j-\gamma_j} \omega_j \; op_M(T^{\gamma_j} h_j) \, \omega_j t^{\gamma_j},$$

where the infinite sum converges in $\mathcal{L}(\mathcal{S}_{\mathfrak{p}}, \mathcal{S}_{\mathfrak{p}+\mathfrak{q}})$ for a suitable choice of the constants c_j, as well as

$$\mathcal{M}^* = \sum_{j=0}^{\infty} t^{j-\delta_j} \omega_j \; op_M(T^{\delta_j} \; m_j) \, \omega_j t^{\delta_j}$$

in $\mathcal{L}(\mathcal{S}_{\mathfrak{p}}, \mathcal{S}_{\mathfrak{p}+\mathfrak{r}})$, $\mathfrak{p} \in \underline{\mathfrak{p}}$ arbitrary as mentioned. Operators in $\mathcal{L}_M$ are called Mellin operators. The function

$$h_j(z) = \sigma_M^{-j}(\mathcal{M})(z)$$

is called conormal symbol of $\mathcal{M}$ of order $-j$.

Now we shall show that the sequence of conormal symbols of a sum of Mellin and Green operators is uniquely determined.

For any given $\eta_0 \in \mathbb{R}$ we define a family of unitary operators in $L^2(\mathbb{R}_+)$

$$S'_\lambda u(t) = \lambda^{-1/2} t^{(1-\lambda)/2\lambda - i\eta_0} u(t^{1/\lambda}), \quad \lambda \in \mathbb{R}_+.$$

Then

$$M(S'_\lambda u)(1/2+i\eta) = \lambda^{1/2} Mu(1/2+i\lambda(\eta-\eta_0)), \quad \eta \in \mathbb{R},$$

and therefore $S'_\lambda u$ weakly tends to zero for $\lambda \longrightarrow \infty$.

10. Proposition. Let

$$\mathfrak{a} = \omega \, op_M(h) \omega + \mathcal{K},$$

$h \in \tilde{C}^\infty_{\mathfrak{q}, \mathfrak{q}'}$ and $\mathcal{K}$ compact in $L^2(\mathbb{R}_+)$. Then, for given η_0 and $u \in C_0^\infty(\mathbb{R}_+)$, supp $u \subset [0,1)$, we have

$$\lim_{\lambda \to \infty} \| S_\lambda'^{-1} \mathfrak{a} S_\lambda' u - h(1/2+i\eta_0)u \|_{L^2(\mathbb{R}_+)} = 0.$$

<u>Proof.</u> Since $\mathcal{k}$ is compact, it follows

$$\lim_{\lambda \to \infty} \| \mathcal{k} S_\lambda' u \|_{L^2(\mathbb{R}_+)} = 0.$$

Moreover,

$$\| S_\lambda'^{-1} \operatorname{op}_M(h) S_\lambda' u - h(1/2+i\eta_0)u \|^2_{L^2(\mathbb{R}_+)}$$

$$= \| \operatorname{op}_M(h) S_\lambda' u - h(1/2+i\eta_0) S_\lambda' u \|^2_{L^2(\mathbb{R}_+)}$$

$$= \frac{1}{2\pi} \| h(z) M(S_\lambda' u)(z) - h(1/2+i\eta_0) M(S_\lambda' u)(z) \|^2_{L^2(\operatorname{Re} z = 1/2)}$$

$$= \frac{1}{2\pi} \int_{-\infty}^{\infty} | h(1/2+i\eta_0+i\lambda^{-1}\eta) - h(1/2+i\eta_0)|^2 |\tilde{u}(1/2+i\eta)|^2 d\eta$$

Lebesgue's theorem on bounded convergence implies

$$\lim_{\lambda \to \infty} \| S_\lambda'^{-1} \operatorname{op}_M(h) S_\lambda' u - h(1/2+i\eta_0)u \|^2_{L^2} = 0.$$

For an arbitrary cut-off function ω we have

$$\| (1-\omega) S_\lambda' u \|^2_{L^2} \leq \lambda^{-1} \int_{\varepsilon}^{1} | t^{(1-\lambda)/2\lambda} u(t^{1/\lambda}) |^2 dt,$$

where $1 - \omega = 0$ for $t \leq \varepsilon < 1$. The substituion $t' = t^{1/\lambda}$ implies

$$\| (1-\omega) S_\lambda' u \|^2_{L^2} = \int_{\varepsilon^{1/\lambda}}^{1} | u(t) |^2 dt \longrightarrow 0 \quad \text{for } \lambda \longrightarrow \infty .$$

Hence we get the assertion. $\square$

11. <u>Theorem.</u> Let $\mathfrak{m} \in \mathcal{L}_M$, $\mathfrak{b} \in \mathcal{L}_G$. Then the sequence of conormal symbols $\sigma_M^{-j}(\mathfrak{m})$ of $\mathfrak{m}$ can be recovered from $\mathfrak{m}+\mathfrak{b}$.

<u>Proof.</u> First we note that Green operators and Mellin operators of the form $t^{j-\gamma_j} \omega \operatorname{op}_M(T^{\gamma_j} h_j) \omega t^{\gamma_j}$ are compact in $L^2(\mathbb{R}_+)$.

In fact, for $j - \gamma_j > 0$ and any given $\varepsilon > 0$ we find a cut-off function $\tilde{\omega}$ such that

$$\| t^{j-\gamma_j} \omega_1 \operatorname{op}_M(T^{\gamma_j} h_j) \omega t^{\gamma_j} \|_{\mathcal{L}(L^2(\mathbb{R}_+))} < \varepsilon .$$

The decomposition $t^{j-\gamma_j}\omega = t^{j-\gamma_j}\tilde{\omega} + t^{j-\gamma_j}(\omega - \tilde{\omega})$ shows that $t^{j-\gamma_j}\omega$ $op_M(T^{\gamma_j}h_j)\,\omega\, t^{\gamma_j}$ is a sum of a compact operator (cf. Lemma 7) and an operator with norm less than $\delta > 0$ arbitrary. Hence it is compact. If $j-\gamma_j = 0$ but $\gamma_j > 0$, the same conclusion applies to $\omega t^{\gamma_j} = \tilde{\omega} t^{\gamma_j} + (\omega - \tilde{\omega})t^{\gamma_j}$. For a given operator $\mathcal{M} + \mathcal{G} \in \mathcal{L}_M + \mathcal{L}_G$ the principal Mellin symbol h_o is uniquely determined according to Proposition 10. Consider now

$$\mathcal{M}_1 + \mathcal{G}_1 := \mathcal{M} + \mathcal{G} - \omega op_M(h_o)\omega = \sum_{j=1}^{\infty} t^{j-\gamma_j}\omega_j op_M(T^{\gamma_j}h_j)\omega_j\, t^{\gamma_j} + \mathcal{G}_1$$

$$= t^{1-\gamma_1}\Big(\sum_{k=0}^{\infty} t^{k-\gamma_k'}\omega_{k+1}\, op_M(T^{\gamma_k'+\gamma_1}h_{k+1})t^{\gamma_k'}\Big)t^{\gamma_1} + \mathcal{G}_2,$$

$\mathcal{G}_2$ being defined by the last equality. Let $v_1, \ldots, v_N$ be a basis of the finite-dimensional space $\mathcal{K}_{1-\gamma_1}(\mathcal{S}_{p+q})$ (for the notation cf. 1.2.1) orthonormal with respect to the $L^2(\mathbb{R}_+)$ scalar product. For the Green operator $\mathcal{G}_3$ with finite-dimensional range $\mathcal{G}_3 u := \sum_{j=1}^{N}((\mathcal{M}_1 + \mathcal{G}_1)u, v_j)v_j$ we have $\mathcal{G}_2 - \mathcal{G}_3 = t^{1-\gamma_1}\mathcal{G}_2'$, $\mathcal{G}_2' \in \mathcal{L}_G$. The same conclusion for the adjoints yields a Green operator $\mathcal{G}_4$ with finite-dimensional range, $\mathcal{G}_4$ being uniquely determined by $\mathcal{M}_1 + \mathcal{G}_1$, such that $\mathcal{G}_2 - \mathcal{G}_3 - \mathcal{G}_4 \in t^{1-\gamma_1}\mathcal{L}_G t^{\gamma_1}$. Then $\mathcal{M}_1 + \mathcal{G}_1 - \mathcal{G}_3 - \mathcal{G}_4 \in t^{1-\gamma_1}(\mathcal{L}_M + \mathcal{L}_G)t^{\gamma_1}$, and h_1 can be recovered as in the first step. Repeating this conclusion we may recover h_j for each j. □

By $\mathcal{L}_M^{-j}$ we denote the subspace of $\mathcal{L}_M$ of Mellin operators with vanishing conormal symbols of order $> -j$. Let $j > 0$, then the sequence

$$0 \longrightarrow \mathcal{L}_M^{-j-1} + \mathcal{L}_G \xrightarrow{\ i\ } \mathcal{L}_M^{-j} + \mathcal{L}_G \xrightarrow{\ \sigma_M^{-j}\ } \mathfrak{M} \longrightarrow 0$$

is exact, where i denotes the natural embedding, and $\sigma_M^{-j}(\mathcal{M} + \mathcal{G}) = \sigma_M^{-j}(\mathcal{M})$. For $j = 0$ we have the same with $\mathfrak{M}_o$ instead of $\mathfrak{M}$, which is the subspace of those $h \in \mathfrak{M}$ which have no poles on $\mathrm{Re}\, z = 1/2$.

12. <u>Proposition.</u> Let $a_j : \mathcal{S}_p \longrightarrow C^{\infty}_{p+q,q'}$, $j \in \mathbb{Z}_+$, be a sequence of operators satisfying $\pi_\alpha'(a_j u) \leq d_j \pi_{\beta(\alpha)}(u)$, with $\beta(\alpha)$ independent of j ((π_α') denotes the system of semi-norms belonging to $C^{\infty}_{p+q,q'}$). Then there are cut-off functions ω_j such that $\sum_{j=0}^{\infty} t^j \omega_j a_j$ converges in $\mathcal{L}(\mathcal{S}_p, \mathcal{S}_{p+q})$ for every $p \in \underline{p}$.

Proof. We set $\omega_j(t) = \omega(c_j t)$, ω a fixed cut-off function, and the constants c_j will be chosen in such a way that $\sum_{j=0}^{\infty} t^j \omega_j a_j$ converges. In view of Lemma 1 we have

$$\pi_\alpha(t^j\omega_j a_j u) \le c \sum_{k=0}^{k(\alpha)} \pi_{\alpha_k}(t^j\omega_j)\pi'_{\alpha'_k}(a_j u)$$

with a constant c independent of j. The assumption (3) yields

$$\pi_\alpha(t^j\omega_j a_j u) \le d_j \sum_{k=0}^{k(\alpha)} \pi_{\alpha_{k'}}(t^j\omega_j)\,\pi_{B(\alpha'_k)}(u).$$

As in the proof of 1.2.1, Theorem 14, we can choose c_j such that

$$\sum_{j=0}^{\infty} d_j \Big(\sum_{k=0}^{k(\alpha)} \pi_{\alpha_k}(t^j\omega_j)\Big) < \infty$$

for all α. □

The above results will be applied to Mellin operators.

13. Proposition. Let $h_j \in \mathfrak{M}$ be an arbitrary sequence of Mellin symbols. Then there exists an operator $\mathcal{M} \in \mathcal{L}_M + \mathcal{L}_G$ with the given sequence of conormal symbols and $\mathcal{M}$ is uniquely determined modulo $\mathcal{L}_G$.

Proof. The estimate in Proposition 3 shows that the assumption of Proposition 12 is valid, and we find constants c_j such that $\sum_{j=0}^{\infty} t^{j-\gamma_j}\omega_j \, op_M(T^{\gamma_j}h_j)\,\omega_j t^{\gamma_j}$ converges as well as the series of the adjoints. The uniqueness modulo $\mathcal{L}_G$ follows from Theorem 11. □

14. Corollary. The sequence

$$0 \longrightarrow \mathcal{L}_G \xrightarrow{\ i\ } \mathcal{L}_M + \mathcal{L}_G \xrightarrow{\bigoplus_{j\in\mathbb{Z}_+} \sigma_M^{-j}} \mathfrak{M}_0 \times \mathop{\times}_{j=1}^{\infty} \mathfrak{M} \longrightarrow 0$$

is exact.

Proof. The exactness in the middle follows from

$$\bigcap_{j=0}^{\infty} \mathcal{L}_M^{-j} + \mathcal{L}_G = \mathcal{L}_G.$$

The surjectivity of the symbol map at the end was proved in Proposition 13. □

1.3.2. Compositions and Fredholm property

First discuss the composition of Mellin and Green operators.

1. Proposition. Let $\mathfrak{m} \in \mathcal{L}_M$, $\mathfrak{b} \in \mathcal{L}_G$. Then $\mathfrak{m}\mathfrak{b}$ and $\mathfrak{b}\mathfrak{m}$ belong to $\mathcal{L}_G$, i.e. $\mathcal{L}_G$ is a two-sided ideal in $\mathcal{L}_M + \mathcal{L}_G$.

Proof. Let $\mathfrak{b} = op_G(b)$,

$$b(t,s) = \sum_{j=0}^{\infty} f_j(t)g_j(s) \in \mathcal{S}_q \hat{\otimes}_\pi \mathcal{S}_{\bar{q}} .$$

Then

$$\mathfrak{m}\mathfrak{b} = op_G\Big(\sum_{j=0}^{\infty} \mathfrak{m}(f_j)(t)g_j(s)\Big),$$

and the sum remains convergent in view of the continuity of $\mathfrak{m}$ in $\mathcal{L}(\mathcal{S}_q, \mathcal{S}_{q+\varepsilon})$. The composition $\mathfrak{b}\mathfrak{m}$ is treated by taking adjoints. □

Now we consider the composition of two Mellin operators. Let $h_1, h_2 \in \mathfrak{M}$, $T^\gamma h_1$ and $T^\lambda h_2$ be without poles at $\mathrm{Re}\, z = 1/2$ and $j, k \in \mathbb{Z}_+$. Applying 1.3.1. Lemma 7 and Proposition 8 we obtain

$$\begin{aligned} t^{j-\gamma}\omega op_M(T^\gamma h_1)\omega t^\gamma t^{k-\lambda}\omega op_M(T^\lambda h_2)\omega t^\lambda \\ = t^{j+k-\eta}\omega op_M(T^{-k+\eta}h_1)op_M(T^\eta h_2)\omega t^\eta \qquad (1) \\ + t^{j+k-\eta}\omega\, op_M(T^{-k+\eta}h_1)(1-\omega^2)op_M(T^\eta h_2)\omega t^\eta \\ + op_G(b). \end{aligned}$$

with a suitable Green operator $op_G(b)$ and η such that $T^{-k+\eta}h_1$ and $T^\eta h_2$ have no poles on $\mathrm{Re}\, z = 1/2$.

2. Lemma. Let $h_1 \in \tilde{C}^\infty_{q_1, q_1'}$, $h_2 \in \tilde{C}^\infty_{q_2, q_2'}$. Then

$$op_M(h_1)op_M(h_2) = op_M(h_1h_2) \qquad (2)$$

and $h_1h_2 \in \tilde{C}^\infty_{q_1+q_2, q_1'+q_2'}$.

Proof. From the characterization of $\tilde{C}^\infty_{q,q'}$ it follows that h_1h_2 is again a Mellin symbol. (2) is an immediate consequence of the definition of op_M. □

Hence the first item in (1) is a Mellin operator of order $-j-k$. For the second one we get from 1.3.1, Lemma 7(i), (ii) that it is a Green operator, too. It follows that

$$t^{j-\gamma}\omega\, op_M(T^{\gamma}h_1)\,\omega\, t^{\gamma}\, t^{k-\lambda}\omega\, op_M(T^{\lambda}h_2)\,\omega\, t^{\lambda}$$
$$= t^{j+k-\eta}\omega\, op_M(T^{\eta}((T^{-k}h_1)h_2))\,\omega\, t^{\eta} \mod \mathcal{L}_G \tag{3}$$

and therefore,

$$\sigma_M^{-j-k}(t^{j-\gamma}\omega\, op_M(T^{\gamma}h_1)\omega t^{\gamma+k-\lambda}\omega\, op_M(T^{\lambda}h_2)\omega t^{\lambda}) = (T^{-k}h_1)h_2 \tag{4}$$

3. Proposition. Let $\mathcal{m}_i+\mathcal{b}_i \in \mathcal{L}_M + \mathcal{L}_G$, $i = 1,2$, with the sequence of conormal symbols $\sigma_M^{-j}(\mathcal{m}_i)$. Then $(\mathcal{m}_1+\mathcal{b}_1)(\mathcal{m}_2+\mathcal{b}_2) \in \mathcal{L}_M + \mathcal{L}_G$, and the sequence of conormal symbols is given by

$$\sigma_M^{-j}(\mathcal{m}_1\mathcal{m}_2) = \sum_{k+l=j}(T^{-l}\sigma_M^{-k}(\mathcal{m}_1))\,\sigma_M^{-l}(\mathcal{m}_2) \tag{5}$$

Proof. We have

$$\mathcal{m}_i + \mathcal{b}_i = \sum_{j=0}^{\infty} t^{j-\gamma_j}\omega_j\, op_M(T^{\gamma_j}\sigma_M^{-j}(\mathcal{m}_i))\,\omega_j t^{\gamma_j} + op_G(b_i).$$

From (3) it follows that $\sigma_M^{-j}(\mathcal{m}_1\mathcal{m}_2)$ only depends on $\sigma_M^{-k}(\mathcal{m}_1)$, $\sigma_M^{-k}(\mathcal{m}_2)$, $0 \le k \le j$, and (5) is valid in view of (4). Then the assertion follows from Proposition 1. □

In particular, we get that $\mathcal{L}_M^{-j} + \mathcal{L}_G$, $j \in \mathbb{Z}_+$, are subalgebras of $\mathcal{L}_M + \mathcal{L}_G$ and also two-sided ideals. Moreover,

$$(\mathcal{L}_M^{-j} + \mathcal{L}_G)(\mathcal{L}_M^{-k} + \mathcal{L}_G) \subset \mathcal{L}_M^{-j-k} + \mathcal{L}_G,$$

i.e. $\mathcal{L}_M + \mathcal{L}_G$ is an algebra filtered by the conormal orders. Section 1.2.2 and 1.3.1, Lemma 6, show that $\mathcal{L}_M + \mathcal{L}_G$ and $\mathcal{L}_M^{-j} + \mathcal{L}_G$ are closed under taking L^2 adjoints.

Now let us discuss the Fredholm property of operators of the form

$$1 + \mathcal{m} + \mathcal{b} \in 1 + \mathcal{L}_M + \mathcal{L}_G.$$

In virtue of the specific features of the Mellin transform we may expect that spaces of square integrable functions

$$L^{2,\eta}(\mathbb{R}_+) := \left\{t^{-\eta}u : u \in L^2(\mathbb{R}_+)\right\}, \qquad \eta \in \mathbb{R},$$

are particularly adapted to this situation.

Let $h \in \mathfrak{M}$ and define a linear continuous operator

$$t^{\eta} \, op_M(T^{-\eta} h) t^{-\eta} \; : L^{2,\eta} \longrightarrow L^{2,\eta}$$

for each $\eta \in \mathbb{R}$ such that $T^{-\eta} h$ has no poles on Re z = 1/2. By the diagram

$$\begin{array}{ccc} L^{2,\eta} & \xrightarrow{\mathcal{L}^{\eta}_M} & L^{2,\eta} \\ \downarrow t^{-\eta} & & \downarrow t^{-\eta} \\ L^2 & \xrightarrow{\mathcal{L}_M} & L^2 \end{array}$$

we get a class of Mellin operators $\mathcal{L}^{\eta}_M$. In the same way we define $\mathcal{L}^{\eta}_G$. Obviously, $\mathcal{L}^{\eta}_M + \mathcal{L}^{\eta}_G$ is an algebra.

Consider a system

$$\mathfrak{a} = 1 + \sum_{j=0}^{\infty} t^{j-\gamma_j} \omega op_M(T^{\gamma_j} h_j) \omega t^{\gamma_j} + t^{\gamma} \; op_G(b) t^{-\gamma} \; : L^{2,\gamma}(\mathbb{R}_+, \mathbb{C}^p) \longrightarrow$$

$$L^{2,\gamma} \; (\mathbb{R}_+, \mathbb{C}^p), \; j+\gamma \geq \gamma_j \geq -\gamma, \; b \in \mathcal{S}_{\gamma} \otimes_{\pi} \mathcal{S}_{\gamma} \otimes \mathbb{C}^{p^2}.$$

4. <u>Theorem</u>. The operator $\mathfrak{a} \in 1 + \mathcal{L}^{\gamma}_M + \mathcal{L}^{\gamma}_G$

$$\mathfrak{a} \; : L^{2,\gamma} \; (\mathbb{R}_+, \mathbb{C}^k) \longrightarrow L^{2,\gamma} \; (\mathbb{R}_+, \mathbb{C}^k), \quad \gamma \in \mathbb{R},$$

is Fredholm iff

$$\det(1 + h_0(z)) \neq 0 \text{ on Re } z = 1/2 - \gamma.$$

Then there exists a parametrix $\mathfrak{a}^{(-1)} \in 1 + \mathcal{L}^{\gamma}_M + \mathcal{L}^{\gamma}_G$ such that

$$1 - \mathfrak{a}\mathfrak{a}^{(-1)}, \; 1 - \mathfrak{a}^{(-1)}\mathfrak{a} \in \mathcal{L}^{\gamma}_G.$$

There are finite-dimensional spaces $\mathcal{N}_1 \subset \mathcal{S}_{T^{\gamma}\varphi}$, $\mathcal{N}_2 \subset \mathcal{S}_{T^{\gamma}\varphi}$ with

$$\ker_{L^{2,\gamma}} \mathfrak{a} = \mathcal{N}_1, \; \operatorname{im}_{L^{2,\gamma}} \mathfrak{a} \oplus \mathcal{N}_2 = L^{2,\gamma} \; (\mathbb{R}_+, \mathbb{C}^k).$$

<u>Proof</u>. The assertions are equivalent to the corresponding statements about

$$\mathfrak{a}^{\gamma} = t^{-\gamma} \mathfrak{a} \, t^{\gamma} \; : L^2(\mathbb{R}_+, \mathbb{C}^p) \longrightarrow L^2(\mathbb{R}_+, \mathbb{C}^p).$$

The equivalence of the Fredholm property to $\det(1+h_0(z-\gamma)) \neq 0$ on Re z = 1/2 is proved in [R 2]. For constructing the parametrix we find the sequence of conormal symbols h_k' by successively solving

$$1 = (1+h_0)\, h_0', \quad 0 = \sum_{k+l=j} (T^{-l}\sigma_M^{-k}(\alpha^\gamma))\, h_l', \quad j > 0,$$

which is a similar procedure as in the parametrix construction for usual elliptic pseudo-differential operators.

We obtain

$$h_0' = (1 + h_0)^{-1}.$$

This is a meromorphic function with poles exactly at the points $\{z \in \mathbb{C} : \det(1+h_0(z)) = 0\}$. Since $h_0(z)$ uniformly decreases in each strip $c \leq \operatorname{Re} z \leq c'$, there is only a finite number of zeros of $\det(1+h_0(z))$ in each strip. Moreover, we have

$$h_0'(z) = \sum_{j=0}^{\infty} (-1)^j h_0(z)^j \quad \text{for } c \leq \operatorname{Re} z \leq c',\ |\operatorname{Im} z| \text{ large}$$

and therefore, $h_0' - 1 \in \mathfrak{M}$. From

$$h_j'(z) = -(1 + h_0(z-l))^{-1} \sum_{\substack{k+l=j \\ l<j}} (T^{-l}\sigma_M^{-k}(\alpha^\gamma)(z))\, h_l'(z)$$

we get $h_j' \in \mathfrak{M}$.

The ellipticity of α implies that there is a well-defined Mellin operator $\alpha^{(-1)}$ in $\mathcal{L}_M^\gamma$ with the sequence h_l' as the sequence of conormal symbols. It follows that

$$1 - \alpha\alpha^{(-1)},\ 1 - \alpha^{(-1)}\alpha \in \mathcal{L}_G^\gamma.$$

For $u \in \ker_{L^{2,\gamma}} \alpha$ we have

$$u = (1 - \alpha^{(-1)}\alpha)\, u \subset \operatorname{im}(1 - \alpha^{(1)}\alpha) \subset \mathcal{S}_T^{\gamma_\varphi}$$

The cokernel $\mathcal{N}_2$ can be represented as the kernel of the adjoint operator. □

Theorem 4 yields an explanation of the nature of the conditions of the Fredholm property in spaces with weights. For a given principal conormal symbol h_0 there is a discrete set Ξ of exceptional values $\lambda \in \mathbb{R}$, such that $\operatorname{Re} z = 1/2 - \lambda$ contains poles of $1 + h_0$. The set Ξ precisely contains those weights, where the Mellin operator is not bounded. The construction in the proof gives a sequence of conormal symbols, where no special condition is used. The corresponding Mellin operator is the desired parametrix in $L^{2,\lambda}$ if its action on $L^{2,\lambda}$ can be defined, i.e. if no zeros of det $(1+h_0)$ are on Re $z = 1/2 - \lambda$.

We have a well-known index formula, cf. e.g. [R2],

$$\text{ind } \mathfrak{a} = (2\pi)^{-1} \det (1+h_0(1/2+\lambda+iy)) \Big|_{y=-\infty}^{\infty} .$$

Since $\det(1+h_0(z))$ is a meromorphic function with regular point at ∞, the argument principle (cf. [L1], p. 90) gives

5. Proposition. Let $\mathfrak{a}$, $\mathfrak{a}'$ have the same principal conormal symbol h_0 and $\mathfrak{a} \in 1 + \mathcal{L}^{\gamma}_{M} + \mathcal{L}^{\gamma}_{G}$, $\mathfrak{a}' \in 1 + \mathcal{L}^{\eta}_{M} + \mathcal{L}^{\eta}_{G}$, $\gamma < \eta$, be elliptic. Then

$$\text{ind } \mathfrak{a}' - \text{ind } \mathfrak{a} = N - P,$$

where N is the number of zeros, and P is the number of poles of $\det(1+h_0(z))$ in $1/2 - \gamma < \text{Re } z < 1/2 - \eta$, counted with their multiplicities.

6. Corollary. Let $\mathfrak{a} \in 1 + \mathcal{L}^{\gamma}_{M} + \mathcal{L}^{\gamma}_{G}$,

$$\mathfrak{a} : L^{2,\gamma}(\mathbb{R}_+, \mathbb{C}^p) \longrightarrow L^{2,\gamma}(\mathbb{R}_+, \mathbb{C}^p),$$

be bijective. Then $\mathfrak{a}^{-1} \in 1 + \mathcal{L}^{\gamma}_{M} + \mathcal{L}^{\gamma}_{G}$.

Proof. According to Theorem 4 we find a parametrix $\mathfrak{a}^{(-1)} \in 1 + \mathcal{L}^{\gamma}_{M} + \mathcal{L}^{\gamma}_{G}$ and

$$\text{ind } \mathfrak{a}^{(-1)} = - \text{ind } \mathfrak{a} = 0.$$

Then we can find a Green operator $\mathfrak{b}' \in \mathcal{L}^{\gamma}_{G}$ such that $\mathfrak{a}^{(-1)} + \mathfrak{b}'$ is an isomorphism and

$$(\mathfrak{a}^{(-1)} + \mathfrak{b}')\mathfrak{a} = 1 + \mathfrak{b}'' \in 1 + \mathcal{L}^{\gamma}_{G}$$

has this property, too. By 1.2.2, Theorem 8, we find $(1+\mathfrak{b}'')^{-1} \in 1+\mathcal{L}^{\gamma}_{G}$ and

$$\mathfrak{a}^{-1} = (1+\mathfrak{b}'')^{-1} (\mathfrak{a}^{(-1)} + \mathfrak{b}') \in 1 + \mathcal{L}^{\gamma}_{M} + \mathcal{L}^{\gamma}_{G}. \ \square$$

1.4. Pseudo-differential operators

1.4.1. Mellin expansion and conormal symbols

This and the following sections consider the action of pseudo-differential operators in spaces of functions with discrete conormal singularity on the half axis. The results have consequences for actions in the half space. For this reason we admit the symbols to be functions of $\xi = (\xi', \tau) \in \mathbb{R}^n \setminus 0$, $\xi' \in \mathbb{R}^{n-1}$, $\tau \in \mathbb{R}^1$. In the final higher dimensional calculus we also admit symbols depending on x, cf. Chapter 2.

Denote by $S^{(0)}(\mathbb{R}^n \setminus 0)$ the space of all functions $a(\xi) \in C^\infty(\mathbb{R}^n \setminus 0)$ that are positively homogeneous of order zero, i.e.

$$a(\lambda\xi) = a(\xi) \quad \text{for } \lambda > 0, \xi \neq 0. \tag{1}$$

Let, as above, r^+ be the restriction of functions to $\mathbb{R}_+ = \{t \in \mathbb{R} : t > 0\}$ and e^+ be the operator that extends functions on $\mathbb{R}_+$ by zero to $\mathbb{R}$. Thus, in particular,

$$r^+ : L^2(\mathbb{R}) \longrightarrow L^2(\mathbb{R}_+), \quad e^+ : L^2(\mathbb{R}_+) \longrightarrow L^2(\mathbb{R}).$$

Set

$$op(a)(\xi')v(t) := (2\pi)^{-1} \int e^{it\tau} a(\xi', \tau)\hat{v}(\tau)d\tau,$$

the hat indicates the Fourier transform, $v \in L^2(\mathbb{R})$, and define

$$op_\psi(a)(\xi')u(t) := r^+ op(a)(\xi')e^+u(t),$$

$u \in L^2(\mathbb{R}_+)$. Then $a(\xi) \in S^{(0)}(\mathbb{R}^n \setminus 0)$ induces a family of symbols in $S^0_{cl}(\mathbb{R})$ (being the space of classical symbols on $\mathbb{R}$) and a family of operators

$$op_\psi(a)(\xi') : L^2(\mathbb{R}_+) \longrightarrow L^2(\mathbb{R}_+), \quad \xi' \neq 0. \tag{2}$$

$op(a)$ and $op_\psi(a)$ may also be defined for $a(\xi', \tau)$ which is not smooth in (ξ', τ), but only a multiplier in $L^2(\mathbb{R})$ with respect to τ. This will tacitly be used below.

1. __Lemma.__ Let $a \in S^0_{cl}(\mathbb{R})$ and ω_1, ω_2 be cut-off functions with $\omega_1 = \omega_2\omega_1$. Then

$$(1-\omega_2)op_\psi(a)\omega_1 = op_G(b_1), \quad \omega_1 op_\psi(a)(1-\omega_2) = op_G(b_2),$$

where $b_1 \in \mathcal{S}_0 \otimes_\pi \mathcal{S}_{\bar{0}}$, $b_2 \in \mathcal{S}_{\bar{0}} \otimes_\pi \mathcal{S}_0$.

Proof. The kernel of $op_\psi(a)$ is given by

$$e^{i(t-t')\tau} a(\tau)d\tau, \qquad t,t' \in \mathbb{R}_+ .$$

One easily checks that

$$P_\alpha(t)P'_\beta(t')(1-\omega_2(t)) \int e^{i(t-t')\tau} a(\tau)d\tau\, \omega_1(t') \in L^2(\mathbb{R}_+ \times \mathbb{R}_+),$$

where P_α, P'_β are the corresponding operators for $\mathcal{S}_{\varphi_0}$ and $\mathcal{S}_0$, defined in Section 1.2.2. Then the assertion follows from 1.2.2, Proposition 5. The second assertion follows in a similar way. □

The proof of Lemma 1 shows that the mappings

$$S^0_{cl} \ni a \longrightarrow b_1 \in \mathcal{S}_0 \otimes_\pi \mathcal{S}_{\varphi_0}, \quad S^0_{cl} \ni a \longrightarrow b_2 \in \mathcal{S}_{\varphi_0} \otimes_\pi \mathcal{S}_0$$

are continuous. For $a \in S^{(0)}(\mathbb{R}^n \setminus 0)$ the family $a(\xi',.) \in S^0_{cl}(\mathbb{R})$ induces smooth families $op_G(b_1)(\xi')$ and $op_G(b_2)(\xi')$.

2. Lemma. Let $a_{-N}(\tau) \in L^\infty(\mathbb{R})$ and $c_N := \| (1+|\tau|)^N a_{-N}(\tau) \|_{L^\infty} < \infty$. Then there exists a $k = k(N) \in \mathbb{Z}_+$ and for any cut-off function ω some $b(a_{-N}) \in \mathcal{S}^{(k)}_{\varphi_0} \otimes_\pi \mathcal{S}^{(k)}_{\varphi_0}$ such that

$$\omega\, op_\psi(a_{-N})\omega = op_G(b(a_{-N})).$$

Here $k(N) \longrightarrow \infty$ as $N \longrightarrow \infty$.

Proof. The kernel $\omega(t) \int e^{i(t-s)\tau} a_{-N}(\tau)d\tau\, \omega(s)$ has N-2 continuous derivatives up to the boundary in $\overline{\mathbb{R}}_+ \times \overline{\mathbb{R}}_+$. Then we find an integer $k = k(N)$ as stated in the assertion. □

Consider the Taylor expansion of $a(\xi',\tau)$ for $\tau \longrightarrow \pm\infty$, $\xi' \neq 0$. The homogeneity of $a(\xi)$ implies

$$a(\xi',\tau) = a(\xi'/|\tau| , \pm 1),$$

and for $\tau \longrightarrow \pm\infty$ we obtain

$$a(\xi',\tau) = \sum_{k=0}^{N-1} \frac{1}{k!} \frac{\partial^k}{\partial\rho^k} a(\rho\xi', \pm 1)\big|_{\rho=0} |\tau|^{-k}$$

$$+ \frac{1}{N!} \int_0^{|\tau|^{-1}} \frac{\partial^N}{\partial\gamma^N} a(\gamma\xi', \pm 1)(|\tau|^{-1} - \gamma)^N d\gamma.$$

Set

$$a_k^{\pm}(\xi') = \frac{1}{k!} (\pm i)^k \frac{\partial^k}{\partial \rho^k} a(\rho \xi', \pm 1)\Big|_{\rho=0} \quad . \tag{3}$$

$a_k^{\pm}(\xi')$ is a homogeneous polynomial in ξ' of order k. In particular, $a_0^{\pm}$ is independent of ξ' and

$$a_0^{\pm} = \lim_{\tau \to \pm\infty} a(\xi', \tau) = a(0, \pm 1). \tag{4}$$

Let $\theta^{\pm}(\tau)$ be the characteristic function of $\mathbb{R}_+$ and $\mathbb{R}_-$, respectively. Introduce the functions

$$g^-(z) = (1-e^{2\pi iz})^{-1}, \; g^+(z) = 1-g^-(z) \tag{5}$$

$z \in \mathbb{C}$. It is well known (cf. [R2], p.56) that

$$op_{\psi}(\theta^{\pm}) = op_M(g^{\pm}) \tag{6}$$

with the generalized Mellin symbols $g^{\pm}(z)$. Operators of the form

$$u \longrightarrow M^{-1}(h(z)Mu(z)) := op_M(h)u \; ,$$

where $h(z)\big|_{\mathrm{Re}\, z = 1/2}$ is a multiplier in $L^2(\mathrm{Re}\, z = 1/2)$, are also considered as Mellin operators, although they do not belong to the algebra $\mathcal{L}_M$. Of course, the Mellin symbols that we have in mind are meromorphic. $g^{\pm}(z)$ are meromorphic functions with simple poles at $z_j = j$, $j \in \mathbb{Z}$.

Set

$$1_+^{j}(\delta, \tau) = (\delta + i\tau)^j, \; j \in \mathbb{Z},$$

$\delta > 0$, fixed.

3. <u>Proposition</u>. There exists a sequence of polynomials $A_k^{\pm}$ in (ξ', δ), $k \in \mathbb{Z}_+$, of the form

$$A_0^{\pm} = a_0^{\pm} \; ,$$

$$A_k^{\pm}(\xi', \delta) = \sum_{j=1}^{k} \rho_{kj} \, a_j^{\pm}(\xi')\delta^{k-j}, \; k \geq 1,$$

such that for every $N \in \mathbb{Z}_+$

$$op_{\psi}(a)(\xi') = \sum_{j=0}^{N} \left\{ A_j^+(\xi', \delta) op_M(g^+) + A_j^-(\xi', \delta) op_M(g^-) \right\} op_{\psi}(1_+^{-j})(\delta) + op_G(b_{N,\delta})(\xi') \tag{7}$$

with some function

$$b_{N,\delta}(t,s,\xi') \in C^\infty(\mathbb{R}^{n-1}_{\xi'} \setminus 0,\ C^m(\overline{\mathbb{R}}_+ \times \overline{\mathbb{R}}_+)) \tag{8}$$

$m = m(N) \longrightarrow \infty$ as $N \longrightarrow \infty$.

Proof. Consider the Taylor expansion of $a(\xi',\tau)$ near $\tau = \pm\infty$,

$$a(\xi',\tau) \sim \sum_{k=0}^{\infty} (i\tau)^{-k} \quad a_k^{\pm}(\xi') \quad ,$$

and substitute

$$(i\tau)^{-1} \sim 1_+^{-1}(\delta,\tau) \sum_{k=0}^{\infty} \delta^k\ 1_+^{-k}(\delta,\tau),\ (i\tau)^{-2} \sim 1_+^{-2}(\delta,\tau) \sum_{k=0}^{\infty} (k+1)\,\delta^k \times$$
$$\times\ 1_+^{-k}(\delta,\tau),$$

$$(i\tau)^{-3} \sim 1_+^{-3}(\delta,\tau) \sum_{k=0}^{\infty} \tfrac{1}{2}(k+1)(k+2)\,\delta^k\ 1_+^{-k}(\delta,\tau), \ldots$$

Then we obtain

$$a(\xi',\tau) = \sum_{j=0}^{N} \left\{A_j^+(\xi',\delta)\theta^+(\tau) + A_j^-(\xi',\delta)\theta^-(\tau)\right\} 1_+^{-j}(\delta,\tau) + R_{N,\delta}(\xi',\tau)$$

and hence the expression (7), since $op_\psi(\theta^{\pm}\ 1_+^{-j}) = op_M(g^{\pm})\ op_\psi(1_+^{-j})$, and (8) is an obvious consequence of Lemma 2. □

In particular

$$A_0^{\pm} = a_0^{\pm},\quad A_1^{\pm} = a_1^{\pm},$$

$$A_2^{\pm} = \delta a_1^{\pm} + a_2^{\pm},\quad A_3^{\pm} = \delta^2 a_1^{\pm} + 2\delta a_2^{\pm} + a_3^{\pm},\ \ldots$$

In a similar way we can prove

3'. Proposition. For

$$1_-^{-j}(\delta,\tau) = (\delta - i\tau)^{-j},\quad j \in \mathbb{Z}_+$$

$\delta > 0$, fixed, and every $N \in \mathbb{Z}_+$ there is a decomposition

$$op_\psi(a)(\xi') = \sum_{j=0}^{N} op_\psi(1_-^{-j})(\delta)\left\{B_j^+(\xi',\delta)op_M(g^+) + B_j^-(\xi',\delta)op_M(g^-)\right\}$$
$$+\ op_G(b_{N,\delta})(\xi')$$

with polynomials $B_j^{\pm}(\xi',\delta)$ of a similar form as in Proposition 3 and a kernel $b_{N,\delta}$ with the property (8), $m = m(N) \longrightarrow \infty$ as $N \longrightarrow \infty$.

4. Proposition. Let $j \in \mathbb{Z}_+$, $N \in \mathbb{Z}_+$. Then we have

$$\omega\, op_\psi(1_+^{-j})(\delta)\,\omega = \omega t^j \sum_{k=0}^{N} (t\delta)^k\ op_M(h_{kj})\omega + op_G(b_{N,\delta}),$$

where

$$h_{kj}(z) = \begin{cases} (-1)^k \binom{j+k-1}{k} \frac{1}{1-z} \cdot \ldots \cdot \frac{1}{j+k-z}, & j > 0, \\ \delta_{jk}. & j = 0, \end{cases} \tag{9}$$

$b_{N,\delta} \in \mathcal{S}_0^{(k)} \otimes_\pi \mathcal{S}_{\varphi_0}^{(k)}$, $k = k(N) \longrightarrow \infty$ as $N \longrightarrow \infty$.

Proof. For $u \in C_0^\infty(\mathbb{R}_+)$ we have

$$\begin{aligned} op_\psi(l_+^{-j})(\delta)u(t) &= (2\pi)^{-1} r^+ \int_{-\infty}^{\infty}\Big\{\int_0^{\infty} e^{i(t-s)\tau}(\delta+i\tau)^{-j}u(s)ds\Big\}d\tau \\ &= (2\pi)^{-1} r^+ \int_0^{\infty} \Big\{ \int_{-\infty}^{\infty} e^{i(t-s)\tau}(\delta+i\tau)^{-j}d\tau\Big\}u(s)ds. \end{aligned}$$

Then

$$\int_{-\infty}^{\infty} e^{it\tau}(\delta+i\tau)^{-j}d\tau = \begin{cases} \frac{2\pi}{\Gamma(j)} t^{j-1}e^{-\delta t} & \text{for } t > 0, \\ 0 & \text{for } t \leq 0 \end{cases}$$

implies for $t > 0$

$$op_\psi(l_+^{-j})(\delta)u(t) = \frac{1}{\Gamma(j)} \int_0^t (t-s)^{j-1}e^{-\delta(t-s)}u(s)ds.$$

Now we insert the expansion

$$e^{-\delta(t-s)} = \sum_{k=0}^{N} \frac{(-\delta)^k}{k!} (t-s)^k + R_{N,\delta}(t,s),$$

where $R_{N,\delta}(t,s)$ has a zero of order $N+1$ at $s = t$. Then with $r = t/s$, $ds = -tr^{-2}\,dr$,

$$\begin{aligned} \int_0^t (t-s)^{j-1}(t-s)^k u(s)ds &= \int_1^{\infty} (t-t/r)^{k+j-1} t r^{-2} u(t/r)dr \\ &= t^{k+j} \int_1^{\infty} (s-1)^{k+j-1} s^{-(k+j)} u(t/s)\frac{ds}{s} \end{aligned}$$

$u \in C_0^\infty(\mathbb{R}_+)$. Thus

$$\int_0^t (t-s)^{k+j-1}u(s)ds = t^{k+j} \int_0^{\infty} \mu_{kj}(s)u(t/s)\,\frac{ds}{s}$$

with

$$\mu_{kj}(s) := \begin{cases} (s-1)^{k+j-1}s^{-(k+j)} & \text{for } s > 1, \\ 0 & \text{for } s \leq 1, \end{cases}$$

and we may write

$$\int_0^t (t-s)^{k+j-1} u(s)ds = t^{k+j} \, op_M(m_{kj})u \tag{10}$$

with

$$m_{kj}(z) = M\mu_{kj}(z) = \int_1^\infty s^{z-1-k-j}(s-1)^{k+j-1} ds.$$

A classical integral formula says

$$m_{kj}(z) = \frac{\Gamma(1-z)\,\Gamma(j+k)}{\Gamma(1-z+j+k)} = \frac{(j+k-1)!}{(j+k-z)\ldots(1-z)} .$$

Hence (9) is proved.

The kernel of the operator associated to $R_{N,\delta}$ is $\theta^+(t-s)(t-s)^{j-1}R_{N,\delta}(t,s)$ which has $N + j - 1$ continuous derivatives in $\overline{\mathbb{R}}_+ \times \overline{\mathbb{R}}_+$ up to the boundary. Hence this kernel is in $\mathcal{S}_0^{(k)} \otimes_\pi \mathcal{S}_{\gamma_0}^{(k)}$ for some $k = k(N)$, $k \longrightarrow \infty$, if $N \longrightarrow \infty$. $\square$

Similar conclusions are valid for $\omega\, op_+(1_+^{-\alpha})(\delta)\omega$ for $\alpha \in \mathbb{C}$, $\operatorname{Re}\alpha > 0$. The expansion of Proposition 4 remains valid with

$$h_{k\alpha} = (-1)^k \frac{\Gamma(1-z)\,\Gamma(\alpha+k)}{k!\,\Gamma(1-z+\alpha+k)\,\Gamma(\alpha)} = \binom{-\alpha}{k} \frac{\Gamma(1-z)}{\Gamma(1-z+\alpha+k)} \tag{11}$$

Set

$$\sigma_M^0(a)(z) := a_0^+ \, g^+(z) + a_0^- \, g^-(z), \tag{12}$$

$$\sigma_M^{-1}(a)(\xi',z) := \left\{ a_1^+(\xi') \, g^+(z) + a_1^-(\xi') g^-(z) \right\} \prod_{j=1}^{1} (j-z)^{-1}, \tag{13}$$

$1 \geq 1$. Below we also write

$$\sigma_M^{-1}(a) =: \sigma_M^{-1}(op_\psi(a))$$

and call $\sigma_M^{-1}(a)$ the <u>conormal symbol</u> of a or $op_\psi(a)$, <u>of order</u> -1, $1 \in \mathbb{Z}_+$.

Now we derive the Mellin expansion of the pseudo-differential action, where the dependence on δ in the expansion of Proposition 3 is cancelled out.

5. <u>Theorem</u>. Let $a \in S^{(0)}(\mathbb{R}^n \setminus 0)$ and ω be a cut-off function. For every $N \in \mathbb{Z}_+$ there is a $k=k(N)$, $k(N) \longrightarrow \infty$ as $N \longrightarrow \infty$, such that

$$\omega\, op_\psi(a)(\xi')\omega = \sum_{j=0}^{N} t^j \omega\, op_M(\sigma_M^{-j}(a))(\xi')\omega + op_G(b_N(a))(\xi') \tag{14}$$

with some kernel $b_N(a) = b_N(a)(t,s,\xi') \in C^\infty(\mathbb{R}^{n-1}_{\xi'} \setminus 0, \mathcal{S}^{(k)}_{\gamma_0} \hat{\otimes}_\pi \mathcal{S}^{(k)}_{\gamma_0})$.

Proof. Let ω_1 be another cut-off function with $\omega_1\omega = \omega$. In view of Lemma 1 and Proposition 3 we have

$$\omega op_\psi(a)(\xi')\omega = \omega\sum_{j=0}^{N} op_M(A_j^+(\xi',\delta)g^+ + A_j^-(\xi',\delta)g^-)op_\psi(1_+^{-j})(\delta)\omega$$

$$\mod op_G(\mathcal{S}^{(k)}_{\gamma_0} \hat{\otimes}_\pi \mathcal{S}^{(k)}_{\gamma_0})$$

$$= \omega\sum_{j=0}^{N} op_M(A_j^+(\xi',\delta)g^+ + A_j^-(\xi',\delta)g^-)\,\omega_1 op_\psi(1_+^{-j})(\delta)\omega$$

$$\mod op_G(\mathcal{S}^{(k)}_{\gamma_0} \hat{\otimes}_\pi \mathcal{S}^{(k)}_{\gamma_0}).$$

Inserting the decomposition of Proposition 4 we obtain modulo $op_G(\mathcal{S}^{(k)}_{\gamma_0} \hat{\otimes}_\pi \mathcal{S}^{(k)}_{\gamma_0})$

$$\begin{aligned}\omega op_\psi(a)(\xi')\omega = &\sum_{j=0}^{N}\sum_{k=0}^{N}\omega op_M(A_j^+(\xi',\delta)\,g^+ + A_j^-(\xi',\delta)g^-)\,\omega_1 t^j t^k \delta^k op_M(h_{kj})\omega\\ &+ op_G(b_{N,\delta})\sum_{k=0}^{N}\omega_1 t^{j+k}\delta^k op_M(h_{kj})\omega \qquad (15)\\ &+ \sum_{j=0}^{N}\omega op_M(A_j^+(\xi',\delta)g^+ + A_j^-(\xi',\delta)g^-)\,\omega_1 op_G(b).\end{aligned}$$

The second and the third term on the right of (15) are of the class $op_G(\mathcal{S}^{(k)}_{\gamma_0} \hat{\otimes}_\pi \mathcal{S}^{(k)}_{\gamma_0})$. Furthermore, we have

$$\omega op_M(g^\pm)t^j\,\omega_1 = \omega\, t^j op_M(g^\pm)\,\omega_1 \mod op_G(\mathcal{S}_{\gamma_0} \hat{\otimes}_\pi \mathcal{S}_{\gamma_0}),$$

$$(1-\omega_1)\,op_M(h_{kj})\omega \in op_G(\mathcal{S}_0 \hat{\otimes}_\pi \mathcal{S}_{\gamma_0}).$$

Thus (15) implies

$$\omega op_\psi(a)(\xi')\omega = \sum_{j=0}^{N}\sum_{k=0}^{N} t^{j+k}\delta^k\,\omega\, op_M((A_j^+(\xi',\delta)g^+ + A_j^-(\xi',\delta)g^-)h_{kj})\omega \qquad (16)$$

$$\mod op_G(\mathcal{S}^{(k)}_{\gamma_0} \hat{\otimes}_\pi \mathcal{S}^{(k)}_{\gamma_0}).$$

Now it is obvious that

$$op_M(\sum_{j,k}\delta^k\{A_j^+(\xi',\delta)g^+ + A_j^-(\xi',\delta)g^-\}h_{kj})$$

$$= op_M(\sum_j\{a_j^+(\xi')g^+ + a_j^-(\xi')g^-\}h_{oj}).$$

Thus Theorem 5 is proved. □

Note that $a \longrightarrow b_N(a)$ is continuous.

1.4.2. The action in $\mathcal{S}_{\mathcal{P}}(\mathbb{R}_+)$

The expansion of a ψDO on the half-axis in 1.4.1, Theorem 5, suggests that $op_\psi(a)$ induces a mapping $\mathcal{S}_{\mathcal{P}} \longrightarrow \mathcal{S}_{\mathcal{P}+\mathcal{P}_0}$ for any $\mathcal{P} \in \underline{\mathcal{P}}$ satisfying $T^{-j}\mathcal{P} \subseteq \mathcal{P}$ for each $j \in \mathbb{Z}_+$. In fact, the increase by $\mathcal{P}_0$ is caused by the simple poles of $g^+(z)$ and $g^-(z)$ at the points of $\mathbb{Z} \subset \mathbb{C}$.

1. Theorem. Let $a \in S^{(0)}(\mathbb{R}^n \setminus 0)$. Then, for every $\mathcal{P} \in \underline{\mathcal{P}}$ with $T^{-j}\mathcal{P} \subseteq \mathcal{P}$ for $j \in \mathbb{Z}_+$ and $\xi' \neq 0$ fixed

$$op_\psi(a)(\xi') : \mathcal{S}_{\mathcal{P}} \longrightarrow \mathcal{S}_{\mathcal{P}+\mathcal{P}_0} \tag{1}$$

is continuous, and

$$op_\psi(\cdot) : S^0_{cl}(\mathbb{R}) \longrightarrow \mathcal{L}(\mathcal{S}_{\mathcal{P}}, \mathcal{S}_{\mathcal{P}+\mathcal{P}_0})$$

is continuous, $\mathcal{L}(\ldots)$ being considered in the topology of uniform convergence on bounded sets.

Proof. Consider an arbitrary norm on $\mathcal{S}_{\mathcal{P}+\mathcal{P}_0}$ given by

$$\| t^{\eta-\delta} \partial_t^l \mathcal{H}_\delta^{\mathcal{P}+\mathcal{P}_0} v \|^2_{L^2(\mathbb{R}_+)} + \| v \|^2_{L^2(\mathbb{R}_+)} .$$

For suitable η', l' and a constant c we have

$$\| Q_\beta v \|^2_{L^2(\mathbb{R}_+)} \leq c(\| t^{\eta'} \partial_t^{l'} v \|^2_{L^2(\mathbb{R}_+)} + \| v \|^2_{L^2(\mathbb{R}_+)}),$$

since for v with finite right hand side we get a Taylor expansion up to order $l'-1$, and $v^{(j)}(0)$, $j = 0,\ldots,l'-1$, is bounded by $c'(\| t^{\eta'} \partial_t^{l'} v \|^2_{L^2(\mathbb{R}_+)} + \| v \|^2_{L^2(\mathbb{R}_+)})^{1/2}$. The operator $\partial_t^{l'}$ is applied to v as function on $\mathbb{R}_+$.

Let $\omega_1, \omega_2, \omega_3$ be cut-off functions with $\omega_1\omega_2 = \omega_2 . \omega_1\omega_3 = \omega_1$. Then

$$\begin{aligned} op_\psi(a) &= (1-\omega_1)\, op_\psi(a) + \omega_1\, op_\psi(a) \\ &= (1-\omega_1)\, op_\psi(a)(1-\omega_2) + (1-\omega_1)\, op_\psi(a)\omega_2 \\ &\quad + \omega_1\, op_\psi(a)(1-\omega_3) + \omega_1\, op_\psi(a)\, \omega_3 . \end{aligned}$$

In view of 1.4.1, Lemma 1, the second and the third item have kernels in $\mathcal{S}_{\varphi_0} \otimes_\pi \mathcal{S}_{\varphi_0}$. The first one is a composition of continuous mappings

$$\mathcal{S}_{\varphi} \xrightarrow{(1-\omega_2)} \mathcal{S}_0 \xrightarrow{e^+} \mathcal{S} \xrightarrow{op(a)} \mathcal{S} \xrightarrow{r^+} \mathcal{S}_{\varphi_0} \xrightarrow{1-\omega_1} \mathcal{S}_0 .$$

and the norms are bounded by semi-norms of $a \in S^0_{cl}$.

Let $u \in \mathcal{S}_\varphi$ be arbitrary and $u = \kappa^\varphi_\varepsilon u + (1- \kappa^\varphi_\varepsilon)u$, where ε will be chosen below. Then

$$t^{\eta'}\partial_t^{l'} \; op(a) \; \kappa^\varphi_\varepsilon u = \sum_{k=0}^{\eta'} op(D_\tau^k a)(t^{\eta'-k}\partial_t^{l'} \kappa^\varphi_\varepsilon u)$$

implies

$$\begin{aligned}\| t^{\eta'}\partial_t^{l'} \; op_\psi(a) \; \kappa^\varphi_\varepsilon u\|_{L^2(\mathbb{R}_+)} &= \| r^+ t^{\eta'}\partial_t^{l'} \; op(a)e^+ \; \kappa^\varphi_\varepsilon u\|_{L^2(\mathbb{R}_+)} \\ &\le \| t^{\eta'}\partial_t^{l'} \; op(a)e^+ \; \kappa^\varphi_\varepsilon u\|_{L^2(\mathbb{R})} \\ &\le \sum_{k=0}^{\eta'} \| op(D_\tau^k a)(t^{\eta'-k}\partial_t^{l'})e^+ \; \kappa^\varphi_\varepsilon u\|_{L^2(\mathbb{R})} .\end{aligned}$$

Taking ε sufficiently large we get

$$(t^{\eta'-k}\partial_t^{l'})e^+ \; \kappa^\varphi_\varepsilon u = e^+(t^{\eta'-k}\partial_t^{l'} \kappa^\varphi_\varepsilon u),$$

and the estimate

$$\begin{aligned}&\| t^{\eta'}\partial_t^{l'} \; op_\psi(a) \; \kappa^\varphi_\varepsilon u\|_{L^2(\mathbb{R}_+)} \\ &\qquad \le \sum_{k=0}^{\eta'} \|op(D_\tau^k a)\|_{\mathcal{L}(L^2(\mathbb{R}))} \|e^+(t^{\eta'-k}\partial_t^{l'} \kappa^\varphi_\varepsilon u)\|_{L^2(\mathbb{R})} \\ &\qquad = \sum_{k=0}^{\eta'} \|op(D_\tau^k a)\|_{\mathcal{L}(L^2(\mathbb{R}))} \|t^{\eta'-k}\partial_t^{l'} \kappa^\varphi_\varepsilon u\|_{L^2(\mathbb{R}_+)} \\ &\qquad \le c\|P_\alpha u\|_{L^2(\mathbb{R}_+)} ,\end{aligned}$$

where $u \longrightarrow \|P_\alpha u\|_{L^2}$ denotes a semi-norm on $\mathcal{S}_\varphi$, and c is bounded by a linear combination of semi-norms of $a \in S^0_{cl}(\mathbb{R})$. The same conclusions are valid for $\omega_1 \; op_\psi(a)\omega_3$.

Consider now $\omega_1 \; op_\psi(a) \; \omega_3(1- \kappa^\varphi_\varepsilon)$. We can apply the decomposition of

$\omega_1 \, op_\psi(a)\omega_3$ from 1.4.1, Theorem 5, and obtain

$$\omega_1 \, op_\psi(a)\omega_3(1-\pi_\varepsilon^\gamma) = \sum_{l=0}^{N} t^l \omega \, op_M(h_l(a))\omega(1-\pi_\varepsilon^\gamma) + op_G \, b_N(a)(1-\pi_\varepsilon^\gamma)$$

with $b_N(a) \in \mathcal{S}_{\gamma_0}^{(k)} \otimes_\pi \mathcal{S}_{\gamma_0}^{(k)}$ and suitably generalized Mellin symbols $h_l(a)$. $(1-\pi_\varepsilon^\gamma)$ is a finite-dimensional Green operator with kernel in $\mathcal{S}_\gamma \otimes_\pi L^2$. Hence

$$op_G b_N(a)(1-\pi_\varepsilon^\gamma) : \mathcal{S}_\gamma \longrightarrow \mathcal{S}_{\gamma_0}^{(k)}$$

is continuous, and the norms are bounded by norms of $b_N(a) \in \mathcal{S}_{\gamma_0}^{(k)} \otimes_\pi \mathcal{S}_{\gamma_0}^{(k)}$. Since $a \longrightarrow b_N(a)$ is continuous, we also have bounds by semi-norms of $a \in S_{cl}^0$. In the same way the results about Mellin operators give that

$$\omega \, op_M(h_l(a))\omega(1-\pi_\varepsilon^\gamma) : \mathcal{S}_\gamma \longrightarrow \mathcal{S}_{\gamma+\gamma_0}$$

is continuous and continuously depends on a. $\square$

The case of symbols $a \in S_{cl}^\mu$, $\mu \in \mathbb{R}$ arbitrary, will be treated in 2.3.5. In particular for $a^{-j}(\xi',\tau) \in C^\infty(\mathbb{R}^n \setminus 0)$ homogeneous of order $-j$, $j \in \mathbb{Z}_+$, we obtain a continuous mapping

$$op_\psi(a^{-j}) : \mathcal{S}_\gamma \longrightarrow \mathcal{S}_{T^{-j}\gamma+\gamma_0} \tag{2}$$

which continuously depends on a^{-j}.

One easily checks that

$$\xi' \longrightarrow op_\psi(a)(\xi') \in \mathcal{L}(\mathcal{S}_\gamma, \mathcal{S}_{\gamma+\gamma_0})$$

is smooth.

An important special case is that of ψDO's with the so-called transmission property, which was introduced by Višik, Eskin [V1] and Boutet de Monvel [B1]. A function $a(\xi',\tau) \in C^\infty(\mathbb{R}^n \setminus 0)$ being positively homogeneous of order zero is said to have the transmission property if

$$a_k^+(\xi') = a_k^-(\xi') \quad \text{for all } k \in \mathbb{Z}_+, \quad \xi' \neq 0. \tag{3}$$

Then the conormal symbols being of the form $a_l^+(\xi')\prod_{j=1}^{l}(j-z)^{-1}$ for $l \geq 1$, a_0^+ for $l = 0$, have no poles at $0,-1,-2,\ldots$. Thus the additional γ_0-

singularity does not occur. As a conclusion we get

2. Proposition. Let a be as above, satisfying the transmission property (3) and $\varphi \in \underline{\Phi}$ such that $\varphi_0 \subseteq \varphi$. Then $op_\psi(a)$ induces a continuous mapping

$$op_\psi(a) : \mathcal{S}_\varphi \longrightarrow \mathcal{S}_\varphi \tag{4}$$

which continuously depends on a.
One can prove the converse, too.

1.5. The algebra $\mathcal{A}(\mathbb{R}^{n-1} \setminus 0)$

1.5.1. Compositions of Mellin and Green operators with pseudo-differential operators

By $\mathcal{A}_\psi(\mathbb{R}^{n-1} \setminus 0)$ we denote the set of all families of operators in $L^2(\mathbb{R}_+)$ of the form $op_\psi(a)(\xi')$, $a \in S^{(0)}(\mathbb{R}^n \setminus 0)$, $\xi' \in \mathbb{R}^{n-1} \setminus 0$. Then

$$op_\psi(a)(\lambda \xi') = \varkappa_\lambda \, op_\psi(a)(\xi') \, \varkappa_\lambda^{-1}, \; \lambda > 0$$

for $(\varkappa_\lambda u)(t) := \lambda^{1/2} u(\lambda t)$, $\lambda > 0$. Moreover, denote by $\mathcal{A}_M(\mathbb{R}^{n-1} \setminus 0)$ the set of all families of operators $\alpha(\xi')$ in $L^2(\mathbb{R}_+)$ with

$$\alpha(\lambda \xi') = \varkappa_\lambda \, \alpha(\xi') \, \varkappa_\lambda^{-1}, \; \lambda > 0 , \tag{1}$$

where $\alpha(\xi')$ is in $\mathcal{A}_M$ for every fixed $\xi' \in \mathbb{R}^{n-1} \setminus 0$ and $\sigma_M^{-j}(\alpha)(\xi')$ is a homogeneous polynomial in ξ' of order j. Finally let $\mathcal{A}_G(\mathbb{R}^{n-1} \setminus 0)$ be the set of all families of Green operators $\alpha(\xi')$ satisfying the homogeneity condition (1), $\alpha(\xi') = op_G(b(\xi'))$, and with $b \in C^\infty(\mathbb{R}^{n-1} \setminus 0, \mathcal{S}_\varphi \hat{\otimes}_\pi \mathcal{S}_q)$ for certain $\varphi, q \in \underline{\Phi}$ independent of ξ'. Denote by $\mathcal{A}(\mathbb{R}^{n-1} \setminus 0)$ the class of families

$$\alpha(\xi') = \alpha_\psi(\xi') + \alpha_M(\xi') + \alpha_G(\xi')$$

where $\alpha_\psi \in \mathcal{A}_\psi(\mathbb{R}^{n-1} \setminus 0)$, $\alpha_M \in \mathcal{A}_M(\mathbb{R}^{n-1} \setminus 0)$, $\alpha_G \in \mathcal{A}_G(\mathbb{R}^{n-1} \setminus 0)$. First note that from α we can recover the so-called interior symbol $a(\xi) =: \sigma_\psi(\alpha)(\xi)$ in the usual way. Let $\chi \in C_0^\infty(\mathbb{R}_+)$, $\chi \equiv 1$ near $t_0 > 0$. Then we have

$$\chi(t) e^{-i\lambda t \tau_0} \alpha(e^{i\lambda t \tau_0} \chi(t))_{t=t_0} \sim \sum_{j=0}^{\infty} a_j(\tau_0) \lambda^{-j},$$

since $\chi(t)(m + h)\ \chi(t)$ have smooth kernels. Having determined a finite sum of homogeneous parts of the symbol to the remaining convolution operator we can apply 2.1.2.3, Lemma 12 of [R1] in order to find the symbol exactly. We get a mapping

$$\sigma_\psi : \mathcal{L}(\mathbb{R}^{n-1} \setminus 0) \longrightarrow S^{(0)}(\mathbb{R}^n \setminus 0).$$

Then the sequence

$$0 \longrightarrow \mathcal{L}_M(\mathbb{R}^{n-1} \setminus 0) + \mathcal{L}_G(\mathbb{R}^{n-1} \setminus 0) \xrightarrow{i} \mathcal{L}(\mathbb{R}^{n-1} \setminus 0) \xrightarrow{\sigma_\psi} S^{(0)}(\mathbb{R}^n \setminus 0) \longrightarrow 0$$

is exact, i the natural embedding, and op_ψ yields a right inverse of σ_ψ. In the sequel we keep ξ' fixed. The smoothness in ξ' of the operators generated in the calculations is obvious. The main goal of this section is the following

1. <u>Theorem</u>. $\mathcal{L}_M(\mathbb{R}^{n-1} \setminus 0) + \mathcal{L}_G(\mathbb{R}^{n-1} \setminus 0)$ is a two-sided ideal in $\mathcal{L}(\mathbb{R}^{n-1} \setminus 0)$. Moreover, for the sequence of conormal symbols of the composition formula 1.3.2.(5) remains valid.

In view of the results of 1.3.2 the proof of the theorem consists of two essential parts:

1. The composition of ψDO's and Green operators, which is a simple consequence of 1.4.2, Theorem 1
2. The composition of ψDO's and Mellin operators, where besides 1. the expansion of the ψDO's in Mellin operators is used.

In the formulation of the corresponding Propositions 2 and 3 we get information about the increase of the conormal singularity.

2. <u>Proposition</u>. Let $a \in S^{-j}_{cl}(\mathbb{R})$, $j \in \mathbb{Z}_+$, and $b \in \mathcal{S}_q \hat{\otimes}_\pi \mathcal{S}_\gamma$. Then

$$op_\psi(a)\ op_G(b) = op_G(b_1),\quad op_G(b)\ op_\psi(a) = op_G(b_2),$$

where $b_1 \in \mathcal{S}_{T^{-j}q+\gamma_0} \hat{\otimes}_\pi \mathcal{S}_\gamma$, $b_2 \in \mathcal{S}_q \hat{\otimes}_\pi \mathcal{S}_{T^{-j}\gamma+\gamma_0}$.

This immediately follows from 1.4.2.(2) and 1.2.2.(17). For $a \in S^{(0)}(\mathbb{R}^n \setminus 0)$ and a smooth family of Green operators $op_G(b)(\xi')$ the families of Green operators $op_G(b_1)(\xi')$ and $op_G(b_2)(\xi')$ are smooth.

3. **Proposition.** Let $a \in S^0_{cl}$, $h \in \mathcal{M}$, $k \in \mathbb{Z}_+$, $0 \le \gamma \le k$ and assume that $T^\gamma h$ has no poles on $\mathrm{Re}\, z = 1/2$. Then

$$op_\psi(a) t^{k-\gamma}\, \omega\, op_M(T^\gamma h)\omega t^\gamma = \sum_{l=0}^{\infty} t^{k+l-\lambda}\, \omega_1 op_M(T^\lambda h_l)\omega t^\lambda + op_G(b_1),$$

$$t^{k-\gamma}\omega\, op_M(T^\gamma h)\omega t^\gamma \quad op_\psi(a) = \sum_{l=0}^{\infty} t^{k+l-\eta}\, \omega_1 op_M(T^\eta h'_l)\omega t^\eta + op_G(b_2)$$

where $\omega_1(t) = \omega(c_1 t)$,

$$h_l(z) = \sigma_M^{-l}(a)(z-k)h(z), \quad h'_l(z) = h(z-l)\,\sigma_M^{-l}(a)(z), \qquad (2)$$

λ, η such that $T^\lambda h_l$, $T^\eta h'_l$ have no poles on $\mathrm{Re}\, z = 1/2$ and $b_1, b_2 \in \mathcal{S}_{\underline{\varphi}} \hat{\otimes}_\pi \mathcal{S}_{\underline{\varphi}}$.

__Proof.__ We consider

$$op_\psi(a) t^{k-\gamma}\omega\, op_M(T^\gamma h)\omega t^\gamma$$

$$= \omega_1 op_\psi(a)\omega t^{k-\gamma} \quad op_M(T^\gamma h)\omega t^\gamma \quad \mathrm{mod}\ op_G(\mathcal{S}_0 \hat{\otimes}_\pi \mathcal{S}_{\varphi_0}),$$

where $\omega = \omega_1\omega$ (cf. 1.4.1, Lemma 1). The second composition can be reduced to this case by taking adjoints. The expansion of $\omega_1 op_\psi(a)\omega$ in Mellin operators from 1.4.1, Theorem 5, yields

$$\omega_1 op_\psi(a)\omega t^{k-\gamma} op_M(T^\gamma h)\omega t^\gamma = \sum_{l=0}^{N} t^l\, \omega\, op_M(\sigma_M^{-l}(a)\omega t^{k-\gamma} op_M(T^\gamma h)\omega t^\gamma$$

$$+ op_G(b_N(a)) t^{k-\gamma}\ op_M(T^\gamma h)\omega t^\gamma.$$

The kernel of $op_G(b_N(a)) t^{k-\gamma}\ op_M(T^\gamma h)\omega t^\gamma$ belongs to $\mathcal{S}^{(k)}_{\varphi_0} \hat{\otimes}_\pi \mathcal{S}_\varphi$ for some $\varphi \in \underline{\varphi}$. By applying the results of 1.3.2. to the composition of Mellin operators $op_M(\sigma_M^{-l}(a))\,\omega\, t^{k-\gamma} op_M(T^\gamma h)\omega$ we we get (2). Since $h(z)$ is rapidly decreasing in all strips $c \le \mathrm{Re}\, z \le c'$ and $\sigma_M^{-l}(a)$ are bounded there, the arising symbols are in $\mathcal{M}$. Since N is arbitrary and $k(N) \longrightarrow \infty$ for $N \longrightarrow \infty$, the assertion follows. $\square$

1.5.2. __Compositions of pseudo-differential operators__

The space of interior symbols $\sigma_\psi(\alpha)$ belonging to $\alpha \in \mathcal{L}(\mathbb{R}^{n-1} \setminus 0)$ is an algebra under the usual multiplication. Now we introduce an algebra of conormal symbols. Denote by $\mathcal{M}^\varphi$ the vector space of all finite linear combinations of the functions

$$g^{\pm}(z),\ g^{\pm}(z)\prod_{l=1}^{k}(l-z)^{-1},\quad k\in\mathbb{Z}_+ .$$

The elements of $\mathfrak{M}^{\psi}$ are meromorphic with poles at most at the real integers $\mathbb{Z}$ in $\mathbb{C}$, and every $f\in\mathfrak{M}^{\psi}$ is bounded for $|\mathrm{Im}\, z|>\varepsilon$ for any $\varepsilon>0$. The vector space $\mathfrak{M}^{\psi}\oplus\mathfrak{M}$ is an algebra with $\{0\}\oplus\mathfrak{M}$ as a two-sided ideal. Denote by $(\mathfrak{M}^{\psi}\oplus\mathfrak{M})[\xi']$ the vector space of polynomials in ξ' with coefficients in $\mathfrak{M}^{\psi}\oplus\mathfrak{M}$ and by $(\mathfrak{M}^{\psi}\oplus\mathfrak{M})^{(k)}[\xi']$ the subspace of homogeneous polynomials of order k. Remember that for $op_{\psi}(a)$

$$\sigma_M^{-j}(op_{\psi}(a))\in\mathfrak{M}^{\psi}[\xi'],\ j\in\mathbb{Z}_+ .$$

Moreover, for $\mathfrak{m}\in\mathcal{L}_M(\mathbb{R}^{n-1}\setminus 0)$,

$$\sigma_M^{-j}(\mathfrak{m})\in\mathfrak{M}[\xi'],\ j\in\mathbb{Z}_+ .$$

The conormal symbol of order $-j$ of $\mathfrak{a}\in\mathcal{L}(\mathbb{R}^{n-1}\setminus 0)$, $\mathfrak{a}=op_{\psi}(a)+\mathfrak{m}+\mathfrak{g}$, is defined by

$$\sigma_M^{-j}(\mathfrak{a})=\sigma_M^{-j}(op_{\psi}(a))+\sigma_M^{-j}(\mathfrak{m}),$$

which is an element in $(\mathfrak{M}^{\psi}\oplus\mathfrak{M})^{(k)}[\xi']$. By $\mathcal{S}(\mathcal{L}(\mathbb{R}^{n-1}\setminus 0))$ we denote the subspace of sequences $(a,(h_k)_{k\in\mathbb{Z}_+})$ of

$$S^{(0)}(\mathbb{R}^{n}\setminus 0)\times\mathop{\times}_{k=0}^{\infty}(\mathfrak{M}^{\psi}\oplus\mathfrak{M})^{(k)}[\xi']$$

for which

$$h_k-\sigma_M^{-k}(a)\in\mathfrak{M}[\xi']\ \text{for all}\ k\in\mathbb{Z}_+\setminus\{0\},\ \in\mathfrak{M}_0[\xi']\ \text{for}\ k=0 .$$

Then we have a mapping

$$\sigma=(\sigma_{\psi},(\sigma_M^{-k})_{k\in\mathbb{Z}_+}):\ \mathcal{L}(\mathbb{R}^{n-1}\setminus 0)\longrightarrow\mathcal{S}(\mathcal{L}(\mathbb{R}^{n-1}\setminus 0))$$

and the sequence

$$0\longrightarrow\mathcal{L}_G(\mathbb{R}^{n-1}\setminus 0)\longrightarrow\mathcal{L}(\mathbb{R}^{n-1}\setminus 0)\xrightarrow{\ \sigma\ }\mathcal{S}(\mathcal{L}(\mathbb{R}^{n-1}\setminus 0))\longrightarrow 0$$

is exact.

The main result of this section is the following

1. Theorem. $\mathcal{L}(\mathbb{R}^{n-1}\setminus 0)$ is an algebra.

Proof. Consider two operators

$$\mathfrak{a}_j = op_\psi(a_j) + \sum_{k=0}^{\infty} t^{k-\gamma_k}\omega_k op_M(h_k^j)\omega_k + op_G(b^j), \; j = 1,2.$$

In view of 1.5.2, Theorem 1, and 1.2.2(17) it remains to consider $op_\psi(a_1)\, op_\psi(a_2)$. For any cut-off function ω we have

$$(1-\omega)r^+op(a_1)e^-op(a_2)e^+ + r^+op(a_1)e^-op(a_2)(1-\omega)e^+ = op_G(b),$$

where $b \in \mathcal{S}_{\varphi_0} \otimes_\pi \mathcal{S}_{\varphi_0}$. This can be proved in the same way as 1.4.1, Lemma 1. Then we obtain

$$op_\psi(a_1)op_\psi(a_2) - op_\psi(a_1a_2) = -r^+op(a_1)e^-op(a_2)e^+$$

$$= \omega\left[op_\psi(a_1)op_\psi(a_2) - op_\psi(a_1a_2)\right]\omega + op_G b.$$

Now apply the decomposition from 1.4.1,

$$\omega op_\psi(a)\omega = \sum_{k=0}^{N} t^k\omega\, op_M(\sigma_M^{-k}(a))\omega + op_G(b_N(a)).$$

Then

$$\omega\, op_\psi(a_1)op_\psi(a_2)\omega = (\omega op_\psi(a_1)\omega)(\omega\, op_\psi(a_2)\omega) + \omega\, op_\psi(a_1)(1-\omega^2)op_\psi(a_2)\omega$$

$$= \left\{\sum_{j=0}^{N} t^j\, \omega op_M(\sigma_M^{-j}(a_1))\omega + op_G(b_N(a_1))\right\}$$

$$\times\left\{\sum_{k=0}^{N} t^k\omega\, op_M(\sigma_M^{-k}(a_2))\omega + op_G(b_N(a_2))\right\} + op_G(b^1)$$

$$= \sum_{l=0}^{N} t^l\omega\, op_M\Big(\sum_{j+k=l}(T^{-k}\sigma_M^{-j}(a_1))\,\sigma_M^{-k}(a_2)\Big)\omega + op_G(b^2),$$

where $b^1, b^2 \in \mathcal{S}^{(k)}_{\varphi_0+\varphi_0} \otimes_\pi \mathcal{S}^{(k)}_{\varphi_0}$, $k = k(N)$. So we have to prove that

$$\sum_{j+k=l}(T^{-k}\sigma_M^{-j}(a_1))\,\sigma_M^{-k}(a_2) - \sigma_M^{-l}(a_1a_2) \in \mathfrak{M}[\xi'] .$$

According to 1.4.1 (3) we may write

$$(a_1a_2)_k^{\pm}$$

$$= \frac{1}{k!}(\pm i)^k\frac{\partial^k}{\partial\varrho^k}\left(a_1(\varrho\xi', \pm 1)a_2(\varrho\xi', \pm 1)\right)\Big|_{\varrho=0}$$

$$= \frac{1}{k!}(\pm i)^k\Big(\sum_{k'=0}^{k}\frac{k!}{k'!(k-k')!}\frac{\partial^{k'}}{\partial\varrho^{k'}}a_1(\varrho\xi', \pm 1)\Big|_{\varrho=0}\frac{\partial^{k-k'}}{\partial\varrho^{k-k'}}a_2(\varrho\xi', \pm 1)\Big|_{\varrho=0}$$

$$= \sum_{k'=0}^{k} a^{\pm}_{1,k'}\, a^{\pm}_{2,k-k'}$$

For any fixed $l \geq 0$ we have

$$\sum_{k+j=l} (T^{-k}\sigma_M^{-j}(a_1))\sigma_M^{-k}(a_2)(z)$$

$$= \sum_{j+k=l} (a^{+}_{1,j}\, g^{+} + a^{-}_{1,j}\, g^{-})(a^{+}_{2,k}\, g^{+} + a^{-}_{2,k}\, g^{-}) \prod_{m=1}^{l} (m-z)^{-1}$$

and

$$\sigma_M^{-l}(a_1 a_2)(z) = ((a_1 a_2)^{+}_{l}\, g^{+} + (a_1 a_2)^{-}_{l}\, g^{-}) \prod_{j=1}^{l} (j-z)^{-1} .$$

Since $g^{+}g^{-} \in \mathfrak{M}$ and because of $(g^{+})^2 - g^{+}$, $(g^{-})^2 - g^{-} \in \mathfrak{M}$ it remains to observe that the coefficients of $(g^{\pm})^2$ and $g^{\pm}$ are equal. $\square$

In the proof of Theorem 1 we showed that $\mathcal{T}(\mathcal{L}(\mathbb{R}^{n-1} \setminus 0))$ is an algebra, where the composition is defined by

$$(a^1, (h^1_k)_{k \in \mathbb{Z}_+})(a^2, (h^2_l)_{l \in \mathbb{Z}_+}) = (a^1 a^2, (h^3_i)_{i \in \mathbb{Z}_+})$$

$$h^3_i = \sum_{k+l=i} (T^{-l} h^1_k) h^2_l \quad .$$

Moreover, $\sigma : \mathcal{L}(\mathbb{R}^{n-1} \setminus 0) \longrightarrow \mathcal{T}(\mathcal{L}(\mathbb{R}^{n-1} \setminus 0))$ is an algebra homomorphism. Let us consider the case of ψDO's on the half-axis with symbols depending on t, too. For simplicity, we assume that $a(t,\tau)$ is the restriction to $t \in \mathbb{R}_+$ of a symbol defined on $\mathbb{R}$. Then we have the Taylor expansion

$$a(t,\tau) = \sum_{j=0}^{N} \frac{t^j}{j!}\; \partial_t^j a(0,\tau) + t^{N+1} a_{N+1}(t,\tau) \; .$$

For given symbols $a_1(t,\tau)$, $a_2(t,\tau)$ denote by $a_1 \circ a_2$ an arbitrary symbol

$$a_1 \circ a_2 \sim \sum_{k=0}^{\infty} \frac{1}{k!}\; \partial_\tau^k a_1 D_t^k a_2 \; .$$

Then $op(a_1 \circ a_2) - op(a_1)op(a_2)$ has a smooth kernel and

$$op_\psi(a_1 \circ a_2) - op_\psi(a_1)op_\psi(a_2) = r^{+}op(a_1 \circ a_2)e^{+} - r^{+}op(a_1)\theta^{+}op(a_2)e^{+}$$

$$= r^{+}op(a_1)\theta^{-}op(a_2)e^{+} + r ,$$

where r has a kernel in $C^\infty(\overline{\mathbb{R}}_+ \times \overline{\mathbb{R}}_+)$. Modulo operators with smooth kernels we may consider $\omega op_\psi(a_1 \circ a_2)\omega - \omega op_\psi(a_1)op_\psi(a_2)\omega$ with an arbitrary cut-off function ω. We shall prove that this difference is in $\mathcal{L}_M(\mathbb{R}^{n-1} \setminus 0) + \mathcal{L}_G(\mathbb{R}^{n-1} \setminus 0)$ modulo operators with smooth kernel. If a_1 or a_2 have a zero of order N at $t = 0$, then the kernel of $r^+op(a_1)e^-$ $op(a_2)e^+$ has N-2 continuous derivatives in $\mathbb{R}_+ \times \mathbb{R}_+$. Therefore, we may assume that

$$a_1(t,\tau) = \sum_{j=0}^{N} t^j a_{1,j}(\tau), \quad a_2(t,\tau) = \sum_{j=0}^{N} t^j a_{2,j}(\tau).$$

Then we have

$$\omega\Big(\sum_{j=0}^{N} t^j op_\psi(a_{1,j})\Big)\Big(\sum_{j'=0}^{N} t^{j'} op_\psi(a_{2,j'})\Big)\omega$$

$$= \omega\Big(\sum_{j,j'=0}^{N} t^j \sum_{k=0}^{j'} \binom{j'}{k} t^{j'-k} op_\psi(D_\tau^k a_{1,j}) op_\psi(a_{2,j'})\Big)\omega$$

and

$$\omega\, op_\psi(a_1 \circ a_2)\omega = \omega\, op_\psi\Big(\sum_k \frac{1}{k!} D_\tau^k a_1 \partial_t^k a_2\Big)\omega$$

$$= \omega\Big(\sum_{j,j'=0}^{N} t^j \sum_{k=0}^{j'} \binom{j'}{k} t^{j'-k} op_\psi(D_\tau^k a_{1,j} a_{2,j'})\Big)\omega$$

In the proof of Theorem 1 it was shown that

$$\omega\Big[op_\psi(D_\tau^k a_{1,j}) op_\psi(a_{2,j'}) - op_\psi(D_\tau^k a_{1,j} a_{2,j'})\Big]\omega \in \mathcal{L}_M(\mathbb{R}^{n-1} \setminus 0) + \mathcal{L}_G(\mathbb{R}^{n-1} \setminus 0).$$

Hence we have proved the following result.

2. Theorem. The space of operators of the form

$$\mathfrak{a} = op_\psi(a) + \mathfrak{m} + \mathfrak{h} + \mathfrak{k}$$

$a(t,\tau) \in S^0_{cl}(\overline{\mathbb{R}}_+ \times \mathbb{R})$, $\mathfrak{m} + \mathfrak{h} \in \mathcal{L}_M(\mathbb{R}^{n-1} \setminus 0) + \mathcal{L}_G(\mathbb{R}^{n-1} \setminus 0)$, $\mathfrak{k}$ an integral operator with kernel in $C^\infty(\overline{\mathbb{R}}_+ \times \overline{\mathbb{R}}_+)$, is an algebra, and the mappings sending $\mathfrak{a}$ into complete ψDO symbols and the sequence of conormal symbols, respectively, are algebra homomorphisms.

Note that $\mathfrak{k}$ can be dropped if all ψDO symbols $a(t,\tau)$ are independent of t outside a bounded set $0 \le t \le T$, $T > 0$.

1.5.3. Ellipticity of operators in $\mathcal{L}(\mathbb{R}^{n-1}\setminus 0)$

The theory of operators in $\mathcal{L}(\mathbb{R}^{n-1}\setminus 0)$ was basically established for studying families of operators of the form

$$\mathcal{C}(\xi') = \begin{pmatrix} \mathcal{A} & \mathrm{op}(k) \\ r'\mathrm{op}(d) & q \end{pmatrix}(\xi') : \begin{matrix} \mathcal{S}_{\Psi}(\mathbb{R}_+,\mathbb{C}^p) \\ \oplus \\ \mathbb{C}^j \end{matrix} \longrightarrow \begin{matrix} \mathcal{S}_{\Psi}(\mathbb{R}_+,\mathbb{C}^r) \\ \oplus \\ \mathbb{C}^g \end{matrix} , \tag{1}$$

$\xi' \in \mathbb{R}^{n-1}\setminus 0$, with

$$\mathcal{A}(\xi') = \mathrm{op}_{\psi}(a)(\xi') + \mathcal{M}(\xi') + \mathrm{op}_G(b)(\xi') \in \mathcal{L}(\mathbb{R}^{n-1}\setminus 0) , \tag{2}$$

a so-called potential boundary symbol

$$\mathrm{op}(k)(\xi') : \mathbb{C}^j \longrightarrow \mathcal{S}_{\Psi}(\mathbb{R}_+,\mathbb{C}^r) , \tag{3}$$

a trace boundary symbol

$$r'\mathrm{op}(d)(\xi') : \mathcal{S}_{\Psi}(\mathbb{R}_+,\mathbb{C}^p) \longrightarrow \mathbb{C}^g , \tag{4}$$

and some pseudo-differential symbol $q(\xi')$ of order zero in $\mathbb{R}^{n-1}$. The operators of the form (1) play the role of boundary symbols of boundary value problems in $\mathbb{R}^n_+$ (or on manifolds with boundary, and may also be x' dependent then). Note that (1) can be extended to a continuous operator

$$\bar{\mathcal{C}}(\xi') : \begin{matrix} L^2(\mathbb{R}_+,\mathbb{C}^p) \\ \oplus \\ \mathbb{C}^j \end{matrix} \longrightarrow \begin{matrix} L^2(\mathbb{R}_+,\mathbb{C}^r) \\ \oplus \\ \mathbb{C}^g \end{matrix} . \tag{5}$$

For simplicity assume that $p = r = 1$. In order to define the action of $\mathrm{op}(k)$ and $r'\mathrm{op}(d)$, respectively, we also may suppose $j = g = 1$.

Let $k(\xi',t) \in C^{\infty}(\mathbb{R}^{n-1}\setminus 0, \mathcal{S}_{\Psi})$ and

$$k(\lambda\xi', \lambda^{-1}t) = \lambda^{1/2}k(\xi',t), \ \lambda > 0.$$

Then $k(\xi',t)$ is called a potential kernel of order zero (in the boundary symbolic sense).

$$\mathrm{op}(k)(\xi')c := ck(\xi',t)$$

defines an action $\mathbb{C} \longrightarrow \mathcal{S}_{\Psi}$ for any fixed $\xi' \neq 0$.
If $\bar{d}(\xi',t)$ is another potential kernel of order zero, we define a trace boundary symbol by duality

$$(r'\mathrm{op}(d)(\xi')u,\ c')_{\mathbb{C}} := (u,\ \mathrm{op}(d)(\xi')c')_{L^2(\mathbb{R}_+)},$$

$u \in L^2(\mathbb{R}_+)$, $c' \in \mathbb{C}$.

The operators $\mathcal{A}(\xi')$ of the form (1) then form an algebra $\mathcal{R}$ in the sense that the composition (when the spaces fit together) is of this type again and

$$\begin{pmatrix} \varkappa_\lambda & 0 \\ 0 & 1 \end{pmatrix} \mathcal{A}(\xi') \begin{pmatrix} \varkappa_\lambda & 0 \\ 0 & 1 \end{pmatrix}^{-1} = \mathcal{A}(\lambda\xi') \tag{6}$$

for $\xi' \neq 0$, $\lambda > 0$. Here

$$(\varkappa_\lambda u)(t) := \lambda^{1/2}u(\lambda t),\ \lambda > 0.$$

It is not our main objective here to study this algebra. We shall return to this question in Chapter 3. Let us only mention that the composition

$$\mathrm{op}(k)(\xi') \circ r'\mathrm{op}(d)(\xi') = \mathrm{op}_G(b)(\xi') \tag{7}$$

is of Green type with $b(\xi',t,s) = k(\xi',t)d(\xi',s)$.

The objects to be studied in $\mathcal{R}(\mathbb{R}^{n-1} \setminus 0)$ such as elliptic operators and inverses (or parametrices) can be reduced to left upper corners by composing with isomorphisms

$$\bar{\mathcal{A}}_1(\xi') : \begin{matrix} L^2(\mathbb{R}_+,\mathbb{C}^r) \\ \oplus \\ \mathbb{C}^g \end{matrix} \longrightarrow L^2(\mathbb{R}_+,\mathbb{C}^r),$$

$$\bar{\mathcal{A}}_2(\xi') : L^2(\mathbb{R}_+,\mathbb{C}^p) \longrightarrow \begin{matrix} L^2(\mathbb{R}_+,\mathbb{C}^p) \\ \oplus \\ \mathbb{C}^j \end{matrix},$$

which induce isomorphisms

$$\mathcal{A}_1(\xi') : \begin{matrix} \mathcal{S}_{\underline{\varphi}}(\mathbb{R}_+,\mathbb{C}^r) \\ \oplus \\ \mathbb{C}^g \end{matrix} \longrightarrow \mathcal{S}_{\underline{\varphi}}(\mathbb{R}_+,\mathbb{C}^r),$$

$$\mathcal{A}_2(\xi') : \mathcal{S}_{\underline{\varphi}}(\mathbb{R}_+,\mathbb{C}^p) \longrightarrow \begin{matrix} \mathcal{S}_{\underline{\varphi}}(\mathbb{R}_+,\mathbb{C}^p) \\ \oplus \\ \mathbb{C}^j \end{matrix},$$

where $\mathcal{A}_i^{-1} \in \mathcal{R}(\mathbb{R}^{n-1} \setminus 0)$, $i = 1,2$. We may assume that $\mathcal{A}_i$ are boundary symbols in Boutet de Monvel's class (cf. [B1],[R1]) and in addition $\sigma_\psi(\mathcal{A}_1) = \sigma_\psi(\mathcal{A}_2^{-1})$ for $r=p$, $j=g$. Thus we only have to deal with $\tilde{\mathcal{A}} = \mathcal{A}_1\mathcal{A}\mathcal{A}_2$.

1. Theorem. Let $\mathfrak{r}(\xi') \in \mathcal{R}(\mathbb{R}^{n-1} \setminus 0)$ and (5) be invertible. Then $\mathfrak{r}^{-1}(\xi') \in \mathcal{R}(\mathbb{R}^{n-1} \setminus 0)$.

As already mentioned, it suffices to prove the assertion in the case $\mathfrak{r} \in \mathcal{L}(\mathbb{R}^{n-1} \setminus 0)$, where $\mathfrak{r}$ is necessarily elliptic, cf. Definition 2, below. The remaining part of this section is devoted to a systematic study of the ellipticity. Then Theorem 1 will be obtained as a corollary.

2. Definition. An operator $\mathfrak{a}(\xi') \in \mathcal{L}(\mathbb{R}^{n-1} \setminus 0)$, $\mathfrak{a}(\xi') : L^2(\mathbb{R}_+, \mathbb{C}^p) \longrightarrow L^2(\mathbb{R}_+, \mathbb{C}^p)$, is called elliptic if both, the principal interior symbol $\sigma_\psi(\mathfrak{a})(\xi) = a_0(\xi)$, $\xi \neq 0$, and the principal conormal symbol $\sigma_M^0(\mathfrak{a})$ on $\mathrm{Re}\, z = 1/2$ (being p x p matrices) are non-singular.

3. Definition. An operator $\mathfrak{b}(\xi') \in \mathcal{L}(\mathbb{R}^{n-1} \setminus 0)$ is called a parametrix of $\mathfrak{a}(\xi')$ if both, $\mathfrak{a}\mathfrak{b}-1$ and $\mathfrak{b}\mathfrak{a}-1$ belong to $\mathcal{L}_G(\mathbb{R}^{n-1} \setminus 0)$.

In view of the previous results (cf. the Sections 1.2, 1.3, 1.4) any p x p system $\mathfrak{a}(\xi') \in \mathcal{L}(\mathbb{R}^{n-1} \setminus 0)$ induces a map

$$\mathfrak{a}(\xi') : \quad \mathcal{S}_{\underline{\varphi}}(\mathbb{R}_+, \mathbb{C}^p) \longrightarrow \mathcal{S}_{\underline{\varphi}}(\mathbb{R}_+, \mathbb{C}^p) \tag{8}$$

possessing a continuous extension

$$\bar{\mathfrak{a}}(\xi') : L^2(\mathbb{R}_+, \mathbb{C}^p) \longrightarrow L^2(\mathbb{R}_+, \mathbb{C}^p) . \tag{9}$$

4. Theorem. Let $\mathfrak{a}(\xi') \in \mathcal{L}(\mathbb{R}^{n-1} \setminus 0)$. Then the following conditions are equivalent:

(i) $\mathfrak{a}(\xi')$ is elliptic,

(ii) (9) is Fredholm.

If $\mathfrak{a}(\xi')$ is elliptic, there exists a parametrix $\mathfrak{b}(\xi') \in \mathcal{L}(\mathbb{R}^{n-1} \setminus 0)$ in the sense of Definition 3 and there are finite-dimensional spaces $\mathcal{N}_i(\xi') \subset \mathcal{S}_{\underline{\varphi}}(\mathbb{R}_+, \mathbb{C}^p)$, i = 1,2, with

$$\mathrm{im}\, \bar{\mathfrak{a}}(\xi') \oplus \mathcal{N}_1(\xi') = L^2(\mathbb{R}_+, \mathbb{C}^p), \ \ker \bar{\mathfrak{a}}(\xi') = \mathcal{N}_2(\xi'), \tag{10}$$

i.e.

$$\mathrm{ind}\, \mathfrak{a}(\xi') = \mathrm{ind}\, \bar{\mathfrak{a}}(\xi') . \tag{11}$$

Moreover,

$$\bar{\mathfrak{a}}(\xi')u \in \mathcal{S}_{\underline{\varphi}}(\mathbb{R}_+, \mathbb{C}^p) \Longrightarrow u \in \mathcal{S}_{\underline{\varphi}}(\mathbb{R}_+, \mathbb{C}^p) . \tag{12}$$

Proof. The equivalence of (i) and (ii) was already proved in [R2]. Now we construct a parametrix $\mathcal{b}(\xi') \in \mathcal{L}(\mathbb{R}^{n-1} \setminus 0)$ of $\mathcal{a}(\xi')$.

Let $\sigma_\psi(\mathcal{a})(\xi) = a(\xi)$ be the interior symbol of $\mathcal{a}$ which is positively homogeneous of order zero and invertible for $\xi \neq 0$, and $\sigma_N^0(\mathcal{a})(z) = a_0^+ g^+(z) + a_0^- g^-(z) + h_0(z)$ be the principal conormal symbol which is meromorphic and nonsingular on Re $z = 1/2$. Then we find an operator $\mathcal{b}_1 \in \mathcal{L}(\mathbb{R}^{n-1} \setminus 0)$ with

$$\sigma_\psi(\mathcal{b}) = \sigma_\psi(\mathcal{a})^{-1}, \quad \sigma_M^0(\mathcal{b}_1)\Big|_{\mathrm{Re}\, z = 1/2} = \Big[\sigma_M^0(\mathcal{a})\Big|_{\mathrm{Re}\, z = 1/2}\Big]^{-1} .$$

In fact, the inverse of $\sigma_M^0(\mathcal{a})(z)$ is again meromorphic with a system of poles inherited from the zeros of $\sigma_M^0(\mathcal{a})(z)$ and $\mathfrak{z} = (\Delta, \Delta')$ with $\Delta \in \underline{\varphi}$, $\Delta' \in \underline{\varphi}'$. Then $\mathcal{b}_1\mathcal{a} = \mathcal{a}_1 \in \mathcal{L}(\mathbb{R}^{n-1} \setminus 0)$ is of the form $\mathcal{a}_1 = 1 + \mathcal{m} + \mathcal{g}$ with $\sigma_\psi(\mathcal{a}_1) = 1$, $\sigma_M^0(\mathcal{a}_1) = 1$, a Mellin operator $\mathcal{m}$ of conormal order ≤ -1 and a Green operator $\mathcal{g}$. Clearly, $\mathcal{a}_1$ is elliptic, too, and it suffices to have a parametrix $\mathcal{a}_1^{(-1)}$ of $\mathcal{a}_1$ in $\mathcal{L}(\mathbb{R}^{n-1} \setminus 0)$. Such a parametrix was constructed in Section 1.3.2, Theorem 4, by step by step calculating the lower order conormal symbols. Then $\mathcal{a}_1^{(-1)} \mathcal{b}_1 = \mathcal{b} \in \mathcal{L}(\mathbb{R}^{n-1} \setminus 0)$ is a parametrix of $\mathcal{a}$.

It remains to prove (12). The parametrix $\mathcal{b}$ has the property $\mathcal{b}\mathcal{a} - 1 = \mathrm{op}_G(w)$ with some $w \in \mathcal{S}_{\underline{\varphi}} \hat{\otimes}_\pi \mathcal{S}_{\underline{\varphi}}$. Thus $-u = \mathrm{op}_G(w)u$ for $u \in \ker \bar{\mathcal{a}}$. Then $u \in \mathcal{S}_{\underline{\varphi}}$ follows from 1.2.2, Lemma 9, and hence $\mathcal{N}_2 := \ker \bar{\mathcal{a}} \subset \mathcal{S}_{\underline{\varphi}}$. On the other hand, we also have $\bar{\mathcal{a}}\bar{\mathcal{b}} = 1 + \mathrm{op}_G(w')$ with some $w' \in \mathcal{S}_{\underline{\varphi}} \hat{\otimes}_\pi \mathcal{S}_{\underline{\varphi}}$, and therefore,

$$\mathrm{im}(1 + \mathrm{op}_G(w')) \subseteq \mathrm{im}\, \bar{\mathcal{a}} \ . \tag{13}$$

Again in view of 1.2.2, Lemma 9, there is a finite-dimensional subspace $\mathcal{N}_1' \subset \mathcal{S}_{\underline{\varphi}}$ with

$$\mathrm{im}(1 + \mathrm{op}_G(w')) \oplus \mathcal{N}_1' = L^2(\mathbb{R}_+, \mathbb{C}^p).$$

Then (13) shows the second part of (12). Finally, $\bar{\mathcal{a}}u \in \mathcal{S}_{\underline{\varphi}}$ implies $\bar{\mathcal{b}}\bar{\mathcal{a}}u = (1 + \mathrm{op}_G(w'))u \in \mathcal{S}_{\underline{\varphi}}$ and again $u \in \mathcal{S}_{\underline{\varphi}}$. □

Note that the construction of $\mathcal{b}$ and of the Green operators in the latter proof is explicit in the sense of the involved singularities.

It is a simple exercise to calculate the concrete $\mathfrak{r} \in \underline{\mathfrak{P}}$ for which

$$\bar{\alpha}(\xi')u \in \mathcal{S}_{\mathfrak{q}} \Longrightarrow u \in \mathcal{S}_{\mathfrak{r}} \qquad (14)$$

with given $\mathfrak{q} \in \underline{\mathfrak{P}}$, where $\mathfrak{r}$ depends on $\mathfrak{q}$. Below we discuss an example. Roughly speaking, we obtain that $\mathfrak{r} = \mathfrak{q} + \mathfrak{n}$ with some extra system $\mathfrak{n} \in \underline{\mathfrak{P}}$ associated with the zeros in $\mathbb{C}$ of the principal conormal symbol of $\alpha(\xi')$ (which are independent of ξ').

Remember that for operators in Boutet de Monvel's class (cf. [B1])

$$\alpha(\xi') = \mathrm{op}_{\psi}(a)(\xi') + \mathrm{op}_G(w)(\xi'),$$

where $a(\xi)$ has the transmission property, and $w \in \mathcal{S}_{\mathfrak{P}_0} \hat{\otimes}_{\pi} \mathcal{S}_{\mathfrak{P}_0}$ ($\mathcal{S}_{\mathfrak{P}_0} = \mathcal{S}(\overline{\mathbb{R}}_+)$), the finite-dimensional spaces $\mathcal{N}_1, \mathcal{N}_2$ belong to $\mathcal{S}_{\mathfrak{P}_0}$, and that (14) has the form

$$\bar{\alpha}(\xi')u \in \mathcal{S}_{\mathfrak{P}_0} \Longrightarrow u \in \mathcal{S}_{\mathfrak{P}_0}. \qquad (15)$$

The latter property is known as the C^{∞} regularity of elliptic boundary value problems up to the boundary (in the case of the transmission property). In the general case (14) expresses a conormal regularity near the boundary, and we may regard $\mathcal{S}_{\underline{\mathfrak{P}}}$ as a natural substitute of $\mathcal{S}(\overline{\mathbb{R}}_+)$ from the classical theory.

This is a consequence of the complete conormal symbolic calculus and related to the L^2 context with principal symbols (cf.[R2]) in a similar manner as the complete symbolic calculus with C^{∞} regularity in the interior of a domain to the Sobolev space regularity and a principal symbolic calculus.

5. Proposition. Let (9) be invertible, $\alpha \in \mathcal{L}(\mathbb{R}^{n-1} \setminus 0)$. Then $\bar{\alpha}^{-1} = \bar{\mathfrak{b}}$ for some $\mathfrak{b} \in \mathcal{L}(\mathbb{R}^{n-1} \setminus 0)$.

Proof. In view of Theorem 4 we find a parametrix $\mathfrak{b}_1 \in \mathcal{L}(\mathbb{R}^{n-1} \setminus 0)$ of α. Then ind $\mathfrak{b}_1 = 0$. Recall that we are talking about families of operators parametrized by $\xi' \in \mathbb{R}^{n-1} \setminus 0$. In view of the homogeneity with respect to ξ' and the operators κ_λ we may restrict ourselves to $S^{n-2} = \{\xi' : |\xi'| = 1\}$, which is a compact parameter space. Hence there is a finite dimensional subspace $L \subset \mathcal{S}_{\mathfrak{P}_0}(\mathbb{R}_+)$ such that $L + \mathrm{im}\, \mathfrak{b}_1 = L^2(\mathbb{R}_+)$ for all

$\xi' \in S^{n-2}$. Indeed, first we find a finite-dimensional $L_1 \subset L^2(\mathbb{R}_+)$ with that property (cf. [R1], 1.1.1.2, Proposition 10) and then we get L by approximating a base of L_1 by functions in $\mathcal{S}(\overline{\mathbb{R}}_+) = \mathcal{S}_{\varphi_0}$. Let $N = \dim L$

and choose an isomorphism $k : \mathbb{C}^N \longrightarrow L$. Then $(b_1, k) : L^2(\mathbb{R}_+, \mathbb{C}^p) \oplus \mathbb{C}^N \longrightarrow L^2(\mathbb{R}_+, \mathbb{C}^p)$ is a surjective Fredholm family and $M_1 = \ker (b_1, k)$ is a smooth finite-dimensional vector bundle over S^{n-2} with fibre dimension N. Since our assertion is local with respect to the dependence on the parameter ξ', we may restrict ourselves to small open sets on S^{n-2} that form a covering. Let U be such an open set, U simply connected. Then $M = M_1|_U$ is trivial and we find an isomorphism (t_1, q): $M \longrightarrow \mathbb{C}^N$. Then

$$\mathcal{C}_1 = \begin{pmatrix} b_1 & k \\ t_1 & q \end{pmatrix} : \begin{matrix} L^2(\mathbb{R}_+, \mathbb{C}^p) \\ \oplus \\ \mathbb{C}^N \end{matrix} \longrightarrow \begin{matrix} L^2(\mathbb{R}_+ \otimes \mathbb{C}^p) \\ \oplus \\ \mathbb{C}^N \end{matrix}$$

is bijective for all $\xi' \in U$. For diam U small enough we may assume q to be bijective. Otherwise q can be modified by a small perturbation that turns it into an isomorphism of $\mathbb{C}^N$ (the matrices with non-vanishing determinant form an open dense subset in the space of all matrices). By approximating the components of t_1 being in $L^2(\mathbb{R}_+)$ by functions in $\mathcal{S}_{\varphi_0}$ we get a bijection $\mathcal{C}$ consisting of entries b_1, k, t, q, where t has components in $\mathcal{S}_{\varphi_0}$. Then $b_2 := b_1 - kq^{-1}t$: $L^2(\mathbb{R}_+, \mathbb{C}^p) \longrightarrow L^2(\mathbb{R}_+, \mathbb{C}^p)$ is bijective and belongs to $\mathcal{L}(\mathbb{R}^{n-1} \setminus 0)$. Since $kq^{-1}t \in \mathcal{L}_G(\mathbb{R}^{n-1} \setminus 0)$, b_2 is a bijective parametrix of α. Thus $b_2\alpha = 1 + g_1$ is also bijective, $g_1 \in \mathcal{L}_G(\mathbb{R}^{n-1} \setminus 0)$. We already know (cf. 1.2.2, Theorem 8) that $(1+g_1)^{-1} = 1+g_2$, $g_2 \in \mathcal{L}_G(\mathbb{R}^{n-1} \setminus 0)$. Thus $\alpha^{-1} = (1+g_2)\, b_2 \in \mathcal{L}(\mathbb{R}^{n-1} \setminus 0)$. □

Note that at the same time we have proved Theorem 1.

As an example consider a scalar elliptic symbol $a(\xi)$ with the factorization index $\varkappa$, $\alpha(\xi') = \mathrm{op}_\psi(a)(\xi')$. Let us calculate the zeros of

$$\sigma_M^0(\alpha)(z) = a_0^+ g^+(z) + a_0^- g^-(z)$$

being the poles of the inverse. Without loss of generality, we may assume that $a_0^- = 1$. Then

$$\sigma_M^0(\alpha)(z) = (1 - e^{2\pi iz})^{-1}\{1 - a_0^+ e^{2\pi iz}\} .$$

Since $(1 - e^{2\pi iz})^{-1} \neq 0$. we have to find the values of z for which $a_0^+ e^{2\pi iz} = 1$, which leads to the condition

$$\log|a_0^+| + i \arg a_0^+ + 2\pi iz = 2\pi ik, \quad k \in \mathbb{Z} ,$$

i.e.

$$z = \frac{1}{2\pi i} (\log |a_0^+| + i \arg a_0^+) + k = -\varkappa + k, \quad k \in \mathbb{Z}.$$

In this way we recover a particular result of [E2], Section 3.9, for scalar elliptic operators with constant coefficients, which was proved there by completely different methods.

A special case is that $|\varkappa| < 1/2$, such that $op_\psi(a) : L^2(\mathbb{R}_+) \longrightarrow L^2(\mathbb{R}_+)$ is an isomorphism and $op_\psi(a)u \in \mathcal{S}_{\mathcal{P}_0}(\mathbb{R}_+) = \mathcal{S}(\overline{\mathbb{R}}_+)$ then implies $u \in \mathcal{S}_{\mathcal{Q}}(\mathbb{R}_+)$ with $\mathcal{Q} = \mathcal{P}_0 + \mathcal{P}_1$, $\mathcal{P}_1 = \{(-\varkappa-k, k) : k \in \mathbb{Z}_+\}$. Of course, we may replace $\mathcal{P}_0$ by some other $\mathcal{P} \in \underline{\mathcal{P}}$ and then $\mathcal{Q} = \mathcal{P} + \mathcal{P}_1$. Since the methods in [E 2] make essential use of the factorization of the symbol a, that proof has no generalization to systems.

1.6. Notes

Let us give some further comments to the material in Chapter 1 and to the literature. The list of references here cannot be complete, but it may easily be enlarged by the bibliographies in the mentioned papers.

The elementary theory of the Mellin transform in distribution spaces and the aspect of distributions with asymptotics can be found in [J1]. Of course, the Mellin transform is entirely classical and was already employed in the traditional mathematical physics in the end of the last century, cf. [M 1], [M 2].

In the past two decades the Mellin transform was used by many authors for solving special questions of the sort as mentioned in the Introduction.

There also have been made attempts to develop a unification of the calculus of Mellin and pseudo-differential operators, cf. Cordes, Herman [C1], Collela, Cordes [C3], with a certain kind of principal symbols obtained as the Gelfand representation of the Chalkin algebra. The results of Gohberg, Krupnik [G1], [G2] may also be interpreted in this context as well as the algebra of Eskin [E 2], which is rather close to that in [G1], [G2].

The basic difficulty is that the pseudo-differential operators and the Mellin operators associated with the Fourier and the Mellin transform, respectively, do not fit together at a first glance. So the generated algebra may be expected to be unpleasant in the sense of some symbolic level.

In [E 2] everything is formulated for argument functions in $C_0^\infty(\mathbb{R}_+)$, although the operators have to be interpreted as continuous operators in $L^2(\mathbb{R}_+)$. The negligible operators are the compact ones.

In [R2] the authors have developed a boundary symbolic calculus with principal symbols which extends the algebra of [E2] by trace, potential and Green operators, and it was shown that the inverses of bijective elements belong to this class, again. The main objective in [R2] is the higher dimensional theory for operators without the transmission property, in particular the index theory, where the Mellin symbols play a specific role in the K theoretic classification of elliptic boundary value problems (cf. also Section 2.3.).

The 'conormal role' of the Mellin symbols, which is explained in [R2], is a source of our notations; it is very natural to identify $\{\operatorname{Re} z = 1/2\}$ with the conormal direction to the boundary. On the other hand 'conormal distributions' have been studied by Melrose [M 3], Melrose, Mendoza [M 4] and others. So we found it justified systematically to talk about the conormal singularity (or the conormal asymptotics) whenever an expansion like 1.2.1.(3) holds. In Chapter 2 this notion will considerably be extended.

The invertibility of a bijective boundary symbol within the given class is a crucial point also in Boutet de Monvel's theory of operators with the transmission property, cf. [B1], which refers to $\mathcal{S}(\overline{\mathbb{R}}_+)$. Unfortunately this approach has no immediate generalization to the case without the transmission property for the spaces $\mathcal{S}_{\mathfrak{P}}(\mathbb{R}_+)$, cf. Section 1.5.3. The functional analysis of the $\mathcal{S}_{\mathfrak{P}}$ spaces is more complicated than that of $\mathcal{S}(\overline{\mathbb{R}}_+) = \mathcal{S}_{\mathfrak{P}_0}(\mathbb{R}_+)$. Though any $\mathcal{S}_{\mathfrak{P}}$ contains a countable set with dense linear span, there is no obvious condition to the coefficients of the series for the convergence in $\mathcal{S}_{\mathfrak{P}}$.

The authors could not find in the literature any systematic treatment of the interesting class of $\mathcal{S}_{\mathfrak{P}}$ spaces. For this reason the structure of these spaces was investigated in Section 1.2. from the very beginning. The reader interested in functional analysis may find a number of questions that are worth to be answered. For instance, it would be useful to know whether there is a natural isomorphism between $\mathcal{S}_{\mathfrak{P}}$ and $\mathcal{S}_{\mathfrak{Q}}$ for different $\mathfrak{P}, \mathfrak{Q} \in \underline{\mathfrak{P}}$, which moves, roughly speaking, the positions of the singularities into each other. Then the isomorphism to $\mathcal{S}_{\mathfrak{P}_0}$ would give a base in each $\mathcal{S}_{\mathfrak{P}}$.

The Green operators in Section 1.2.2. are understood in terms of kernels. In this sense they are a realization of a well-known abstract situation which is employed once again in Section 2.1.4.

A novelty in our algebra are the Mellin operators $\mathcal{M}$ associated with the sequences of conormal symbols $\sigma_M^{-1}(\mathcal{M})$. The principal parts correspond to the Mellin operators in [E2], but they are not quite identical with them. In order to generate a $*$ algebra in spaces with conormal singularity it is important to have a cut-off function also from the right. The structure of the lower order terms is a bit more complicated. The distribution of the powers of t on both sides of the Mellin action is needed because of possible poles of the conormal symbols at points $z \in \mathbb{C}$ with $\mathrm{Re}\, z \in \mathbb{Z} + 1/2$ which might move to $\mathrm{Re}\, z = 1/2$ under composition, cf. formula 1.3.2.(5). The lower order Mellin operators are Hilbert-Schmidt in $L^2(\mathbb{R}_+)$. They are included in the Green operators in [R2].

The results of the Sections 1.5.1, 1.5.2 show that these Green operators have the structure $\mathcal{M}+\mathcal{G}$, where $\mathcal{M}$ is a Mellin operator in the sense of 1.3.1. with $\sigma_M^0(\mathcal{M}) = 0$, and $\mathcal{G}$ a Green operator in the sense of 1.2.2.

This result is a consequence of the structure of the algebra $\mathcal{L}$, in particular of the important Mellin expansion of the pseudo-differential action in 1.4.1. Another corollary of this expansion is the continuity of ψDO's in $\mathcal{S}_\varphi$.

The action of a ψDO $op_\psi(a)$ adds a typical singularity to the argument function (namely φ_0) except when a has the transmission property. The Mellin action also adds singularity and carries over the original one, whereas the Green operators forget the singularity and produce another one. In this way the operators in $\mathcal{L}$ transform the conormal singularity in a well-defined manner. Since the algebra is closed under parametrix construction for elliptic elements, we obtain the conormal singularity of the solutions in terms of the singularity of the right hand side and the zeros of the principal conormal symbol in the complex plane. Hence the symbolic calculus in $\mathcal{L}$ explains how (and why) the singularities of the solutions are produced, cf. the remarks at the end of 1.5.3.

A feature of the approach with Mellin expansions of operators is the precise localization at the exceptional point $t = 0$. The meaning of that aspect will be explained in detail in Chapter 3, where we deal with operators that degenerate at $t = 0$.

Our theory emphasizes very much the poles and their multiplicities of meromorphic functions. In the parametrix construction we have to pass to inverse functions, where zeros and poles change their role. Moreover, the zeros of $(Mu)(z)$ for argument functions u may compensate the poles of Mellin symbols h under the action $op_M(h)u$. There exists a more symmetric version of classes with asymptotics, where the theory keeps in mind both poles and zeros of the involved objects. Then, in the sequences $\varphi = \{(p_j, m_j)\}_{j\in\mathbb{Z}_+}$ are also admitted integers $m_j < 0$.

The calculus in Chapter 1 is the simplest variant of a variety of modifications. A number of them will be discussed in the following chapters. The generalizations are suggested also by special investigations of operators on the half line.

It is obvious that our theory contains as a special case the singular integral operators with discontinuous symbols. So we may ask a lot which is custom in that part of the analysis, for instance, with respect to other function spaces (weighted spaces, the L^p case, Hölder spaces and so on). Furthermore, there are typical applications, e.g. to degenerate ordinary differential equations and two-dimensional mixed problems.

Let us finish this discussion with some references that are related to the mentioned problems.

The Mellin transform as one of the classical integral transforms has been studied in a great number of papers and monographs, see, for instance Titchmarsh [T 1], Fung [F 3], Zemanian [Z3], Komeč [K8], Bryčkov Prudnikov [B4], Jeanquartier [J1].

Mellin operators in the context of symbols and certain relations to ψDOs were investigated by Cordes, Herman [C1], Plamenevskiĭ [P3],[P4], Eskin [E2], Nourrigat [N 2], Collela, Cordes [C3], Lewis, Parenti [L2], [L2], Schulze [S4], Rempel, Schulze [R2],[R8].

Applications to boundary problems in plane non-smooth domains were given for instance, by Eskin [E3], Costabel [C5] and Costabel, Stephan, Wendland [C 6].

Aspects of the conormal asymptotics of solutions for degenerate ordinary differential equations were discussed in Elschner [E 1]. For solutions of ψDO's on the half axis (and also in the higher dimensional case) the asymptotics extends the concept of smoothness up to the boundary which is satisfied in the calculus of operators with the transmission property, cf. Višik, Eskin [V1], Boutet de Monvel [B1].

2. Continuous asymptotics and higher order operators

2.1. Spaces of functions with continuous conormal singularity

2.1.1. The spaces $\mathcal{S}_\Lambda(\mathbb{R}_+)$ and $C^\infty_{\Lambda,\Lambda'}(\mathbb{R}_+)$

For solving boundary problems in spaces of functions with conormal singularity we have to admit some x' dependence of the exponents $p_j \in \mathbb{C}$. This is caused by the x' dependence of the pseudo-differential symbols. Then the positions of poles and zeros of the conormal symbols may vary with x' (cf. the example at the end of 1.5.3.). This makes it necessary to introduce the notion of continuous asymptotics, which is more general than the discrete one of Chapter 1. Roughly speaking we associate the spaces $\mathcal{S}_\Lambda$, $C^\infty_{\Lambda,\Lambda'}$ with certain given subsets $\Lambda, \Lambda' \subset \mathbb{C}$. Here Λ and Λ' are unions of disjoint compact sets $\Lambda_j \subset \{\mathrm{Re}\ z < 1/2\}$ and $\Lambda'_j \subset \{\mathrm{Re}\ z > 1/2\}$, respectively, and $\{\Lambda_j\}_{j\in\mathbb{Z}_+}$, $\{\Lambda'_j\}_{j\in\mathbb{Z}_+}$ are the substitutes of the systems of exponents $\{p_j\}_{j\in\mathbb{Z}_+}$, $\{p'_j\}_{j\in\mathbb{Z}_+}$ in the discrete asymptotics. The more complicated general situation with non-compact connected components is necessary, too, and will be studied in Section 2.1.5.

In order to motivate our definitions let us reinterprete the situation of the discrete conormal singularity in terms of analytic functionals. By $\mathcal{A}(\Omega)$, $\Omega \subseteq \mathbb{C}$ open, we denote the space of all holomorphic functions in Ω, equipped with the topology of uniform convergence on compact sets.

The topological dual $\mathcal{A}'(\Omega)$ of $\mathcal{A}(\Omega)$ is called the space of <u>analytic functionals</u> on Ω. For basic properties of $\mathcal{A}'(\Omega)$ we refer to [S2], [H3]. Let us remind here of the following

1. <u>Definition</u>. Let $K \subset \mathbb{C}$ be a compact set. Then $\lambda \in \mathcal{A}'(\mathbb{C})$ is said to be <u>carried by K</u> if for every open neighbourhood U of K there is a constant C_U such that

$$|\langle\lambda,h\rangle| \leq C_U \sup_{z\in U} |h(z)| \quad \text{for all } h \in \mathcal{A}(\mathbb{C}) . \tag{1}$$

By $\mathcal{A}'(K)$ we denote the subspace of all $\lambda \in \mathcal{A}'(\mathbb{C})$ carried by K.

$\mathcal{A}'(K)$ is a Fréchet space with the best constants C_U in (1) as semi-norms, namely

$$\psi_U(\lambda) := \sup_{h \in \mathcal{A}(\mathbb{C})} \left\{ \left(\sup_{z \in U} |h(z)| \right)^{-1} \, |\langle \lambda, h \rangle| \right\} . \tag{2}$$

In view of the Hahn-Banach theorem for every open neighbourhood U of K there is a measure μ with supp $\mu \subset U$ and

$$\langle \lambda, h \rangle = \int h \, d\mu \quad \text{for all } h \in \mathcal{A}(\mathbb{C})$$

(Below we shall construct special representations of μ).

In this way we get extensions of λ to holomorphic functions h in U. In particular if $\Phi(z,w)$ is a fundamental solution of the Cauchy-Riemann operator $\bar{\partial}$ we can consider the potentials $\langle \lambda, \Phi \rangle (z)$ as functions of $z \in U$, where λ acts on the variable w.

By

$$\langle \frac{d^k}{dw^k} \delta_{w-p}, h \rangle := (-1)^k \left(\frac{d^k}{dw^k} h\right)(p),$$

$p \in \Omega$ fixed, we get an element in $\mathcal{A}'(\mathbb{C})$ carried by p.

Let $\wp = \{(p_j, m_j)\}_{j \in \mathbb{Z}_+}$ be as in the definition of the discrete conormal singularity. For $\gamma \geq 0$ define $N(\gamma) \in \mathbb{Z}_+$ as the smallest j for which $\operatorname{Re} p_k < 1/2 - \gamma$ for $k > j$. If $\xi_{jk} \in \mathbb{C}$ are given constants

$$\xi_\gamma := \sum_{j=0}^{N} \sum_{k=0}^{m_j} \xi_{jk} \frac{d^k}{dw^k} \delta_{w-p_j} \tag{3}$$

$N = N(\gamma)$, belongs to $\mathcal{A}'(\mathbb{C})$. Let

$$\pi_{\mathbb{C}} \wp := \{ w \in \mathbb{C} \; : \; w = p_j, \; j \in \mathbb{Z}_+ \} \tag{4}$$

and

$$\Gamma(\gamma)\wp := \pi_{\mathbb{C}} \wp \cap \{ \operatorname{Re} w \geq 1/2 - \gamma \} .$$

The function

$$F(t,w) := t^{-w} \omega(t) \tag{5}$$

belongs to $\mathcal{A}(\mathbb{C})$ for any fixed $t \in \mathbb{R}_+$. For $\gamma \geq 0$ we obviously have

$$\kappa_\gamma u(t) = u(t) - \langle \xi_\gamma, F \rangle (t)$$

and $u \in \mathcal{S}_\wp$ is equivalent to the existence of some $\xi_\gamma \in \mathcal{A}'(\Gamma(\gamma)\wp)$

of the form (3) such that

$$t^{-\gamma+\eta}\partial^1 r_\gamma u = t^{-\gamma+\eta}(\partial^1 u - \langle \zeta_\gamma, \partial^1 F\rangle)(t) \in L^2(\mathbb{R}_+) \tag{6}$$

for all $\alpha = (\eta, \gamma, 1) \in \underline{\alpha}$.

Define the function

$$\phi(z,w) = \int_0^\infty t^{z-1} F(t,w)dt \tag{7}$$

which is a fundamental solution of $\bar{\partial}$.

$\phi(z,w)$ is a meromorphic function with a simple pole at $z=w$. Instead of (6) we may write

$$z^1(\tilde{u} - \langle \zeta_\gamma, \phi\rangle)(z) \in T^{\gamma-\eta}\ \mathbb{L}_0^2 \tag{8}$$

for all $\alpha \in \underline{\alpha}$, $\mathbb{L}_0^2$ being defined as the space of all functions of the form $f(z) = \tilde{v}_1(z) + \tilde{v}_2(z)$, $v_1 \in L^2(\mathbb{R}_+)$, supp v_1 bounded, $v_2 \in \mathcal{S}(\overline{\mathbb{R}}_+)$.

Remember that the Mellin transform u of every $u \in L^2(\mathbb{R}_+)$ belongs to $L^2(\{\text{Re } z = 1/2\})$. It may happen, for instance when $u \in \mathcal{S}_\varphi$, that $\tilde{u}$ has an extension to some domain $\Omega \subset \mathbb{C}$ as a single valued holomorphic function. This will also be denoted by $\tilde{u}$ and called the Mellin transform, again.

Now we extend the concept of the discrete conormal asymptotics to a so-called continuous asymptotics associated with more general sets $\Lambda \subset \mathbb{C}$ and analytic functionals carried by the compact components. Consider a set $\Lambda \subset \mathbb{C}$,

$$\Lambda = \bigcup_{j \in \mathbb{Z}_+} \Lambda_j \tag{9}$$

being the union of disjoint compact sets $\Lambda_j \subset \mathbb{C}$,

$$\text{Re } w < 1/2 \quad \text{for } w \in \Lambda \tag{10}$$

and a sequence

$$d = (\delta_j)_{j \in \mathbb{Z}_+}, \quad \delta_j \geq 0, \tag{11}$$

$\delta_0 = 0$, $\delta_j < \delta_{j+1}$, $\delta_j \longrightarrow \infty$ as $j \longrightarrow \infty$ and

$$\sup_{z \in \Lambda_j} \text{Re } z < 1/2 - \delta_j < \inf_{z \in \Lambda_{j-1}} \text{Re } z \tag{12}$$

for all $j \in \mathbb{Z}_+$, $j \geq 1$. Set

$$\underline{\alpha} := \overline{\mathbb{R}}_+ \times d \times \mathbb{Z}_+ \quad .$$

The function (5) may be considered as element

$$F(t,w) \in \mathcal{A}(\{\mathrm{Re}\, w < 1/2 - \delta\},\ L^2(\mathbb{R}_+, t^{-2\delta}dt))$$
$$= \mathcal{A}(\{\mathrm{Re}\, w < 1/2 - \delta\}) \otimes_\pi L^2(\mathbb{R}_+, t^{-2\delta}dt) \ .$$

Thus, for any $\zeta_\delta \in \mathcal{A}'(\Lambda_{(\delta)})$,

$$\Lambda_{(\delta)} := \Lambda \cap \{\mathrm{Re}\, z > 1/2 - \delta\} \ ,$$

we obtain $< \zeta_\delta , F > \in L^2(\mathbb{R}_+)$. Now, for (7) we have

$$\phi(z,w) = (M_t F)(z,w) \in \mathcal{A}(\{\mathrm{Re}\, w < 1/2 - \delta\}) \otimes_\pi L^2(\{\mathrm{Re}\, z = 1/2 - \delta\})$$

and hence

$$< \zeta_\delta , \phi > (z) = M < \zeta_\delta , F > \in L^2(\{\mathrm{Re}\, z = 1/2\}) \ .$$

2. <u>Definition</u>. By $\mathcal{S}_\Lambda(\mathbb{R}_+)$ we denote the subspace of all $u \in C^\infty(\mathbb{R}_+)$ such that for every $\delta \in d$ there exists some $\zeta_\delta \in \mathcal{A}'(\mathbb{C})$ carried by $\Lambda_{(\delta)}$ and

$$t^{-\delta+\eta} \vartheta^1 (u - < \zeta_\delta , F >)(t) \in L^2(\mathbb{R}_+) \tag{13}$$

for all $\alpha = (\eta, \delta, 1) \in \underline{\alpha}$.

Again (13) may be replaced by

$$z^1 (\tilde{u} - < \zeta_\delta , \phi >)(z) \in T^{\delta - \eta} \mathbb{L}^2_0 \ . \tag{14}$$

Below we shall see that another sequence (11) with the mentioned properties with respect to Λ leads to the same space. For the moment d is chosen and then fixed.

The intuitive interpretation of Definition 2 is that Λ represents some <u>singularity type</u> of $u \in C^\infty(\mathbb{R}_+)$ near t=0, whereas the ζ_δ describe the <u>quantities of singularity</u> concentrated on $\Lambda_{(\delta)}$. This picture will also be discussed in Section 2.1.3 from another point of view, where we refer to more general sets Δ in the complex plane and holomorphy outside of Δ .

3. <u>Lemma</u>. Let ζ'_δ be another functional carried by $\Lambda_{(\delta)}$ and

$$t^{-\delta+\eta} \vartheta^1 (u - < \zeta'_\delta , F >) \in L^2(\mathbb{R}_+)$$

for at least one $\eta \in \overline{\mathbb{R}}_+$, $1 \in \mathbb{Z}_+$. Then $\zeta_\delta = \zeta'_\delta$.

In other words, the correspondence $u \longrightarrow \xi_\delta = \xi_\delta(u)$ of Definition 2 is a well-defined linear mapping

$$\xi_\delta : \mathcal{S}_\Lambda(\mathbb{R}_+) \longrightarrow \mathcal{A}'(\Lambda_{(\delta)}) .$$

<u>Proof</u>: The definition implies that

$$\tilde{u} - \langle \xi_\delta , \Phi \rangle \quad , \quad \tilde{u} - \langle \xi'_\delta , \Phi \rangle$$

have analytic extensions to $\operatorname{Re} z > 1/2 - (\delta+\varepsilon)$, $\varepsilon > 0$ sufficiently small. Thus

$$\langle \xi_\delta - \xi'_\delta , \Phi \rangle \tag{15}$$

also has an analytic extension to $\operatorname{Re} z > 1/2 - (\delta+\varepsilon)$. Since both ξ_δ, ξ'_δ are carried by the compact set $\Lambda_{(\delta)} \subset \{\operatorname{Re} z > 1/2 - (\delta+\varepsilon)\}$ the function (15) is entire. From this we shall conclude that $\xi_\delta = \xi'_\delta$.

Let

$$F_0(t,w) = t^{-w} \varepsilon(t) \tag{16}$$

ε being the characteristic function of $(0,1)$ and

$$\Phi_0(z,w) = \int_0^\infty t^{z-1} F_0(t,w)dt = \frac{1}{z-w} . \tag{17}$$

Then

$$\Psi(z,w) = \Phi(z,w) - \Phi_0(z,w)$$

is entire separately in z and w. Then the function $\langle \lambda , \Psi \rangle(z)$, $\lambda := \xi_\delta - \xi'_\delta$ is also entire and $\langle \lambda , \Phi_0 \rangle(z)$ is entire, too, since $\langle \lambda , \Phi \rangle$ is entire. In view of

$$h(w) = \frac{1}{2\pi i} \int_B h(z') \frac{1}{z'-w} dz', \quad h \in \mathcal{A}(\mathbb{C}) ,$$

for any curve B surrounding $\Lambda_{(\delta)}$, we obtain

$$\langle \lambda , h \rangle = \frac{1}{2\pi i} \int_B h(z') \langle \lambda , \Phi \rangle (z') dz' = 0$$

and hence $\lambda = 0$. □

If $K \subsetneq \mathbb{C}$ is a closed set, by $c(\mathbb{C} \setminus K)$ we denote the union of the connection components of the unbounded part of $\mathbb{C} \setminus K$. Then

$$K^c := \mathbb{C} \setminus c(\mathbb{C} \setminus K) \tag{18}$$

is a closed set with $K \subseteq K^c$ and $(K^c)^c = K^c$. Obviously $K = K^c$ iff K is simply connected.

By $\underline{\Lambda}$ we denote the set of all closed subsets $\Lambda \subset \mathbb{C}$ with the following properties

(i) $\Lambda^c = \Lambda$,

(ii) there is a sequence of compact subsets $\Lambda_j \subseteq \Lambda$, $j \in \mathbb{Z}_+$ with $\Lambda = \bigcup_{j \in \mathbb{Z}_+} \Lambda_j$ and $\sup\{\mathrm{Re}\, z : z \in \Lambda_{j+1}\} < \inf\{\mathrm{Re}\, z : z \in \Lambda_j\}$ for all $j \in \mathbb{Z}_+$, $\sup\{\mathrm{Re}\, z : z \in \Lambda_j\} \to -\infty$ as $j \to \infty$,

(iii) $\Lambda \subset \{z \in \mathbb{C} : \mathrm{Re}\, z < 1/2\}$,

(iv) for any $\eta, \delta \in \mathbb{R}_+$ the set of points

$$B_{\delta\eta} := B_{\delta\eta}(\Lambda) := \{z \in \mathbb{C} \setminus \Lambda : \mathrm{dist}(z, \Lambda) = \tfrac{1}{1+\eta}\} \cap \{z : \mathrm{Re}\, z \geq \tfrac{1}{2} - \delta\} \tag{19}$$

is a piecewise smooth curve.

To every $\Lambda \in \underline{\Lambda}$ there exists a system $d = (\delta_j)_{j \in \mathbb{Z}_+}$ of non-negative reals with the properties mentioned above, cf. Fig. 2.

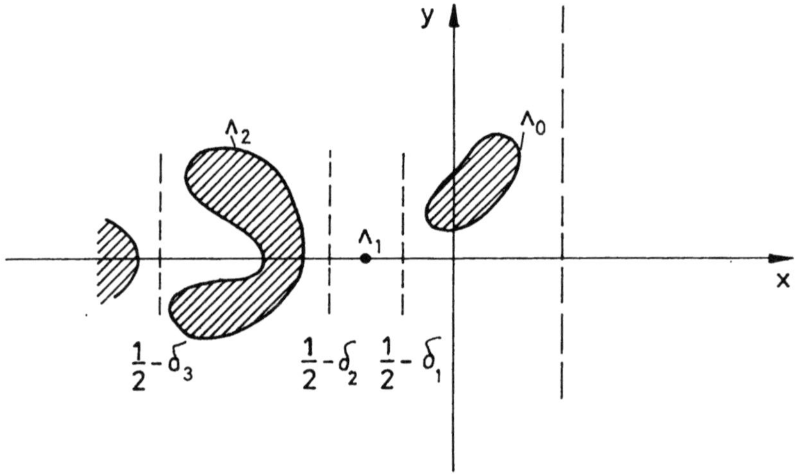

Fig. 1

The property (iv) is not essential for the basic properties of the $\mathcal{S}_\Lambda$ spaces. It was introduced for convenience when we deal with operators of some order, cf. Definition 13 below.

If $K \subset \mathbb{C}$ is compact and $K^c = K$, by $\mathcal{U}(K)$ we denote a countable system of open relatively compact sets $U \supset K$ with smooth boundaries,

$$\overline{U}^c = \overline{U}, \quad U \subseteq U' \text{ or } U' \subseteq U$$

for any $U, U' \in \mathcal{U}(K)$ and

$$K = \bigcap \{ U : U \in \mathcal{U}(K) \} .$$

In particular for $K = \Lambda_{(\delta)}$ we set

$$\mathcal{U}_\delta = \mathcal{U}(\Lambda_{(\delta)})$$

and assume in addition in this case $U \subset \{\operatorname{Re} z > 1/2 - \delta\}$ for all $U \in \mathcal{U}_\delta$.

4. Remark. The analytic functional $\xi_\delta(u)$ describing the singularity quantities can be represented by the values of the analytic extension of $\tilde{u}(z)$, in other words, let $\lambda = \xi_\delta(u)$, $u \in \mathcal{S}_\Lambda$, and $B = \partial U$, $U \in \mathcal{U}_\delta$. Then

$$\langle \lambda, h \rangle = \frac{1}{2\pi i} \int_B \tilde{u}(z)\, h(z)\, dz$$

for every $h \in \mathcal{A}(V)$, $V \supset \overline{U}$ an open neighbourhood, $V \subset \{\operatorname{Re} z > 1/2 - \delta\}$. In fact, we know that $\tilde{u}(z) - \langle \lambda, \phi_o \rangle(z)$ is holomorphic in a neighbourhood of $\overline{U}$, then

$$\int_B \tilde{u}(z)\, h(z)\, dz = \int_B \langle \lambda, \phi_o \rangle (z)\, h(z)\, dz$$

$$= \int_B \langle \lambda_w, \frac{1}{z-w} \rangle\, h(z)\, dz = 2\pi i \langle \lambda, h \rangle .$$

Now we shall formulate further semi-norm systems on $\mathcal{A}'(\Lambda_{(\delta)})$, equivalent to ψ_U. They will be given in terms of $\langle \xi, \phi \rangle$, which differs from the potential of ξ with respect to the Cauchy kernel ϕ_o by an entire function. For any $U \in \mathcal{U}_\delta$ we have a Hermitean form on $\mathcal{A}'(\Lambda_{(\delta)})$

$$\varphi_B(\xi, \lambda) := \int_B \langle \xi, \phi \rangle (z)\, \overline{\langle \lambda, \phi \rangle (z)}\, d\sigma , \tag{20}$$

where $d\sigma$ denotes the measure on $\partial U = B$ induced by the metric of $\mathbb{C}$.

Set

$$\varphi_B(\xi, \xi)^{1/2} =: \varphi_B(\xi) . \tag{21}$$

$\varphi_B(\cdot)$ is a norm on $\mathcal{A}'(\Lambda_{(\delta)})$, since $\varphi_B(\xi) = 0$ implies $\langle\xi,\Phi\rangle(z) = 0$ for all $z \in B$. Then

$$\langle \xi ,h \rangle = \frac{1}{2\pi i} \int_B \langle\xi,\Phi\rangle(z)\; h(z)\; dz \tag{22}$$

vanishes for all $h \in \mathcal{A}(\mathbb{C})$, i.e. $\xi = 0$. Thus (20) is a scalar product on $\mathcal{A}'(\Lambda_{(\delta)})$.

5. **Lemma.** The topologies on $\mathcal{A}'(\Lambda_{(\delta)})$ defined by the systems of seminorms $\{\psi_U : U \in \mathcal{U}_\delta\}$ and $\{\varphi_B : B = \partial U,\ U \in \mathcal{U}_\delta\}$ are equivalent. $\mathcal{A}'(\Lambda_{(\delta)})$ is a nuclear Fréchet space.

Proof. By definition of ψ_U we have

$$|\langle \xi(w),\ \Phi(w,z)\rangle| \leq \psi_U(\xi) \sup_w |\Phi(w,z)|$$

for all $z \in B$, $B = \partial U$. Then

$$\varphi_B(\xi)^2 = \int_B |\langle\xi,\Phi\rangle(z)|^2\, d\sigma$$

$$\leq c\ \psi_U(\xi)^2 \int_B \sup_{w\in\Lambda_{(\delta)}} |\Phi(w,z)|^2\, d\sigma .$$

Conversely

$$|\langle \xi,h\rangle| = \left|\frac{1}{2\pi i}\int \langle \xi,\Phi\rangle(z)\; h(z)\; dz\right|$$

$$\leq c \sup_{z\in B} |h(z)| \int |\langle \xi,\Phi\rangle(z)|\, d\sigma$$

$$\leq c \sup_{z\in U} |h(z)| \left\{\int |\langle \xi,\Phi\rangle(z)|^2\, d\sigma\right\}^{1/2}$$

and hence

$$\psi_U(\xi) \leq c\ \varphi_B(\xi) \tag{23}$$

for all $\xi \in \mathcal{A}'(\Lambda_{(\delta)})$ with some $c > 0$, $B = \partial U$. The nuclearity of $\mathcal{A}'(\Lambda_{(\delta)})$ follows in the same way as for standard spaces of holomorphic functions. In fact, according to (22) each functional ξ can be represented by the holomorphic function $\langle\xi,\Phi\rangle(z)$ in the exterior of $\Lambda_{(\delta)}$. □

Let $U \in \mathcal{U}_\delta$, $B = \partial U$, $\xi \in \mathcal{A}'(\Lambda_{(\delta)})$.

$$\gamma_B(\xi) = \sup_{z\in B} |\langle \xi,\Phi\rangle(z)| . \tag{24}$$

6. <u>Lemma</u>. The topologies on $\mathcal{A}'(\Lambda_{(\delta)})$ defined by the systems of semi-norms $\{\psi_U : U \in \mathcal{U}_\delta\}$ and $\{\gamma_B : B = \partial U,\ U \in \mathcal{U}_\delta\}$ are equivalent.

<u>Proof</u>. The topology of $\mathcal{A}'(\Lambda_{(\delta)})$ defined by $\{\psi_U\}$ is stronger or equal to that defined by $\{\gamma_B\}$. On the other hand it is obvious that the $\{\gamma_B\}$ topology is stronger or equal to the $\{\varphi_B\}$ topology. Thus the desired equivalence follows from Lemma 5. □

Now we show that the convergence of analytic functionals implies the convergence of their potentials $<\xi, \phi>$ in the space of rapidly decreasing functions on the lines

$$\Gamma_\rho = \{ \mathrm{Re}\ z = \rho \} . \tag{25}$$

7. <u>Lemma</u>. Let $\xi_n \longrightarrow \xi$ in $\mathcal{A}'(\Lambda_{(\delta)})$ and $\Gamma_\rho \cap \Lambda_{(\delta)} = \emptyset$. Then

$$(\rho + iy)^l < \xi_n, \phi > (\rho + iy) \longrightarrow (\rho + iy)^l < \xi, \phi > (\rho + iy) \tag{26}$$

in $L^2(\Gamma_\rho)$ for all $l \in \mathbb{Z}_+$.

<u>Proof</u>: For fixed ρ we find some $U \in \mathcal{U}_\delta$ with $\mathrm{Re}\ z \neq \rho$ for all $z \in \overline{U}$. Then

$$< \xi_n, \phi >(z) = \int_B \lambda_n(w)\ \phi(z, w)\ dw$$

with $\lambda_n \in C^\infty(B)$ representing ξ_n. A similar representation holds for ξ. Then

$$\int_{-\infty}^{\infty} |(\rho + iy)^l < \xi_n - \xi, \phi > (\rho + iy)|^2\ dy$$

$$= \int_{-\infty}^{\infty} |(\rho + iy)^l \int_B (\lambda_n(w) - \lambda(w))\, \phi(\rho + iy, w)\ dw|^2\ dy$$

$$\leq \int |\lambda_n(w) - \lambda(w)|^2 |dw| \int |(\rho + iy)^l|^2 \int |\phi(\rho + iy, w)|^2 |dw|\ dy$$

$\longrightarrow 0$ as $n \longrightarrow \infty$. □

Our next aim is to construct a semi-norm system on $\mathcal{S}_\Lambda$ under which it becomes a Fréchet space. It turns out that the $L^2(\mathbb{R}_+)$-norms of the expressions (13) are not sufficient for a topology on $\mathcal{S}_\Lambda$ such that the operators ξ_δ are continuous. We need an extra term for the functionals. So let us define on $\mathcal{S}_\wedge$ a system of scalar products

$$\pi_\gamma(u,v) := (u,v)_{L^2(\mathbb{R}_+)} +$$

$$+ (t^{-\delta+\eta}\vartheta^l(u-\langle\zeta_\delta(u),F\rangle),\ t^{-\delta+\eta}\vartheta^l(v-\langle\zeta_\delta(v),F\rangle))_{L^2(\mathbb{R}_+)}$$

$$+ (\langle\zeta_\delta(u),\phi\rangle(z),\langle\zeta_\delta(v),\phi\rangle(z))_{L^2(B)} \ .$$

Here $\gamma = (\eta,\delta,l,B)$, where $\eta \in \overline{\mathbb{R}}_+$, $\delta \in d$, $l \in \mathbb{Z}_+$ and $B = \partial U$ for $U \in \mathcal{U}_\delta$. The corresponding index set will be denoted by $\underline{\gamma}$. Set

$$\pi_\gamma(u,u)^{1/2} =: \pi_\gamma(u) \ .$$

Let us define a further system of scalar products on $\mathcal{S}_\wedge$ by the expressions

$$\mu_B(u,v) := (\tilde{u},\tilde{v})_{L^2(\Gamma_{1/2})} + (z^l\tilde{u},\ z^l\tilde{v})_{L^2(\Gamma_\rho)} + (\langle\zeta_\delta(u),\phi\rangle,\langle\zeta_\delta(v),\phi\rangle)_{L^2(B)} \tag{27}$$

and associated norms

$$\mu_B(u) := \mu_B(u,u)^{1/2} \ , \tag{28}$$

$B \in \underline{B}$, where $\underline{B}$ is the set of all tuples $B = (\delta,\rho,l,B)$, $\delta \in d$, $\rho \in \{\frac{1}{2} - \delta : \delta \in d\} \cup (\mathbb{Z}_+ \setminus 0)$, $l \in \mathbb{Z}_+$, B as above. Instead of (27) and (28), respectively, we also can consider the system of scalar products

$$\mu'_B(u,v) := (\tilde{u},\tilde{v})_{L^2(\Gamma_{1/2})} + (z^l\tilde{u},\ z^l\tilde{v})_{L^2(\Gamma_\rho)} + (\tilde{u},\tilde{v})_{L^2(B)} \tag{29}$$

and the associated norms $\mu'_B(u) = (\mu'_B(u,u))^{1/2}$.

Another convenient semi-norm system is

$$\bar{\mu}_B(u) := \max\Big\{ \sup_{y\in\mathbb{R}} |\tilde{u}(1/2+iy)|,\ \sup_{y\in\mathbb{R}} |(\rho+iy)^l\ \tilde{u}(\rho+iy)|,\ \sup_{z\in B} |\tilde{u}(z)| \Big\} \ , \tag{30}$$

$B \in \underline{B}$.

8. <u>Theorem</u>. $\mathcal{S}_\wedge$ equipped with the system of norms $(\pi_\gamma)_{\gamma\in\underline{\gamma}}$ is a nuclear Fréchet space. The topology is equivalent to that defined by the system $(\mu_B)_{B\in\underline{B}}$. Other equivalent systems are $(\mu'_B)_{B\in\underline{B}}$ and

$(\bar{\mu}_{\beta})_{\beta\in\underline{\beta}}$.

Proof. The simple arguments that $\mathcal{S}_{\Lambda}$ is Fréchet are left to the reader. Let us check the equivalence of (π_{γ}) and (μ_{β}).

We have to show that $(\mathcal{S}_{\Lambda},(\pi_{\gamma})) \longrightarrow (\mathcal{S}_{\Lambda},(\mu_{\beta}))$ is continuous with continuous inverse, which means that

(i) to any $\beta\in\underline{\beta}$ there are finitely many $\gamma_k\in\underline{\gamma}$, $k = 1,\dots,k_{\beta}$, with

$$\mu_{\beta}(u) \leq c \sup_k \pi_{\gamma_k}(u),\ u\in\mathcal{S}_{\Lambda} \tag{31}$$

with some constant $c = c(\beta,\gamma_1,\dots,\gamma_{k_{\beta}})$.

(ii) to any $\gamma\in\underline{\gamma}$ there are finitely many $\beta_l\in\underline{\beta}$, $l = 1,\dots,l_{\alpha}$, with

$$\pi_{\gamma}(u) \leq c \sup_l \mu_{\beta_l}(u),\ u\in\mathcal{S}_{\Lambda} \tag{32}$$

with some constant $c = c(\gamma,\beta_1,\dots,\beta_{l_{\alpha}})$. First remark that

$$\pi_{\gamma}(u) = \Big\{\|u\|^2_{L^2(\Gamma_{\nu/2})} + \|z^l(\tilde{u}(z) - \langle\zeta_{\delta}(u),\phi\rangle(z)\|^2_{L^2(\Gamma_{\nu/2-\delta+\eta})}$$
$$+\|\langle\zeta_{\delta}(u),\phi\rangle(z)\|^2_{L^2(B)}\Big\}^{\frac{1}{2}},$$

and that we may restrict ourselves to those δ,η for which $\nu/2-\delta+\eta$ equals one of the l_j, $j\in\mathbb{Z}$. We have

$$\|z^l\tilde{u}(z)\|_{L^2(\Gamma_{\nu/2-\delta+\eta})} \leq \|z^l(\tilde{u}(z) - \langle\zeta_{\delta}(u),\phi\rangle(z))\|_{L^2(\Gamma_{\nu/2-\delta+\eta})}$$
$$+\|z^l\langle\zeta_{\delta}(u),\phi\rangle(z)\|_{L^2(\Gamma_{\nu/2-\delta+\eta})}$$

and (31) is a consequence of Lemma 5 and Lemma 7. Conversely,

$$\|z^l(\tilde{u}(z) - \langle\zeta_{\delta}(u),\phi\rangle(z))\|_{L^2(\Gamma_{\nu/2-\delta+\eta})} \leq \|z^l\tilde{u}(z)\|_{L^2(\Gamma_{\nu/2-\delta+\eta})}$$
$$+\|z^l\langle\zeta_{\delta}(u),\phi\rangle(z)\|_{L^2(\Gamma_{\nu/2-\delta+\eta})}$$

together with Lemma 5 and Lemma 7 implies (32).

The nuclearity follows in a similar way as that for the $\mathcal{S}_{\gamma}$ spaces in Section 1.2.2, since we are dealing with spaces of functions the Mellin images of which are holomorphic outside Λ and rapidly decreasing as $|\mathrm{Im}\, z| \longrightarrow \infty$, cf. also Section 2.1.3.

In order to prove the equivalence with (μ_B) first observe that the norms μ_B' on the functions $u \in \mathcal{S}_\Lambda$ are finite and induce a topology which is stronger or equal to that defined by the μ_B. Indeed, the convergence of $\tilde{u}$ in $L^2(B)$ implies the convergence of $\zeta_\delta(u)$ in $\mathcal{A}'(\Lambda_\delta)$, which follows from

$$|\langle \zeta_\delta(u), h \rangle| = \Big| \frac{1}{2\pi i} \int_B \tilde{u}(z)\, h(z)\, dz \Big|$$

$$\leq c_B \sup_{z \in U} |h(z)| \Big\{ \int_B |\tilde{u}(z)|^2 \, d\sigma \Big\}^{1/2} .$$

Thus $\mathcal{S}_\Lambda$ is Fréchet with respect to (μ_B') and hence both topologies coincide.

The proof of the equivalence of $(\bar{\mu}_B)_{B \in \underline{B}}$ to the other systems follows in a similar way as that of Lemma 6 and 1.2.1, Lemma 15. $\square$

It is easily seen that the particular choice of d with the mentioned properties with respect to Λ does not affect the topology and that a countable subset of numbers $\eta_j \geq 0$ with $\eta_j \longrightarrow \infty$ as $j \longrightarrow \infty$ suffices. Similarly, in (μ_B) we may replace the sequence of the numbers ρ by another one, say $(\rho_k)_{k \in \mathbb{Z}}$, where

$$\Gamma_{\rho_k} \cap \Lambda = \emptyset, \; k \in \mathbb{Z}, \quad \rho_k \longrightarrow \pm\infty \quad \text{as } k \longrightarrow \pm\infty .$$

9. Remark. Let $\Lambda \in \underline{\Lambda}$, $\varphi \in \underline{\varphi}$, $\Lambda(\varphi) \subseteq \Lambda$, cf. (4). Then $\mathcal{S}_\varphi$ is a closed subspace of $\mathcal{S}_\Lambda$ and the topology of $\mathcal{S}_\varphi$ defined in Section 1.2.1. coincides with that of $\mathcal{S}_\Lambda$. In particular $\mathcal{S}_0$ is closed in $\mathcal{S}_\Lambda$.

Let us point out that the semi-norm systems (μ_B), (μ_B') or $(\bar{\mu}_B)$ give no immediate characterization of the Mellin image $M\mathcal{S}_\Lambda$. But the above calculations easily lead to a simple necessary and sufficient condition for $\tilde{u} \in M\mathcal{S}_\Lambda$.

Let $A \subseteq \mathbb{C}$ be a closed set. Then a function $\chi \in C^\infty(\mathbb{C})$ is called an A excision if $0 \leq \chi \leq 1$ and $\chi \equiv 0$ in some open neighbourhood of A.

10. Proposition. The functions $\tilde{u} \in M\mathcal{S}_\Lambda$ are characterized by the properties

(i) $\tilde{u} \in \mathcal{A}(\mathbb{C} \setminus \Lambda)$,

(ii)

$$\int_{-\infty}^{\infty} |(\varrho + iy)^l \chi(\varrho + iy)\tilde{u}(\varrho + iy)|^2 \, dy < \infty \tag{33}$$

for all $\varrho \in \mathbb{R}$, $l \in \mathbb{Z}_+$, and all Λ excision functions χ.

This, of course, leads to a further equivalent semi-norm system on $\mathcal{S}_\Lambda$, namely by the square roots of the integrals (33) and the numbers

$$\sup_{z \in K} |\tilde{u}(z)|, \quad K \subset\subset \mathbb{C} \setminus \Lambda .$$

In view of Theorem 3 we have

$$\mathcal{S}_\Lambda = \varprojlim \mathcal{H}_k \tag{34}$$

with a countable system $(\mathcal{H}_k)_{k \in \mathbb{Z}}$ of Hilbert spaces, where $\mathcal{H}_{k+1} \hookrightarrow \mathcal{H}_k$ is Hilbert-Schmidt for $k \geqq 0$ and $\mathcal{H}_j := L^2(\mathbb{R}_+)$ for $j \leqq 0$.

We want to fix some choice of a system $(\mathcal{H}_k)_{k \in \mathbb{Z}}$. If we keep fixed the Hilbert spaces $\mathcal{H}_k$ that are associated with $\mathcal{S}_\Lambda$, we incidentally write

$$\mathcal{S}_\Lambda^{(k)} := \mathcal{H}_k(\Lambda) := \mathcal{H}_k . \tag{35}$$

There are, of course, many possibilities of choosing $\mathcal{H}_k$ such that (34) holds.

The scale that we take now will be a bit different from the closures of $\mathcal{S}_\Lambda$ with respect to any norms from Theorem 8, since it is desirable to refer to a system of strips

$$S_\delta = \{ z \in \mathbb{C} : 1/2 - \delta < \operatorname{Re} z < 1/2 + \delta \} , \tag{36}$$

$\delta \geqq 0$, independently of Λ. Recall the notation (19). Then the set $B_{\delta\eta} \cap \Gamma_{1/2-\delta}$ consists of a finite set of points $z_0, \ldots, z_N$ and we may assume that $\operatorname{Im} z_{l-1} \leqq \operatorname{Im} z_l$, $1 \leqq l \leqq N$. Let

$$I_0 := \left\{ z \in \Gamma_{1/2-\delta} : \operatorname{Im} z \leq \operatorname{Im} z_0 \right\},$$

$$I_l := \left\{ z \in \Gamma_{1/2-\delta} : \operatorname{Im} z_{l-1} \leq \operatorname{Im} z \leq \operatorname{Im} z_l \right\},\ l = 1,\dots,N\ ,$$

$$I_{N+1} := \left\{ z \in \Gamma_{1/2-\delta} : \operatorname{Im} z \geq \operatorname{Im} z_N \right\}.$$

Set

$$C_{\delta\eta} = \Gamma_{1/2+\delta} \cup I_0 \cup I_N \cup \bigcup_{I_l \cap \Lambda = \emptyset} I_l \ .$$

Then $C_{\delta\eta}$ is a piecewise smooth curve except **for** isolated points that may be neglected, cf. Fig. 2.

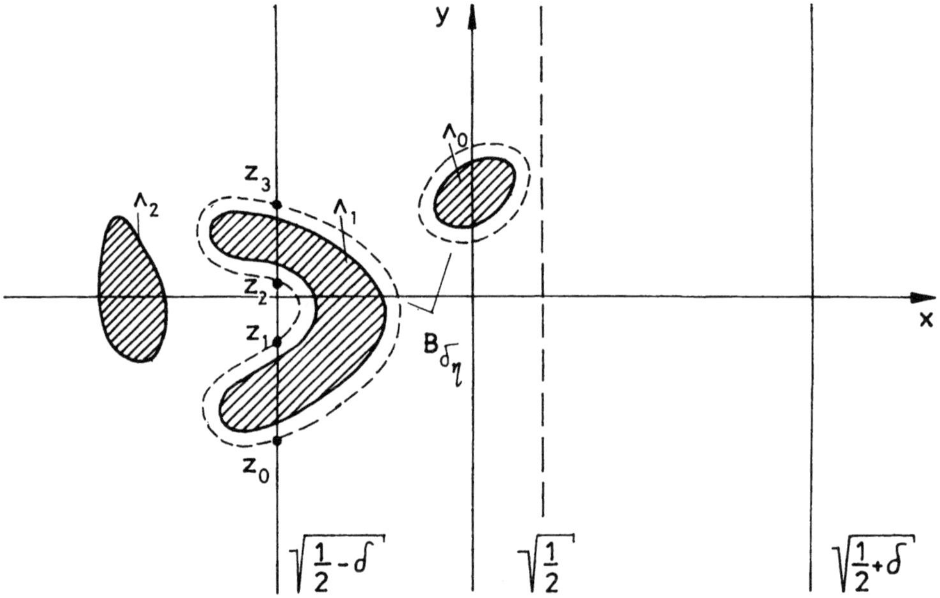

Fig. 2

Now define

$$\lambda_{k\delta\eta}(u)^2 := \sum_{j=0}^{k} \int_{B_{\delta\eta} \cup C_{\delta\eta}} |z^j \tilde{u}(z)|^2 \, d\sigma \quad , \tag{37}$$

$k \in \mathbb{Z}_+$, $\delta, \eta \in \mathbb{R}_+$, where $d\sigma$ is the positive measure on $B_{\delta\eta} \cup C_{\delta\eta}$ induced by the metric of $\mathbb{C}$,

$$\lambda_{k,o,k}(u)^2 := \sum_{j=0}^{k} \| z^j \tilde{u}(z) \|^2_{L^2(\Gamma_{1/2})} \quad , \tag{38}$$

$k \in \mathbb{Z}_+$. Let $\mathcal{H}_{k\delta\eta}(\Lambda)$ be the completion of $\mathcal{S}_\Lambda$ with respect to $\lambda_{k\delta\eta}$. Then it is a simple exercise to check that

$$\mathcal{S}_\Lambda = \varprojlim_{k,\delta,\eta} \mathcal{H}_{k\delta\eta}(\Lambda) .$$

For $\eta = k$ we use the abbreviation

$$\lambda_{k\delta} := \lambda_{k\delta k}, \quad \mathcal{H}_{k\delta}(\Lambda) := \mathcal{H}_{k\delta k}(\Lambda)$$

and for $k = \delta$

$$\lambda_k := \lambda_{kk}, \quad \mathcal{H}_k(\Lambda) := \mathcal{H}_{kk}(\Lambda) .$$

Incidentally we also write $\mathcal{S}_\Lambda^{(k,\delta)}$ and $\mathcal{S}_\Lambda^{(k)}$ instead of $\mathcal{H}_{k\delta}(\Lambda)$ and $\mathcal{H}_k(\Lambda)$, respectively. It is clear that

$$\mathcal{S}_\Lambda = \varprojlim_{k,\delta} \mathcal{H}_{k\delta}(\Lambda) = \varprojlim_{k} \mathcal{H}_k(\Lambda) \tag{39}$$

Denote $\mathcal{H}_k(\Lambda)$ as the standard scale associated with Λ. This notation will be used below also for more general singularity sets $\Delta \subset \mathbb{C}$, cf. 2.1.3. Moreover, by $\mathcal{H}_k(\psi)$, $\psi \in \underline{\Psi}$, we denote the closure of $\mathcal{S}_\psi$ in $\mathcal{H}_k(\nabla\psi)$. Then $\mathcal{S}_\psi = \varprojlim \mathcal{H}_k(\psi)$, and the notion of a standard scale also applies to discrete singularity types in $\underline{\Psi}$. For $\Lambda^1, \Lambda^2 \in \underline{\Lambda}$ and $\Lambda^1 \subseteq \Lambda^2$ we have

$$\mathcal{H}_k(\Lambda^1) \hookrightarrow \mathcal{H}_k(\Lambda^2)$$

with continuous embedding.

Note that

$$\overline{\lambda}_{k\delta\eta}(u) := \max_{0 \le j \le k} \sup_{z \in B_{\delta\eta} \cup C_{\delta\eta}} | z^j \tilde{u}(z) | \tag{40}$$

is another semi-norm system on $\mathcal{S}_\Lambda$ equivalent to the systems in Theorem 8. Similarly we can form equivalent systems $\overline{\lambda}_{k\delta} := \overline{\lambda}_{k\delta k}$, $\overline{\lambda}_k := \overline{\lambda}_{kk}$.

11. <u>Proposition</u>. There exists a countable system $\{\nu_\iota\}_{\iota \in I}$ of norms on $\mathcal{S}_0$ such that for every $\Lambda \in \underline{\Lambda}$ there is a subset $I(\Lambda) \subseteq I$ for which $\mathcal{S}_\Lambda$ has the topology induced by $\{\nu_\iota\}_{\iota \in I(\Lambda)}$. $\Lambda^1 \subseteq \Lambda^2$ implies $I(\Lambda^1) \supseteq I(\Lambda^2)$.

Proof: The semi-norms in $\mathcal{S}_\Lambda$ are defined in terms of sums of integrals over Γ_ρ and curves or pieces of curves in some finite strip $1/2 - \delta <$ $\mathrm{Re}\, z < 1/2 + \delta$, cf. (37), (38). It is obvious that in the definition of $(u,v)_{\mathcal{H}_k}$ we may restrict ourselves to a countable system of δ, say the rational reals.

Thus it suffices to find a countable curve system in the plane which contains a sufficiently rich subset for approximating any given curve involved in the definition of $(u,v)_{\mathcal{H}_k}$. Since any individual curve is piecewise smooth and contained in a compact set (except the infinite pieces on $\Gamma_{1/2\pm\delta}$), we simply may use the system of piecewise smooth curves consisting of intervals $\{x+i\delta : \rho_1 \leq x \leq \rho_2\}$ and $\{\rho + iy :$ $\delta_1 \leq y \leq \delta_2\}$ with rational $\rho, \rho_i, \delta, \delta_i$, $i = 1,2$. $\square$

12. Corollary. Let $\Lambda^1, \Lambda^2 \in \underline{\Lambda}$ and $\Lambda^1 \subseteq \Lambda^2$. Then $\mathcal{S}_{\Lambda^1}$ is a subspace of $\mathcal{S}_{\Lambda^2}$ and the embedding

$$\iota : \mathcal{S}_{\Lambda^1} \longrightarrow \mathcal{S}_{\Lambda^2}$$

is continuous.

Now we shall study continuity properties of linear operators acting between projective limits of Hilbert spaces. First we give the definition of the order of such operators.

13. Definition. Let $\mathcal{S}^i$, $i = 1,2$, be Fréchet spaces which are projective limits of Banach spaces $(\mathcal{H}^i_j)_{j\in\mathbb{Z}}$,

$$\mathcal{S}^i = \varprojlim_j \mathcal{H}^i_j, \quad i = 1,2 \quad . \tag{40}$$

A continuous operator

$$\alpha : \mathcal{S}^1 \longrightarrow \mathcal{S}^2$$

is called of order $\leq k \in \mathbb{Z}$ (with respect to the chosen scales of $\mathcal{H}^i_j$) if α has continuous extensions

$$\alpha^j : \mathcal{H}^1_j \longrightarrow \mathcal{H}^2_{j-k}$$

for all $j \in \mathbb{Z}$.

Clearly the definition refers to the order of the spaces. The $\mathcal{H}^i_j$ are assumed to be fixed in any concrete case and the ambiguity will not affect the essence of our constructions.

Note that for $\Lambda^i \in \underline{\Lambda}$, $\Lambda^1 \subseteq \Lambda^2$, the embedding

$$\iota : \mathcal{S}_{\Lambda^1} \hookrightarrow \mathcal{S}_{\Lambda^2}$$

is of order zero if we fix $\mathcal{H}_k(\Lambda^i)$ as above.

Let $\Lambda^1, \Lambda^2 \in \underline{\Lambda}$ and consider the space $\mathcal{L}(\mathcal{S}_{\Lambda^1}, \mathcal{S}_{\Lambda^2})$ of all continuous operators (41) in the topology of uniform convergence on all bounded sets. Then we have the subclass

$$\mathcal{L}^k(\mathcal{S}_{\Lambda^1}, \mathcal{S}_{\Lambda^2}) \subset \mathcal{L}(\mathcal{S}_{\Lambda^1}, \mathcal{S}_{\Lambda^2}) \tag{42}$$

of all operators of order k, and $\mathcal{L}^k(\mathcal{S}_{\Lambda^1}, \mathcal{S}_{\Lambda^2})$ may be equipped with the following semi-norm system

$$\rho_j(\mathfrak{a}) = \sup\left\{\|\mathfrak{a}u\|_{\mathcal{H}^2_{j-k}} : \|u\|_{\mathcal{H}^1_j} = 1\right\},$$

in which $\mathcal{L}^k(\mathcal{S}_{\Lambda^1}, \mathcal{S}_{\Lambda^2})$ is a Fréchet space. In $\mathcal{L}(\mathcal{S}_{\Lambda^1}, \mathcal{S}_{\Lambda^2})$ we have the semi-norm system

$$\eta_{B,j}(\mathfrak{a}) = \sup\left\{\|\mathfrak{a}u\|_{\mathcal{H}^2_j} : u \in B\right\},$$

where $j \in \mathbb{Z}_+$, and B runs through all bounded sets in $\mathcal{S}_{\Lambda^1}$. A set $B \subset \mathcal{S}_{\Lambda^1}$ is bounded iff $\|u\|_{\mathcal{H}^1_j} \leq c^j$ for all $u \in B$ with constants c^j, $j \in \mathbb{Z}_+$. Then it is obvious that the embedding (42) is continuous.

For $\gamma \in \mathbb{C}$ and any subset $\Lambda \subseteq \mathbb{C}$ we put

$$T^{-\gamma}\Lambda = \{z \in \mathbb{C} : z + \gamma \in \Lambda\} . \tag{43}$$

If $T^{-\gamma}\Lambda \in \underline{\Lambda}$ we define $\mathcal{S}_{T^{-\gamma}\Lambda}$ by means of $T^{-\gamma}(d \setminus \{0\}) \cup \{0\}$, when $\mathcal{S}_\Lambda$ is defined by using d.

As mentioned above, when we talk about orders of operators, we fix the Hilbert space scales, in particular for $\mathcal{S}_\Lambda$ given by (38) we set

$$\mathcal{S}_{\Lambda \setminus \Lambda_{(\delta)}} = \varprojlim \mathcal{H}_k(\Lambda \setminus \Lambda_\delta) .$$

In view of Lemma 5 we also have a representation

$$\mathcal{A}'(\Lambda_{(\delta)}) = \varprojlim \mathcal{A}'(\Lambda_{(\delta)})_{(k)} , \tag{44}$$

$\mathcal{A}'(\Lambda_{(\delta)})_{(k)}$ being the completion of $\mathcal{A}'(\Lambda_{(\delta)})$ with respect to φ_{B_k}, cf.

(21), $B_k = \partial U_k$, $U_k \in \mathcal{U}_\delta$.

14. Proposition. The operator

$$\zeta_\delta : \mathcal{S}_\Lambda \longrightarrow \mathcal{A}'(\Lambda_\delta) \tag{45}$$

defined by $u \longrightarrow \zeta_\delta(u)$ is continuous for every $\delta \in d$.

The operator

$$F_\delta : \mathcal{A}'(\Lambda_{(\delta)}) \longrightarrow \mathcal{S}_\Lambda \tag{46}$$

defined by $F_\delta : \lambda \longrightarrow \langle \lambda, F \rangle (t)$, $\lambda \in \mathcal{A}'(\Lambda_{(\delta)})$, is continuous, too, $\delta \in d$, where $\zeta_\delta \circ F_\delta$ = id on $\mathcal{A}'(\Lambda_\delta)$.

Thus

$$\mathcal{r}_\delta : \mathcal{S}_\Lambda \longrightarrow \mathcal{S}_{\Lambda \setminus \Lambda_{(\delta)}} \tag{47}$$

defined by

$$\mathcal{r}_\delta u = u - \langle \zeta_\delta, F \rangle \tag{48}$$

is a continuous projection, the operator

$$\vartheta^1 : \mathcal{S}_\Lambda \longrightarrow \mathcal{S}_\Lambda \tag{49}$$

is continuous for all $1 \in \mathbb{Z}_+$, and the operator of multiplication by t^γ

$$t^\gamma : \mathcal{S}_\Lambda \longrightarrow \mathcal{S}_{T^{-\gamma}\Lambda} \tag{50}$$

is a topological isomorphism for all $\gamma \in \mathbb{C}$ with $\mathrm{Re}(z - \gamma) < 1/2$ for all $z \in \Lambda$. The order of $\mathcal{r}_\delta$ is zero and ϑ^1 has order 1. The order of $\omega t^\gamma : \mathcal{S}_\Lambda \longrightarrow \mathcal{S}_{T^{-\gamma}\Lambda}$, $\mathrm{Re}\,\gamma \leqq 0$, is zero.

Proof: $u_n \longrightarrow u$ in $\mathcal{S}_\Lambda$ implies $\varphi_B(\zeta_\delta(u_n - u)) \longrightarrow 0$ for $n \longrightarrow \infty$ and in view of Lemma 5 $\zeta_\delta(u_n - u) \longrightarrow 0$ in $\mathcal{A}'(\Lambda_{(\delta)})$. Thus (45) is proved.

For proving (46) consider $\mathcal{S}_\Lambda$ with the semi-norm system $(\mu_B)_{B \in \underline{B}}$, cf. Theorem 8. Lemma 5 shows that $\lambda_n \longrightarrow \lambda$ in $\mathcal{A}'(\Lambda_{(\delta)})$ implies $\varphi_B(\lambda_n - \lambda) \longrightarrow 0$, and a trivial argument in the proof of Lemma 5 (replace $h(z)$ by $z^1 h(z)$) shows that

$$(\langle \lambda_n - \lambda, \phi \rangle (z), \langle \lambda_n - \lambda, \phi \rangle (z))_{L^2(B)} \longrightarrow 0$$

as $n \longrightarrow \infty$. Then (46) follows from this and Lemma 7.

Next we show (49). Let $\delta \in d$ be arbitrarily fixed and $\lambda = \zeta_\delta(u)$. Assume that we have checked that

$$\vartheta^{1} \langle \lambda, F \rangle (t) = \langle \lambda', F \rangle (t) \tag{51}$$

for some $\lambda' \in \mathcal{A}'(\Lambda_{(\delta)})$ and that $\lambda_n \longrightarrow \lambda$ in $\mathcal{A}'(\Lambda_{(\delta)})$ implies $\lambda_n' \longrightarrow \lambda'$ in $\mathcal{A}'(\Lambda_{(\delta)})$. Then the assertion immediately follows from (27). In fact, replacing $\tilde{u}$, $\tilde{v}$ by $z\tilde{u}$, $z\tilde{v}$ in the first two terms on the right of (27) only defines a shift of the index B. Thus we have to discuss the third term and to remark that $\lambda = \zeta_{\delta}(u) = \zeta_{\delta}(\langle \lambda, F \rangle)$. Hence it remains to show (51). For $U \in \mathcal{U}_{\delta}$, $B = \partial U$, $h \in \mathcal{A}(\mathbb{C})$ we define $\lambda' \in \mathcal{A}'(\mathbb{C})$ by $\langle \lambda', h \rangle = \int_B (-z)^{1} \langle \lambda, \Phi \rangle (z)\, h(z)\, dz$.
Then it is clear (cf. the proof of Lemma 5) that $\lambda \longrightarrow \lambda'$ is continuous and that $\Lambda_{(\delta)}$ carries λ'.
The proof of (50) is straightforward and left to the reader. The order properties of $\mathcal{V}_{\delta}$ and ϑ^{1} are obvious. For ωt^{γ}, ω some cut-off function, we mainly have to compare the terms in the $\mathcal{H}_k$ norms and to prove that

$$\sum_{j=0}^{k} \frac{1}{i} \int_{\Gamma_{\frac{1}{2}+k}} | z^{j} M(\omega t^{\gamma} u)(z) |^{2} dz \leqq c \sum_{j=0}^{k} \frac{1}{i} \int_{\Gamma_{\frac{1}{2}+k}} | z^{j} M(u)(z) |^{2}\, dz$$

for some constant $c > 0$, $u \in \mathcal{S}_{\Lambda}$. This is true, indeed, since we may apply the identity

$$\frac{1}{2\pi i} \int_{\Gamma_{\frac{1}{2}+k}} | \tilde{v}(z) |^{2} dz = \int_{\mathbb{R}_{+}} | t^{k} v(t) |^{2}\, dt$$

and 1.1.1.(9). $\square$

Consider a fixed $\Lambda = \bigcup \Lambda_i \in \underline{\Lambda}$. For the operator ζ_{δ}, $\delta \in d$, we then have

$$\zeta_{\delta} = \sum_{i=0}^{N(\delta)} \zeta_i \quad , \tag{52}$$

where

$$\langle \zeta_i(u), h \rangle = \frac{1}{2\pi i} \int_{B_i} \tilde{u}(z)\, h(z)\, dz \tag{53}$$

with a curve B_i surrounding Λ_i and $B_i \cap \Lambda = \emptyset$.

For the $\mathcal{S}_\Lambda$ spaces we have an analogue of 1.2.1, Theorem 14, which is again a generalization of Borel's theorem and essentially states that for any given sequence of functionals $\zeta_i \in \mathcal{A}'(\Lambda_i)$, $i \in \mathbb{Z}_+$, there is a function $u \in \mathcal{S}_\Lambda$, $\Lambda = \bigcup_i \Lambda_i$, such that $\zeta_i(u) = \zeta_i$.

15. Theorem. Let $\mathcal{Z}$ be the direct product $\bigtimes_{i \in \mathbb{Z}_+} \mathcal{A}'(\Lambda_i)$ equipped with the topology of component-wise convergence, and

$\zeta : \mathcal{S}_\Lambda \longrightarrow \mathcal{Z}$ the unique morphism induced by ζ_δ, $\delta \longrightarrow \infty$. Then the sequence of topological vector spaces

$$0 \longrightarrow \mathcal{S}_0 \xrightarrow{\ \iota\ } \mathcal{S}_\Lambda \xrightarrow{\ \zeta\ } \mathcal{Z} \longrightarrow 0$$

is exact.

Proof: We have to show that ζ is surjective, since the remaining assertions follow immediately from Remark 9, Proposition 14 and Lemma 3. Let $\{\zeta_i\}_{i \in \mathbb{Z}_+}$, $\zeta_i \in \mathcal{A}'(\Lambda_i)$ be arbitrarily given. We consider the sequence of continuous norms $\mu_j = \sum_{|\beta| \leq j} \mu_\beta$ on $\mathcal{S}_\Lambda$, which obviously define the topology of $\mathcal{S}_\Lambda$. Set

$$F_c(t,w) = t^{-w}\omega(ct), \; c > 0,$$

and

$$\phi_c(z,w) = \int_0^\infty t^{z-1} F_c(t,w)\, dt.$$

Assume that we have proved that for any given $k \in \mathbb{Z}_+$ there is an $i_0 = i_0(k) \in \mathbb{Z}_+$ such that for all $i \geq i_0$

$$\mu_k(\langle \zeta_i, \phi_c(z,\cdot)\rangle) \longrightarrow 0 \text{ as } c \longrightarrow \infty \quad . \tag{54}$$

Then for each $i \in \mathbb{Z}_+$ we find a c_i such that

$$\mu_k(\langle \zeta_i, \phi_{c_i}(z,\cdot)\rangle) < \frac{1}{2^i} \quad \text{for all } k \text{ with } i_0(k) \leq i \; .$$

Indeed, for choosing c_i we only have to fulfill a finite number of conditions. Then

$$u := \sum_{i=0}^\infty \langle \zeta_i, F_{c_i}(t,\cdot)\rangle$$

converges in $\mathcal{S}_\Lambda$, since we have for every $k \in \mathbb{Z}_+$

$$\mu_k\Big(\sum_{i=0}^{\infty}<\zeta_i,\phi_{c_i}(z,\cdot)>\Big) = \sum_{i<i_o(k)} \mu_k(<\zeta_i,\phi_{c_i}(z,\cdot)>)$$

$$+ \sum_{i\geq i_o(k)} \mu_k(<\zeta_i,\phi_{c_i}(z,\cdot)>) < \infty .$$

Thus it remains to prove (54).

The norm μ_k contains a finite number of expressions

$$I_\rho := \| z^l M(<\zeta_i,t^{-w}> \omega(ct))(z) \|^2_{L^2(\Gamma_\rho)}$$

and

$$I_B := \| M(<\zeta_i,t^{-w}> \omega(ct))(z) \|^2_{L^2(B)}$$

and we shall see that $i_o(k)$ has simply to be chosen so large that

$$\sup \{ \operatorname{Re} z : z \in \Lambda_{i_o(k)} \}$$

is less than the infimum over Re z for $z \in \Gamma_\rho$ or $z \in B$ for all Γ_ρ and B that are involved in μ_k. Denote this infimum by m_k. Let B_i be a curve surrounding Λ_i, $B_i \cap \Lambda = \emptyset$, $\sup \{ \operatorname{Re} z : z \in B_i \} = n_i < m_k$. For I_ρ we have

$$\int |(\rho+iy)^l \int t^{\rho+iy-1}<\zeta_i,t^{-w}> \omega(ct)dt|^2 dy$$

$$= \int |(\rho+iy)|^{2l} | \int (\tfrac{t}{c})^{\rho+iy-1} <\zeta_i,(\tfrac{t}{c})^{-w}> \omega(t)\; c^{-1}\, dt|^2\, dy$$

$$\leq \int |(\rho+iy)|^{2l} \int |(\tfrac{t}{c})^{\rho+iy-1} <\zeta_i,(\tfrac{t}{c})^{-w}> \omega(t)\; c^{-1}\, dt|^2\, dy$$

$$= c^{-\rho} \int |(\rho+iy)|^{2l} \int t^{\rho-1}|<\zeta_i,(\tfrac{t}{c})^{-w}>|\, \omega(t)\, dt|^2\, dy,$$

and

$$|<\zeta_i,(\tfrac{t}{c})^{-w}| = |\int_{B_i} <\zeta_i,\phi(z,\cdot)> (\tfrac{t}{c})^{-z}\, dz|$$

$$\leq (\tfrac{t}{c})^{-n_i} \int_{B_i} |<\zeta_i,\phi(z,\cdot)>| d\sigma = C_i (\tfrac{t}{c})^{-n_i} .$$

Thus

$$I_\rho \leq c^{-\rho+n_i}\, C_i \int |(\rho+iy)|^{2l} \int t^{-n_i+\rho-1}\, \omega(t)\, dt|^2\, dy$$

which is correct for $\rho - n_i - 1 > -1$, i.e. $\rho > n_i$. Thus $I_\rho \leq c^{-\rho+n_i} C_i'$, i.e. I_ρ has the desired property. The discussion of I_B is similar and left to the reader. $\square$

Note that $\mathcal{S}_o$ has a topological complement in $\mathcal{S}_\Lambda$ unless Λ is compact. In the latter case we have

$$\mathcal{S}_\Lambda \cong \mathcal{S}_o \oplus \mathcal{A}'(\Lambda),$$

where $\{<\zeta(w), F(t,w)> : \zeta \in \mathcal{A}'(\Lambda)\}$ is a topological complement of $\mathcal{S}_o$.

The principle of the proof of Theorem 15 will often be applied below for proving the convergence of infinite sums of Mellin operators. The basic observation is that $t^j \omega(ct)$ has a norm $< \varepsilon$ in $\mathcal{L}(\mathcal{H}_k(\Lambda),$ $\mathcal{H}_k(T^{-j}\Lambda))$ for $j > k$ and c large enough. We have $\mathcal{H}_k(T^{-j}\Lambda) = \mathcal{H}_k(\emptyset)$ and

$$\| t^j \omega(ct) u \|_{\mathcal{H}_k(\emptyset)} \leq \text{const} \| u \|_{\mathcal{H}_k(\Lambda)}$$

with a constant independent of Λ. Thus we may choose the constants c_i above independently of $\Lambda \in \underline{\Lambda}$.

Recall 1.2.1, Theorem 14, which is a sharper version of the Mittag-Leffler theorem, cf. [H2]. For the given sequence of Laurent expansions at the points $z_j = p_j$ we obtained a meromorphic function f in $\mathbb{C}$ possessing these z_j as poles with the given Laurent coefficients at the negative powers. In addition our construction yields a rapid decrease of f as $|\text{Im } z| \to \infty$. Theorem 15 admits a similar interpretation. The sequence of local data $<\zeta_i, \phi>(z)$, being holomorphic outside Λ_i, may be glued together to a function f being holomorphic in $\mathbb{C} \setminus \Lambda$ and rapidly decreasing for $|\text{Im } z| \to \infty$.

A special case is $\Lambda_i = \{z_i\}$ and ζ_i defined by

$$<\zeta_i, h> = \sum_{k=0}^{\infty} \frac{1}{k!} a_k \left(\frac{d^k}{dz^k} h\right)(z_i), \ h \in \mathcal{A}(\mathbb{C}) \tag{55}$$

$|a_k| \leq c_\varepsilon \varepsilon^k$ for every $\varepsilon > 0$ with some constant c_ε. In this case ζ_i is carried by $\{z_i\}$. The space $\mathcal{S}_\Lambda$ that belongs to $\Lambda = \bigcup \{z_i\}$ consists of functions with an asymptotic behaviour of the form

$$u(t) \sim \sum_{j=0}^{\infty} \sum_{k=0}^{\infty} \zeta_{jk}\, t^{-z_j} \log^k t \quad \text{as } t \longrightarrow 0,$$

where ζ_{jk} satisfies a growth condition for $k \longrightarrow \infty$.

16. Remark. $u(t) \in \mathcal{S}_\Lambda$ implies $u(t) \log^k t \in \mathcal{S}_\Lambda$ for arbitrary $k \in \mathbb{Z}_+$, $\Lambda \in \underline{\Lambda}$.

In fact, $u \in \mathcal{S}_\Lambda$ implies $u(t) \log^k t \in L^2(\mathbb{R}_+) \cap C^\infty(\mathbb{R}_+)$ and

$$M(u \log^k .)(z) = \frac{d^k}{dz^k} M(u)(z)$$

cf. 1.1.1(11). Thus $z^l \frac{d^k}{dz^k} M(u)(z)$ belongs to $L^2(\Gamma_\rho)$ for all Γ_ρ involved in the semi-norm system $(\mu_\beta)_{\beta \in \underline{\beta}}$ and $\frac{d^k}{dz^k} M(u)(z)$ is holomorphic in $\mathbb{C} \setminus \Lambda$. Hence it represents a sequence of analytic functionals

$$\langle \zeta_i(u \log^k .), h \rangle = \frac{1}{2\pi i} \int_{B_i} \frac{d^k}{dz^k} M(u)(z)\, h(z)\, dz,$$

$h \in \mathcal{A}(\mathbb{C})$.

The latter remarks show that the concept of $\mathcal{S}_\Lambda$ spaces only generalizes that of $\mathcal{S}_\varphi$ spaces with respect to the singularity type represented by a set in the complex plane but not with respect to certain fixed 'orders' of the functionals. The applications in Section 2.1.2. below will show that the $\mathcal{S}_\Lambda$ spaces are natural. On the other hand, there are spaces of a mixed type in the following sense.

Let $\Lambda \in \underline{\Lambda}$, $\varphi \in \underline{\varphi}$, $\varphi = \{(p_j, m_j) : j \in \mathbb{Z}_+\}$, $\pi_\mathbb{C}\varphi \cap \Lambda = \emptyset$. By $\mathcal{S}_{\Lambda+\varphi}$ we denote the subspace of all $u \in \mathcal{S}_{\Lambda+\pi_\mathbb{C}\varphi}$ for which $(Mu)(z)$ has poles at $z_j = p_j$ of order $\leq m_j+1$ for all $j \in \mathbb{Z}_+$, equipped with a Fréchet topology that corresponds to the $\mathcal{S}_\Lambda$ definition near Λ and to the $\mathcal{S}_\varphi$ definition near $\pi_\mathbb{C}\varphi$. If $\pi_\mathbb{C}\varphi \cap \Lambda \neq \emptyset$, we have a **maximal** $\varphi_1 \subseteq \varphi$ with $\pi_\mathbb{C}\varphi_1 \cap \Lambda = \emptyset$ and then $\mathcal{S}_{\Lambda+\varphi} := \mathcal{S}_{\Lambda+\varphi_1}$. Similarly as $\mathcal{S}_\varphi$ is closed in $\mathcal{S}_{\pi_\mathbb{C}\varphi}$, $\mathcal{S}_{\Lambda+\varphi}$ is a closed subspace of $\mathcal{S}_{\Lambda+\pi_\mathbb{C}\varphi}$.

It is our next goal to study the Mellin convolution between functions in $\mathcal{S}_\Lambda$ spaces.

17. Definition. A couple Λ^1, Λ^2 of sets in $\underline{\Lambda}$ is called admissible if

$$\Lambda = \Lambda^1 + \Lambda^2 := (\Lambda^1 \cup \Lambda^2)^c$$

belongs to $\underline{\Lambda}$, again (cf. the notation (18)).

18. Theorem. Let $\Pi, \Xi \in \underline{\Lambda}$ be admissible, and $op_M[f]u := M^{-1}(\tilde{f}\,\tilde{u})$, $f \in \mathcal{S}_{\Xi}$. Then $op_M[f]$ induces a continuous operator

$$op_M[f] \;:\; \mathcal{S}_{\Pi} \longrightarrow \mathcal{S}_{\Pi+\Xi}$$

of order zero. Moreover,

$$op_M[\cdot] \;:\; \mathcal{S}_{\Xi} \longrightarrow \mathcal{L}^0(\mathcal{S}_{\Pi}, \mathcal{S}_{\Pi+\Xi})$$

is continuous.

Proof: The operator $op_M[f]$ may be regarded as the composition of the embedding $\iota : \mathcal{S}_{\Pi} \longrightarrow \mathcal{S}_{\Lambda}$, $\Lambda := \Pi + \Xi$ and $op_M[f] : \mathcal{S}_{\Lambda} \longrightarrow \mathcal{S}_{\Lambda}$. As noted above ι is continuous and of order zero. Then it suffices to observe that for $\mathcal{H}_k(\Lambda)$ defined by the norms (38) we have

$$\| op_M[f]u \|_{\mathcal{H}_k(\Lambda)} \leq c \, \| u \|_{\mathcal{H}_k(\Lambda)}$$

with some $c > 0$. This estimate is obvious with

$$c = \sup \{ |\tilde{f}(z)| : z \in B_{kk} \cup C_{kk} \} .$$

It also implies the asserted continuity of $op_M[\cdot]$. $\square$

We also have an analogue of the spaces of the type $C^{\infty}_{\mathcal{P},\mathcal{P}'}$, cf. Section 1.2.1, here in the setting of analytic functionals. Denote by $\underline{\Lambda}'$ the set of all closed subsets $\Lambda' \subset \mathbb{C}$ with

$$\{ z : 1-z \in \Lambda' \} \in \underline{\Lambda} .$$

To any $\Lambda' \in \underline{\Lambda}'$ there exists a system $d' = (\delta'_j)_{j \in \mathbb{Z}_+}$ of non-negative numbers with $\delta'_0 = 0$, $\delta'_j < \delta'_{j+1}$, $\delta'_j \longrightarrow \infty$ as $j \longrightarrow \infty$, $d' \cap \Lambda' = \emptyset$. Set

$$\Lambda'_{(\delta')} = \{ \operatorname{Re} z \leq 1/2 + \delta' \} \cap \Lambda' .$$

19. Definition. Let $\Lambda \in \underline{\Lambda}$, $\Lambda' \in \underline{\Lambda}'$ and d, d' be associated systems of non-negative reals. By $C^{\infty}_{\Lambda,\Lambda'}(\mathbb{R}_+)$ we denote the subspace of all $u \in C^{\infty}(\mathbb{R}_+)$ such that for every $\delta \in d$, $\delta' \in d'$ there exist elements $\zeta_{\delta} \in \mathcal{A}'(\Lambda_{(\delta)})$, $\zeta'_{\delta'} \in \mathcal{A}'(\Lambda'_{(\delta')})$ such that

$$t^{\gamma} \, \vartheta^1 (u - \langle \zeta_{\delta}, F \rangle - \langle \zeta'_{\delta'}, F' \rangle)(t) \in L^2(\mathbb{R}_+)$$

for all $\gamma \in [-\delta, \delta']$, $l \in \mathbb{Z}_+$. Here

$$F'(t,w) = t^{-w}(1-\omega(t))$$

for some cut-off function ω.

The theory of the spaces $C^\infty_{\Lambda,\Lambda'}$ is entirely analogous to the $\mathcal{S}_\Lambda$ theory. In particular, $C^\infty_{\Lambda,\Lambda'}$ is a nuclear Fréchet space and we have equivalent semi-norm systems expressed in terms of

$$M\, C^\infty_{\Lambda,\Lambda'} \subset \mathcal{A}(\mathbb{C} \setminus (\Lambda \cup \Lambda')) \quad .$$

There are obvious analogues of the semi-norm systems λ_k and $\tilde{\lambda}_k$ defined above. Let us use the same letters here, also for the analogues of the curves $B_k = B_k(\Lambda,\Lambda')$, $C_k = C_k(\Lambda,\Lambda')$. Then

$$C^\infty_{\Lambda,\Lambda'} = \varprojlim_k \mathcal{H}_k(\Lambda,\Lambda') \,,$$

$\mathcal{H}_k(\Lambda,\Lambda')$ being the closure of $C^\infty_{\Lambda,\Lambda'}$ with respect to λ_k.

$\Lambda'^{1}, \Lambda'^{2} \in \underline{\Lambda}'$ are called admissible if $\Lambda'^{1} + \Lambda'^{2} := (\Lambda'^{1} \cup \Lambda'^{2})^c \in \underline{\Lambda}'$.

20. Theorem. Let both $\Pi, \Xi \in \underline{\Lambda}$ and $\Pi', \Xi' \in \underline{\Lambda}'$ be admissible and $f \in C^\infty_{\Xi,\Xi'}$. Then $op_M[f]$ induces a continuous operator

$$op_M[f] : C^\infty_{\Pi,\Pi'} \longrightarrow C^\infty_{\Pi+\Xi,\Pi'+\Xi'}$$

of order zero. Moreover,

$$op_M[\cdot] : C^\infty_{\Xi,\Xi'} \longrightarrow \mathcal{L}^0(C^\infty_{\Pi,\Pi'}, C^\infty_{\Pi+\Xi,\Pi'+\Xi'})$$

is continuous.

The proof is similar as for Theorem 18. With $\Lambda := \Pi + \Xi$, $\Lambda' = \Pi' + \Xi'$ we have

$$\| op_M[f]u \|_{\mathcal{H}_k(\Lambda,\Lambda')} \le c \, \| u \|_{\mathcal{H}_k(\Lambda,\Lambda')},$$

$$c = \sup \left\{ |\tilde{f}(z)| \quad : z \in B_{kk} \cup C_{kk} \right\} .$$

Observe that a larger class of functions $f \in C^\infty(\mathbb{C} \setminus (\Lambda \cup \Lambda'))$ induces continuous operators in $C^\infty_{\Lambda,\Lambda'}$. Let us explicitly describe some classes of multipliers in $MC^{\infty}_{\Lambda,\Lambda'}$.

21. Definition. Let μ be a function on $\mathbb{R} \ni x = \mathrm{Re}\, z = 1/2$ which is bounded on compact sets. Denote by $\mathcal{M}^{\mu}_{\Lambda,\Lambda'}$ the space of functions $h \in \mathcal{A}(\mathbb{C} \setminus (\Lambda \cup \Lambda'))$ such that for every $\rho_1, \rho_2 \in \mathbb{R}$, $\rho_1 \leq \rho_2$, and any $\Lambda \cup \Lambda'$ excision function χ

$$\sup\left\{ (1+|z|)^{-\mu} h(z)\, \chi(z) \; : \; \rho_1 \leq \mathrm{Re}\, z \leq \rho_2 \right\} < \infty . \tag{56}$$

The semi-norms (56) together with

$$\sup_{z \in K} |h(z)| , \quad K \subset \mathbb{C} \setminus (\Lambda \cup \Lambda') \text{ compact},$$

induce a Fréchet topology on $\mathcal{M}^{\mu}_{\Lambda,\Lambda'}$.

Set

$$\mathrm{op}_M(h)\, u = M^{-1}(h\, \tilde{u}).$$

22. Theorem. Any $h \in \mathcal{M}^{\mu}_{\Lambda,\Lambda'}$ induces a continuous operator

$$\mathrm{op}_M(h) \; : \; C^{\infty}_{\Lambda,\Lambda'} \longrightarrow C^{\infty}_{\Lambda,\Lambda'} . \tag{57}$$

If $\mu = 0$, the operator (57) is of order zero in the sense of Definition 13 and $h \longrightarrow \mathrm{op}_M(h)$ induces a continuous operator

$$\mathrm{op}_M \; : \; \mathcal{M}^{0}_{\Lambda,\Lambda'} \longrightarrow \mathcal{L}^{0}(C^{\infty}_{\Lambda,\Lambda'}, C^{\infty}_{\Lambda,\Lambda'}) .$$

The straightforward proof is left to the reader.

Basically we are interested in operators of the form $\omega\, \mathrm{op}_M(h)\omega$, ω a cut-off function.

23. Proposition. Let $\Lambda \in \underline{\Lambda}$, $\Lambda' \in \underline{\Lambda}'$. The operator of multiplication by ω induces a continuous operator

$$\omega \; : \; C^{\infty}_{\Lambda,\Lambda'} \longrightarrow \mathcal{S}_{\Lambda}$$

of order zero.

The proof is an immediate consequence of the definitions of $\mathcal{S}_{\Lambda}$ and $C^{\infty}_{\Lambda,\Lambda'}$.

In this way we finally obtain

24. Theorem. Any $h \in \mathcal{M}^{\mu}_{\Lambda,\Lambda'}$ induces a continuous operator

$$\omega\, \mathrm{op}_M(h)\omega \; : \; \mathcal{S}_{\Lambda} \longrightarrow \mathcal{S}_{\Lambda} . \tag{58}$$

If $\mu = 0$ the operator (58) is of order zero and $h \longrightarrow \omega\, op_M(h)\omega$ induces a continuous operator

$$\omega\, op_M(\cdot)\omega \quad : \mathcal{M}^0_{\Lambda,\Lambda'} \longrightarrow \mathcal{L}^0(\mathcal{S}_\Lambda, \mathcal{S}_\Lambda) \ .$$

2.1.2. Spaces of the type $C^\infty(\Omega, \mathcal{S}_\Lambda)$, $C^\infty(\Omega, \mathcal{S}_{\Lambda'})$, ...

Let $\Omega \subset \mathbb{R}^d$ be an open set and E a locally convex vector space. Then the topological vector spaces $C^\infty(\Omega,E)$ and $C_0^\infty(\Omega,E)$ are defined in a standard way. There are canonical isomorphisms

$$C^\infty(\Omega,E) \cong C^\infty(\Omega) \otimes_\pi E, \ C_0^\infty(\Omega,E) \cong C_0^\infty(\Omega) \otimes_\pi E,$$

where $\otimes_\pi$ denotes the projective tensor product. In view of the nuclearity of $C^\infty(\Omega)$ and $C_0^\infty(\Omega)$ we can replace $\otimes_\pi$ by the inductive tensor product $\otimes_\varepsilon$. For an arbitrary $u \in C^\infty(\Omega,E)$ we obtain a (non-unique) representation

$$u = \sum_{j=1}^{\infty} f_j e_j \qquad , \ f_j \in C^\infty(\Omega), \ e_j \in E,$$

which converges in $\otimes_\pi$. Conversely, each converging series of this form gives a function in $C^\infty(\Omega,E)$. Similar statements hold for $C_0^\infty(\Omega,E)$ with $C_0^\infty(\Omega)$ instead of $C^\infty(\Omega)$.

This will be used in particular for

$$E = \mathcal{S}_\Lambda(\mathbb{R}_+), \quad E = C^\infty_{\Lambda,\Lambda'}(\mathbb{R}_+)$$

or a tensor product of spaces of this type. As a prototype in our discussions we restrict ourselves to $C^\infty(\Omega, \mathcal{S}_\Lambda)$ and the associated spaces $C^\infty(\Omega, \mathcal{A}'(K))$ for compact $K \subset\subset \mathbb{C}$. Operations on $\mathcal{S}_\Lambda$ induce in a natural way operations on $C^\infty(\Omega) \otimes_\pi \mathcal{S}_\Lambda$ and $C_0^\infty(\Omega) \otimes_\pi \mathcal{S}_\Lambda$. Let $\Lambda = \bigcup_{j=0}^{\infty} \Lambda_j$ and

$$\zeta_j : \mathcal{S}_\Lambda \longrightarrow \mathcal{A}'(\Lambda_j)$$

be the linear continuous mapping which can be represented by

$$\langle \zeta_j(u), h \rangle = \frac{1}{2\pi i} \int_{B_j} \tilde{u}(z)\, h(z)\, dz, \quad h \in \mathcal{A}(\mathbb{C}) \ ,$$

where B_i is a smooth positively oriented closed curve in $\mathbb{C}$, $B_j \cap \Lambda = \emptyset$ surrounding Λ_j and no other Λ_i, $j \neq i$. Then

$$1 \otimes \zeta_j : C^\infty(\Omega) \otimes_\pi \mathcal{S}_\Lambda \longrightarrow C^\infty(\Omega) \otimes_\pi \mathcal{A}'(\Lambda_j)$$

is also a linear continuous mapping which can be represented by the formula

$$<(1 \otimes \zeta_j)u(y,t),h> = \frac{1}{2\pi i} \int_{B_j} \tilde{u}(y,z)h(z)dz, \qquad h \in \mathcal{A}(\mathbb{C}) .$$

For an arbitrary multi-index $\alpha \in \mathbb{Z}_+^d$ the diagram

$$\begin{array}{ccc} C^\infty(\Omega) \otimes_\pi \mathcal{S}_\Lambda & \xrightarrow{1 \otimes \zeta_i} & C^\infty(\Omega) \otimes_\pi \mathcal{A}'(\Lambda_i) \\ D^\alpha \otimes 1 \downarrow & & \downarrow D^\alpha \otimes 1 \\ C^\infty(\Omega) \otimes_\pi \mathcal{S}_\Lambda & \xrightarrow{1 \otimes \zeta_i} & C^\infty(\Omega) \otimes_\pi \mathcal{A}'(\Lambda_i) \end{array}$$

consists of linear continuous mappings and commutes. If F_i denotes the linear continuous mapping

$$F_i : \mathcal{A}'(\Lambda_i) \longrightarrow \mathcal{S}_\Lambda$$

defined by $\zeta \longrightarrow F_i(\zeta)(t) = <\zeta_i(w), F(t,w)>$, cf. 2.1.1.(5), which has the property $\zeta_i \circ F_j = \delta_{ij}$, then we get the linear continuous mapping

$$1 \otimes F_i : C^\infty(\Omega) \otimes_\pi \mathcal{A}'(\Lambda_i) \longrightarrow C^\infty(\Omega) \otimes_\pi \mathcal{S}_\Lambda$$

and $(1 \otimes \zeta_i) \circ (1 \otimes F_j) = \delta_{ij}$.
Below we shall write simply ζ_j instead of $1 \otimes \zeta_j$, D^α instead of $D^\alpha \otimes 1$ if no confusion is possible.
Let $\mathfrak{a}$ be a smooth family of operators in $\mathcal{L}(\mathcal{S}_{\Lambda 1}, \mathcal{S}_{\Lambda 2})$, i.e. $\mathfrak{a} \in C^\infty(\Omega, \mathcal{L}(\mathcal{S}_{\Lambda 1}, \mathcal{S}_{\Lambda 2})) \cong C^\infty(\Omega) \otimes_\pi \mathcal{L}(\mathcal{S}_{\Lambda 1}, \mathcal{S}_{\Lambda 2})$. Then $\mathfrak{a}$ induces a linear continuous operator $\mathfrak{a} : C^\infty(\Omega) \otimes_\pi \mathcal{S}_{\Lambda^1} \longrightarrow C^\infty(\Omega) \otimes_\pi \mathcal{S}_{\Lambda^2}$.

Now let us define a class $\mathcal{P}(\Omega, \underline{\varphi})$, $\Omega \subseteq \mathbb{R}^d$ open, as the space of all mappings $\varphi : \Omega \longrightarrow \underline{\varphi}$ that are continuous at each $y_o \in \Omega$ in the following sense.
First denote by $\pi_{\mathbb{C}}(\varphi(y_o))_\delta$, $\delta > 0$, the finite set of points

$$\pi_{\mathbb{C}}(\varphi(y_o))_\delta := \pi_{\mathbb{C}}(\varphi(y_o)) \cap \{\operatorname{Re} z \geq \tfrac{1}{2} - \delta\} .$$

An element $U \in \underline{\Lambda}$ is called a closed neighbourhood of $\pi_{\mathbb{C}}(\varphi(y_o))_\delta$ if $\pi_{\mathbb{C}}(\varphi(y_o))_\delta \subset \operatorname{int} U$. Then $\varphi : \Omega \longrightarrow \underline{\varphi}$ is called continuous at $y_o \in \Omega$ if for each $\delta > 0$ and each closed neighbourhood U of $\pi_{\mathbb{C}}(\varphi(y_o))_\delta$ there is a closed set $K \subset \Omega$, $y_o \in \operatorname{int} K$, such that $\pi_{\mathbb{C}}(\varphi(y))_\delta \subset \operatorname{int} U$ for all $y \in K$. Denote by $N(y,\delta,\varphi)$, $y \in \Omega$, the number of points of $\varphi(y)$ in $\{\operatorname{Re} z \geq \frac{1}{2} - \delta\}$ counted with their multiplicities. Set for $K \subset\subset \Omega$

$$N(K,\delta,\varphi) = \sup_{y \in K} N(y,\delta,\varphi) \qquad .$$

Then we define $\mathcal{P}(\Omega,\underline{\varphi})$ as the class of all mappings $\varphi : \Omega \longrightarrow \underline{\varphi}$ that are continuous at each $y \in \Omega$ and satisfy

$$N(K,\delta,\varphi) < \infty \qquad \text{for each } K \subset\subset \Omega\,,\ \delta \in \mathbb{R}_+ \ .$$

Let Ω' be a subset with compact closure $\overline{\Omega'} \subset\subset \Omega$. If we are given for each $y \in \overline{\Omega'}$ a closed neighbourhood $U(y)$ of $\pi_{\mathbb{C}}(\varphi(y))_\delta$, we get a covering of $\overline{\Omega'}$ by open sets int $K(y)$, from which we can select a finite covering. If $U(y)$ is given for each $y \in \Omega$, we find a locally finite covering $\{K_j\}_{j \in J}$ of Ω, where each K_j corresponds to some $y_j \in \Omega$ and $U_j = U(y_j) \in \underline{\Lambda}$, $\pi_{\mathbb{C}}(\varphi(y))_\delta \subset \operatorname{int} U_j$ for all $y \in K_j$. Denote by $\mathcal{K}_\delta(\varphi)$ the system of all sequences $\{K_j, U_j\}$ with these properties, associated with $\varphi \in \mathcal{P}(\Omega,\underline{\varphi})$. In $\mathcal{K}_\delta(\varphi)$ we have a semi-ordering and every two elements have a common refinement.

Note that the covering of Ω by int K_j cannot be arbitrary. If some K_j is too large, it may be impossible to find a neighbourhood of $\pi_{\mathbb{C}}(\varphi(y))$ for all $y \in K_j$ that belong to $\underline{\Lambda}$. Conversely, it is clear that on compact parts of Ω a sufficiently fine covering is always possible.

Use the notation $\mathcal{S}_\Lambda^{(1,\delta)}$ for the closure of $\mathcal{S}_\Lambda$, $\Lambda \in \underline{\Lambda}$, with respect to the Hilbert-norm belonging to the strip $\{\frac{1}{2} - \delta \leq \operatorname{Re} z \leq \frac{1}{2} + \delta\}$ and $\mathcal{J}$ derivatives of orders ≤ 1.

Let $\varphi \in \mathcal{P}(\Omega,\varphi)$ be given and denote by $C^\infty(\Omega,\mathcal{S}_\varphi)$ the space of all functions $u(y,t) \in C^\infty(\Omega \times \mathbb{R}_+)$ with the properties

(i) $u(y_o,\cdot) \in \mathcal{S}_{\varphi(y_o)}$ for each $y_o \in \Omega$,

(ii) for all $\delta > 0$, $\{K_j, U_j\} \in \mathcal{K}_\delta(\varphi)$, $l \in \mathbb{Z}_+$, we have for every j

$$\rho_{K_j} u(y,t) \in C^\infty(\operatorname{int} K_j, \mathcal{S}_{U_j}^{(l,\delta)})$$

ρ_{K_j} being the operator of restriction to int K_j.

By definition we have a system of operators

$$\rho_{\{y\}} : C^\infty(\Omega, \mathcal{S}_\varphi) \longrightarrow \mathcal{S}_{\varphi(y)},$$

$$\rho_{K_j} : C^\infty(\Omega, \mathcal{S}_\varphi) \longrightarrow C^\infty(\operatorname{int} K_j, \mathcal{S}_{U_j}^{(l,\delta)}) ,$$

where $C^\infty(\operatorname{int} K_j, \mathcal{S}_{U_j}^{(l,\delta)})$ is a nuclear Fréchet space. It is obvious that

$$\bigcap \rho_{K_j}^{-1} (\{0\}) = \{0\} ,$$

where the intersection is taken over all $\mathcal{K}_\delta(\varphi), \delta, j, l$. Consider $C^\infty(\Omega, \mathcal{S}_\varphi)$ in the projective limit topology with respect to the mappings $\rho_{\{y\}}$, $y \in \Omega$, and the ρ_{K_j} for all $\mathcal{K}_\delta(\varphi), \delta, j, l$ (the weakest topology under which all operators $\rho_{\{y\}}$, ρ_{K_j} are continuous). Then $C^\infty(\Omega, \mathcal{S}_\varphi)$ is a nuclear locally convex space.

In view of 2.1.1, Remark 9, the system of maps $\rho_{\{y\}}$, $y \in \Omega$, is superfluous for introducing the topology in $C^\infty(\Omega, \mathcal{S}_\varphi)$, i.e.

$$C^\infty(\Omega, \mathcal{S}_\varphi) = \varprojlim_{\substack{\{K_j, U_j\} \in \mathcal{K}_\delta(\varphi) \\ \delta \in \mathbb{R}_+, l \in \mathbb{Z}_+}} C^\infty(\operatorname{int} K_j, \mathcal{S}_{U_j}^{(l,\delta)}) .$$

Then, by choosing rational approximations of the singularity types in finite strips, rational approximations of the K_j, and $\delta \in \mathbb{Z}_+$, we get a countable semi-norm system in $C^\infty(\Omega, \mathcal{S}_\varphi)$ under which it is obviously complete and hence a Fréchet space.

Now introduce some further notations. Let $\mathcal{P}_\Delta(\Omega, \underline{p})$, $\Delta \subseteq \mathbb{C}$, be the set of all $\varphi \in \mathcal{P}(\Omega, \underline{p})$ for which $\pi_{\mathbb{C}}(\varphi(y)) \subseteq \Delta$ for all $y \in \Omega$, and define

$$C^\infty(\Omega, \mathcal{S}_\Delta) = \varinjlim_{\varphi \in \mathcal{P}_\Delta(\Omega, \underline{p})} C^\infty(\Omega, \mathcal{S}_\varphi),$$

being a nuclear space, again.

Observe that the continuity of some $\varphi(y)$ does not refer to the multiplicities involved in the discrete singularity types. So we also may extend the notion and introduce, in an analogous manner, the class $\mathcal{P}(\Omega, \pi_{\mathbb{C}}\underline{\varphi})$ as the set of all mappings $\pi_{\mathbb{C}}\varphi : \Omega \longrightarrow \pi_{\mathbb{C}}\underline{\varphi}$ **that** are continuous at each $y_0 \in \Omega$. In the same way as $C^\infty(\Omega, \mathcal{S}_\varphi)$ we can introduce the space $C^\infty(\Omega, \mathcal{S}_{\pi_{\mathbb{C}}\varphi})$, where (i) is replaced by (i') $u(y_0, \cdot) \in \mathcal{S}_{\pi_{\mathbb{C}}\varphi(y_0)}$ for each $y_0 \in \Omega$.

This simply means that point-wise we allow functionals with discrete support and possibly infinite order.

The definition of spaces of the types $C^\infty(\Omega, \mathcal{S}_\varphi \otimes_\pi \mathcal{S}_\eta)$, $C^\infty(\Omega, \mathcal{S}_\Lambda \otimes_\pi \mathcal{S}_\eta)$ ...., for $\varphi, \eta \in \mathcal{P}(\Omega, \underline{\varphi})$ is straightforward and left to the reader. We also use notations of the sort $C^\infty(\Omega, \mathcal{S}_{\Delta^1} \otimes_\pi \mathcal{S}_{\Delta^2 \cdot})$, $C^\infty(\Omega, \mathcal{S}_{\Delta^1 \cdot} \otimes_\pi \mathcal{S}_{\Delta^2 \cdot})$ for subsets $\Delta^i \subset \mathbb{C}$, $i = 1,2$, in an analogous meaning as above.

Let us remark that the notion of continuity of some $\varphi \in \mathcal{P}(\Omega, \underline{\varphi})$ does not imply that we always find a system of functions $p_j : \Omega \longrightarrow \mathbb{C}$, $j \in \mathbb{Z}_+$, with $\pi_{\mathbb{C}}\varphi(y) = \{p_j(y)\}_{j \in \mathbb{Z}_+}$. In general we have to expect branching points and the multiplicities may have jumps.

Consider a simple example of a smooth family of discrete functionals which illustrates the situation. Consider an arbitrary non-constant polynomial in z

$$g(y,z) = \sum_{j=0}^{m} g_j(y)\, z^j$$

with $g_j \in C^\infty(\Omega)$, $|g_j(y)| \leqq \text{const.}$, $y \in \Omega$, $j = 0,\dots,m$. Then there is a compact set $K \subset \mathbb{C}$ such that $g(y,z) \neq 0$ for all $y \in \Omega$, $z \notin K$.

Let B be a positively oriented connected curve surrounding K and define the family of functionals in $\mathcal{A}'(\mathbb{C})$ by the formula

$$\langle \xi(y), v \rangle = \frac{1}{2\pi i} \int_B \frac{1}{g(y,z)}\, v(z)\,dz, \quad v \in \mathcal{A}(\mathbb{C}).$$

Then we have a smooth function $\xi : \Omega \longrightarrow \mathcal{A}'(\mathbb{C})$. The derivatives can be calculated by differentiating g^{-1} with respect to y. Denote by

$$r^0(y), \dots, r^{M(y)}(y)$$

the different poles of the meromorphic function $g^{-1}(y,z)$, y fixed, and let

$$\sum_{k=0}^{m_j(y)} \xi_k^j(y)\ (z - r^j(y))^{-(k+1)}$$

be the principal part of the Laurent expansion of g^{-1} at $r^j(y)$, $j = 0,1,\ldots,M(y)$. Then, for $v \in \mathcal{A}(\mathbb{C})$

$$\langle \xi(y), v\rangle = \frac{1}{2\pi i}\sum_{j=0}^{M(y)} \int_B \Big\{ \sum_{k=0}^{m_j(y)} \xi_k^j(y)(z-r^j(y))^{-(k+1)} \sum_{l=0}^{m_j(y)} \frac{1}{l!}\frac{\partial^l v}{\partial z^l}(r^j(y))(z-r^j(y))^l \Big\}\, dz$$

$$= \frac{1}{2\pi i}\sum_{j=0}^{M(y)} \int_B \Big\{ \sum_{l=0}^{m_j(y)} \frac{1}{l!}\, \xi_l^j(y) \frac{\partial^l v}{\partial z^l}(r^j(y))\Big\} (z-r^j(y))^{-1} dz$$

$$= \sum_{j=0}^{M(y)} \sum_{l=0}^{m_j(y)} \frac{1}{l!}\, \xi_l^j(y) \frac{\partial^l v}{\partial z^l}(r^j(y))$$

This shows that for fixed y the functional ξ is discrete and belongs to

$$\varphi(y) = \{ (-r^j(y), m_j(y)) : j = 0,\ldots,M(y) \} .$$

It is a well-known result that the zeros $r^j(y)$ of the family of polynomials $g(y,z)$ defining the positions of carriers of $\xi(y)$ satisfy the condition (ii) of the definition of $C^\infty(\Omega_y, \mathcal{S}_{\varphi(y)})$ (cf. Kato [K 1]). Of course, $M(y)$ and the $r^j(y)$ are no smooth functions of y in general. For instance, if

$$g(y,z) = (a(y)-z)(b(y)-z) \tag{1}$$

and $N = \{ y \in \Omega : a(y) = b(y)\}$, we have

$$M(y) = \begin{cases} 1 & \text{for } y \in N, \\ 0 & \text{for } y \notin N . \end{cases}$$

Any $u(y,t) \in C^\infty(\Omega, \mathcal{S}_\Lambda)$ which is associated with the functional ξ belonging to (1) (e.g. $u(y,t) = \langle \xi(y), t^{-z}\omega(t)\rangle$) has an asymptotics of the form

$$\xi_1(y)\ t^{a(y)} + \xi_2(y)\ t^{b(y)} \qquad \text{for } y \notin N , \tag{2}$$

$$\xi_3(y)\ t^{a(y)} \log t \qquad \text{for } y \in N . \tag{3}$$

The asymptotics of $D_y^\alpha u(y,t)$ cannot be calculated by differentiating (2), (3) with respect to y unless the exponents and the coefficients are differentiable. If they are smooth, then, of course, we may differentiate the asymptotic expansion and we observe that the powers of the logarithm increase.

These remarks show that the systems of y depending exponents p(y) together with the multiplicities (powers of the logarithm +1) generated by smooth functions

$$\xi : \Omega \longrightarrow \mathcal{A}'(\mathbb{C}) \tag{4}$$

can be very capricious. Since such a behaviour automatically occurs in the asymptotics of boundary problems, we want to study the $C^\infty(\Omega, \mathcal{S}_{\Lambda}^{\bullet})$ spaces in more detail, $\Lambda \in \underline{\Lambda}$.

For $u \in C^\infty(\Omega, \mathcal{S}_{\Lambda}^{\bullet}) \subset C^\infty(\Omega, \mathcal{S}_{\Lambda})$, $\Lambda = \bigcup \Lambda_i$, we obviously have $\xi_i(u) \in C^\infty(\Omega) \otimes_\pi \mathcal{A}'(\Lambda_i)$ and for each fixed $y \in \Omega$ the analytic functional $\xi_i(u)(y) \in \mathcal{A}'(\Lambda_i)$ is a finite sum of derivatives of the Dirac functional. Since neither the positions of these Dirac functionals nor the coefficients smoothly depend on y, it is not obvious that $D_y^\alpha \xi_i(u)(y)$ is a family of discrete functionals of finite order, again. However, we can prove the following

1. <u>Theorem</u>. For every $\varphi \in \mathcal{P}(\Omega, \underline{\varphi})$ and any $\alpha \in \mathbb{Z}_+^d$ there is a $\sigma_\alpha = \sigma_\alpha(\varphi) \in \mathcal{P}(\Omega, \underline{\varphi})$ such that the operator

$$D_y^\alpha \; : \; C^\infty(\Omega, \mathcal{S}_\varphi) \longrightarrow C^\infty(\Omega, \mathcal{S}_{\sigma_\alpha}) \tag{5}$$

is continuous, and $\pi_{\mathbb{C}}\varphi(y) \subseteq \pi_{\mathbb{C}}\sigma_\alpha(\varphi)(y)$, $y \in \Omega$,

$$N(y, \delta, \sigma_\alpha(\varphi)) \leq 2^{|\alpha|} N(y, \delta, \underline{\varphi}), \; y \in \Omega, \; \delta \in \mathbb{R}_+ .$$

Moreover,

$$D_y^\alpha \; : \; C^\infty(\Omega, \mathcal{S}_{\Lambda}^{\bullet}) \longrightarrow C^\infty(\Omega, \mathcal{S}_{\Lambda}^{\bullet})$$

is continuous, $\Lambda \in \underline{\Lambda}$.

The proof will be given in several steps. First we may restrict ourselves to $|\alpha| = 1$, then it follows inductively. For $|\alpha| = 1$, $u \in C^\infty(\Omega, \mathcal{S}_\varphi)$ we have

$$\rho_{K_j} D_y^\alpha u(y,t) \in C^\infty(\text{int } K_j, \mathcal{S}_{U_j}^{(1,\delta)})$$

for any collection $\{K_j, U_j\} \in \mathcal{K}_\delta(\varphi)$, $\delta > 0$, $l \in \mathbb{Z}_+$.

Thus, the function $t \longrightarrow D_y^\alpha u(y,t)$ has a discrete singularity type for each fixed $y \in \Omega$ but possibly with infinite multiplicities. So we have to establish the estimate for $N(y, \delta, \eta_\alpha)$.

For the proof we need a Lemma about special families of meromorphic functions. Let $K \subset \mathbb{C}$ be a compact simply connected set. Let $y \longrightarrow f(y,z)$ be a family of meromorphic functions in z, $y \in \Omega$, with the following properties:

(i) $f|_{\Omega \times (\mathbb{C} \setminus K)} \in C^\infty(\Omega \times (\mathbb{C} \setminus K))$,

if $w_1(y), \ldots, w_N(y)$, $N = N(y,f)$ are the poles of $f(y,\cdot)$ counted with their multiplicities, then

(ii) $N(f) = \sup\limits_{y \in \Omega} N(y,f) < \infty$.

Then we have

2. __Lemma.__ For an arbitrary multi-index $\alpha \in \mathbb{Z}_+^d$ the function $D_y^\alpha(f(y,z)_{\Omega \times (\mathbb{C} \setminus K)})$ can be extended to a family of meromorphic functions satisfying (i), (ii) and

$$N(D_y^\alpha f) \leq 2^{|\alpha|} N(f) \quad . \tag{6}$$

The proof of Lemma 2 follows by induction from the following

3. __Lemma.__ Let $\{f_k\}_{k \in \mathbb{Z}_+}$ be a sequence of meromorphic functions with poles in K, $N(f_k) \leq N$ for all k and assume that $f_k|_{\mathbb{C} \setminus K}$ converges to a function f' in $\mathcal{A}(\mathbb{C} \setminus K)$ as $k \longrightarrow \infty$. Then f' can be extended to a meromorphic function f, i.e. $f|_{\mathbb{C} \setminus K} = f'$, and $N(f) \leq N$.

__Proof.__ Let $w_1^k, \ldots, w_{N(f_k)}^k$ be the poles of f_k counted with their multiplicities. Then we define a sequence of polynomials

$$g_k(z) = (z - w_1^k) \ldots (z - w_{N(f_k)}^k) \quad .$$

Since all w_j^k belong to the compact set K and $N(f_k) \leq N$, the set of these polynomials belongs to a bounded subset of the space of all polynomials of degree $\leq N$. Then we find a subsequence $\{g_{k_l}\}_{l \in \mathbb{Z}_+}$ of the sequence $\{g_k\}_{k \in \mathbb{Z}_+}$ which converges to the polynomial g of degree $\leq N$. Then the sequence of functions $\{h_l\}_{l \in \mathbb{Z}_+}$, $h_l(z) = f_{k_l}(z)\, g_{k_l}(z) \in \mathcal{A}(\mathbb{C})$ tends to a limit $h \in \mathcal{A}(\mathbb{C})$. If we set $f = h/g$, the assertion follows. $\square$

Proof of Lemma 2. It suffices to consider $|\alpha| = 1$. For example let $\alpha = (1,0,\dots 0)$. Then Lemma 4 can be applied to the sequence

$$f_k(z) = (y_1 - y_1^k)^{-1} \, (f(y_1, y', z) - f(y_1^k, y', z))$$

for fixed $y = (y_1, y') \in \Omega$ and $y_1^k \longrightarrow y_1$ as $k \longrightarrow \infty$. (6) follows from $N(f_k) \leq 2N(f)$. $\square$

Now we complete the proof of Theorem 2. The inclusion $\pi_{\mathbb{C}}\varphi \supseteq \pi_{\mathbb{C}}\varphi_\alpha$ for all y is obvious. Now we prove the estimate for $N(y, \delta, \varphi_\alpha)$.

For fixed $\delta > 0$ let us choose a closed neighbourhood U of $\pi_{\mathbb{C}}(\varphi(y))_\delta$ for all y in a compact set $K \subset \Omega$ and assume U to be compact, $U = U^c$, $B = \partial U$ smooth. Define a function $\operatorname{int} K \ni y \longrightarrow \mathcal{A}'(U)$ by

$$\langle \zeta(\rho_K u), h \rangle = \frac{1}{2\pi i} \int_B u(y, \tilde{z}) \, h(z) dz,$$

$h \in \mathcal{A}(\mathbb{C})$, $u \in C^\infty(\Omega, \mathcal{S}_\varphi)$. Then

$$\zeta(\rho_K u) \in C^\infty(\operatorname{int} K, \mathcal{A}'(U)),$$

where $\zeta(\rho_K u)(y)$ is discrete with finite multiplicities. In this way we get a continuous operator

$$\zeta \circ \rho_K : C^\infty(\Omega, \mathcal{S}_\varphi) \longrightarrow C^\infty(\operatorname{int} K, \mathcal{A}'(U)),$$

commuting with D_y^α. The function

$$f(y,z) := \langle \zeta \circ \rho_K(y,w), \phi(z,w) \rangle$$

satisfies the assumptions of Lemma 2 (with obvious substitution of notations). Hence $D_y^\alpha(\zeta \circ \rho_K)$ is of finite order, again, for every $y \in \operatorname{int} K$, with the desired estimates for the multiplicities. In particular this consideration implies $\pi_{\mathbb{C}}\varphi(y) \supseteq \pi_{\mathbb{C}}\varphi_\alpha(\varphi)(y)$, $y \in \Omega$, and $\varphi_\alpha(\varphi) \in \mathcal{P}(\Omega, \underline{\varphi})$. The continuity of $D_y^\alpha : C^\infty(\Omega, \mathcal{S}_\varphi) \longrightarrow C^\infty(\Omega, \mathcal{S}_{\varphi_\alpha})$ follows by arguments of the following sort. Let, in general, E,F be projective limits

$$E = \varprojlim_{\gamma \in \Gamma} (\rho_\gamma, E_\gamma), \quad F = \varprojlim_{\eta \in H} (\lambda_\eta, F_\eta),$$

where, for instance, F has the weakest topology, for which $\lambda_\eta : F \longrightarrow F_\eta$ is continuous for all $\eta \in H$. Then an operator $A : E \longrightarrow F$ is continuous iff $\lambda_\eta \circ A : E \longrightarrow F_\eta$ is continuous for all $\eta \in H$. The latter is the case iff for every $\eta \in H$ and any semi-norm f_η in F_η there is a finite set of semi-norms $e^1,\ldots,e^{N(f_\eta)}$ in E such that

$$f_\eta(u) \leq C \sup_{1\leq j\leq N(f_\eta)} e^j_\eta(\lambda_\eta \circ A\, u)$$

with some $C > 0$ for all $u \in E$. Now the semi-norms in E in our example $E = C^\infty(\Omega, \mathcal{S}_\varphi)$ are defined by the mappings ρ_K and semi-norms in $C^\infty(\text{int } K, \mathcal{S}_U^{(1,\delta)})$ for certain $K \subset \Omega$, $U \supset \pi_{\mathbb{C}}(\varphi(y))_\delta$, $y \in K$. The λ_η are interpreted in an analogous manner. Then the continuity of $A = D_y^\alpha$ is rather obvious by choosing e_η^j in terms of the underlying geometric picture of localizations of singularity types in the δ strip and locally finite coverings of Ω.

Though elementary the consideration is clumsy and will not always be carried out explicitely in the calculus below. Let us quote it as the localization argument whenever we deal with the spaces with branching asymptotics and establish continuities that way. So the localization argument gives us the continuity of D_y^α. This also implies the continuity in $C^\infty(\Omega, \mathcal{S}_{\Lambda^\bullet})$. $\square$

Point out once again that, roughly speaking, the localization argument reduces assertions for the spaces with branching asymptotics to assertions for families of $C^\infty(G, \mathcal{S}_U)$ spaces by taking the open sets G and the singularity types U small.

2.1.3. Other spaces with asymptotics

In this section we return to functions $u(t)$ on the half axis $\mathbb{R}_+$ and discuss the asymptotic behaviour for $t \longrightarrow 0$ from another point of view. Namely, instead of the analytic functionals we look at the sets Λ and also admit more general configurations.

The interpretation of a subset $\Delta \subset \mathbb{C}$ is that it represents some type of asymptotics of $u \in C^\infty(\mathbb{R}_+) \cap L^2(\mathbb{R}_+)$ for $t \longrightarrow 0$ $(t \longrightarrow \infty)$ when $\tilde{u}(z)$ is in $\mathcal{A}(\mathbb{C} \setminus \Delta)$ and satisfies some growth condition for $|\text{Im } z| \longrightarrow \infty$. The precise definitions are as follows.

Denote by $\underline{\Delta}$ the system of all closed subsets $\Delta \subset \mathbb{C}$ contained in $\{\operatorname{Re} z < \frac{1}{2}\}$, satisfying $\Delta^c = \Delta$, $\{z \in \Delta : \rho_1 \leqq \operatorname{Re} z \leqq \rho_2\}$ compact for all $\rho_1, \rho_2 \in \mathbb{R}$, and $B_{\delta\eta}(\Delta)$ (defined in the same way as 2.1.1. (19)) is piecewise smooth, $\eta, \delta \in \mathbb{R}_+$. Let $\Delta \in \underline{\Delta}$ be fixed and $\chi \in C^\infty(\mathbb{C})$ a Δ excision function, i.e. $\chi \equiv 0$ in an open neighbourhood of Δ, $\chi \equiv 1$ outside some other open neighbourhood. Denote by $\mathcal{S}_{(\Delta)}$ the space of all $u \in C^\infty(\mathbb{R}_+)$ for which $(Mu)(z) = \tilde{u}(z)$ is holomorphic in $\mathbb{C} \setminus \Delta$ and satisfies the estimates

$$\Big| \int_{\Gamma_\rho} |z^l \chi(z) \tilde{u}(z)|^2 dz \Big| < \infty \quad , \; l \in \mathbb{Z}_+,$$

for every $\rho \in \mathbb{R}$, cf. 2.1.1, Proposition 10. Define a system of norms $\lambda_k = \lambda_{kk}$ by the same scheme as in 2.1.1. (38) and denote by $\mathcal{H}_k(\Delta)$ the closure of $\mathcal{S}_{(\Delta)}$ with respect to λ_k. The Hilbert spaces $\mathcal{H}_k(\Delta)$ are regarded as the standard scale of spaces for the singularity type Δ. By similar arguments as in Section 2.1.1. we get the

1. <u>Theorem</u>. $\mathcal{S}_{(\Delta)}$ equipped with the semi-norm system $(\lambda_k)_{k \in \mathbb{Z}_+}$ is a nuclear Fréchet space, $\mathcal{S}_{(\Delta)} = \varprojlim \mathcal{H}_k(\Delta)$. For $\Delta \in \underline{\Lambda}$ we have $\mathcal{S}_\Delta = \mathcal{S}_{(\Delta)}$.

The basic aspects of the theory of $\mathcal{S}_\Lambda$ spaces developed in 2.1.1. can be generalized to the case of $\mathcal{S}_{(\Delta)}$ spaces. We return to this question in Section 2.1.5., where we introduce spaces $\mathcal{S}_\Delta$ associated with Δ in another way, and it will be proved that the two definitions are equivalent (for the moment let us still use the paranthesis at the subscript Δ).

It is easy to introduce the concept of Δ-asymptotics near $t = 0$ also for extendable distributions on $\mathbb{R}_+$ (i.e. distributions $u \in \mathcal{D}'(\mathbb{R}_+)$ with $u = v|_{\mathbb{R}_+}$ for some $v \in \mathcal{D}'(\mathbb{R})$).

Until now we have assumed that Δ is contained in $\{\operatorname{Re} z < \frac{1}{2}\}$. It is useful also to define appropriate spaces of holomorphic functions in $\mathbb{C} \setminus Z$, where $Z \subset \mathbb{C}$ is an arbitrary closed subset with $Z^c = Z$ and $\{z \in Z : \rho_1 \leqq \operatorname{Re} z \leqq \rho_2\}$ compact for arbitrary $\rho_1, \rho_2 \in \mathbb{R}$, $\rho_1 \leqq \rho_2$. In general we do not assume that $Z \cap \Gamma_\rho = \emptyset$ for some $\rho \in \mathbb{R}$.

The scheme is exactly the same as for the $\mathcal{S}_{(\Delta)}$ spaces. In particular, for $\Gamma_{\frac{1}{2}} \cap Z = \emptyset$

$$\Delta = \left\{ z \in Z : \operatorname{Re} z < \tfrac{1}{2} \right\}, \quad \Delta' = \left\{ z \in Z : \operatorname{Re} z > \tfrac{1}{2} \right\}$$

we get an analogue of the $C^\infty_{\Lambda,\Lambda'}$ spaces, here denoted by $C^\infty_{(\Delta,\Delta')}$. For more general Z we prefer to talk about subspaces in $\mathcal{A}(\mathbb{C} \setminus Z)$. Denote by $\mathcal{M}_{(Z)}$ the space of all holomorphic functions in $\mathbb{C} \setminus Z$ for which

$$\Big| \int_{\Gamma_\rho} | z^l \chi(z)\, h(z)|^2 \, dz \Big| < \infty \ , \ l \in \mathbb{Z}_+ \ ,$$

for all $\rho \in \mathbb{R}$ and any Z excision function χ. Then we get a natural Fréchet space structure on $\mathcal{M}_{(Z)}$ for which $\mathcal{M}_{(Z)}$ is nuclear and $\mathcal{M}_{(Z)} = \varprojlim \mathcal{H}_k(Z)$, $\mathcal{H}_k(Z)$ being the standard scale to Z, defined by the procedure of 2.1.1.

Incidentally we also consider the Hilbert spaces $\mathcal{H}_{k\delta}(Z)$ defined by the norm $\lambda_{k\delta}$ in a similar manner as in 2.1.1. and then we use the notation

$$\mathcal{M}^{(k,\delta)}_{(Z)} = \mathcal{H}_{k\delta}(Z) \ . \qquad (1)$$

2.2. Operators with continuous singularity

2.2.1. The action of operators on $\mathcal{S}_\Lambda(\mathbb{R}_+)$

In this section we extend the theory of Green, Mellin and pseudo-differential operators in $\mathcal{S}_\gamma$ to the case of $\mathcal{S}_\Lambda$ spaces and also discuss the dependence on parameters. The operators on the half axis are part of the boundary symbols to be used in Section 2.3.

The theory of Green operators can be formulated in terms of general nuclear Fréchet spaces. First let us study the abstract situation of scales of complex Hilbert spaces $(\mathcal{H}^i_j)_{j \in \mathbb{Z}}$, $i = 1,2,\ldots$, with $\mathcal{H}^i_{j+1} \hookrightarrow \mathcal{H}^i_j$ nuclear for $j \geq 0$, $\mathcal{H}^i_j = \mathcal{H}^i_0$ for $j \leq 0$ and the projective limits

$$\mathcal{S}^i = \varprojlim_j \mathcal{H}^i_j, \ i = 1,2,\ldots$$

The theory applies to $\mathcal{H}^i_0 = L^2(X^i, d\mu^i)$, X^i a locally compact space, μ^i a positive measure on X^i, $i = 1,2$. In this Chapter we have $\mathcal{H}^i_0 = L^2(\mathbb{R}_+)$. In Section 4.2. we set $\mathcal{H}^i_0 = L^2(\mathbb{R}_+ \times M)$ for some closed compact

manifold M. For convenience let us assume here $\mathcal{H}_o^i = L^2(\mathbb{R}_+)$. Denote by $J : L^2(\mathbb{R}_+) \longrightarrow L^2(\mathbb{R}_+)$ the anti-linear involution $u \longrightarrow \bar{u}$. For a subspace $\mathcal{H} \subset L^2(\mathbb{R}_+)$ we get another subspace $J\mathcal{H} \subset L^2(\mathbb{R}_+)$.
Let $w \in L^2(\mathbb{R}_+) \otimes_H L^2(\mathbb{R}_+) = L^2(\mathbb{R}_+ \times \mathbb{R}_+)$ be a given function and denote by

$$op_G(w) : L^2(\mathbb{R}_+) \longrightarrow L^2(\mathbb{R}_+)$$

the associated linear continuous operator

$$op_G(w)\, u(t) = \int_{\mathbb{R}_+} w(t,s)\, u(s)\, ds = (w(t,s), Ju(s))_{L^2(\mathbb{R}_+)},$$

$u \in L^2(\mathbb{R}_+)$. Obviously, $op_G(w)$ is a Hilbert-Schmidt operator. Denote by $op_G(L^2(\mathbb{R}_+) \otimes_H L^2(\mathbb{R}_+))$ the space of all those operators and by $op_G(\mathcal{T})$ the subspace of operators with kernels in $\mathcal{T} \subset L^2(\mathbb{R}_+) \otimes_H L^2(\mathbb{R}_+)$. Then the adjoint $op_G(w)^*$ defined by

$$(op_G(w)u,v)_{L^2(\mathbb{R}_+)} = (u, op_G(w)^* v)_{L^2(\mathbb{R}_+)}, \quad u,v \in L^2(\mathbb{R}_+)$$

possesses the kernel $w^*(t,s) = \overline{w(s,t)}$, i.e. $op_G(w)^* = op_G(w^*)$.

The composition $op_G(w_1)\, op_G(w_2)$ of two Hilbert-Schmidt operators is again Hilbert-Schmidt, where

$$op_G(w_1)\, op_G(w_2) = op_G(w_3),$$

$$w_3(t,s) = \int_{\mathbb{R}_+} w_1(t,s')\, w_2(s',s)\, ds'$$

1. <u>Lemma</u>. Let $\mathcal{H}_j^1 \otimes_H \mathcal{H}_k^2$ be the Hilbert space tensor product of $\mathcal{H}_j^1$ and $\mathcal{H}_k^2$. Then

$$\mathcal{S}^1 \otimes_\pi \mathcal{S}^2 = \varprojlim_{j,k} \mathcal{H}_j^1 \otimes_H \mathcal{H}_k^2 .$$

In a similar way we obtain

$$\mathcal{S}^1 \otimes_\pi \mathcal{H}_o^2 = \varprojlim_{j} \mathcal{H}_j^1 \otimes_H \mathcal{H}_o^2 .$$

The proof is basically the same as that of 1.2.2. (13), (14).
The spaces $\mathcal{L}(\mathcal{H}_o^1, \mathcal{S}^2)$, $\mathcal{L}(\mathcal{S}^1, \mathcal{S}^2)$,... are considered in the topology

of uniform convergence on all bounded sets. In particular in $\mathcal{L}(\mathcal{H}_0^1, \mathcal{S}^2)$ we have the semi-norm system

$$\lambda_\alpha(\sigma) = \sup_{u \in \mathcal{H}_0^1} \frac{\mu_\alpha(\sigma u)}{\|u\|_{\mathcal{H}_0^1}}, \quad \sigma \in \mathcal{L}(\mathcal{H}_0^1, \mathcal{S}^2),$$

where μ_α runs through a semi-norm system in $\mathcal{S}^2$.

2. <u>Proposition</u>. op_G induces a topological isomorphism

$$op_G : \mathcal{S}^2 \otimes_\pi \mathcal{H}_0^1 \longrightarrow \mathcal{L}(\mathcal{H}_0^1, \mathcal{S}^2).$$

<u>Proof</u>: First we show that op_G is continuous. $\mathcal{S}^2 \otimes_\pi \mathcal{H}_0^1$ is defined as the closure of the set of the finite sums $w = \sum f_i \otimes g_i$, $f_i \in \mathcal{S}^2$, $g_i \in \mathcal{H}_0^1$ under the semi-norm system

$$\tau_\alpha(w) = \inf \sum \mu_\alpha(f_i) \, \| g_i \|_{\mathcal{H}_0^1} ,$$

where $(\mu_\alpha)_{\alpha \in \underline{\alpha}}$ defines the topology in $\mathcal{S}^2$. For these w we have

$$\sigma u = op_G(w)u = \sum f_i < g_i, u > = \sum f_i(g_i, Ju)$$

and hence

$$\mu_\alpha(\sigma u) \leq \sum \mu_\alpha(f_i) \, \| g_i \|_{\mathcal{H}_0^1} \| u \|_{\mathcal{H}_0^1} ,$$

which implies $\lambda_\alpha(\sigma) \leq \tau_\alpha(w)$. Extension by continuity leads to the continuity of op_G.

Now assume that $\sigma \in \mathcal{L}(\mathcal{H}_0^1, \mathcal{S}^2)$. Then $\sigma \in \mathcal{L}(\mathcal{H}_0^1 \mathcal{H}_k^2)$ for every $k \in \mathbb{Z}_+$. Since $\mathcal{H}_{k+1}^2 \hookrightarrow \mathcal{H}_k^2$ is Hilbert-Schmidt, σ has a kernel in $\mathcal{H}_k^2 \otimes_H \mathcal{H}_0^1$. Thus the kernel also belongs to $\varprojlim_k \mathcal{H}_k^2 \otimes_H \mathcal{H}_0^1 = \mathcal{S}^2 \otimes_\pi \mathcal{H}_0^1$. Thus we have the surjectivity of op_G. The injectivity is clear, since a Hilbert-Schmidt operator has a uniquely determined kernel in $\mathcal{H}_0^2 \otimes_H \mathcal{H}_0^1$. □

Note that Proposition 2 is an abstract version of 1.2.2. Proposition 7.

The operator $J: L^2(\mathbb{R}_+) \longrightarrow L^2(\mathbb{R}_+)$ transforms every vector subspace into another one. In particular, for

$$\mathcal{S}^i = \varprojlim \mathcal{H}_j^i, \quad i = 1,2,\ldots,$$

we have

$$J\mathcal{S}^i = \varprojlim J\mathcal{H}_j^1 ,$$

which is another nuclear Fréchet space. The Hilbert space structure in $J\mathcal{H}^i_j$ is given by the scalar product

$$(u,v)_{J\mathcal{H}^i_j} = (Jv,\ Ju)_{\mathcal{H}^i_j} \quad .$$

For $b \in \mathcal{S}^2 \otimes_\pi J\mathcal{S}^1$ we get $\mathrm{op}_G(b)^* = \mathrm{op}_G(b^*)$ with $b^* \in \mathcal{S}^1 \otimes_\pi J\mathcal{S}^2$. If $b_1 \in \mathcal{S}^2 \otimes_\pi J\mathcal{S}^1$ and $b_2 \in \mathcal{S}^3 \otimes_\pi J\mathcal{S}^2$, then $\mathrm{op}_G(b_2)\ \mathrm{op}_G(b_1) = \mathrm{op}_G(b_3)$, where $b_3 \in \mathcal{S}^3 \otimes_\pi J\mathcal{S}^1$. Proposition 2 shows that

$$\mathrm{op}_G(b) \in \mathcal{L}(\mathcal{H}^1_o, \mathcal{S}^2),\ \mathrm{op}_G(b)^* \in \mathcal{L}(\mathcal{H}^2_o, \mathcal{S}^1) \tag{1}$$

implies

$$b(t,s) \in (\mathcal{S}^2 \otimes_\pi \mathcal{H}^1_o) \cap (\mathcal{H}^2_o \otimes_\pi J\mathcal{S}^1) \quad . \tag{2}$$

It is obvious that we have a continuous embedding

$$\mathcal{S}^2 \otimes_\pi J\mathcal{S}^1 \hookrightarrow (\mathcal{S}^2 \otimes_\pi \mathcal{H}^1_o) \cap (\mathcal{H}^2_o \otimes_\pi J\mathcal{S}^1) \ .$$

If $\mathcal{S}^i = \varprojlim_j \mathcal{H}^i_j$, $i = 1,2$, we have the space $\mathcal{L}^k(\mathcal{S}^1, \mathcal{S}^2)$ of all continuous operators $\alpha : \mathcal{S}_1 \longrightarrow \mathcal{S}_2$ of order k, cf. 2.1.1. Definition 13. For all $b \in \mathcal{S}^2 \otimes_\pi \mathcal{H}^1_o$ we have

$$\mathrm{op}_G(b) \in \bigcap_{k \in \mathbb{Z}_+} \mathcal{L}^{-k}(\mathcal{S}^1, \mathcal{S}^2) = \mathcal{L}^{-\infty}(\mathcal{S}^1, \mathcal{S}^2)$$

Conversely, any operator $\alpha \in \mathcal{L}^{-\infty}(\mathcal{S}^1, \mathcal{S}^2)$ induces a continuous map in $\mathcal{L}(\mathcal{H}^1_o, \mathcal{H}^2_k)$ for arbitrary $k \in \mathbb{Z}_+$, and hence it belongs to $\mathcal{L}(\mathcal{H}^1_o, \mathcal{S}^2)$. In this way we obtain from Proposition 2

3. <u>Corollary</u>. $\mathcal{S}^2 \otimes_\pi \mathcal{H}^1_o \cong \mathcal{L}^{-\infty}(\mathcal{S}^1, \mathcal{S}^2)$.

In the following proposition we set $\mathcal{H}_o = \mathcal{H}^i_o$, $i = 1,2$. Applying 1.2.2. Proposition 8 to $H = \mathcal{H}_o$ and $\mathcal{A} = \mathrm{op}_G(\mathcal{S}^2 \otimes_\pi \mathcal{H}_o)$ and $\mathcal{A}^* = \mathrm{op}_G(\mathcal{S}^1 \otimes_\pi \mathcal{H}_o)$ we get immediately

4. <u>Proposition</u>. Let $b \in (\mathcal{S}^2 \otimes_\pi \mathcal{H}_o) \cap (\mathcal{H}_o \otimes_\pi \mathcal{S}^1)$. Then

$$1 + \mathrm{op}_G(b) : \mathcal{H}_o \longrightarrow \mathcal{H}_o \tag{3}$$

is a Fredholm operator with index zero. Moreover, $\ker(1+\mathrm{op}_G(b))$ is a finite-dimensional subspace of $\mathcal{S}^2$ and $\mathrm{coker}(1+\mathrm{op}_G(b)) = (\ker(1+\mathrm{op}_G(b)^*))^\perp$, $\ker(1+\mathrm{op}_G(b)^*) \subset \mathcal{S}^1$. If $u \in \mathcal{H}_o$ satisfies the equation $(1+\mathrm{op}_G(b))u = f \in \mathcal{S}^2$, then $u \in \mathcal{S}^2$.

In the same way we get

5. Proposition. Let $b \in (\mathcal{S}^2 \otimes_\pi \mathcal{H}_0) \cap (\mathcal{H}_0 \otimes_\pi J\mathcal{S}^1)$ and $1+op_G(b) : \mathcal{H}_0 \to \mathcal{H}_0$ be bijective. Then $(1+op_G(b))^{-1} = 1+op_3(b_1)$, where $b_1 \in (\mathcal{S}^2 \otimes_\pi \mathcal{H}_0) \cap (\mathcal{H}_0 \otimes_\pi J\mathcal{S}^1)$.

This concept will be applied to $\mathcal{S}^i = \mathcal{S}_{\wedge^i} = \lim\limits_{\leftarrow k} \mathcal{H}^i_k$, cf. 2.1.1.(34) for $\wedge = \wedge^i$.

Now we discuss the parameter-depending case.

For proving an analogue of Proposition 5 we need some extra arguments that ensure the smooth dependence on y of the inverse.

6. Proposition. Let

$$1 + op_G(b)(y) : \mathcal{H}_0 \longrightarrow \mathcal{H}_0$$

be bijective for every $y \in \Omega$, $b \in C^\infty(\Omega, \mathcal{S}^2 \otimes_\pi \mathcal{H}_0 \cap \mathcal{H}_0 \otimes_\pi J\mathcal{S}^1)$. Then

$$(1 + op_G(b)(y))^{-1} = 1 + op_G(w)(y)$$

for some $w \in C^\infty(\Omega, \mathcal{S}^2 \otimes_\pi \mathcal{H}_0 \cap \mathcal{H}_0 \otimes_\pi J\mathcal{S}^1)$.

Proof: First observe that

$$\begin{aligned} b(y,t,s) &\in C^\infty(\Omega, \mathcal{S}^2 \otimes_\pi \mathcal{H}_0 \cap \mathcal{H}_0 \otimes_\pi J\mathcal{S}^1) \\ &= C^\infty(\Omega) \otimes_\pi (\mathcal{S}^2 \otimes_\pi \mathcal{H}_0 \cap \mathcal{H}_0 \otimes_\pi J\mathcal{S}^1) \\ &= C^\infty(\Omega) \otimes_\pi (\mathcal{S}^2 \otimes_\pi \mathcal{H}_0) \cap C^\infty(\Omega) \otimes_\pi (\mathcal{H}_0 \otimes_\pi J\mathcal{S}^1) \end{aligned}$$

cf. 2.1.5. Lemma 3 . In particular $b \in C^\infty(\Omega) \otimes_\pi \mathcal{S}^2 \otimes_\pi \mathcal{H}_0$ has a representation

$$b(y,t,s) = \sum_{i=0}^{\infty} c_i(y)\ f_i(t)\ g_i(s),$$

where $c_i \in C^\infty(\Omega)$, $f_i \in \mathcal{S}^2$, $g_i \in \mathcal{H}_0$, and the series converges in the tensor product topology. Write

$$b_M(y,t,s) = \sum_{i=0}^{M} c_i(y)\ f_i(t)\ g_i(s)$$

and

$$b_M'(y,t,s) = b(y,t,s) - b_M(y,t,s).$$

Then $\| b'_M(y,\ldots)\|_{\mathcal{H}_0 \otimes_H \mathcal{H}_0} \longrightarrow 0$ as $M \to \infty$ uniformly with all derivatives on each compact $K \subset\subset \Omega$. Thus, for $M = M(K)$ sufficiently large $1 + op_G(b_M)(y) : \mathcal{H}_0 \longrightarrow \mathcal{H}_0$ is invertible for all $y \in K$. Since $op_G(b_M)(y)$ is a finite-dimensional operator, we obtain

$$(1 + op_G(b_M)(y))^{-1} = 1 + op_G(c_M)(y),$$

where $op_G(c_M)(y)$ is a smooth family of finite-dimensional operators. Then we have

$$(1+op_G(c_M)(y))(1+op_G(b)(y)) = 1+(1+op_G(c_M)(y))op_G(b'_M)(y) = 1+op_G(b''_M)(y),$$

where $op_G(b''_M)(y)$ is a smooth family of operators with $\| b''_M(y,\ldots)\|_{\mathcal{H}_0 \otimes_H \mathcal{H}_0} \longrightarrow 0$ as $M \to \infty$ uniformly on K. Then we can represent $(1+op_G(b''_M)(y))^{-1}$ by a Neumann series which converges uniformly for $y \in K$. For the derivatives with respect to y one concludes in the same way as well as for the remaining semi-norms $\| \cdot \|_{\mathcal{H}_k \otimes_H \mathcal{H}_0}$. The analogous conclusion may be applied for the adjoint.

7. Proposition. Let $\Omega \subseteq \mathbb{R}^N$ be contractible and consider the family of Fredholm operators

$$1 + op_G(b)(y) : \mathcal{H}_0 \longrightarrow \mathcal{H}_0,\ y \in \Omega\ ,$$

$b \in C^\infty(\Omega, \mathcal{S}^2 \otimes_\pi \mathcal{H}_0 \cap \mathcal{H}_0 \otimes_\pi \mathcal{J}\mathcal{S}^1)$. Then, if dim ker$(1+op_G(b)(y))$ is independent of y, there are C^∞ mappings $\mathcal{N}_1 : \Omega \longrightarrow \mathcal{S}^1$, $\mathcal{N}_2 : \Omega \longrightarrow \mathcal{S}^2$ with (3) for every fixed $y \in \Omega$.

The simple proof is straightforward and left to the reader. If Ω is not contractible, we have vector bundles over Ω as kernel and cokernel. The theory of Green operators will be applied to Frêchet spaces of the type $\mathcal{S}^2 = \mathcal{S}_\Lambda$, $\mathcal{S}^1 = \mathcal{S}_\Xi$ for certain singularity types $\Lambda, \Xi \in \underline{\Delta}$ and in 2.1.5. to more general singularity types. Introduce the following

8. Notation.

$$\mathcal{S}_\Lambda \otimes_\Gamma \mathcal{S}_\Xi = (\mathcal{S}_\Lambda \otimes_\pi L^2(\mathbb{R}_+)) \cap (L^2(\mathbb{R}_+) \otimes_\pi \mathcal{S}_\Xi). \tag{4}$$

Our next task is to generalize the concept of Mellin operators with symbols that are associated with singularity types in $\underline{\Delta}$ and $\underline{\Delta}'$, respectively.

We shall have to deal systematically with certain sets in $\mathbb{C}$ and various operations between them. Recall the notation

$$\Gamma_\rho = \{z \in \mathbb{C} : \operatorname{Re} z = \operatorname{Re} \rho\}, \ \rho \in \mathbb{C} \tag{5}$$

further, for $Z \subseteq \mathbb{C}$

$$T^\gamma Z = \{z \in \mathbb{C} : z + \gamma \in Z\}, \ \gamma \in \mathbb{C} . \tag{6}$$

If $Z \subset \mathbb{C}$ is some set with $\Gamma_{\frac{1}{2}} \cap T^{\gamma} Z = \emptyset$, $\gamma \in \mathbb{C}$ fixed, define

$$Z_{(\gamma)} := \{z \in Z : \operatorname{Re} (z+\gamma) < \tfrac{1}{2}\} , \tag{7}$$

$$Z'_{(\gamma)} := Z \setminus Z_{(\gamma)} . \tag{8}$$

Then

$$Z_{(\gamma)} = T^{-\gamma}(T^{\gamma}Z)_{(o)}, \ Z'_{(\gamma)} = T^{-\gamma}(T^{\gamma}Z)'_{(o)} . \tag{9}$$

Set

$$Z^* := \{z \in \mathbb{C} : 1 - \bar{z} \in Z\} , \tag{10}$$

$$Z^{\vee} := \{z \in \mathbb{C} : 1 - z \in Z\} . \tag{11}$$

Then, obviously,

$$(T^{\gamma}Z)^* = T^{\bar{\gamma}}Z^*, \quad (T^{\gamma}Z)^{\vee} = T^{-\gamma}Z^{\vee}, \tag{12}$$

$$(Z_{(\gamma)})^* = (Z^*)_{(-\gamma)}, \quad (Z'_{(\gamma)})^* = (Z^*)'_{(-\gamma)} \tag{13}$$

and similarly for the v-operation. If Z^1, Z^2 are two closed subsets of $\mathbb{C}$, we define

$$Z^1 + Z^2 := (Z^1 \cup Z^2)^c$$

or more generally for $Z^j \subseteq \mathbb{C}$, $j = 1,\dots,N \leq \infty$

$$\sum_{j=1}^{N} Z^j := (\bigcup_{j=1}^{N} Z^j)^c .$$

9. <u>Definition.</u> By $\underline{Z}^d$ we denote the set of all closed subsets $Z \subset \mathbb{C}$ with the following properties

(i) $Z^c = Z$,

(ii) there is a sequence of compact subsets $Z_j \subseteq Z$, $j \in \mathbb{Z}$, with $Z = \bigcup_{j \in \mathbb{Z}} Z_j$ and

$$\sup\{\operatorname{Re} z : z \in Z_j\} < \inf\{\operatorname{Re} z : z \in Z_{j+1}\}$$

for all $j \in \mathbb{Z}_+$, $\sup\{\operatorname{Re} z : z \in Z_j\} \longrightarrow -\infty$

as $j \longrightarrow -\infty$, $\inf\{\operatorname{Re} z : z \in Z_j\} \longrightarrow +\infty$

as $j \longrightarrow \infty$,

(iii) for any $\eta, \delta \in \mathbb{R}_+$, the set of points

$$B_{\delta\eta}(Z) := \{z \in \mathbb{C} \setminus Z : \operatorname{dist}(z,Z) = \frac{1}{1+\eta}\} \cap \{z : \frac{1}{2} - \delta \leq \operatorname{Re} z \leq \frac{1}{2}+\delta\}$$

is a piecewise smooth curve.

$Z^1, Z^2 \in \underline{Z}^d$ are called <u>admissible</u> if $Z^1 + Z^2 \in \underline{Z}^d$.

The assumption of admissibility in this section has only technical reasons. Nevertheless, it is often automatically satisfied. In Section 2.2.2 this assumption will completely be dropped and the results remain valid. The general case may be reduced to the present one. So, for the moment we have to be careful with the geometric conditions.

Any $Z \in \underline{Z}^d$ may be written in the form

$$Z = T^{\gamma}(\Lambda + \Lambda')$$

for certain $\gamma \in \mathbb{R}$, and $\Lambda \in \underline{\Delta}$, $\Lambda' \in \underline{\Delta}'$. Then we define

$$\mathfrak{M}_Z := \{T^{-\gamma}h : h \in MC^{\infty}_{\Lambda,\Lambda'}\} .$$

$\mathfrak{M}_Z$ is equipped with a natural nuclear Fréchet topology, cf. Section 2.1.1.

The topology in $\mathfrak{M}_Z$ may also be defined in terms of supremum norms. The definition of $\mathfrak{M}_Z$ implies

$$h \in \mathfrak{M}_Z \Longleftrightarrow T^{-\gamma} h \in \mathfrak{M}_{T^{\gamma}Z},$$

$\gamma \in \mathbb{R}$.

Remember that in 2.1.1. Definition 21 we introduced the classes $\mathfrak{M}^{\mu}_{\Lambda,\Lambda'}$ of not strongly decreasing Mellin symbols. In a similar sense we also can define $\mathfrak{M}^{\mu}_Z := T^{\gamma}\mathfrak{M}^{\mu}_{T^{\gamma}Z}$, $\gamma \in \mathbb{R}$, such that $(T^{\gamma}Z)_{(o)} \in \underline{\Delta}$, $(T^{\gamma}Z)'_{(o)} \in \underline{\Delta}'$, $Z \in \underline{Z}^d$ arbitrary. Later on we restrict ourselves to more precise classes of not strongly decreasing Mellin symbols (as $|\operatorname{Im} z| \longrightarrow \infty$), which are necessary for solving the problems in the chapters below. For this reason we neglect the more general classes for the moment and derive the calculus for $\mathfrak{M}_Z$. Extensions to $\mathfrak{M}_Z^r$ are obvious in most cases and left to the reader.

Let $h \in \mathfrak{M}_Z$, $\Gamma_{\frac{1}{2}} \cap Z = \emptyset$, and set as usual

$$op_M(h)\, u = M^{-1}(h\, Mu) \; . \tag{14}$$

$op_M(h)$ represents a continuous operator in $L^2(\mathbb{R}_+)$ and for the adjoint we have

$$(op_M(h))^* = op_M(h^*) \text{ with } h^*(z) = \bar{h}(1-\bar{z}),$$

such that $h^* \in \mathfrak{M}_{Z^*}$. Let

$$Iu(z) := t^{-1}\, u(t^{-1}),\; u \in L^2(\mathbb{R}_+),$$

I being an isometric involutive isomorphism of $L^2(\mathbb{R}_+)$. Then $MI = \tilde{I}M$, where $(\tilde{I}f)(z) := f(1-z) := f^{\vee}(z)$. In this way

$$I^{-1}\, op_M(h) I = op_M(h^{\vee}),$$

and

$$h \in \mathfrak{M}_Z \Longrightarrow h^{\vee} \in \mathfrak{M}_{Z^{\vee}} \; .$$

The proof of 1.3.1. Proposition 12 is valid also for Mellin operators (14). It follows that the Mellin symbol h is uniquely determined by the operator.

For any $h \in \mathfrak{M}_Z$, $\Gamma_{\frac{1}{2}} \cap Z = \emptyset$, we have

$$op_M(h) : L^2(\mathbb{R}_+) \longrightarrow L^2(\mathbb{R}_+) \cap C^{\infty}(\mathbb{R}_+) \; .$$

Let $\Lambda \in \underline{\Lambda}$ and Λ, $Z_{(o)}$ be admissible. Then

$$\omega\, op_M(h)\omega \; : \mathcal{S}_{\Lambda} \longrightarrow \mathcal{S}_{\Lambda + Z_{(o)}} \tag{15}$$

is continuous for any cut-off function ω, cf. 2.1.1. Theorem 20. In view of the continuity of $\mathcal{S}_{\Lambda^1} \hookrightarrow \mathcal{S}_{\Lambda^2}$ for $\Lambda^1 \subseteq \Lambda^2$, in (15) we may pass to the maximal singularity type $\Xi = \Lambda + Z_{(o)}$, for which

$$\omega\, op_M(h)\omega \;:\; \mathcal{S}_{\Xi} \longrightarrow \mathcal{S}_{\Xi} \tag{16}$$

is continuous. Remember that (15), (16) are of order zero, cf. 2.1.1. Definition 13, 2.1.1. Theorem 22.

10. <u>Lemma</u>. Let $h \in \mathfrak{M}_Z$, $Z \in \underline{Z}^d$, $\Gamma_{\frac{1}{2}} \cap Z = \emptyset$, $\Lambda := Z_{(o)}$, $\Lambda' := Z'_{(o)}$, and

ω_i be cut-off functions, i = 1,2,3,4. Then

(i) $\omega_1 \, op_M(h)(1-\omega_2) \in op_G(\mathcal{S}_\Lambda \otimes_\Gamma \mathcal{J}C^\infty_{o,\Lambda^*})$,

(ii) $(1-\omega_1) \, op_M(h)\omega_2 \in op_G(C^\infty_{o,\Lambda'} \otimes_\Gamma \mathcal{J}\mathcal{S}_{\Lambda'^*})$,

(iii) $\omega_1 \, op_M(h)(\omega_2-\omega_3) \in op_G(\mathcal{S}_\Lambda \otimes_\Gamma \mathcal{S}_o)$,

(iv) $(\omega_1-\omega_2) \, op_M(h)\omega_3 \in op_G(\mathcal{S}_o \otimes_\Gamma \mathcal{J}\mathcal{S}_{\Lambda'^*})$,

(v) $\omega_1 \, op_M(h)\omega_2 - \omega_3 \, op_M(h)\omega_4 \in op_G(\mathcal{S}_\Lambda \otimes_\Gamma \mathcal{J}\mathcal{S}_{\Lambda'^*})$.

<u>Proof</u>. First consider (i). In view of Proposition 2 and Theorem 8 it suffices to show that

$$\omega_1 \, op_M(h)(1-\omega_2) \in \mathcal{L}(L^2(\mathbb{R}_+), \mathcal{S}_\Lambda) \qquad (17)$$

and

$$(1-\omega_2) \, op_M(h)^* \omega_1 \in \mathcal{L}(L^2(\mathbb{R}_+), C^\infty_{o,\Lambda^*}) \; . \qquad (18)$$

In order to obtain (17) let $u \in L^2(\mathbb{R}_+)$. Then $(1-\omega_2)u$ has a support that does not contain the origin, hence $f(z) = M((1-\omega_2)u)(z)$ is holomorphic in $\mathrm{Re}\, z < \frac{1}{2}$ and f is characterized in terms of $L^2(\Gamma_\rho)$ norms, $\rho \leq \frac{1}{2}$, $\Gamma_\rho = \{\mathrm{Re}\, z = \rho\}$, cf. 1.1.2. In addition hf(z) is the Mellin image of some C^∞ function because of the decrease of h on $\Gamma_{\frac{1}{2}}$, further hf is holomorphic in $\{\mathrm{Re}\, z < \frac{1}{2}\} \setminus \Lambda$. Thus, roughly speaking, $\omega_1 M^{-1}(hf)$ belongs to $\mathcal{S}_\Lambda$. These heuristic arguments can easily be made exact. First we can conclude in the same way for all Γ_ρ, $\rho < \frac{1}{2}$, with $\Gamma_\rho \cap \Lambda = \emptyset$ and also observe the $L^2(B)$ norms of hf(z), B a curve surrounding a component of Λ. This immediately leads to

$$\mu'_B(v) \leq c \, \| u \|_{L^2(\mathbb{R}_+)} , \quad v = \omega_1 \, op_M(h)(1-\omega_2)u,$$

for every B with a constant c = c(j), cf. 2.1.1. (29). (18) and the other relations (ii) - (v) follow in precisely the same way. $\square$

Now we pass to a slight extension of the concept of operators of the form (15). Let $\alpha, \gamma \in \mathbb{C}$, $\mathrm{Re}\,\alpha \geq \mathrm{Re}\,\gamma \geq 0$, $h \in \mathcal{M}_Z$, $Z \in \underline{Z}^d$, $\Gamma_{\frac{1}{2}} \cap T^{-\gamma}Z = \emptyset$. Then

$$t^{\alpha-\gamma}\omega \, op_M(T^\gamma h)\omega t^\gamma : \mathcal{S}_\Lambda \longrightarrow \mathcal{S}_{T^{-\alpha}(\Lambda + Z_{(\gamma)})} \qquad (19)$$

is continuous (for $\Lambda \in \underline{\Lambda}$ and $\Lambda, Z_{(\gamma)}$ admissible).

Moreover, (19) is of order zero in the sense of the representations

$$\mathcal{S}_\Lambda = \underleftarrow{\lim}\, \mathcal{H}_k(\Lambda),\quad \mathcal{S}_{T^{-\alpha}(\Lambda + Z_{(\gamma)})} = \underleftarrow{\lim}\, \mathcal{H}_k(T^{-\alpha}(\Lambda + Z_{(\gamma)}))\,,$$

fixed in Section 2.1.1., cf. 2.1.1. Proposition 14.

11. <u>Proposition</u>. Let $\alpha, \gamma, \lambda \in \mathbb{C}$,

$$0 \leqq \operatorname{Re}\gamma \leqq \operatorname{Re}\lambda \leqq \operatorname{Re}\alpha\,,$$

$h \in \mathfrak{M}_Z$, $Z \in \underline{Z}^d$, $\Gamma_{\frac{1}{2}} \cap T^{-\gamma}Z = \Gamma_{\frac{1}{2}} \cap T^{-\lambda}Z = \emptyset$. Then for every cut-off function ω we have

$$t^\alpha \omega \left\{ t^{-\gamma} \operatorname{op}_M(T^\gamma h) t^\gamma - t^{-\lambda} \operatorname{op}_M(T^\lambda h) t^\lambda \right\} \omega = \operatorname{op}_G(b)$$

for some $b \in \mathcal{S}_{\Lambda^1} \otimes_\Gamma J\mathcal{S}_{\Lambda^2}$,

$$\Lambda^1 = T^{-\alpha}(Z_{(\lambda)} \setminus Z_{(\gamma)}),\quad \Lambda^2 = T^{-\bar{\alpha}}((Z^*)_{(\bar{\gamma})} \setminus (Z^*)_{(\bar{\lambda})})\,. \tag{20}$$

<u>Proof</u>: Let $u \in L^2(\mathbb{R}_+)$. Then $v = u\omega$ has a bounded support and $\tilde{v}(z)$ is holomorphic for $\operatorname{Re} z > \frac{1}{2}$. We have

$$\begin{aligned} &t^{\alpha-\gamma} \omega \operatorname{op}_M(T^\gamma h) t^\gamma \\ &\quad = \frac{1}{2\pi i} \omega \int_{\Gamma_{\frac{1}{2}}} t^{\alpha-\gamma-z} h(z+\gamma) \tilde{v}(z+\gamma) dz \\ &\quad = \frac{1}{2\pi i} \omega \int_{\Gamma_{\frac{1}{2}+\gamma_1-\alpha_1}} t^{-z} h(z+\alpha) \tilde{v}(z+\alpha)\, dz, \end{aligned}$$

$\alpha_1 = \operatorname{Re}\alpha$, $\gamma_1 = \operatorname{Re}\gamma$. A similar formula holds for the second term. Thus

$$\begin{aligned} f(t) &:= t^{\alpha-\gamma} \omega \operatorname{op}_M(T^\gamma h) t^\gamma v - t^{\alpha-\lambda} \omega \operatorname{op}_M(T^\lambda h) t^\lambda v \\ &= \frac{1}{2\pi i} \omega(t) \int_{B_1} t^{-z} h(z+\alpha) \tilde{v}(z+\alpha)\, dz \\ &= \frac{1}{2\pi i} \omega(t) \int_B t^{-z} h(z+\alpha) \tilde{v}(z+\alpha)\, dz\,. \end{aligned} \tag{21}$$

Here $B_1 = \Gamma_{\frac{1}{2}+\gamma_1-\alpha_1} \cup \Gamma_{\frac{1}{2}+\lambda_1-\alpha_1}$ $(\lambda_1 = \operatorname{Re}\lambda)$ is positively oriented and B is a smooth positively oriented closed contour around Λ^1. In the second equation of (21) we have used that $h(z+\alpha)\,\tilde{v}(z+\alpha)$ is holomorphic in

$$\left\{z \in \mathbb{C} : \tfrac{1}{2} + \gamma_1 - \alpha_1 < \operatorname{Re} z < \tfrac{1}{2} + \lambda_1 - \alpha_1\right\} \setminus \Lambda^1 .$$

We also have the strong-decrease of $h(z+\alpha)\tilde{v}(z+\gamma)$ as $|\operatorname{Im} z| \longrightarrow \infty$, i.e. $f \in C^\infty(\mathbb{R}_+)$. Moreover, supp f is bounded. For the Mellin image of (21) we get

$$\tilde{f}(w) = \frac{1}{2\pi i}\int_B \phi(w,z)\, h(z+\alpha)\, \tilde{v}(z+\alpha)\, dz,$$

cf. 2.1.1. (7), which is the potential with respect to ϕ of an analytic functional carried by Λ^1. Thus $f \in \mathcal{S}_{\Lambda^1}$. It is again immediately seen from the semi-norm systems that $u \longrightarrow f$ defines a continuous operator $L^2(\mathbb{R}_+) \longrightarrow \mathcal{S}_{\Lambda^1}$. Applying Proposition 2 we obtain that the kernel of this operator belongs to $\mathcal{S}_{\Lambda^1} \otimes_\pi L^2(\mathbb{R}_+)$. A similar argument holds for the adjoints. We have for $\delta := \bar{\alpha} - \bar{\gamma}$

$$\begin{aligned}(t^{\alpha-\gamma}\omega\, op_M(T^\gamma h)t^\gamma \omega)^* &= t^{\bar\gamma}\omega(op_M(T^\gamma h))^*\, t^{\bar\alpha-\bar\gamma}\omega\\ &= t^{\bar\gamma}\omega\, op_M(T^{-\bar\gamma}h^*)t^{\bar\alpha-\bar\gamma}\omega\\ &= t^{\bar\alpha-\delta}\omega\, op_M(T^\delta(T^{-\bar\alpha}h^*))t^\delta\omega ,\end{aligned}$$

$h^* \in \mathfrak{M}_{Z^*}$, $T^{-\bar\alpha}h^* \in \mathfrak{M}_{T^{\bar\alpha}Z^*}$, and a similar formula for $\eta := \bar\alpha - \bar\lambda$. Thus

$$\begin{aligned}&(t^{\alpha-\gamma}\omega\, op_M(T^\gamma h)t^\gamma\omega) - (t^{\alpha-\lambda}\omega\, op_M(T^\lambda h)t^\lambda\omega)^*\\ &= t^{\bar\alpha}\omega\left\{t^{-\eta}\, op_M(T^\eta(T^{-\bar\alpha}h^*))t^\eta - t^{-\delta}\, op_M(T^\delta(T^{-\bar\alpha}h^*))t^\delta\right\}\omega\end{aligned}$$

has a kernel in $\mathcal{S}_{\Lambda^2} \otimes_\Gamma L^2(\mathbb{R}_+)$. $\Lambda^2 := T^{-\bar\alpha}\left\{(T^{\bar\alpha}Z^*)_{(\eta)} \setminus (T^{\bar\alpha}Z^*)_{(\delta)}\right\}$ just coincides with Λ^2 in (20). $\square$

12. <u>Corollary</u>. Let $h \in \mathfrak{M}_Z$, $Z \in \underline{Z}^d$, $\Gamma_{\frac{1}{2}} \cap Z = \Gamma_{\frac{1}{2}} \cap T^\eta Z = \emptyset$ for $\eta \in \mathbb{C}$, $\operatorname{Re} \eta \geqq 0$. Then

$$t^\eta\omega\, op_M(T^{-\eta}h)\omega - \omega\, op_M(h)\omega\, t^\eta = op_G(b)$$

for some $b \in \mathcal{S}_{\Pi^1} \otimes_\Gamma J\mathcal{S}_{\Pi^2}$,

$$\Pi^1 = T^{-\eta}((T^\eta Z)_{(\eta)} \setminus (T^\eta Z)_{(0)}),$$

$$\Pi^2 = T^{-\bar\eta}((T^\eta Z)^*_{(0)} \setminus (T^\eta Z)^*_{(\bar\eta)}).$$

Now let us consider operators of the form

$$\mathcal{M} := t^\gamma\omega\, op_M(T^\delta h)t^\eta\omega , \tag{22}$$

$\gamma, \delta, \eta \in \mathbb{C}$, $\operatorname{Re} \gamma \geq 0$, $\operatorname{Re} \eta \geq 0$, $h \in \mathfrak{M}_Z$, $\Gamma_{\frac{1}{2}} \cap T^{-\delta} Z = \emptyset$. (22) induces a continuous operator

$$\mathfrak{M} : \mathcal{S}_\Lambda \longrightarrow \mathcal{S}_{T^{-\gamma}(T^{-\eta}\Lambda + (T^{-\delta}Z)_{(o)})}, \tag{23}$$

$\Lambda \in \underline{\Lambda}$.

$$\mathfrak{M}^* = t^{\bar{\eta}}\, \omega\, \mathrm{op}_M(T^{-\bar{\delta}} h^*) t^{\bar{\gamma}} \omega \tag{24}$$

induces a continuous operator

$$\mathfrak{M}^* : \mathcal{S}_\Pi \longrightarrow \mathcal{S}_{T^{-\bar{\eta}}(T^{-\bar{\gamma}}\Pi + (T^{\bar{\delta}}Z^*)_{(o)})}, \tag{25}$$

$\Pi \in \underline{\Lambda}$.

Everywhere in this section we assume once and for all that all occurring couples of singularity types to be added are admissible. Moreover, we suppose that whenever $\mathrm{op}_M(T^\gamma h)$ is formed, $\gamma \in \mathbb{C}$, $h \in \mathfrak{M}_Z$, the condition $\Gamma_{\frac{1}{2}} \cap T^{-\gamma} Z = \emptyset$ holds.

13. <u>Proposition</u>. Let $\mathfrak{M}$ be given by (22) and $\mathcal{G} = \mathrm{op}_G(b)$, $b \in \mathcal{S}_\Lambda \otimes_\Gamma J\mathcal{S}_\Pi$, $\Lambda, \Pi \in \underline{\Lambda}$. Then

$$\mathfrak{M}\mathcal{G} = \mathrm{op}_G(b_1), \quad \mathcal{G}\mathfrak{M} = \mathrm{op}_G(b_2)$$

with certain

$$b_1 \in \mathcal{S}_{T^{-(\gamma+\eta)}\Lambda + T^{-\gamma}(T^{-\delta}Z)_{(o)}} \otimes_\Gamma J\mathcal{S}_\Pi ,$$

$$b_2 \in \mathcal{S}_\Lambda \otimes_\Gamma J\mathcal{S}_{T^{-(\bar{\eta}+\bar{\gamma})}\Pi + T^{-\bar{\eta}}(T^{\bar{\delta}}Z^*)_{(o)}} .$$

<u>Proof</u>. $\mathfrak{M}\mathcal{G}$ is continuous as operator from $L^2(\mathbb{R}_+)$ to $\mathcal{S}_{T^{-(\gamma+\eta)}\Lambda + T^{-\gamma}(T^{-\delta}Z)_{(o)}}$. The adjoint is continuous as operator from $L^2(\mathbb{R}_+)$ to $\mathcal{S}_\Pi$. Thus the singularity types of the kernel b_1 follow from Proposition 2 applied both to $\mathfrak{M}\mathcal{G}$ and $(\mathfrak{M}\mathcal{G})^* = \mathcal{G}^*\mathfrak{M}^*$ and the continuities (23) and (25). The second assertion follows in the same way. $\square$

14. <u>Proposition</u>. Let $Z^1, Z^2 \in \underline{Z}^d$, $Z = Z^1 + Z^2 \in \underline{Z}^d$. Then the multiplication of functions in $\mathbb{C} \setminus Z$ induces a mapping

$$\mathfrak{M}_{Z^1} \times \mathfrak{M}_{Z^2} \longrightarrow \mathfrak{M}_Z \tag{26}$$

which is continuous separately in the arguments.

The assertion is a straightforward generalization of 2.1.1. Theorem 20.

So we may drop the details. Now we want to express compositions of operators of the form (22). Let

$$\mathfrak{m}_i = t^{\gamma_i} \omega \, op_M(T^{\delta_i} h_i) t^{\eta_i} \omega \tag{27}$$

with $h_i \in \mathfrak{M}_{Z^i}$, $Z^i \in \underline{Z}^d$, $\gamma_i, \delta_i, \eta_i \in \mathbb{C}$, $\operatorname{Re}\gamma_i \geqq 0$, $\operatorname{Re}\eta_i \geqq 0$, $i = 1,2$.
Then

$$\mathfrak{m}_1 \mathfrak{m}_2 = t^{\gamma_1} \omega \, op_M(T^{\delta_1} h_1) \omega t^{\alpha} t^{\beta} \omega \, op_M(T^{\delta_2} h_2) \omega t^{\eta_2}$$

with

$$\alpha + \beta = \gamma_1 + \gamma_2, \quad \operatorname{Re}\alpha \geqq 0 \quad \operatorname{Re}\beta \geqq 0. \tag{28}$$

15. <u>Proposition</u>. Assume that $\mathfrak{m}_i$, $i = 1,2$, are given as above and

$$\Gamma_{\frac{1}{2}} \cap T^{-\delta_1 + \alpha} Z^1 = \emptyset, \quad \Gamma_{\frac{1}{2}} \cap T^{-\delta_2 - \beta} Z^2 = \emptyset ,$$

α, β as in (28). Then

$$\mathfrak{m}_1 \mathfrak{m}_2 = t^{\gamma} \omega \, op_M(T^{\delta_1 - \alpha} h_1 \, T^{\delta_2 + \beta} h_2) \omega t^{\eta} + \mathcal{G} \tag{29}$$

with

$$\gamma := \gamma_1 + \alpha, \quad \eta := \eta_2 + \beta$$

and a Green operator $\mathcal{G}$. In particular for $\delta_1 = \eta_1$, $\delta_2 = \eta_2$ we obtain

$$\mathfrak{m}_1 \mathfrak{m}_2 = t^{\gamma} \omega \, op_M(T^{\eta}(\{T^{-(\gamma_2+\eta_2)} h_1\} h_2)) \omega t^{\eta} + \mathcal{G} . \tag{30}$$

<u>Proof</u>: Applying Corollary 12 we obtain

$$t^{\gamma_1} \omega op_M(T^{\delta_1} h_1) \omega t^{\alpha} = t^{\gamma_1 + \alpha} \omega \, op_M(T^{\delta_1 - \alpha} h_1) \omega + \mathcal{G}_1,$$

$$\omega t^{\beta} \, op_M(T^{\delta_2} h_2) \omega t^{\eta_2} = \omega \, op_M(T^{\delta_2 + \beta} h_2) \omega t^{\beta + \eta_2} + \mathcal{G}_2$$

with Green operators $\mathcal{G}_1, \mathcal{G}_2$. Applying Proposition 13 we thus obtain the assertion. ▯ The singularity type of $\mathcal{G}$ in (29) can easily be expressed in terms of the given ones in $\mathfrak{m}_1$ and $\mathfrak{m}_2$ by corresponding iterations of the resulting singularity types in Corollary 12 and Proposition 13.
Consider a sequence $h_j \in \mathfrak{M}_{Z^j}$, $Z^j \in \underline{Z}^d$, $j \in \mathbb{Z}_+$, and a sequence of $\gamma_j \in \mathbb{R}$, $0 \leqq \gamma_j \leqq j$, where

$$j - \gamma_j \longrightarrow \infty , \quad \gamma_j \longrightarrow \infty \text{ as } j \longrightarrow \infty .$$

We have continuous operators

$$\mathfrak{m}_j := t^{j-\gamma_j}\,\omega(c_j t)\,\mathrm{op}_M(T^{\gamma_j} h_j)\,\omega(c_j t) t^{\gamma_j}, \tag{31}$$

$$\mathfrak{m}_j : \mathcal{S}_\Lambda \longrightarrow \mathcal{S}_{T^{-j}(\Lambda + Z_{(-\gamma_j)})}$$

for $\Lambda \in \underline{\Lambda}$. The constants $c_j > 0$ are arbitrary for the moment. Set

$$h(\Lambda) = \sum_{j=0}^{\infty} T^{-j}(\Lambda + Z_{(-\gamma_j)}) \ . \tag{32}$$

At the same time we also want to observe the adjoint operators

$$\mathfrak{m}_j^* = t^{j-\delta_j}\,\omega(c_j t)\,\mathrm{op}_M(T^{\delta_j} h_j^{(*)})\,\omega(c_j t) t^{\delta_j} \ ,$$

where $\delta_j := j - \gamma_j$, $h_j^{(*)} := T^{-j} h_j^* \in \mathfrak{M}_{T^j z^{j*}}$, such that

$$\mathfrak{m}_j^* : \mathcal{S}_\Pi \longrightarrow \mathcal{S}_{T^{-j}(\Pi + (T^j z^{j*})_{(-\delta_j)})}$$

for any $\Pi \in \underline{\Lambda}$. Set

$$h^*(\Pi) = \sum_{j=0}^{\infty} T^{-j}(\Pi + (T^j z^{j*})_{(-\delta_j)}).$$

16. <u>Theorem</u>. Let $h_j \in \mathfrak{M}_{z^j}$, $z^j \in \underline{z}^d$, $j \in \mathbb{Z}_+$, be a sequence and γ_j, $\mathfrak{m}_j$ as mentioned. Then there is a choice of constants $c_j > 0$ such that

$$\mathfrak{m} = \sum_{j=0}^{\infty} \mathfrak{m}_j \tag{33}$$

converges in $\mathcal{L}^0(\mathcal{S}_\Lambda, \mathcal{S}_{h(\Lambda)})$ for all $\Lambda \in \underline{\Lambda}$ with $h(\Lambda) \in \underline{\Lambda}$ and that in addition

$$\mathfrak{m}^* = \sum_{j=0}^{\infty} \mathfrak{m}_j^* \tag{34}$$

converges in $\mathcal{L}^0(\mathcal{S}_\Pi, \mathcal{S}_{h^*(\Pi)})$ for all $\Pi \in \underline{\Lambda}$ with $h^*(\Pi) \in \underline{\Lambda}$. If $\tilde{\mathfrak{m}}$ is defined in an analogous manner by another choice of ω and constants c_j, γ_j, then $\mathfrak{m} - \tilde{\mathfrak{m}}$ is a Green operator.

<u>Proof</u>: First remember the representations $\mathcal{S}_\Lambda = \varprojlim \mathcal{H}_k(\Lambda)$, $\Lambda \in \underline{\Lambda}$, fixed in Section 2.1.1., where the norm of $\mathcal{H}_k(\Lambda)$ refers to a strip $S_k = \{\frac{1}{2} - k < \mathrm{Re}\ z < \frac{1}{2} + k\}$. Then we have to prove that $\mathfrak{m}$ converges as operator

$$\mathfrak{m} : \mathcal{H}_k(\Lambda) \longrightarrow \mathcal{H}_k(h(\Lambda))$$

for every k and the same for the adjoint, uniformly for $\Lambda \in \underline{\Lambda}$. Write $\mathfrak{M} = \mathfrak{M}^{(o)} + \mathfrak{M}^{(1)}$,

$$\mathfrak{M}^{(o)} = \sum_{j=0}^{N} \mathfrak{M}_j, \quad \mathfrak{M}^{(1)} = \sum_{j=N+1}^{\infty} \mathfrak{M}_j \quad .$$

$\mathfrak{M}^{(o)}$ as a finite sum of zero order operators has order zero. Thus, it suffices to deal with $\mathfrak{M}^{(1)}$. For $k \in \mathbb{Z}_+$ fixed and N large enough we have $-(j-\gamma_j) < \frac{1}{2} - k$ for all $j \geq N + 1$. Thus $\mathfrak{M}_j$ does not affect the singularity types in $\{\mathrm{Re}\ z > \frac{1}{2} - k\}$ for $j \geq N + 1$, and hence it suffices to show that $\mathfrak{M}^{(1)}$ converges as a sum of operators

$$\mathfrak{M}^{(1)} : \mathcal{H}_k(\Lambda) \longrightarrow \mathcal{H}_k(\Lambda) ,$$

since then it may be composed with the embedding $\mathcal{H}_k(\Lambda) \hookrightarrow \mathcal{H}_k(h\ (\Lambda))$. $\| \mathfrak{M}_j u \|^2_{\mathcal{H}_k(\Lambda)}$ is a finite sum of expressions of the form

$$I_\Gamma^2 = \| z^1\ n_j(z) \|^2_{L^2(\Gamma)} \quad , \qquad (35)$$

$$I_B^2 = \| z^1\ n_j(z) \|^2_{L^2(B)} \quad , \qquad (36)$$

B being a piecewise smooth curve of finite length, Γ finitely many lines or half lines parallel to the imaginary axis

$$n_j(z) = M\left[t^{j-\gamma_j}\ \omega(c_j t)\ \ M^{-1}\{ h_j(z+\gamma_j)\ M(\omega(c_j t) t^{\gamma_j} u\} \right] .$$

Set for abbreviation $\gamma = \gamma_j$, $c = c_j$, $h = h_j$. Then

$$M(\omega(cs)s^{\gamma}u)(w) = c^{-\gamma}\ c^{1-w} \int s^{w-1+\gamma}\ \omega(s)u(\tfrac{s}{c})\ \frac{ds}{c},$$

$$M^{-1}\{ h(w+\gamma)M(\omega(cs)s^{\gamma}u)(w)\}(t)$$

$$= c^{-\gamma} \int_{\Gamma_{\frac{1}{2}}} t^{-w}\ h(w+\gamma)\ c^{1-w} \left[\int s^{w-1+\gamma}\ \omega(s)u(\tfrac{s}{c})\ \frac{ds}{c} \right] đw \quad , \quad đw = \frac{1}{2\pi i}\ dw ,$$

$$n_j(z) = c^{-\gamma} \int t^{z-1+j-\gamma}\ \omega(ct) \Big\{ \int_{\Gamma_{\frac{1}{2}}} t^{-w}\ h(w+\gamma)\ c^{1-w}$$

$$\left[\int s^{w-1+\gamma}\ \omega(s)u(\tfrac{s}{c})\ \frac{ds}{c} \right] đw \Big\} dt$$

$$= c^{-j} \int c^{-z}\ t^{z-1+j-\gamma} \omega(t) \Big\{ \int_{\Gamma_{\frac{1}{2}}} t^{-w}\ h(w+\gamma)$$

$$\left[\int s^{w-1+\gamma}\ \omega(s)u(\tfrac{s}{c})\ ds \right] đw \Big\} dt$$

The proof of the following simple Lemma is given below.

17. **Lemma**. For every $k_o \in \mathbb{Z}_+$ there is an $m \in \mathbb{R}_+$ such that

$$\delta_c^m : u(s) \longrightarrow c^{-m} u(\tfrac{s}{c})$$

defines a family of operators

$$\delta_c^m : \mathcal{H}_k \longrightarrow \mathcal{H}_k$$

which are uniformly bounded in $c \in [1,\infty)$, $0 \leq k \leq k_o$.

It is now obvious that

$$H_c^m : u \longrightarrow t^{j-\gamma}\omega(t) \int_{\Gamma_{\frac{1}{2}}} t^{-w} h(w+\gamma) \left[\int s^{w-1+\gamma}\omega(s)(\delta_c^m u)(s)ds\right] dw$$

induces a family of continuous operators $H_c^m : \mathcal{H}_k(\Lambda) \longrightarrow \mathcal{H}_k(\Lambda)$ uniformly bounded in $c \in [1,\infty)$. Thus

$$n_j(z) = c^{-j+m+m'} \int c^{-m'-z} t^{z-1}(H_c^m u)(t)\, dt$$

for arbitrary $m' \in \mathbb{R}$. Since both

$$\left\| z^1 \left[\int c^{-m'-z} t^{z-1}(H_c^m u)(t)\, dt\right] \right\|_{L^2(\Gamma)}$$

and

$$\left\| z^1 \left[\int c^{-m'-z} t^{z-1}(H_c^m u)(t)\, dt\right] \right\|_{L^2(B)}$$

remain finite uniformly in $c \in [1,\infty)$ for m' sufficiently large and fixed, we find that

$$\| \mathcal{M}_j u \|^2_{\mathcal{H}_k(\Lambda)} \leq c^{-j+m+m'}\, C\, \| u \|^2_{\mathcal{H}_k(\Lambda)}$$

with some constant $C > 0$. For $j = N+1$, N large enough, we get

$$\| \mathcal{M}_j \|_{\mathcal{H}_k(\Lambda)} \longrightarrow 0 \text{ as } c \longrightarrow \infty .$$

Thus, an appropriate choice of the constants c_j implies the convergence of $\mathcal{M}^{(1)}$ in $\mathcal{L}(\mathcal{H}_k(\Lambda))$. A standard diagonal argument yields the convergence for all k. Another diagonal argument yields the convergence for every $\Lambda \in \underline{\Lambda}$, since any Λ may be shrinked a bit in the strip S_k such that the $\mathcal{H}_k(\Lambda)$ norms of functions are still finite. Thus it suffices to show the statement for a countable system of norms for every k (cf. also 2.1.1. Proposition 12).

For the series of adjoints we may conclude in the same way. Then, by taking the maximum of the occurring constants for every j we get the desired convergence for $\mathcal{M}$ and $\mathcal{M}^*$.

Let $\tilde{\mathcal{M}} = \sum_{j=0}^{\infty} \tilde{\mathcal{M}}_j$ be defined like (33) by another choice of the data ω, c_j, γ_j. Then

$$\mathcal{G} := \sum_{j=0}^{\infty} (\mathcal{M}_j - \tilde{\mathcal{M}}_j)$$

also converges in $\mathcal{L}^0(\mathcal{S}_\Lambda, \mathcal{S}_{h(\Lambda)})$ for all Λ as well as the series of the adjoints. Set

$$\mathcal{G}_N := \sum_{j=0}^{\infty} (\mathcal{M}_j - \tilde{\mathcal{M}}_j), \quad \mathcal{G}_N' = \mathcal{G} - \mathcal{G}_N .$$

In view of Lemma 10 and Proposition 11 $\mathcal{G}_N$ is a Green operator. Further it is clear that for every $k \in \mathbb{Z}_+$ there is an $N = N(k)$ such that $\mathcal{G}_N'$ is continuous as operator from $L^2(R_+)$ to $\mathcal{H}_k$. Thus $\mathcal{G} \in \mathcal{L}(L^2(\mathbb{R}_+), \mathcal{H}_k)$ for all k and hence, in view of Proposition 2 $\mathcal{G}$ has a kernel in $\mathcal{S}_\Lambda \hat{\otimes}_\pi L^2(\mathbb{R}_+)$. The same argument for the adjoint operators yields the second assertion of Theorem 16. $\square$

<u>Proof of Lemma 17</u>. Obviously

$$M(\delta_c^m u)(z) = c^{z-m}(Mu)(z),$$

and hence we have to insert $c^{z-m}\tilde{u}(z)$ into the norms λ_k. They are obviously uniformly finite in c for $0 \leq k \leq k_o$ and $m = \frac{1}{2} + k_o$. $\square$

18. <u>Definition</u>. If $\mathcal{M}$ is given by (33), we set

$$\sigma_M^{-j}(\mathcal{M}) = h_j, \quad j \in \mathbb{Z}_+ \tag{37}$$

and call h_j the conormal symbol of $\mathcal{M}$ of order $-j$.

Consider two operators

$$\mathcal{M} := \mathcal{M}^1 := \sum_{j=0}^{\infty} t^{j-\gamma_j} \omega^1(c_j^1 t) \, op_M(T^{\gamma_j} h_j) \, \omega^1(c_j^1 t) t^{\gamma_j},$$

$$\mathcal{M}^2 := \sum_{k=0}^{\infty} t^{k-\mu_k} \omega^2(c_k^2 t) \, op_M(T^{\mu_k} m_k) \, \omega^2(c_k^2 t) t^{\mu_k},$$

$h_j \in \mathfrak{M}_{z^j}$, $m_k \in \mathfrak{M}_{w^k}$, z^j, $w^k \in \underline{Z}^d$, which exist in the sense of Theorem 16. Set

$$f_l(z) = \sum_{j+k=l} (T^{-k} h_j)(z)\, m_k(z) \tag{38}$$

(cf. Proposition 14) and assume that $f_l \in \mathfrak{M}_{v^l}$ for certain $v^l \in \underline{Z}^d$. With the sequence $(f_l)_{l\in Z_+}$ and some fixed choice of constants ϑ_l, $0 \leq \vartheta_l \leq 1$, $\vartheta_l \leq$ const. independent of l, $\Gamma_{\frac{1}{2}} \cap T^{-\vartheta_l} v^l = \emptyset$, we can form $\Lambda \longrightarrow f(\Lambda)$ in a similar way as (32) related to $(h_j, \gamma_j)_{j\in Z_+}$, and for every $\Lambda \in \underline{\Lambda}$ with $f(\Lambda) \in \underline{\Lambda}$ we can apply Theorem 16 to the sequence $(f_l, \vartheta_l)_{l\in Z_+}$. This yields an operator

$$\mathcal{M} := \sum_{l=0}^{\infty} \omega(c_l^3 t)\, t^{1-\vartheta_l}\, op_M(T^{\vartheta_l} f_l)\, \omega(c_l^3 t)\, t^{\vartheta_l} \tag{39}$$

and we get the following

19. Proposition. Under the above assumptions we have

$$\mathcal{M}^1 \mathcal{M}^2 = \mathcal{M} + \mathcal{G}$$

with some Green operator $\mathcal{G}$.

Proof: The proof follows by completely the same arguments as at the end of the proof of Theorem 16, where the Green property of $\mathcal{M} - \tilde{\mathcal{M}}$ was established. Here we have to apply Proposition 15 to the finite partial sums. $\square$

20. Corollary. Let $\mathcal{M}^i$, $i = 1,2$, be as in Proposition 19. Then

$$\sigma_M^{-l}(\mathcal{M}^1 \mathcal{M}^2) = \sum_{j+k=l} \left\{ T^{-k} \sigma_M^{-j}(\mathcal{M}^1) \right\} \sigma_M^{-k}(\mathcal{M}^2) \quad . \tag{40}$$

Another application of the arguments at the end of the proof of Theorem 16 and of Proposition 13 yields

21. Proposition: Let $\mathcal{M}$ be as in Theorem 16 and $\mathcal{G}$ a Green operator. Then $\mathcal{M}\mathcal{G}$ and $\mathcal{G}\mathcal{M}$ are of Green type.

Point out once again that in all occurring constructions we pose the geometric conditions to the patterns of the involved singularity types that are needed for applying all auxiliary statements. But the assertions remain true in the general setting of Section 2.2.2. without any geometric restrictions to the singularity types.

The next point to be discussed is the action of pseudo-differential operators on spaces of the type $\mathcal{S}_\Lambda$. Here we refer to the notations in the Sections 1.1.4. and 1.1.5. Let

$$\Lambda^{o} = \{-j \ : \ j \in \mathbb{Z}_{+}\}. \tag{41}$$

22. Proposition. Let $\Lambda \in \underline{\Lambda}$, $T^{-j}\Lambda \subseteq \Lambda$ for all $j \in \mathbb{Z}_{+}$, and $a \in S^{(o)}(\mathbb{R}^{n} \setminus 0)$. Then $op_{\psi}(a)(\xi')$ induces a continuous operator

$$op_{\psi}(a)(\xi') \ : \mathcal{S}_{\Lambda} \longrightarrow \mathcal{S}_{\Lambda+\Lambda^{o}} \tag{42}$$

for every $\xi' \neq 0$, and

$$op_{\psi}(\cdot) \ : S^{o}_{cl}(R) \longrightarrow \mathcal{L}^{o}(\mathcal{S}_{\Lambda}, \mathcal{S}_{\Lambda+\Lambda^{o}}) \tag{43}$$

induced by (42) is continuous.

Proof: In view of 1.4.1. Theorem 5 we have a decomposition

$$\omega\, op_{\psi}(a)\,\omega = \mathcal{M}_{N} + op_{G}(b_{N}),$$

$$\mathcal{M}_{N} := \sum_{j=0}^{N} t^{j}\, \omega\, op_{M}(\sigma_{M}^{-j}(a))\,\omega .$$

In order to show the asserted continuity we set

$$\mathcal{S}_{\Lambda} = \varprojlim \mathcal{H}_{k}(\Lambda),\ \mathcal{S}_{\Lambda+\Lambda^{o}} = \varprojlim \mathcal{H}_{1}(\Lambda+\Lambda^{o}).$$

In view of the concrete definition of the scales $\mathcal{H}_{k}(\Lambda)$, $\mathcal{H}_{k}(\Lambda+\Lambda^{o})$ for $N = N(k)$ sufficiently large we get that both

$$\mathcal{M}_{N} : \mathcal{H}_{k}(\Lambda) \longrightarrow \mathcal{H}_{k}(\Lambda+\Lambda^{o})$$

and

$$op_{G}(b_{N}) : \mathcal{H}_{k}(\Lambda) \longrightarrow \mathcal{H}_{k}(\Lambda+\Lambda^{o})$$

are continuous. Thus

$$\omega\, op_{\psi}(a)\,\omega : \mathcal{H}_{k}(\Lambda) \longrightarrow \mathcal{H}_{k}(\Lambda+\Lambda^{o})$$

is continuous. Now we have

$$op_{\psi}(a) = \omega\, op_{\psi}(a)\,\omega + (1-\omega)\, op_{\psi}(a)\,\omega$$
$$+ \omega\, op_{\psi}(a)(1-\omega) + (1-\omega)\, op_{\psi}(a)(1-\omega).$$

The continuity of

$$(1-\omega)\, op_{\psi}(a)(1-\omega) : \mathcal{S}_{\Lambda} \longrightarrow \mathcal{S}_{\Lambda+\Lambda^{o}}$$

is obvious as well as that of

$$\omega\, op_{\psi}(a)(1-\omega) : \mathcal{S}_{\Lambda} \longrightarrow \mathcal{S}_{\Lambda+\Lambda^{o}}$$

because of 1.4.2. Theorem 1 and the continuity of $\mathcal{S}_{\gamma_{0}} \hookrightarrow \mathcal{S}_{\Lambda^{o}}$. Further let ω' be another cut-off function with $\omega = \omega'\omega$. Then

$$(1-\omega)\, op_\psi(a)\omega = (1-\omega')\, op_\psi(a)\omega + (\omega-\omega')\, op_\psi(a)\omega .$$

For $(\omega-\omega')\, op_\psi(a)\omega$ we argue as in the beginning, whereas for $(1-\omega')\, op_\psi(a)\omega$ we may apply 1.4.1. Lemma 1, which shows that $(1-\omega')\, op_\psi(a)\omega : \mathcal{S}_\Lambda \longrightarrow \mathcal{S}_{\Lambda+\Lambda^o}$ is continuous, too. The continuity of (43) may be proved by obvious estimates of the involved semi-norms. $\square$

Note that $\Lambda^o = \Lambda(\gamma_o)$, cf. 1.2.1. (36), 2.1.1. (4), and that we also have a continuous operator

$$op_\psi(a)(\xi') : \mathcal{S}_\Lambda \longrightarrow \mathcal{S}_{\Lambda+\gamma_o} .$$

23. Proposition. Let $\mathcal{G} \in op_G(\mathcal{S}_\Gamma \otimes_r \mathcal{S}_\Delta)$, $\Gamma, \Delta \in \underline{\Lambda}$, $T^{-j}\Gamma \subseteq \Gamma$, $T^{-j}\Delta \subseteq \Delta$, $j \in \mathbb{Z}_+$, $a \in S^{(o)}(\mathbb{R}^n \setminus 0)$. Then, for every fixed $\xi' \neq 0$

$$op_\psi(a)(\xi')\,\mathcal{G} \in op_G(\mathcal{S}_{\Gamma+\Lambda^o} \otimes_r \mathcal{S}_\Delta), \tag{44}$$

$$\mathcal{G}\, op_\psi(a)(\xi') \in op_G(\mathcal{S}_\Gamma \otimes_r \mathcal{S}_{\Delta+\Lambda^o}) . \tag{45}$$

Proof: We show (44), then (45) follows by duality. In view of Proposition 2 and Theorem 8 it suffices to know that

$$op_\psi(a)\,\mathcal{G} \in \mathcal{L}(L^2(\mathbb{R}_+), \mathcal{S}_{\Gamma+\Lambda^o}) , \tag{46}$$

$$\mathcal{G}^* op_\psi(a)^* \in \mathcal{L}(L^2(\mathbb{R}_+), J\mathcal{S}_\Delta) . \tag{47}$$

(46) is obvious because of $\mathcal{G} \in \mathcal{L}(L^2(\mathbb{R}_+), \mathcal{S}_\Gamma)$ and Proposition 22. (47) is obvious, too, since $op_\psi(a)^* \in \mathcal{L}(L^2(\mathbb{R}_+), L^2(\mathbb{R}_+))$ and $\mathcal{G}^* \in \mathcal{L}(L^2(\mathbb{R}_+), J\mathcal{S}_\Delta)$. $\square$

Note that we have even

$$op_\psi(a)\,\mathcal{G} \in op_G(\mathcal{S}_{\Gamma+\gamma_o} \otimes_r \mathcal{S}_\Delta)$$

and similarly for $\mathcal{G}\, op_\psi(a)$.

Let $\mathcal{M}$ be as in Theorem 16 and $a \in S^{(o)}(\mathbb{R}^n \setminus 0)$. Replacing m_k in formula (38) by $\sigma_M^{-k}(a)$ we get a function $f_1(z)$ which is holomorphic outside $\left(\sum_{j+k=1} T^k z^j\right) \cup \{j : j \in Z\} =: V^1$. Assume that $V^1 \in \underline{Z}^d$. The properties of $\sigma_M^{-k}(a)$ for $|\mathrm{Im}\, z| \longrightarrow \infty$ show that then $f_1(z) \in \mathfrak{M}_{V^1}$. With $(f_1, \nu_1)_{1 \in \mathbb{Z}_+}$, ν_1 constants as above, we can pass to some operator (39). Then we have the following

24. **Theorem.** Under the described conditions we have

$$\mathcal{M}\ \mathrm{op}_\psi(a) = \mathcal{M} + \mathcal{G}$$

with some Green operator $\mathcal{G}$, and an analogous composition rule with $\mathrm{op}_\psi(a)$ as the left factor.

Proof: Applying (47) we have to discuss the items

$$\mathcal{M}\ \omega\ \mathrm{op}_\psi(a)\omega\,, \tag{48}$$

$$\mathcal{M}(1-\omega)\ \mathrm{op}_\psi(a)\,\omega\,, \tag{49}$$

$$\mathcal{M}\ \omega\ \mathrm{op}_\psi(a)(1-\omega)\,, \tag{50}$$

$$\mathcal{M}(1-\omega)\ \mathrm{op}_\psi(a)(1-\omega)\,. \tag{51}$$

The mapping properties of $\mathcal{M}(1-\omega)$ show that $\mathcal{M}(1-\omega)$ is a Green operator (cf. also Lemma 10). Hence (49), (51) are Green operators, in view of Proposition 23.

Choose another cut-off function ω' with $\omega' = \omega\omega'$. Then

$$\omega\mathrm{op}_\psi(a)(1-\omega) = \omega'\ \mathrm{op}_\psi(a)(1-\omega) + (\omega-\omega')\ \mathrm{op}_\psi(a)(1-\omega),$$

and

$$\omega'\mathrm{op}_\psi(a)(1-\omega) = \mathrm{op}_G(b)$$

with $b \in \mathcal{S}_{\gamma_0} \otimes_\pi \mathcal{S}_{\gamma_0}$, cf. 1.4.1. Lemma 1. Thus

$$\mathcal{M}\omega\ \mathrm{op}_\psi(a)(1-\omega) = \mathcal{M}\mathrm{op}_G(b) + \mathcal{M}(\omega-\omega')\ \mathrm{op}_\psi(a)(1-\omega)$$

is of Green type, since $\mathcal{M}(\omega-\omega')$ is Green and, hence we may apply Proposition 21 and Proposition 23. Finally (48) equals $\mathcal{M} + \mathcal{G}_1$ with some Green operator $\mathcal{G}_1$. This follows in the same way as Proposition 19 when we apply the Mellin expansion of the pseudo-differential action and the property that the f_l are Mellin symbols in our class. □

2.2.2. Spaces of the type $\sum \mathcal{S}_{\Lambda^i}$, and the algebra $\mathcal{G}(\mathbb{R}^d)$

The calculus of ψDO's, Mellin and Green operators requires an extension of the concept of $\mathcal{S}_\Lambda$ spaces, $\Lambda \in \underline{\Lambda}$. If, for instance, $\mathcal{M}_i : \mathcal{S}_{\Lambda^i} \longrightarrow \mathcal{S}_{\Xi^i}$, $i = 1,2$, are Mellin operators of the form $\mathcal{M}_i = \omega\,\mathrm{op}_M(h_i)\omega$ and $h_i \in \mathfrak{M}_{Z^i}$, $\Xi^i = \Lambda^i + Z^i_{(o)}$, we have no addition $\mathcal{M}_1 + \mathcal{M}_2$ and composition $\mathcal{M}_1\mathcal{M}_2$ in this class unless Λ^i, Z^i satisfy the admissibility conditions of 2.2.1. The general case gives rise to spaces of functions

$h_1 + h_2$ with $h_i \in \mathfrak{M}_{Z^i}$, $i = 1,2$, and $u_1 + u_2$ with $u_i \in \mathcal{S}_{\wedge}^i$. Using the topological sum of Fréchet spaces we will completely drop the admissibility conditions and the calculus of Mellin, Green and ψDO's remains valid.

First let us remind of the notion of sums of Banach spaces. If $\mathcal{E}$ is a complete locally convex vector space and $\mathcal{H}^1, \mathcal{H}^2$ Banach spaces being vector subspaces of $\mathcal{E}$, we denote by $\mathcal{H}^1 + \mathcal{H}^2$ the space of all elements $u = u_1 + u_2$ for $u_1 \in \mathcal{H}^1$, $u_2 \in \mathcal{H}^2$. Then $\mathcal{H}^1 + \mathcal{H}^2$ is a Banach space with the norm

$$\| u \|_{\mathcal{H}^1 + \mathcal{H}^2} = \inf_{u=u_1+u_2} (\| u_1 \|_{\mathcal{H}^1} + \| u_2 \|_{\mathcal{H}^2}) \tag{1}$$

or the equivalent norm

$$\inf_{u=u_1+u_2} (\| u_1 \|^2_{\mathcal{H}^1} + \| u_2 \|^2_{\mathcal{H}^2})^{\frac{1}{2}} . \tag{2}$$

The intersection $\mathcal{H}^1 \cap \mathcal{H}^2$ becomes a Banach space with the norm

$$\| u \|_{\mathcal{H}^1 \cap \mathcal{H}^2} = \max \left\{ \| u \|_{\mathcal{H}^1}, \| u \|_{\mathcal{H}^2} \right\}$$

or the equivalent norm

$$(\| u \|^2_{\mathcal{H}^1} + \| u \|^2_{\mathcal{H}^2})^{\frac{1}{2}} .$$

Consider the direct sum $\mathcal{H}^1 \oplus \mathcal{H}^2$ with the norm

$$\| (u_1, u_2) \|_{\mathcal{H}^1 \oplus \mathcal{H}^2} = (\| u_1 \|^2_{\mathcal{H}^1} + \| u_2 \|^2_{\mathcal{H}^2})^{\frac{1}{2}}, \quad (u_1, u_2) \in \mathcal{H}^1 \oplus \mathcal{H}^2$$

and set

$$\mathcal{N} = \left\{ (u_1, u_2) \in \mathcal{H}^1 \oplus \mathcal{H}^2 : \exists \ u \in \mathcal{H}^1 \cap \mathcal{H}^2 \text{ with } u_1 = u, \ u_2 = -u \right\} .$$

Then $\mathcal{N}$ is a closed subspace of $\mathcal{H}^1 \oplus \mathcal{H}^2$ isomorphic to $\mathcal{H}^1 \cap \mathcal{H}^2$. The canonical mapping $\mathcal{H}^1 \oplus \mathcal{H}^2 \longrightarrow \mathcal{H}^1 + \mathcal{H}^2$ given by $(u_1, u_2) \longrightarrow u_1 + u_2$ induces an isomorphism

$$\mathcal{H}^1 \oplus \mathcal{H}^2 / \mathcal{N} \longrightarrow \mathcal{H}^1 + \mathcal{H}^2 . \tag{3}$$

In fact, the factor norm on the left

$$\inf_{w=(u,-u) \in \mathcal{N}} (\| u_1 - u \|^2_{\mathcal{H}^1} + \| u_2 - u \|^2_{\mathcal{H}^2})^{\frac{1}{2}}$$

coincides with the norm on $\mathcal{H}^1 + \mathcal{H}^2$.

In particular if $\mathcal{H}^1, \mathcal{H}^2$ are Hilbert spaces, (3) induces a Hilbert space structure on $\mathcal{H}^1 + \mathcal{H}^2$.

Obviously, we have the continuous embedding $\mathcal{H}^1 \hookrightarrow \mathcal{H}^1 + \mathcal{H}^2$, since

$$\| u \|_{\mathcal{H}^1 + \mathcal{H}^2} = \inf_{u = u_1 + u_2} (\| u_1 \|_{\mathcal{H}^1} + \| u_2 \|_{\mathcal{H}^2}) \le \| u \|_{\mathcal{H}^1}$$

for $u \in \mathcal{H}^1$. Moreover, $\mathcal{H} + \mathcal{H} = \mathcal{H}$. In fact, the continuous embedding $\mathcal{H} + \mathcal{H} \hookrightarrow \mathcal{H}$ follows from

$$\| w \|_{\mathcal{H}} = \inf_{w = u + v} (\| u \|_{\mathcal{H}} + \| v \|_{\mathcal{H}}) .$$

Then the continuous embedding $\mathcal{H}^2 \hookrightarrow \mathcal{H}^1$ implies $\mathcal{H}^1 + \mathcal{H}^2 = \mathcal{H}^1$, since $\mathcal{H}^1 \hookrightarrow \mathcal{H}^1 + \mathcal{H}^2 \hookrightarrow \mathcal{H}^1 + \mathcal{H}^1 = \mathcal{H}^1$.

Let

$$T^i : \mathcal{H}^i \longrightarrow \tilde{\mathcal{H}}^i, \ i = 1,2,$$

be linear continuous operators satisfying

$$T^1|_{\mathcal{H}^1 \cap \mathcal{H}^2} = T^2|_{\mathcal{H}^1 \cap \mathcal{H}^2} .$$

Then $T^1 \oplus T^2 : \mathcal{H}^1 \oplus \mathcal{H}^2 \longrightarrow \tilde{\mathcal{H}}^1 \oplus \tilde{\mathcal{H}}^2$ induces a unique linear continuous operator

$$T^1 + T^2 : \mathcal{H}^1 + \mathcal{H}^2 \longrightarrow \tilde{\mathcal{H}}^1 + \tilde{\mathcal{H}}^2.$$

We also need sums and intersections of Fréchet spaces. Let $\mathcal{S}^1, \mathcal{S}^2$ be Fréchet spaces which are continuously embedded into $\mathcal{E}$ and denote by $\mathcal{S}^1 + \mathcal{S}^2$ the space of all $u = u_1 + u_2$, $u_1 \in \mathcal{S}^1$, $u_2 \in \mathcal{S}^2$. Set

$$\mathcal{M} = \{ (u,-u) \in \mathcal{S}^1 \oplus \mathcal{S}^2 : u \in \mathcal{S}^1 \cap \mathcal{S}^2 \} .$$

Then the algebraic isomorphism

$$\mathcal{S}^1 \oplus \mathcal{S}^2 / \mathcal{M} \longrightarrow \mathcal{S}^1 + \mathcal{S}^2$$

given by $u_1 \oplus u_2 \longrightarrow u_1 + u_2$ induces on $\mathcal{S}^1 + \mathcal{S}^2$ the factor space topology which is Fréchet, since $\mathcal{M}$ is closed in $\mathcal{S}^1 \oplus \mathcal{S}^2$.

1. Proposition. Let $\mathcal{S}^1, \mathcal{S}^2$ be nuclear. Then $\mathcal{S}^1 + \mathcal{S}^2$ is nuclear, too.

Indeed, it is trivial that the direct sum $\mathcal{S}^1 \oplus \mathcal{S}^2$ is a nuclear Fréchet space. Then the factor space with respect to a closed subspace is nuclear, too.

Note that we also have defined the topological intersection $\mathcal{S}^1 \cap \mathcal{S}^2$ of the Fréchet spaces $\mathcal{S}^1, \mathcal{S}^2$ with the topology induced by $\mathcal{M}$. If $\mathcal{S}^1, \mathcal{S}^2$ are nuclear, the space $\mathcal{M}$ as a closed subspace of $\mathcal{S}^1 \oplus \mathcal{S}^2$ is nuclear, too, i.e. $\mathcal{S}^1 \cap \mathcal{S}^2$ is a nuclear Fréchet space.

If $\mathcal{S}^i$, $i = 1,\ldots,N$, are Fréchet spaces we may iterate the construction and obtain Fréchet spaces

$$(\ldots((\mathcal{S}^1 + \mathcal{S}^2) + \mathcal{S}^3) + \ldots) + \mathcal{S}^N \cong \mathcal{S}^1 + (\ldots(\mathcal{S}^2 + \mathcal{S}^3) + \mathcal{S}^4) + \ldots) + \mathcal{S}^N)$$
$$\cong \ldots \quad ,$$

where the isomorphisms are topological. Thus it is justified to write

$$\mathcal{S} := \mathcal{S}^1 + \mathcal{S}^2 + \ldots + \mathcal{S}^N$$

and

$$\mathcal{S} \cong \mathcal{S}^1 \oplus \mathcal{S}^2 \oplus \ldots \oplus \mathcal{S}^N / \mathcal{M} \ ,$$

where

$$\mathcal{M} := \left\{ (v_1,\ldots,v_N) \in \mathcal{S}^1 \oplus \ldots \oplus \mathcal{S}^N : v_k \in \bigcap_{i=1}^{N} \mathcal{S}^i,\ k = 1,\ldots,N, \sum_{k=1}^{N} v_k = 0 \right\}$$

is a closed subspace of the direct sum.

For

$$\mathcal{S}^i = \varprojlim_j \mathcal{H}^i_j$$

we have

$$\mathcal{S}^1 + \ldots + \mathcal{S}^N = \varprojlim_j (\mathcal{H}^1_j + \ldots + \mathcal{H}^N_j) \ .$$

Let $T^1 : \mathcal{S}^1 \longrightarrow \tilde{\mathcal{S}}^1$ and $T^2 : \mathcal{S}^2 \longrightarrow \tilde{\mathcal{S}}^2$ be linear continuous operators between Fréchet spaces satisfying

$$T^1 \big|_{\mathcal{S}^1 \cap \mathcal{S}^2} = T^2 \big|_{\mathcal{S}^1 \cap \mathcal{S}^2} \ .$$

Then $T^1 \oplus T^2 : \mathcal{S}^1 \oplus \mathcal{S}^2 \longrightarrow \tilde{\mathcal{S}}^1 \oplus \tilde{\mathcal{S}}^2$ induces a unique linear continuous operator

$$T^1 + T^2 : \mathcal{S}^1 + \mathcal{S}^2 \longrightarrow \tilde{\mathcal{S}}^1 + \tilde{\mathcal{S}}^2 \ .$$

Now we prove the distributivity of sum and tensor product of Fréchet spaces.

2. <u>Lemma</u>. Let E,F,G be Fréchet spaces. Then $(E + F) \otimes_\pi G \cong E \otimes_\pi G + F \otimes_\pi G$.

Proof: Each $w \in (E + F) \otimes_\pi G$ admits a representation as absolutely converging series

$$w = \sum_{i=1}^{\infty} \lambda_i u_i \otimes g_i ,$$

where $\sum_{i=1}^{\infty} |\lambda_i| < \infty$, $u_i \longrightarrow 0$ in $E + F$, $g_i \longrightarrow 0$ in G for $i \longrightarrow \infty$. There exist sequences $e_i \longrightarrow 0$ in E, $f_i \longrightarrow 0$ in F such that $u_i = e_i + f_i$. Then

$$w = \sum_{i=1}^{\infty} \lambda_i (e_i+f_i) \otimes g_i = \sum_{i=1}^{\infty} \lambda_i e_i \otimes g_i + \sum_{i=1}^{\infty} \lambda_i f_i \otimes g_i$$

and the continuous embedding $(E + F) \otimes_\pi G \hookrightarrow E \otimes_\pi G + F \otimes_\pi G$ is proved. Conversely, the continuous embeddings $E \hookrightarrow E + F$, $F \hookrightarrow E + F$ induce continuous embeddings $E \otimes_\pi G \hookrightarrow (E+ F) \otimes_\pi G$ and $F \otimes_\pi G \hookrightarrow (E + F) \otimes_\pi G$, and therefore $E \otimes_\pi G + F \otimes_\pi G \hookrightarrow (E + F) \otimes_\pi G$. $\square$

3. Lemma. Let E,F,G be Fréchet spaces that may be written as projective limits

$$E = \varprojlim E^j , \quad F = \varprojlim F^k , \quad G = \varprojlim G^l$$

of separable Hilbert spaces, $E^{j+1} \hookrightarrow E^j$, $F^{k+1} \hookrightarrow F^k$, $G^{l+1} \hookrightarrow G^l$ continuous. Let either E and F or G be nuclear. Then

$$(E \cap F) \otimes_\pi G = (E \otimes_\pi G) \cap (F \otimes_\pi G)$$

with the corresponding intersection Fréchet topologies.

Proof: First observe that we have continuous embeddings

$$(E \cap F) \otimes_\pi G \hookrightarrow E \otimes_\pi G, \quad (E \cap F) \otimes_\pi G \hookrightarrow F \otimes_\pi G .$$

Then we also get a continuous embedding

$$(E \cap F) \otimes_\pi G \hookrightarrow (E \otimes_\pi G) \cap (F \otimes_\pi G) .$$

So the non-trivial part of the assertion is the converse. If E,F are nuclear, we may assume that the embeddings $E^{j+1} \hookrightarrow E^j$, $F^{k+1} \hookrightarrow F^k$ are nuclear and, similarly, $G^{l+1} \hookrightarrow G^l$ is nuclear if G is nuclear. We have

$$E \cap F = \varprojlim_j (E^j \cap F^j) ,$$

which is an immediate consequence of the definition of the intersection of Fréchet spaces. Thus

$$(E \cap F) \otimes_\pi G = \varprojlim_j (E^j \cap F^j) \otimes_\pi \varprojlim_l G^l = \varprojlim_j (E^j \cap F^j) \otimes_\pi G^j ,$$

where the latter equation follows from the definition of the projective tensor product. Under the assumptions on E,F,G we get

$$\varprojlim (E^j \cap F^j) \otimes_\pi G^j = \varprojlim (E^j \cap F^j) \otimes_H G^j .$$

We shall prove that

$$(E^j \cap F^j) \otimes_H G^j = (E^j \otimes_H G^j) \cap (F^j \otimes_H G^j) .$$

Then

$$\begin{aligned}(E \cap F) \otimes_\pi G &= \varprojlim \left[(E^j \otimes_H G^j) \cap (F^j \otimes_H G^j)\right] \\ &= \varprojlim E^j \otimes_H G^j \cap \varprojlim F^j \otimes_H G^j \\ &= \varprojlim E^j \otimes_\pi G^j \cap \varprojlim F^j \otimes_\pi G^j \\ &= (E \otimes_\pi G) \cap (F \otimes_\pi G),\end{aligned}$$

where we have used the assumptions on E,F,G again.

It remains to show the following. Let $\mathcal{E}, \mathcal{F}, \mathcal{G}$ be separable Hilbert spaces. Then

$$(\mathcal{E} \cap \mathcal{F}) \otimes_H \mathcal{G} = (\mathcal{E} \otimes_H \mathcal{G}) \cap (\mathcal{F} \otimes_H \mathcal{G}),$$

with the Hilbert space structures of the intersections that are induced by the diagonals of $\mathcal{E} \oplus \mathcal{F}$ and $(\mathcal{E} \otimes_H \mathcal{G}) \oplus (\mathcal{F} \otimes_H \mathcal{G})$, respectively. Let $h \in (\mathcal{E} \otimes_H \mathcal{G}) \cap (\mathcal{F} \otimes_H \mathcal{G})$. Then, for an orthogonal base $(g_j)_{j \in \mathbb{Z}_+}$ in $\mathcal{G}$ we have convergence

$$\sum_{j=0}^{\infty} (h, g_j)_{\mathcal{G}} \, g_j \longrightarrow h$$

both in $\mathcal{E} \otimes_H \mathcal{G}$ and in $\mathcal{F} \otimes_H \mathcal{G}$ and, hence, in the intersection. Since $(h, g_j)_{\mathcal{G}} \in \mathcal{E} \cap \mathcal{F}$ for all $j \in \mathbb{Z}_+$, we have

$$\sum_{j=0}^{N} (h, g_j)_{\mathcal{G}} \, g_j \in (\mathcal{E} \cap \mathcal{F}) \otimes_H \mathcal{G}$$

for all $N \in \mathbb{Z}_+$ and the only point is to ensure the convergence for $N \longrightarrow \infty$ in $(\mathcal{E} \cap \mathcal{F}) \otimes_H \mathcal{G}$. This follows, since the scalar product in $(\mathcal{E} \cap \mathcal{F}) \otimes_H \mathcal{G}$ is induced by the diagonal of $(\mathcal{E} \oplus \mathcal{F}) \otimes_H \mathcal{G} = (\mathcal{E} \otimes_H \mathcal{G}) \oplus (\mathcal{F} \otimes_H \mathcal{G})$, and this induces at the same time the above $(\mathcal{E} \otimes_H \mathcal{G}) \cap (\mathcal{F} \otimes_H \mathcal{G})$-scalar product. $\square$

As an example let $\Lambda^1, \Lambda^2 \in \underline{\Lambda}$ (which does not necessarily satisfy the compatibility condition that $\Lambda^1 + \Lambda^2 \in \underline{\Lambda}$). Then we have the sum of Fréchet spaces

$$\mathcal{S}_\Delta := \mathcal{S}_{\Lambda^1} + \mathcal{S}_{\Lambda^2}, \quad \Delta := \Lambda^1 + \Lambda^2 ,$$

$\mathcal{S}_\Delta \subset L^2(\mathbb{R}_+)$.

This notation will be justified when we show that the space $\mathcal{S}_\Delta$ does not depend on the decomposition of Δ into the sum of elements in $\underline{\Lambda}$ (cf. Proposition 4). On the $\mathcal{S}_\Lambda$ spaces we have an additional structure from the representation as projective limit of Hilbert spaces

$$\mathcal{S}_{\Lambda^i} = \varprojlim_j \mathcal{H}^i_j .$$

Then

$$\mathcal{S}_{\Lambda^1} + \mathcal{S}_{\Lambda^2} = \varprojlim_{l=j+k} (\mathcal{H}^1_j + \mathcal{H}^2_k) .$$

Since the embeddings

$$e^i_j : \mathcal{H}^i_j \longrightarrow \mathcal{H}^i_{j-1}, \; j \in \mathbb{Z}, \; i = 1,2,$$

which are induced by the inclusion $\mathcal{H}^i_j \subset L^2(\mathbb{R}_+)$, satisfy

$$e^1_j|_{\mathcal{H}^1_j \cap \mathcal{H}^2_j} = e^2_j|_{\mathcal{H}^1_j \cap \mathcal{H}^2_j} ,$$

we get

$$e^1_j + e^2_j : \mathcal{H}^1_j + \mathcal{H}^2_j \longrightarrow \mathcal{H}^1_{j-1} + \mathcal{H}^2_{j-1} \quad \text{for all } j \in \mathbb{Z}$$

and

$$\mathcal{S}_{\Lambda^1} + \mathcal{S}_{\Lambda^2} = \varprojlim_j (\mathcal{H}^1_j + \mathcal{H}^2_j) .$$

The same is true of arbitrary finite sums $\sum_{i=1}^{p} \mathcal{S}_{\Lambda^i}$. We shall need also infinite sums of $\mathcal{S}_{\Lambda^i}$. One reason is that each $\Lambda \in \underline{\Lambda}$ is an infinite sum of <u>compact</u> sets Λ_j.

Consider an arbitrary sequence $(\Lambda^i)_{i \in \mathbb{Z}_+}$, $\Lambda^i \in \underline{\Lambda}$ satisfying

$$\sup \{ \operatorname{Re} z : z \in \Lambda^i \} \longrightarrow -\infty \quad \text{as } i \longrightarrow \infty .$$

Then we can represent $\mathcal{S}_{\Lambda^i}$ as $\mathcal{S}_{\Lambda^i} = \varprojlim \mathcal{H}^i_j$ in such a way that for each j there exists $i_o = i_o(j)$ such that $\mathcal{H}^i_j \hookrightarrow \sum_{k=1}^{i_o} \mathcal{H}^k_j$ for all $i > i_o$.

In fact, the norm defining $\mathcal{H}_j^i$ for fixed j takes care only about a fixed strip $c(j) < \mathrm{Re}\, z < \frac{1}{2}$ to which no singularities are contributed for large i. Therefore,

$$\sum_{i=1}^{\infty} \mathcal{H}_j^i = \sum_{i=1}^{i_o} \mathcal{H}_j^i$$

and we set

$$\sum_{i=1}^{\infty} \mathcal{S}_{\Lambda^i} = \varprojlim_{j} \left(\sum_{i=1}^{\infty} \mathcal{H}_j^i\right) .$$

Note that $\sum_{i=1}^{\infty} \mathcal{S}_{\Lambda^i}$ is a nuclear Fréchet space, again. It is easy to check that for $\Lambda = \sum_{i=1}^{\infty} \Lambda_i \in \underline{\Lambda}$, Λ_i compact, $\sum_{i=1}^{\infty} \mathcal{S}_{\Lambda_i} = \mathcal{S}_{\Lambda}$ (cf. 2.1.1. Definition 2).

Every $\Delta \in \underline{\Delta}$ (cf. the definition of $\underline{\Delta}$ in 2.1.3.) can be represented as $\Delta = \sum_{j=1}^{\infty} \Delta_j$, where $\Delta_j \in \underline{\Lambda}$ compact and

$$\sup \{\mathrm{Re}\, z : z \in \Delta_j\} \longrightarrow -\infty \quad \text{as } j \longrightarrow \infty .$$

Then we set

$$\mathcal{S}_{\Delta} = \sum_{j=1}^{\infty} \mathcal{S}_{\Delta_j} .$$

4. <u>Proposition</u>. The Fréchet space $\mathcal{S}_{\Delta}$ does not depend on the choice of the decomposition $\Delta = \sum_{j=1}^{\infty} \Delta_j$ with the above properties, i.e. if $\Delta = \sum_{k=1}^{\infty} \Delta'_k$ is another such decomposition, then

$$\sum_{j=1}^{\infty} \mathcal{S}_{\Delta_j} = \sum_{k=1}^{\infty} \mathcal{S}_{\Delta'_k} ,$$

and the topologies are equivalent.

<u>Proof</u>: Denote by $\tilde{\Delta}_l$ another covering of Δ by compact sets which is a common refinement of Δ_j and Δ'_k, i.e. $\Delta = \sum_{l=1}^{\infty} \tilde{\Delta}_l$, $\tilde{\Delta}_l$ compact and

for each j there is an $l(j)$ such that $\tilde{\Delta}_{l(j)} \subset \Delta_j$ and for each k there is an $l'(k)$ such that $\tilde{\Delta}_{l'(k)} \subset \Delta'_k$. Then $\sup \{\mathrm{Re}\, z : z \in \tilde{\Delta}_l\} \longrightarrow -\infty$ as $l \longrightarrow \infty$ and $\mathcal{S}_{\tilde{\Delta}_{l(j)}} \hookrightarrow \mathcal{S}_{\Delta_j}$, $\mathcal{S}_{\tilde{\Delta}_{l'(k)}} \hookrightarrow \mathcal{S}_{\Delta'_k}$ and therefore we obtain continuous embeddings

$$\sum_{l=1}^{\infty} \mathcal{S}_{\tilde{\Delta}_l} \hookrightarrow \sum_{j=1}^{\infty} \mathcal{S}_{\Delta_j}, \quad \sum_{l=1}^{\infty} \mathcal{S}_{\tilde{\Delta}_l} \hookrightarrow \sum_{k=1}^{\infty} \mathcal{S}_{\Delta'_k} .$$

The assertion is proved if we show the converse embeddings. Consider a fixed $j \geq 1$ and set $L(j) = \{ l \geq 1 : \tilde{\Delta}_l \cap \Delta_j \neq \emptyset \}$, which is a finite set of indices. Then $\Delta_j \subset \sum_{l \in L(j)} \tilde{\Delta}_l$. Now the task is to decompose an arbitrary $u \in \mathcal{S}_{\Delta_j}$ into a sum of functions in $\mathcal{S}_{\tilde{\Delta}_l}$, $l \in L(j)$. The main point is the decomposition of the functional $\zeta(u) \in \mathcal{A}'(\Delta_j)$, which is possible according to the following

5. <u>Lemma</u>. Let $K, K_1, K_2 \subset \mathbb{C}$ be compact, simply connected sets, $K = K_1 + K_2$. Then

$$\mathcal{A}'(K) = \mathcal{A}'(K_1) + \mathcal{A}'(K_2)$$

with equivalent topologies of the Fréchet spaces.

The proof of Lemma 5 will be given below. Obviously the assertion is valid for an arbitrary finite union of compact sets. Therefore we have

$$\zeta(u) = \sum_{l \in L(j)} \zeta_l, \quad \zeta_l \in \mathcal{A}'(\tilde{\Delta}_l),$$

and

$$u_l := \langle \zeta_l, F \rangle \in \mathcal{S}_{\tilde{\Delta}_l} .$$

It follows that $u - \sum_{l \in L(j)} u_l \in \mathcal{S}_0$. Thus we have proved that

$$\mathcal{S}_{\Delta_j} \subset \sum_{l \in L(j)} \mathcal{S}_{\tilde{\Delta}_l}$$

and therefore

$$\sum_{j=1}^{\infty} \mathcal{S}_{\Delta_j} \subset \sum_{l=1}^{\infty} \mathcal{S}_{\tilde{\Delta}_l} . \quad \square$$

The proof of Lemma 5 relies on the well-known fact that the Cousin problem is solvable in the complex plane.

Observe that it is not always true that $\mathcal{A}'(K_1 \cap K_2) = \mathcal{A}'(K_1) \cap \mathcal{A}'(K_2)$. But this equality holds if $h_i \in \mathcal{A}(\mathbb{C} \setminus K_i)$ with $h_1 = h_2$ over $\mathbb{C} \setminus (K_1 + K_2)$ always implies the existence of some $h \in \mathcal{A}(\mathbb{C} \setminus (K_1 \cap K_2))$ with $h_i = h|_{\mathbb{C} \setminus K_i}$, $i = 1,2$. This is true **if** K_1, K_2 are convex but not if K_1, K_2 are different curves connecting different points $z_1, z_2 \in \mathbb{C}$.

Proof of Lemma 5: It is obvious that the canonical embedding

$$\iota : \mathcal{A}'(K_1) + \mathcal{A}'(K_2) \hookrightarrow \mathcal{A}'(K)$$

is continuous. Since both $\mathcal{A}'(K_1) + \mathcal{A}'(K_2)$ and $\mathcal{A}(K)$ are Fréchet spaces, it suffices to prove that ι is surjective. Set $\Omega_i = \mathbb{C} \setminus K_i$, $i = 1,2$, $\Omega = \Omega_1 \cap \Omega_2$. Then $K \subseteq \mathbb{C} \setminus \Omega$ and every $\zeta \in \mathcal{A}'(K)$ can be represented by a function $h \in \mathcal{A}(\Omega)$, namely $h = \langle \zeta, F \rangle$. In view of the solvability of the Cousin problem there exist functions $h_i \in \mathcal{A}(\Omega_i)$ such that $h = h_1 - h_2$ over Ω, and h_i represent elements $\zeta_i \in \mathcal{A}'(K_i)$. Thus $\zeta = \zeta_1 - \zeta_2$ and hence ι is surjective. $\square$

Obviously, each $\Delta \in \underline{\Delta}$ admits a decomposition $\Delta = \Lambda^1 + \Lambda^2$, $\Lambda^i \in \underline{\Lambda}$, $i = 1,2$. Then we have

$$\mathcal{S}_{\Lambda^i} = \sum_{j=0}^{\infty} \mathcal{S}_{\Lambda^i_j} \quad \text{if} \quad \Lambda^i = \sum_{j=0}^{\infty} \Lambda^i_j, \quad i = 1,2,$$

and therefore

$$\mathcal{S}_{\Lambda^1} + \mathcal{S}_{\Lambda^2} = \sum_{j=0}^{\infty} \mathcal{S}_{\Lambda^1_j} + \sum_{j=0}^{\infty} \mathcal{S}_{\Lambda^2_j} = \mathcal{S}_\Delta \quad ,$$

since $\sum_{j=0}^{\infty} \Lambda^1_j + \sum_{j=0}^{\infty} \Lambda^2_j = \Delta$.

The decomposition of singularity types and of the associated spaces into sums will be used systematically in the following calculus. For references below we then talk of the <u>decomposition principle</u>.

Any singularity type $\Delta \in \underline{\Delta}$ may be decomposed into a sum $\Delta = \Delta^1 + \Delta^2$ with $\Gamma_{\frac{1}{2}+\gamma^i} \cap \Delta^i = \emptyset$, $i = 0,1$, for arbitrary $\gamma^1, \gamma^2 \in \mathbb{R}$, $\gamma^1 \neq \gamma^2$. It turns out that then the associated spaces $\mathcal{S}_\Delta$ and $\mathcal{S}_{\Delta^1} + \mathcal{S}_{\Delta^2}$ coincide. Similar arguments hold for the spaces of Mellin symbols below.

Now let us show that the spaces $\mathcal{S}_{(\Delta)}$ (defined in 2.1.3.) and $\mathcal{S}_\Delta$ actually coincide.

6. <u>Proposition</u>. For each $\Delta \in \underline{\Delta}$ we have

$$\mathcal{S}_\Delta = \mathcal{S}_{(\Delta)}$$

and the topologies are equivalent.

Proof: Let $\Lambda^1, \Lambda^2 \in \underline{\Lambda}$ be such that $\Lambda^1 + \Lambda^2 = \Delta$. Then

$$\mathcal{S}_\Delta = \mathcal{S}_{\Lambda^1} + \mathcal{S}_{\Lambda^2} \hookrightarrow \mathcal{S}_{(\Delta)}$$

and it is sufficient to show surjectivity. Let $u \in \mathcal{S}_{(\Delta)}$. Then we can decompose $\tilde{u} \in \mathcal{A}(\mathbb{C} \setminus \Delta)$

$$\tilde{u} = \tilde{v}_1 + \tilde{w}_1, \ \tilde{v}_1 \in \mathcal{A}(\mathbb{C} \setminus \Lambda^1), \ \tilde{w}_1 \in \mathcal{A}(\mathbb{C} \setminus \Lambda^2),$$

according to the solvability of the Cousin problem. The holomorphic functions $\tilde{v}_1$ and $\tilde{w}_1$ represent sequences of analytic functionals carried by Λ^1_j and Λ^2_j, respectively, $\Lambda^i = \sum_{j=1}^{\infty} \Lambda^i_j$. Applying 2.1.1. Theorem 15 we find $v \in \mathcal{S}_{\Lambda^1}$ and $w \in \mathcal{S}_{\Lambda^2}$ representing the same sequences of analytic functionals. Then $\tilde{u} - \tilde{v} - \tilde{w} \in M(\mathcal{S}_0)$. Hence we get the desired decomposition of u. □

From Proposition 6 and the semi-norm systems in 2.1.3. it immediately follows that for $\Delta^1, \Delta^2 \in \underline{\Delta}$ and $\Delta^1 \subseteq \Delta^2$ there is a canonical continuous embedding $\mathcal{S}_{\Delta^1} \hookrightarrow \mathcal{S}_{\Delta^2}$. Moreover, we have an analogue of 2.1.1. Proposition 11, namely

7. **Proposition**. There exists a countable system $\{\nu_\iota\}_{\iota \in I}$ of norms on $\mathcal{S}_0$ such that for every $\Delta \in \underline{\Delta}$ there is a subset $I(\Delta) \subseteq I$ for which $\mathcal{S}_\Delta$ has the topology induced by $\{\nu_\iota\}_{\iota \in I(\Delta)}$ and $\Delta^1 \subseteq \Delta^2$ implies $I(\Delta^1) \supseteq I(\Delta^2)$.

Incidentally we use the notation

$$\mathcal{S}_{\underline{\Delta}} = \varinjlim \mathcal{S}_\Delta$$

for the inductive limit over all $\Delta \in \underline{\Delta}$. We also have an analogue of the spaces $C^\infty_{\Lambda,\Lambda'}$ of 2.1.1., here for $\Lambda \in \underline{\Delta}$ and $\Lambda' \in \underline{\Delta}'$,

$$\underline{\Delta}' := \{\Delta' \subseteq \mathbb{C} : \Delta'^{\vee} \in \underline{\Delta}\}.$$

The obvious details may be dropped here. We return to these spaces below under the aspect of general Mellin symbols.

Let us now mention a certain counterpart of the representation of $\mathcal{S}_\Delta$, $\Delta \in \underline{\Delta}$, as $\mathcal{S}_{\Lambda^1} + \mathcal{S}_{\Lambda^2}$, $\Delta = \Lambda^1 + \Lambda^2$.

Proposition 8. Let Δ, $\Delta^1, \Delta^2 \in \underline{\Delta}$ and $\Delta = \Delta^1 \cap \Delta^2$. Assume that for each $\delta \in \mathbb{R}_+$ there exist decompositions $\Delta^i = \Delta^i_\delta + \Lambda^i_\delta$, $i = 1,2$, with $\Delta^i_\delta \subset \{\operatorname{Re} z < \frac{1}{2} - \frac{\delta}{2}\}$ and $\mathcal{A}'(\Lambda^1_\delta \cap \Lambda^2_\delta) = \mathcal{A}'(\Lambda^1_\delta) \cap \mathcal{A}'(\Lambda^2_\delta)$. Then

$$\mathcal{S}_\Delta = \mathcal{S}_{\Delta^1} \cap \mathcal{S}_{\Delta^2} \quad .$$

The simple proof is left to the reader. Remember that the conditions of Proposition 8 are satisfied if Λ^i_δ are convex.

With each function in $\mathcal{S}_\Lambda$, $\Lambda \in \underline{\Lambda}$, a sequence of singularity quantities is associated, namely the sequence of analytic functionals $\zeta_i \in \mathcal{A}'(\Lambda_i)$ $\Lambda = \bigcup_i \Lambda_i$, which generalizes the information about the coefficients ζ_{jk} for functions in $\mathcal{S}_\varphi$. In the situation of the $\mathcal{S}_\Delta$-spaces, in general, there is no analogue of the singularity quantities. More precisely, singularity quantities are associated only with the bounded connected components of Δ, in contrast to the unbounded ones. In order to show this we may restrict ourselves to such $\Delta \in \underline{\Delta}$ without bounded components.

9. Proposition. Let $\Delta \in \underline{\Delta}$ be without bounded components. Then $\mathcal{S}_0$ is dense in $\mathcal{S}_\Delta$.

Proof: Let $\varphi \in C_0^\infty(\mathbb{R}_+)$, $\varphi \equiv 1$ near 1, and set $\varphi_c(t) = \varphi(t^c)$, $0 < c \leq 1$. Obviously

$$\tilde{\varphi}_c(z) = \int_0^\infty t^z \varphi(t^c) \frac{dt}{t} = \frac{1}{c} \int_0^\infty t^{\frac{z}{c}} \varphi(t) \frac{dt}{t} = \frac{1}{c} \tilde{\varphi}\left(\frac{z}{c}\right) .$$

For each $u \in \mathcal{S}_\Delta$ we have

$$(\varphi_c u)^\sim(z) = \frac{1}{2\pi i} \int_{\Gamma_{\frac{1}{2}}} \tilde{\varphi}_c(z-w)\tilde{u}(w)dw = \frac{1}{2\pi i} \int_{\Gamma} \tilde{\varphi}_c(z-w)\tilde{u}(w)dw \in \tilde{\mathcal{S}}_0 ,$$

where Γ is a contour occurring in the semi-norms of $\mathcal{S}_\Delta$, i.e. in a compact part of a strip $\frac{1}{2} + \gamma < \operatorname{Re} z < \frac{1}{2}$ it approximates Δ and outside this compact part it coincides with $\Gamma_{\frac{1}{2}+\gamma}$. We show now that $(\varphi_c u)^\sim(z) \longrightarrow \tilde{u}(z)$ for $c \longrightarrow 0$ and all $z \in \Gamma$ and

$$\sup \left\| \int_\Gamma \tilde{\varphi}_c(z-w)u(w)dw \right\|_{L^2(\Gamma)} \| \tilde{u}(z) \|^{-1}_{L^2(\Gamma)} < M$$

where the constant M may depend on Γ, but not on c. Then it follows by Lebesgue's theorem that

$$\left\| \frac{1}{2\pi i} \int_{\Gamma} \tilde{\varphi}_c(z-w)\tilde{u}(w)dw - \tilde{u}(z) \right\|_{L^2(\Gamma)} \longrightarrow 0 \quad \text{for } c \longrightarrow 0 .$$

From

$$\frac{1}{2\pi i} \int_{\Gamma} \tilde{\varphi}_c(z-w)dw = \frac{1}{2\pi i c} \int_{\Gamma} \tilde{\varphi}\left(\frac{z-w}{c}\right) dw$$

$$= \frac{1}{2\pi i} \int_{\Gamma_{\frac{1}{2}}} \tilde{\varphi}\left(\frac{z}{c}-w\right) dw = \frac{1}{2\pi i} \int_{\Gamma_{\frac{1}{2}}} \tilde{\varphi}(w)\, dw = 1$$

it follows that

$$\frac{1}{2\pi i} \int_{\Gamma} \tilde{\varphi}_c(z-w)\tilde{u}(w)dw - \tilde{u}(z) = \frac{1}{2\pi i} \int_{\Gamma} \tilde{\varphi}_c(z-w)\left[\tilde{u}(w)-\tilde{u}(z)\right]dw$$

$$= \frac{1}{2\pi i c} \int_{\Gamma} \tilde{\varphi}\left(\frac{z-w}{c}\right) \left[\tilde{u}(w) - \tilde{u}(z)\right] dw \longrightarrow 0 \quad \text{for } c \longrightarrow 0,$$

since $\tilde{u}(w) - \tilde{u}(z)$ has a zero at $w = z$. From the estimates

$$\left\| \int_{\Gamma} \tilde{\varphi}_c(z-w)\tilde{u}(w)dw \right\|^2_{L^2(\Gamma)} \leqq \left\| \int_{\Gamma} |\tilde{\varphi}_c(z-w)|^{\frac{1}{2}} (|\tilde{\varphi}_c(z-w)|^{\frac{1}{2}} |\tilde{u}(w)|)\, |dw| \right\|^2_{L^2(\Gamma)} \leqq$$

$$\leqq \left\| \int_{\Gamma} |\tilde{\varphi}_c(z-w)||dw| \int_{\Gamma} |\tilde{\varphi}_c(z-w)|\, |u(w)|^2 |dw| \right\|^2_{L^2(\Gamma)}$$

$$\leqq \sup_{z\in\Gamma} \int_{\Gamma} |\tilde{\varphi}_c(z-w)|\, |dw| \quad \sup_{w\in\Gamma} \int_{\Gamma} |\tilde{\varphi}_c(z-w)|\, |dz|\, \|\tilde{u}\|^2_{L^2(\Gamma)}$$

and

$$\sup_{z\in\Gamma} \int_{\Gamma} |\tilde{\varphi}_c(z-w)|\, |dw| = \sup_{z\in\Gamma} \frac{1}{c} \int_{\Gamma} \left|\tilde{\varphi}\left(\frac{z-w}{c}\right)\right| |dw| = \sup_{z\in\Gamma} \int_{\frac{z}{c}-\frac{\Gamma}{c}} |\tilde{\varphi}(w)|\, |dw| \leqq M'$$

where M' is independent of c we obtain the desired estimate for

$$\sup \left\| \int_{\Gamma} \tilde{\varphi}_c(z-w)\tilde{u}(w)dw \right\|_{L^2(\Gamma)} \|\tilde{u}(z)\|^{-1}_{L^2(\Gamma)} .$$

The same arguments apply to $z^k\tilde{u}(z)$ for every k. Now let $(\Gamma_j)_{j\in\mathbb{Z}_+}$ be a system of curves of the quality of Γ such that the semi-norms $u \longrightarrow \|z^k\tilde{u}(z)\|_{L^2(\Gamma_j)}$, $j,k\in\mathbb{Z}_+$, define the topology of $\mathcal{S}_\Delta$. Then our arguments give

$$z^k(\varphi_c u)^\sim(z) \longrightarrow z^k\tilde{u}(z) \quad \text{in} \quad L^2(\Gamma_j)$$

as $c \longrightarrow 0$ for all j,k, and hence the desired approximation. $\square$

This result suggests some further discussion on the intuitive interpretation of asymptotic types represented by subsets $\Delta \subset \mathbb{C}$. Remember that $\mathcal{S}_0$ is closed in $\mathcal{S}_\Lambda$ for $\Lambda \in \underline{\Lambda}$, cf. 2.1.1. Remark 9. For $\Delta \in \underline{\Delta} \setminus \underline{\Lambda}$ the space $\mathcal{S}_0$ is never closed in $\mathcal{S}_\Delta$ and $\mathcal{S}_0$ is dense in $\mathcal{S}_\Delta$ iff $\mathbb{C} \setminus \Delta$ is simply connected.

Let $\mathcal{S} \subset C^\infty(\mathbb{R}_+)$ be an arbitrary vector subspace with a Fréchet structure, and $e^+(1-\omega)\mathcal{S} \subset \mathcal{S}(\mathbb{R})$, ω a cut-off function. Assume that $\mathcal{S}_0$ is closed in $\mathcal{S}$. Then $\mathcal{S}/\mathcal{S}_0$ is a Fréchet space in the quotient topology. Intuitively $\mathcal{S}/\mathcal{S}_0$ might be interpreted as the space of asymptotic quantities of functions in $\mathcal{S}$.

For $\Delta \in \underline{\Delta} \setminus \underline{\Lambda}$ this picture is not adequate. Nevertheless, it is justified to talk of an asymptotics associated with Δ. Indeed, let us choose a decomposition $\Delta = \Delta_\eta + \Theta_\eta$, $\eta \geq 0$, where $\Theta_\eta \subseteq \Delta$ is compact $\Theta_\eta \in \underline{\Lambda}$,

$$\Theta_\eta \subseteq \left\{ z : \operatorname{Re} z \geq \tfrac{1}{2} - \eta \right\},$$

$$\Delta_\eta \subseteq \left\{ z : \operatorname{Re} z \leq \tfrac{1}{2} - \eta \right\}.$$

Then any $u \in \mathcal{S}_\Delta$ admits a decomposition

$$u = u_\eta + v_\eta ,$$

where $u_\eta \in \mathcal{S}_{\Delta_\eta}$, $v_\eta \in \mathcal{S}_{\Theta_\eta}$, and hence

$$\partial_t^k(u - v_\eta) \in t^{\eta-\varepsilon} L^2(\mathbb{R}_+)$$

for all $k \in \mathbb{Z}_+$ and any $\varepsilon > 0$. The decomposition of u follows by arguments of solving a Cousin problem, similarly as in the proof of Lemma 5. Then

$$v_\eta(t) = \omega(t) \langle \zeta_\eta(w), t^{-w} \rangle + r_\eta(t) ,$$

where $\zeta_\eta \in \mathcal{A}'(\Theta_\eta)$, $r_\eta \in \mathcal{S}_0$. Thus ζ_η represents an asymptotic quantity of u. In contrast to the $\mathcal{S}_\Lambda$ theory here the choice of ζ_η is not unique, since in general $\Delta_\eta \cap \Theta_\eta \neq \emptyset$.

Now we are in a position to extend the calculus of 2.1.4. to the spaces $\mathcal{S}_\Delta$, $\Delta \in \underline{\Delta}$. It is clear that the abstract description of Green operators applies. We denote by $\mathcal{G}_G$ the class of all $op_G(b)$

$$b \in \mathcal{S}_{\underline{\Delta}} \otimes_\Gamma \mathcal{S}_{\underline{\Delta}} = \{\mathcal{S}_{\underline{\Delta}} \otimes_\pi L^2(\mathbb{R}_+)\} \cap \{L^2(\mathbb{R}_+) \otimes_\pi \mathcal{S}_{\underline{\Delta}}\}$$

acting on $L^2(\mathbb{R}_+)$ or $\mathcal{S}_{\underline{\Delta}}$.

The equality

$$\mathcal{S}_\varphi \otimes_\pi \mathcal{S}_\psi = \mathcal{S}_\varphi \otimes_\Gamma \mathcal{S}_\psi$$

established in 1.2.2. Proposition 13 reflects a special property of the $\mathcal{S}_\varphi$ spaces. The proof is based on the fact that for each fixed weight $\mathcal{S}_\varphi$ is only a finite-dimensional modification of $\mathcal{S}_o$. For the $\mathcal{S}_\Delta$ spaces, $\Delta \in \underline{\Delta}$, we have $\otimes_\pi \neq \otimes_\Gamma$, more precisely the proper inclusion

$$\mathcal{S}_{\Delta^1} \otimes_\pi \mathcal{S}_{\Delta^2} \subsetneqq \mathcal{S}_{\Delta^1} \otimes_\Gamma \mathcal{S}_{\Delta^2} = \mathcal{S}_{\Delta^1} \otimes_\pi L^2 \cap L^2 \otimes_\pi \mathcal{S}_{\Delta^2} .$$

Assume that $\mathcal{S}_{\Delta^1} \otimes_\pi \mathcal{S}_{\Delta^2} = \mathcal{S}_{\Delta^1} \otimes_\Gamma \mathcal{S}_{\Delta^2}$ as sets for $\Delta^j \in \underline{\Delta}$. Then the topologies have to coincide and each norm on $\mathcal{S}_{\Delta^1} \otimes_\pi \mathcal{S}_{\Delta^2}$ can be estimated by a sufficiently strong norm on $\mathcal{S}_{\Delta^1} \otimes_\pi L^2 \cap L^2 \otimes_\pi \mathcal{S}_{\Delta^2}$. In other words, for given piece-wise smooth curves C_1, C_2 which coincide with $\Gamma_{\frac{1}{2}-\delta_1}$, $\Gamma_{\frac{1}{2}-\delta_2}$ outside a compact set and $k_1, k_2 \in \mathbb{Z}_+$ there are C_1', C_2' which coincide with $\Gamma_{\frac{1}{2}-\delta_1'}$, $\Gamma_{\frac{1}{2}-\delta_2'}$, outside a compact set (cf. Fig. 3) such that

$$\int_{C_1}\int_{C_2} |\tilde{u}(z_1)\tilde{v}(z_2)|^2 (1+|z_1|)^{2k_1} (1+|z_2|)^{2k_2} |dz_1|\,|dz_2|$$

$$\leqq c\Big\{\int_{C_1'}\int_{\Gamma_{\frac{1}{2}}} |\tilde{u}(z_1)\tilde{v}(z_2)|^2 (1+|z_1|)^{2k_1'} |dz_1|\,|dz_2|$$

$$+ \int_{\Gamma_{\frac{1}{2}}}\int_{C_2'} |\tilde{u}(z_1)\tilde{v}(z_2)|^2 (1+|z_2|)^{2k_2'} |dz_1|\,|dz_2|\Big\}$$

$u \in \mathcal{S}_{\Delta^1}$, $v \in \mathcal{S}_{\Delta^2}$ with a constant c independent of u,v.

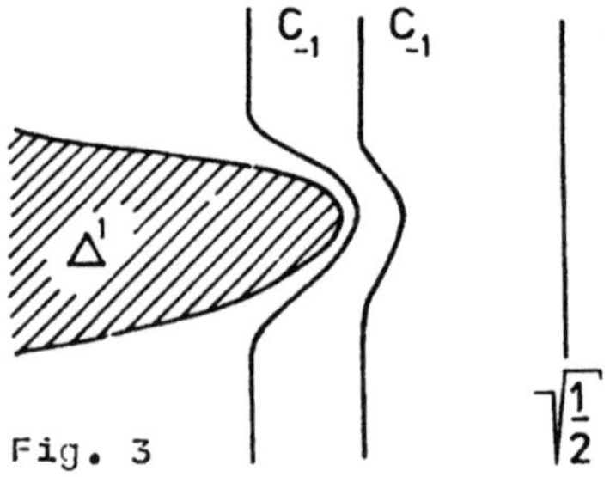

Fig. 3

We shall get a contradiction by inserting families $u_\lambda(t)$, $v_\lambda(t)$, to be specified later, and if we make a special choice of C_1, C_2. This choice is based on a result from the theory of univalent functions. Let $C(\delta, \Delta^j)$ be the preimage of $\Gamma_o = \{\mathrm{Re}\, z = 0\}$ of the conformal mapping Φ_j of the unbounded part Ω_j of the domain $\{z \in \mathbb{C} \setminus \Delta_j : \frac{1}{2} - \delta < \mathrm{Re}\, z < \frac{1}{2}\}$ to the strip $\{-\frac{1}{2} < \mathrm{Re}\, z < \frac{1}{2}\}$, $\Phi_j(\pm\infty) = \pm\infty$. Then there exists a constant K such that

$$\frac{1}{K} \leq \frac{\mathrm{dist}\,(z, \Gamma_{1/2})}{\mathrm{dist}\,(z, \partial\Omega_j \setminus \Gamma_{\frac{1}{2}})} \leq K$$

for all $z \in C(\delta, \Delta^j)$ (cf. Pommerenke [P5], ex. 1, p. 318). This means that the 'middle stream line' stays roughly in the middle between the boundary parts. Now we choose C_j in such a way that a part of it is so close to Δ_j that this part lies to the left of $C(\delta, \Delta^j)$ for all $\delta > 0$. Now we have

$$\int_{C_1}\int_{C_2} |\tilde{u}_\lambda(z_1)\,\tilde{v}_\lambda(z_2)|^2 (1+|z_1|)^{2k_1} (1+|z_2|)^{2k_2} \,|dz_1|\,|dz_2|$$

$$= \int_{\Phi_1(C_1)} |\tilde{u}_\lambda \circ \Phi_1^{-1}(w_1)|^2 (1+|\Phi_1^{-1}(w_1)|)^{2k_1} \left|\frac{Dz_1}{Dw_1}\right| |dw_1| \times$$

$$\int_{\Phi_2(C_2)} |\tilde{v}_\lambda \circ \Phi_2^{-1}(w_2)|^2 (1+|\Phi_2^{-1}(w_2)|)^{2k_2} \left|\frac{Dz_2}{Dw_2}\right| |dw_2| \,.$$

Choosing u_λ, v_λ such that

$$|\tilde{u}_\lambda \circ \Phi_1^{-1}(w_1)| = \lambda^{1/2 - \mathrm{Re}\, w_1} |\tilde{u}_1 \circ \Phi_1^{-1}(w_1)| \,,$$

$$|\tilde{v}_\lambda \circ \Phi_2^{-1}(w_2)| = \lambda^{1/2 - \mathrm{Re}\, w_2} |\tilde{v}_1 \circ \Phi_2^{-1}(w_2)| \,, \quad \lambda \geq 1,$$

we get an estimate from below by $c\lambda^{2+2\varepsilon}$, $\varepsilon > 0$, where

$$c = \int_{\Phi_1^{\varepsilon}(C_1)} |\tilde{u}_1 \circ \Phi_1^{-1}(w_1)|^2 \, (1+|\Phi_1^{-1}(w_1)|)^{2k_1} \left|\frac{Dz_1}{Dw_1}\right| |dw_1| \int_{\Phi_2^{\varepsilon}(C_2)} |(\tilde{v}_1 \circ \Phi_2^{-1}(w_2)|^2$$

$$(1+|\Phi_2^{-1}(w_2)|)^{2k_2} \left|\frac{Dz_2}{Dw_2}\right| |dw_2| \, , \; \Phi_j^{\varepsilon}(C_j) = \{ w \in \Phi_j(C_j) \; : \; \mathrm{Re}\, w < -\varepsilon \} \, .$$

For the right-hand side of the inequality in question we get the estimate from above, which gives the desired contradiction for $\lambda \to \infty$

$$\int_{C_1'} \int_{\Gamma_{\frac{1}{2}}} |\tilde{u}_\lambda(z_1)|^2 \, |\tilde{v}_\lambda(z_2)|^2 \, (1+|z_1|)^{2k_i} \, |dz_1| \, |dz_2| =$$

$$= \int_{\Gamma_{-\frac{1}{2}}} |\tilde{u}_\lambda \circ \Phi_1^{-1}(w_1)|^2 \, (1+|\Phi_1^{-1}(w_1)|)^{2k_i} \left|\frac{Dz_1}{Dw_1}\right| |dw_1| \int_{\Gamma_{\frac{1}{2}}} |\tilde{v}_\lambda \circ \Phi_2^{-1}(w_2)|^2 \left|\frac{Dz_2}{Dw_2}\right| |dw_2|$$

$$\leq \lambda^2 \int_{\Gamma_{-\frac{1}{2}}} |\tilde{u}_1 \circ \Phi_1^{-1}(w_1)|^2 (1+|\Phi_1^{-1}(w_1)|)^{2k_i} \left|\frac{Dz_1}{Dw_1}\right| |dw_1| \int_{\Gamma_{\frac{1}{2}}} |\tilde{v}_1 \circ \Phi_2^{-1}(w_2)|^2 \left|\frac{Dz_2}{Dw_2}\right| |dw_2|$$

10. Definition. Denote by $\underline{Z}$ the set of all closed $Z \subset \mathbb{C}$ with the following properties

(i) $Z^c = Z$

(ii) $\{ z \in Z \; : \; \rho_1 \leq \mathrm{Re}\, z \leq \rho_2 \}$ is compact for all $\rho_1, \rho_2 \in \mathbb{R}$

(iii) for any $\eta, \delta \in \mathbb{R}_+$ the set of points $B_{\delta\eta}(Z)$ defined as in 2.4.1. Definition 9 is a piecewise smooth curve.

It is obvious that every $Z \in \underline{Z}$ can be written in the form

$$Z = Z^1 + Z^2, \; Z^i \in \underline{Z}^d, \; i = 1,2.$$

11. Definition. Denote by $\mathfrak{M}_Z$ the space $\mathfrak{M}_Z = \mathfrak{M}_{Z^1} + \mathfrak{M}_{Z^2}$ for $Z^1, Z^2 \; \underline{Z}^d, \; Z^1 + Z^2 = Z$.

Proposition 12 below shows that the definition is correct. In view of Proposition 1 $\mathfrak{M}_Z$ is a nuclear Fréchet space, since $\mathfrak{M}_{Z^i}$, $i = 1,2$, are nuclear Fréchet spaces. Write $\mathfrak{M}_{\underline{Z}} = \varinjlim \mathfrak{M}_Z$ for the inductive limit over all $Z \in \underline{Z}$. In 2.1.3. was associated a space $\mathfrak{M}_{(Z)}$ with each $Z \in \underline{Z}$.

12. Proposition. For each $Z \in \underline{Z}$ we have $\mathfrak{M}_Z = \mathfrak{M}_{(Z)}$ and the topologies are equivalent. In particular $\mathfrak{M}_Z$ is independent of the choice of the decomposition of Z.

The proof is based on the decomposition principle. The arguments of the proof of Proposition 6 are repeated with obvious modifications. For arbitrary $Z^i \in \underline{Z}$, $i = 1,\dots,p$, $Z := \sum_{i=1}^{p} Z^i$ we have

$$\mathfrak{M}_Z = \sum_{i=1}^{p} \mathfrak{M}_{Z^i} \quad .$$

Now consider Mellin operators with symbols in $\mathfrak{M}_Z$, $Z \in \underline{Z}$. First we omit the assumption that $Z \in \underline{Z}^d$ but only suppose that $\Gamma_{\frac{1}{2}+\gamma} \cap Z = \emptyset$ for some $\gamma \in \mathbb{C}$, Re $\gamma \geqq 0$. In a second step the latter condition will also be dropped. The second step is based on the first one by applying the decomposition principle in the sense of Definition 11 and Proposition 12.

Let $h \in \mathfrak{M}_Z$, $Z \in \underline{Z}$, $\Gamma_{\frac{1}{2}} \cap T^{-\gamma} Z = \emptyset$, Re $\gamma \geqq 0$. Fix some $\alpha \in \mathbb{C}$, $0 \leqq$ Reγ $\leqq$ Re α , and define the operator

$$\mathcal{M} := \omega\, t^{\alpha-\gamma} \;\; op_M(T^{\gamma} h) t^{\gamma} \omega \; , \tag{4}$$

say on $L^2(\mathbb{R}_+)$.

For the continuity of (4) in spaces with asymptotics we need

13. __Proposition.__ The multiplication induces mappings

$$M(\mathcal{S}_\Delta) \times \mathfrak{M}_Z \longrightarrow \mathfrak{M}_{Z+\Delta} \, , \, \mathfrak{M}_Z \times \mathfrak{M}_{\tilde{Z}} \longrightarrow \mathfrak{M}_{Z+\tilde{Z}}$$

$\Delta \in \underline{\Delta}$, $Z, \tilde{Z} \in \underline{Z}$, which are continuous separately in the arguments.

__Proof:__ In view of Proposition 6 and Proposition 12 $M(\mathcal{S}_\Delta)$ and $\mathfrak{M}_Z$ consist of holomorphic functions outside Δ and Z which are rapidly decreasing for $|\text{Im } z| \longrightarrow \infty$ in each strip $c_1 \leq$ Re $z \leq c_2$. Then the product is holomorphic outside $\Delta + Z$ and also rapidly decreasing. The continuity in each argument is obvious. The second assertion follows in the same way. □

Remark that another proof can be given by using the decomposition principle and 2.2.1. Proposition 14. Then we immediately obtain

14. __Proposition.__ Let $Z \in \underline{Z}$, $\Gamma_{\frac{1}{2}} \cap T^{-\gamma} Z = \emptyset$, $0 \leqq$ Re $\gamma \leqq$ Re α and $h \in \mathfrak{M}_Z$. Then

$$\omega\, t^{\alpha-\gamma} op_M(T^{\gamma} h) \; t^{\gamma} \omega \;\; : \;\; \mathcal{S}_\Delta \longrightarrow \mathcal{S}_{T^{-\alpha}(\Delta + Z_{(-\gamma)})}$$

is continuous for every $\Delta \in \underline{\Delta}$.

Let $Z \in \underline{Z}$ be arbitrary and $h \in \mathfrak{M}_Z$. For given $\alpha \in \mathbb{C}$, Re $\alpha > 0$, choose $\gamma_i \in \mathbb{C}$, $0 \leq$ Re $\gamma_i \leq$ Re α, $Z^i \in \underline{Z}$, $i = 1, \ldots, p$ such that $\Gamma_{\frac{1}{2}} \cap T^{-\gamma_i} Z^i = \emptyset$, $Z = \sum_{i=1}^{p} Z^i$. Then

$$h = \sum_{i=1}^{p} h_i, \text{ where } h_i \in \mathfrak{M}_{Z^i}$$

and

$$\mathcal{M}_i := \omega t^{\alpha-\gamma_i} op_M(T^{\gamma_i} h_i) t^{\gamma_i} \omega \qquad (5)$$

define continuous mappings

$$\mathcal{M}_i : \mathcal{S}_\Delta \longrightarrow \mathcal{S}_{T^{-\alpha}(\Delta + Z^i_{(-\gamma_i)})}, \quad \Delta \in \underline{\Delta} .$$

Then

$$\mathcal{M} := \sum_{i=1}^{p} \mathcal{M}_i : \mathcal{S}_\Delta \longrightarrow \sum_{i=1}^{p} \mathcal{S}_{T^{-\alpha}(\Delta + Z^i_{(-\gamma_i)})} = \mathcal{S}_{T^{-\alpha}(\Delta + \sum_{i=1}^{p} Z^i_{(-\gamma_i)})}$$

is continuous for each $\Delta \in \underline{\Delta}$. We call $\mathcal{M}$ a <u>Mellin operator of conormal order $-\alpha$</u> associated with $h \in \mathfrak{M}_Z$. The choice of the decomposition $h = \sum_{i=1}^{p} h_i$ with the above properties does not affect the resulting operator modulo $\mathcal{G}_G$. In order to show this we use the following

15. <u>Lemma</u>. Let $Z \in \underline{Z}$, $\alpha, \gamma, \lambda \in \mathbb{C}$, $0 \le \mathrm{Re}\,\gamma \le \mathrm{Re}\,\lambda \le \mathrm{Re}\,\alpha$, $\Gamma_{\frac{1}{2}} \cap T^{-\gamma} Z = \emptyset$, $\Gamma_{\frac{1}{2}} \cap T^{-\lambda} Z = \emptyset$ and $n \in \mathfrak{M}_Z$. Then

$$\omega t^{\alpha-\gamma} op_M(T^{\gamma} n) t^{\gamma} \omega - \omega t^{\alpha-\lambda} op_M(T^{\lambda} n) t^{\lambda} \omega \in \mathcal{G}_G,$$

where $\mathcal{G}_G = op_G(\mathcal{S}_{\underline{\Delta}} \otimes_\Gamma \mathcal{S}_{\underline{\Delta}})$.

The proof immediately follows from 2.1.4. Corollary 12 and the decomposition principle. We refer to Lemma 15 as to the <u>translation principle</u>. The formula for the resulting singularity types of the Green symbol is the same as in 2.2.1.

16. <u>Proposition</u>. Let $h \in \mathfrak{M}_Z$, $Z \in \underline{Z}$, $\alpha \in \mathbb{C}$, $\mathrm{Re}\,\alpha > 0$ and $h = \sum_{j=1}^{q} \tilde{h}_j$, $\tilde{h}_j \in \mathfrak{M}_{\tilde{Z}^j}$, $Z = \sum_{j=1}^{q} \tilde{Z}^j$, $\Gamma_{\frac{1}{2}} \cap T^{-\tilde{\gamma}_j} \tilde{Z}^j = \emptyset$, $0 \le \mathrm{Re}\,\tilde{\gamma}_j \le \mathrm{Re}\,\alpha$ be another decomposition of h and $\tilde{\mathcal{M}}$ the corresponding Mellin operator. Then

$$\mathcal{M} - \tilde{\mathcal{M}} \in \mathcal{G}_G.$$

<u>Proof</u>: Let $\delta_0, \delta_1 \in \mathbb{C}$, $0 \le \mathrm{Re}\,\delta_0 < \mathrm{Re}\,\delta_1 \le \mathrm{Re}\,\alpha$. Then each $h_i \in \mathfrak{M}_{Z^i}$ can be decomposed

$h_i = h_{i,o} + h_{i,1}$, $h_{i,o} \in \mathfrak{M}_{Z^{i,o}}$, $h_{i,1} \in \mathfrak{M}_{Z^{i,1}}$ where $Z^{i,o}, Z^{i,1} \in \underline{Z}$

$Z^i = Z^{i,o} + Z^{i,1}$, $\Gamma_{\frac{1}{2}} \cap T^{-\delta_o} Z^{i,o} = \emptyset$, $\Gamma_{\frac{1}{2}} \cap T^{-\delta_1} Z^{i,1} = \emptyset$.

Then

$$\omega t^{\alpha-\gamma_j} op_M(T^{\gamma_j} h_j) t^{\gamma_j} \omega = \omega t^{\alpha-\delta_o} op_M(T^{\delta_o} h_{j,o}) t^{\delta_o} \omega + \omega t^{\alpha-\delta_1} op_M(T^{\delta_1} h_{j,1}) t^{\delta_1} \omega \mod \mathcal{G}_G$$

according to the translation principle.

Let $\tilde{h}_j = \tilde{h}_{j,o} + \tilde{h}_{j,1}$ be the corresponding decomposition of $\tilde{h}_j$. Then

$$\sum_{j=1}^{p} h_{j,o} - \sum_{k=1}^{q} \tilde{h}_{k,o} = \sum_{k=1}^{q} \tilde{h}_{k,1} - \sum_{j=1}^{p} h_{j,1}$$

and both sides of the latter equation are holomorphic near $\Gamma_{\frac{1}{2}+\delta_i}$, $i=1,2$.

Thus

$$\mathcal{M} - \tilde{\mathcal{M}} = \omega t^{\alpha-\delta_o} op_M(T^{\delta_o}\{\sum_{j=1}^{p} h_{j,o} - \sum_{k=1}^{q} \tilde{h}_{k,o}\}) t^{\delta_o} \omega$$

$$+ \omega t^{\alpha-\delta_1} op_M(T^{\delta_1}\{\sum_{k=1}^{q} \tilde{h}_{k,1} - \sum_{j=1}^{p} h_{j,1}\}) \mod \mathcal{G}_G$$

vanishes modulo $\mathcal{G}_G$ in view of the translation principle. $\square$

17. **Lemma.** Let $h \in \mathfrak{M}_Z$, $Z \in \underline{Z}$, $\Gamma_{\frac{1}{2}} \cap Z = \emptyset$, $\Lambda := Z_{(o)}$, $\Lambda' := Z'_{(o)}$, and let ω_i, $i = 1,2,3,4$, be cut-off functions. Then the relations of 2.2.1. Lemma 10 (i),....,(v) hold.

This can easily be proved by the decomposition argument and 2.2.1. Lemma 10.

Another consequence of decomposition arguments is that

$$\mathcal{M}\mathcal{G}, \mathcal{G}\mathcal{M} \in \mathcal{G}_G \tag{6}$$

for any $\mathcal{G} \in \mathcal{G}_G$ and $\mathcal{M}$ of the form (4). Clearly the singularity types of the Green operators in (6) can easily be expressed in terms of the singularity types in $\mathcal{G}$ and Z, γ.

Often we will use the notation

$$op_M^{-\alpha}(h) := \sum_{i=1}^{p} t^{\alpha-\gamma_i} op_M(T^{\gamma_i} h_i) t^{\gamma_i}$$

with α, γ_i, h_i as in (5). This is a well-defined action on L^2 functions on $\mathbb{R}_+$ with bounded support. Then, in view of Proposition 16, Lemma 17,

$$\mathfrak{m} = \omega\, op_M^{-\alpha}(h)\,\omega$$

is determined by h modulo $\mathcal{G}_G$. For $\alpha = 0$ we simply set $op_M^0(h) = op_M(h)$. We call h the conormal symbol of $\mathfrak{m}$ (of conormal order $-\alpha$) and write

$$\sigma_M^{-\alpha}(\mathfrak{m})(z) = h(z) = \sum_{j=1}^{p} h_j(z) \quad .$$

Denote by $\omega op_M^{-\alpha}(\mathfrak{M}_Z)\omega$ the space of all Mellin operators of conormal order $-\alpha$ with conormal symbol $h \in \mathfrak{M}_Z$.

If $\mathfrak{m} \in \omega op_M^{-\alpha}(\mathfrak{M}_Z)\omega$ and $\mathfrak{m}^*$ is the formal L^2 adjoint, then $\mathfrak{m}^* \in \omega op_M^{-\bar{\alpha}}(\mathfrak{M}_{T^{-\bar{\alpha}}Z^*})\omega$ and $\sigma_M^{-\bar{\alpha}}(\mathfrak{m}^*) = T^{-\bar{\alpha}}\ \sigma_M^{-\alpha}(\mathfrak{m})^*$.

Now let us calculate compositions. First assume $h \in \mathfrak{M}_Z$, $n \in \mathfrak{M}_W$, $\Gamma_{\frac{1}{2}} \cap Z = \Gamma_{\frac{1}{2}} \cap W = \emptyset$ and consider

$$\mathfrak{n} = \omega\, op_M(h)\,\omega\, t^{\varkappa}\ \omega\, op_M(n)\omega,$$

$\varkappa \in \mathbb{C}$, $\operatorname{Re}\varkappa \geq 0$. Set for $\lambda \in \mathbb{C}$, $\varepsilon > 0$

$$S_{\lambda,\varepsilon} = \left\{ z \in \mathbb{C} : \tfrac{1}{2} - \varepsilon < \operatorname{Re}(z-\lambda) < \tfrac{1}{2} + \varepsilon \right\}$$

and define

$$Z^1 = Z \setminus S_{-\frac{\varkappa}{2},\varepsilon} \qquad Z^2 = Z \cap \bar{S}_{-\frac{\varkappa}{2},2\varepsilon}\ ,$$

$$W^1 = W \setminus S_{\frac{\varkappa}{2},\varepsilon} \qquad W^2 = W \cap \bar{S}_{\frac{\varkappa}{2},2\varepsilon}\ ,$$

$\varepsilon > 0$ sufficiently small. Then $Z = Z^1 + Z^2$, $W = W^1 + W^2$, and a decomposition argument yields

$$h = h_1 + h_2,\ h_i \in \mathfrak{M}_{Z^i},$$

$$n = n_1 + n_2,\ n_i \in \mathfrak{M}_{W^i},\quad i = 1,2\ .$$

Thus $\mathfrak{n}$ equals

$$\mathfrak{n} = \mathfrak{n}_1 + \mathfrak{n}_2 + \mathfrak{n}_3 + \mathfrak{n}_4$$

with

$$\mathfrak{n}_1 = \omega\, op_M(h_1)\,\omega\, t^{\varkappa}\ \omega\ op_M(n_1)\,\omega\ ,$$
$$\mathfrak{n}_2 = \omega\, op_M(h_1)\,\omega\, t^{\varkappa}\ \omega\ op_M(n_2)\,\omega\ ,$$
$$\mathfrak{n}_3 = \omega\, op_M(h_2)\,\omega\, t^{\varkappa}\ \omega\ op_M(n_1)\,\omega\ ,$$
$$\mathfrak{n}_4 = \omega\, op_M(h_2)\,\omega\, t^{\varkappa}\ \omega\ op_M(n_2)\,\omega\ .$$

Since

$$\Gamma_{\frac{1}{2}} \cap T^{\frac{\varkappa}{2}} Z^1 = \emptyset, \quad \Gamma_{\frac{1}{2}} \cap T^{-\frac{\varkappa}{2}} W^1 = \emptyset ,$$

a translation argument yields

$$\mathcal{M}\mathcal{O}_1 = \omega t^{\frac{\varkappa}{2}} op_M(T^{-\frac{\varkappa}{2}} h_1) \omega^2 op_M(T^{\frac{\varkappa}{2}} n_1) t^{\frac{\varkappa}{2}} \omega$$

modulo Green operators. Applying Lemma 17 and (6) we may omit ω^2 modulo Green terms and then

$$\mathcal{M}\mathcal{O}_1 = \omega t^{\frac{\varkappa}{2}} op_M(T^{-\frac{\varkappa}{2}} h_1 T^{\frac{\varkappa}{2}} n_1) t^{\frac{\varkappa}{2}} \omega \bmod \mathcal{C}_G .$$

In order to calculate $\mathcal{M}\mathcal{O}_2$ we observe that $\Gamma_{\frac{1}{2}} \cap T^{-\varkappa} W^2 = \emptyset$ for $\varepsilon < \mathrm{Re}\,\frac{\varkappa}{4}$. Then translation yields modulo $\mathcal{C}_G$

$$t^{\varkappa} \omega\, op_M(n_2) \omega = \omega\, op_M(T^{\varkappa} n_2) \omega\, t^{\varkappa}$$

and hence similarly as before

$$\mathcal{M}\mathcal{O}_2 = \omega\, op_M(h_1 T^{\varkappa} n_2) t^{\varkappa} \bmod \mathcal{C}_G .$$

In a similar way we argue for $\mathcal{M}\mathcal{O}_4$, i.e.

$$\mathcal{M}\mathcal{O}_4 = \omega\, op_M(h_2 T^{\varkappa} n_2) \omega\, t^{\varkappa}$$

$$= \omega\, t^{\varkappa} op_M((T^{-\varkappa} h_2) n_2) \omega \bmod \mathcal{C}_G .$$

For $\mathcal{M}\mathcal{O}_3$ observe that $\Gamma_{\frac{1}{2}} \cap T^{\varkappa} Z^2 = \emptyset$ for $\varepsilon < \mathrm{Re}\,\frac{\varkappa}{4}$. Thus

$$\mathcal{M}\mathcal{O}_3 = \omega\, t^{\varkappa} op_M((T^{-\varkappa} h_2) n_1) \omega \bmod \mathcal{C}_G .$$

Thus we have

$$\mathcal{M}\mathcal{O} = \mathcal{M}\mathcal{O}_1 + \mathcal{M}\mathcal{O}_2 + \mathcal{M}\mathcal{O}_3 + \mathcal{M}\mathcal{O}_4 \in \omega\, op_M^{-\varkappa}(\mathcal{M}_{Z+W}) \omega + \mathcal{C}_G$$

and

$$\sigma_M^{-\varkappa}(\mathcal{M}\mathcal{O}) = T^{-\varkappa} h \cdot n .$$

Then we obtain in an obvious way

18. <u>Proposition</u>. Let $\mathcal{M} = \omega\, op_M^{-\alpha}(h_1)\omega$, $h_1 \in \mathcal{M}_X$, $X \in \underline{Z}$, $\mathcal{M}' = \omega\, op_M^{-\beta}(h_2)\omega$, $h_2 \in \mathcal{M}_Y$, $Y \in \underline{Z}$. Then

$$\mathcal{M}\mathcal{M}' = \omega\, op_M^{-\alpha-\beta}(T^{-\beta} h_1 \cdot h_2) \omega \bmod \mathcal{C}_G .$$

This proposition shows that the composition law for conormal symbols from 1.3.2. remains valid in the present more general case.

The final class of Mellin operators with conormal symbols in $\mathfrak{M}_{\underline{Z}}$ consists of infinite sums of Mellin operators of conormal order α_j, $\operatorname{Re}\alpha_j \longrightarrow +\infty$ for $j \longrightarrow \infty$ (cf. 1.3.1.). In the applications to mixed problems we only need $\alpha_j = j$, $j \in \mathbb{Z}_+$. For the same reasons as in 1.3.1. we assume that the numbers γ_{ij}, which are involved in the definition of op_M^{-j}, are real and satisfy

$$\gamma_{ij} \longrightarrow \infty \ , \ j - \gamma_{ij} \longrightarrow \infty \quad \text{as } j \longrightarrow \infty \quad \text{for each } i \ .$$

As shown in the proof of Proposition 14 we may restrict ourselves to decompositions of $Z \in \underline{Z}$ into a sum of <u>two</u> subsets : $Z = Z^o + Z^1$. For arbitrary given $Z^j \in \underline{Z}$, $j \in \mathbb{Z}_+$, where Z^o satisfies $\Gamma_{\frac{1}{2}} \cap Z^o = \emptyset$, we consider an arbitrary sequence $h_j \in \mathfrak{M}_{Z^j}$. Choose two sequences $\{\gamma_{j,o}\}_{j\geq 1}$, $\{\gamma_{j,1}\}_{j\geq 1}$ of real non-negative numbers such that $j - \gamma_{j,i} \geq 0$, $i = 1,2$,

$$j - \gamma_{j,i} \longrightarrow \infty \ , \ \gamma_{j,i} \longrightarrow \infty \quad \text{as } j \longrightarrow \infty \ , \ i = 0,1.$$

For given decompositions $Z^j = Z^{j,o} + Z^{j,1}$, $\Gamma_{\frac{1}{2}} \cap T^{-\gamma_{j,o}} Z^{j,o} = \emptyset$,

$\Gamma_{\frac{1}{2}} \cap T^{-\gamma_{j,1}} Z^{j,1} = \emptyset$ we obtain $h_{j,i} \in Z^{j,i}$, $h_{j,o} + h_{j,1} = h_j$, and

$$\begin{aligned} \omega\, op_M^{-j}(h_j)\omega = {} & \omega t^{j-\gamma_{j,o}} op_M(T^{\gamma_{j,o}} h_{j,o}) t^{\gamma_{j,o}} \omega \ + \\ & + \omega t^{j-\gamma_{j,1}} op_M(T^{\gamma_{j,1}} h_{j,1}) t^{\gamma_{j,1}} \omega \end{aligned}$$

defines a continuous operator

$$\mathcal{S}_\Delta \longrightarrow \mathcal{S}_{T^{-j}(\Delta + (Z^{j,o})_{(-\gamma_{j,o})} + (Z^{j,1})_{(-\gamma_{j,1})})}, \ \Delta \in \underline{\Delta} \ .$$

Since $j - \gamma_{j,i} \longrightarrow \infty$ for $j \longrightarrow \infty$, $i = 0,1$, only a finite number of operators $\omega\, op_M^{-j}(h_j)\omega$ in the sequence contribute to the resulting singularity type in each fixed strip $\frac{1}{2} - c < \operatorname{Re} z < \frac{1}{2}$. The adjoints

$$\begin{aligned} (\omega\, op_M^{-j}(h_j)\omega)^* = {} & \omega t^{\gamma_{j,o}} op_M(T^{\gamma_{j,o}} h_{j,o})^* \ t^{j-\gamma_{j,o}} \omega \ + \\ & + \omega t^{\gamma_{j,1}} op_M(T^{\gamma_{j,1}} h_{j,1})^* \ t^{j-\gamma_{j,1}} \omega \end{aligned}$$

define continuous operators

$$\mathcal{S}_\Delta \longrightarrow \mathcal{S}_{T^{-j}\Delta + (Z^{j,o})^*(\gamma_{j,o}) + (Z^{j,1})^*(\gamma_{j,1})} .$$

Since $\gamma_{j,1} \longrightarrow \infty$ as $j \longrightarrow \infty$, $i = 0,1$, the resulting singularity in each strip $\frac{1}{2} - c < \operatorname{Re} z < \frac{1}{2}$ is affected only by a finite number of operators $(\omega \, op_M^{-j}(h_j)\omega)^*$ in the sequence. The following result can be proved in the same way as 1.3.1. Proposition 15.

19. Proposition. Let $h_j \in \mathfrak{M}_{Z^j}$, $Z \in \underline{Z}$, $\Gamma_{\frac{1}{2}} \cap Z^o = \emptyset$, $j \in \mathbb{Z}_+$ be an arbitrary sequence. If $op_M^{-j}(h_j)$ is defined with $\gamma_{j,i}$ as above, $op_M^0 := op_M$, one can choose constants $c_j > 0$ such that

$$\mathcal{M} = \sum_{j=0}^{\infty} \omega(c_j t) \, op_M^{-j}(h_j) \, \omega(c_j t)$$

converges in $\mathcal{L}^0(\mathcal{S}_\Delta, \mathcal{S}_{\sum_j T^{-j}(\Delta + (Z^{j,o})(\gamma_{j,o}) + (Z^{j,1})(\gamma_{j,1}))})$ for all $\Delta \in \underline{\Delta}$

and

$$\mathcal{M}^* = \sum_{j=0}^{\infty} \omega(c_j t)(op_M^{-j}(h_j))^* \, \omega(c_j t)$$

converges in $\mathcal{L}^0(\mathcal{S}_\Delta, \mathcal{S}_{\sum_j \{T^{-j}\Delta + (Z^{j,o})^*(\gamma_{j,o}) + (Z^{j,1})^*(\gamma_{j,1})\}})$

for all $\Delta \in \underline{\Delta}$. Another choice of ω and c_j preserves $\mathcal{M}$ modulo $\mathcal{G}_G$.

20. Definition. Denote by $\mathcal{G}_M + \mathcal{G}_G$ the space of all operators in $\mathcal{L}(L^2(\mathbb{R}_+))$ of the form $\mathcal{M} + \mathcal{g}$, where $\mathcal{g} \in \mathcal{G}_G$ and $\mathcal{M}$ is a Mellin operator of a form which occurs in Proposition 17. The function

$$\sigma_M^{-j}(\mathcal{M} + \mathcal{g})(z) := h_j(z)$$

is called conormal symbol of order $-j$.

Note that the sequence of conormal symbols is uniquely determined by the operation and does not depend on $\mathcal{g} \in \mathcal{G}_G$. In fact, the conclusions of 1.3.1. Theorem 13 can be applied with obvious modifications. As in 1.3.1. Corollary 16 we have the exact sequence

$$0 \longrightarrow \mathcal{G}_G \longrightarrow \mathcal{G}_M + \mathcal{G}_G \xrightarrow{\bigoplus_{j \in \mathbb{Z}_+} \sigma_M^{-j}} \mathfrak{M}_{\underline{Z}^o} \bigoplus_{j \geq 1} \mathfrak{M}_{\underline{Z}} \longrightarrow 0 ,$$

where $\mathfrak{M}_{\underline{Z}^o}$ denotes the inductive limit over all $\mathfrak{M}_{Z^o}$, $Z^o \in \underline{Z}$, $\Gamma_{\frac{1}{2}} \cap Z^o = \emptyset$.

In the same way as 1.3.2. Proposition 3 we get from 2.2.1. Proposition 2 and Proposition 16

21. <u>Theorem</u>. $\mathcal{G}_M + \mathcal{G}_G$ is a $*$ algebra, $\mathcal{G}_G$ is a two-sided ideal in $\mathcal{G}_M + \mathcal{G}_G$, and

$$\sigma_M^{-j}(\mathcal{M}\mathcal{M}') = \sum_{k+l=j} T^{-l}\sigma_M^{-k}(\mathcal{M})\,\sigma_M^{-l}(\mathcal{M}') .$$

Now we define an algebra $\mathcal{G}(\mathbb{R}^d)$ of families of operators depending on the parameter $\eta \in \mathbb{R}^d$ (cf. the analogous definitions in 1.5.3). By $\mathcal{G}_G(\mathbb{R}^d)$ we denote the set of all families $op_G(b)(\eta)$ with

$$b(t,s,\eta) \in C^\infty(\mathbb{R}^d, \mathcal{S}_{\underline{\Delta}} \otimes_\Gamma \mathcal{S}_{\underline{\Delta}})$$

and

$$op_G(b)(\lambda\eta) = \varkappa_\lambda\, op_G(b)(\eta)\, \varkappa_\lambda^{-1}$$

for $\lambda \geq 1$ and $\eta \in \mathbb{R}^d$, $|\eta| \geq$ const., where the constant may depend on b, $(\varkappa_\lambda u)(t) = \lambda^{1/2} u(\lambda t)$.

Choose a function $\zeta \in C^\infty(\mathbb{R}_+)$, $\zeta > 0$,

$$\zeta(t) = \begin{cases} \frac{1}{2} & \text{for } t > \frac{1}{2}, \\ t & \text{for } t > 1, \end{cases}$$

and set

$$\omega_{|\eta|,j}(t) = \omega(c_j t\, \zeta(|\eta|)),$$

$c_j > 0$ some constant depending on $j \in \mathbb{Z}_+$.

Denote by $\mathfrak{M}_{Z}^{(k)}[\eta]$ the space of homogeneous polynomials of order k in $(\eta_1,\dots,\eta_d)$ with coefficients in $\mathfrak{M}_{\underline{Z}}$. Then Proposition 19 may be extended in the following way.

22. <u>Proposition</u>. Let $h_j \in \mathfrak{M}_{Z^j}^{(j)}[\eta]$, $Z^j \in \underline{Z}$, $j \in \mathbb{Z}_+$, be an arbitrary sequence, $\Gamma_{\frac{1}{2}} \cap Z^o = \emptyset$, and define $op_M^{-j}(h_j)(\eta)$ as above. Then there are constants $c_j > 0$ such that

$$\mathcal{M}(\eta) := \sum_{j=0}^{\infty} \omega_{|\eta|,j}(t)\, op_M^{-j}(h_j)(\eta)\, \omega_{|\eta|,j}(t)$$

converges in $\mathcal{L}^o(\mathcal{S}_\Delta, \mathcal{S}_{h(\Delta)})$ for all $\Delta \in \underline{\Delta}$, $h(\Delta) := \sum_{j=0}^{\infty} T^{-j}(\Delta + (Z^{j,0})_{(\cdot\gamma_{j,0})} + (Z^{j,1})_{(\cdot\gamma_{j,1})})$ uniformly in $\eta \in K \subset\subset \mathbb{R}^d$, and

$$\mathcal{M}^*(\eta) := \sum_{j=0}^{\infty} \omega_{|\eta|,j}(t)(\mathrm{op}_M^{-j}(h_j))^*(\eta)\,\omega_{|\eta|,j}(t)$$

converges in $\mathcal{L}^o(\mathcal{S}_\Delta, \mathcal{S}_{h^*(\Delta)})$ for all $\Delta \in \underline{\Delta}$, $h^*(\Delta) := \sum_{j=0}^{\infty}\{T^{-j}\Delta + (Z^{j,0})^*_{(\gamma_{j,0})} + (Z^{j,1})^*_{(\cdot\gamma_{j,1})}\}$, uniformly in $\eta \in K \subset\subset \mathbb{R}^d$. Another choice of ω and c_j preserves $\mathcal{M}(\eta)$ modulo $\mathcal{G}_G(\mathbb{R}^d)$.

The proof may be given in the same way as in Proposition 19, where obvious arguments yield in addition uniform convergence on compact subsets of $\mathbb{R}^d \ni \eta$.

In Section 3.2.3. the convergence of sums of parameter-depending Mellin operators will be discussed from the point of view of semi-norms of spaces of operator-valued pseudo-differential symbols.

Denote by $\mathcal{G}_M(\mathbb{R}^d)$ the set of all families of operators $\mathcal{M}(\eta)$ as in Proposition 22.

Finally let us define the space $\mathcal{G}_\psi(\mathbb{R}^d)$ of all operators $\mathrm{op}_\psi(a)(\eta)$, where $a(\xi) \in S^o(\mathbb{R}^{d+1})$ is arbitrary, $a(\lambda\xi) = a(\xi)$ for $\lambda \geq 1$, $|\xi| \geq$ const. $\xi = (\eta, \tau)$. Set

$$\mathcal{G}(\mathbb{R}^d) = \mathcal{G}_\psi(\mathbb{R}^d) + \mathcal{G}_M(\mathbb{R}^d) + \mathcal{G}_G(\mathbb{R}^d) .$$

From the Mellin expansion of a ψDO in $\mathcal{G}_\psi(\mathbb{R}^d)$ (cf. 1.4.1. Theorem 5) we obtain the following generalization of 1.4.2. Theorem 1.

23. __Theorem.__ $\mathrm{op}_\psi(a) \in \mathcal{G}_\psi(\mathbb{R}^d)$, $\eta \in \mathbb{R}^d$ fixed, induces a continuous operator

$$\mathrm{op}_\psi(a)(\eta) : \mathcal{S}_\Delta \longrightarrow \mathcal{S}_{\sum_{j=0}^{\infty} T^{-j}\Delta + \Lambda^o} \tag{7}$$

for every $\Delta \in \underline{\Delta}$ of order zero and we have a continuous map $a \longrightarrow \mathcal{L}^o(\mathcal{S}_\Delta, \mathcal{S}_{\sum T^{-j}\Delta + \Lambda^o})$, where the space of all symbols a is considered in the natural Fréchet topology.

$\mathrm{op}_\psi(a)(\eta)$ has a sequence of conormal symbols

$$\sigma_M^{-j}(a)(\eta) \in \mathfrak{M}^{\psi(j)}[\eta] \;, \quad j \in \mathbb{Z}_+ \;.$$

The formulas are the same as in Section 1.4.1. (cf. formulas (12),(13)). For $\alpha = \mathrm{op}_\psi(a) + \mathcal{M} + \mathcal{G} \in \mathcal{Y}(\mathbb{R}^d)$ we define $\sigma_\psi(\alpha)(\xi)$, $\xi = (\eta,\tau)$, as the homogeneous extension of $a \,\big|_{|\xi| \geq \text{const}}$ to $\mathbb{R}^{d+1} \setminus 0$ and

$$\sigma_M^{-j}(\alpha)(\eta,z) = \sigma_M^{-j}(\mathrm{op}_\psi(a))(\eta,z) + \sigma_M^{-j}(\mathcal{M})(\eta,z) \;.$$

Then

$$\sigma_M^{-j}(\alpha)(\eta,z) \in (\mathfrak{M}_{\underline{Z}} + \mathfrak{M}^{\psi})^{(j)}[\eta] \;, \quad j \in \mathbb{Z}_+ \;.$$

24. <u>Theorem.</u> $\mathcal{Y}(\mathbb{R}^d)$ is a $*$ algebra, $(\mathcal{Y}_M + \mathcal{Y}_G)(\mathbb{R}^d)$ and $\mathcal{Y}_G(\mathbb{R}^d)$ are two-sided ideals in $\mathcal{Y}(\mathbb{R}^d)$. For $\alpha_1, \alpha_2 \in \mathcal{Y}(\mathbb{R}^d)$ we have

$$\sigma_\psi(\alpha_1 \alpha_2)(\xi) = \sigma_\psi(\alpha_1)(\xi)\, \sigma_\psi(\alpha_2)(\xi) \;, \tag{8}$$

$$\sigma_M^{-l}(\alpha_1\alpha_2)(\eta,z) = \sum_{j+k=l} (T^{-k}\, \sigma_M^{-j}(\alpha_1)(\eta,z))\, \sigma_M^{-k}(\alpha_2)(\eta,z) \tag{9}$$

$l \in \mathbb{Z}_+$. In particular, if $\mathcal{G}(\eta) \in \mathrm{op}_G(\mathcal{S}_\Gamma \otimes_\pi J\mathcal{S}_\Delta)$, $\Gamma, \Delta \in \underline{\Delta}$,

$$\mathrm{op}_\psi(a)(\eta)\, \mathcal{G}(\eta) \in \mathrm{op}_G(\mathcal{S}_{\sum T^{-j}\Gamma + \Lambda^0} \otimes_\pi J\mathcal{S}_\Delta) \tag{10}$$

$$\mathcal{G}(\eta)\, \mathrm{op}_\psi(a)(\eta) \in \mathrm{op}_G(\mathcal{S}_\Gamma \otimes_\pi J\mathcal{S}_{\sum T^{-j}\Delta + \Lambda^0}) \;. \tag{11}$$

<u>Proof:</u> First observe that (10), (11) is a consequence of (7), 2.2.1. Proposition 2 and

$$C^\infty(\Omega, \mathcal{S}_\Delta \otimes_\pi \mathcal{S}_{\underline{\Xi}}) = C^\infty(\Omega, L_2 \otimes_\pi \mathcal{S}_{\underline{\Xi}}) \cap C^\infty(\Omega, \mathcal{S}_\Delta \otimes_\pi L^2),$$

$\Omega \subseteq \mathbb{R}^d$ open, $\Delta, \Xi \in \underline{\Delta}$, cf. Lemma 3. Applying Theorem 21 we obtain that $\mathcal{Y}_G(\mathbb{R}^d)$ is a two-sided ideal in $\mathcal{Y}(\mathbb{R}^d)$. It remains to check that

$$\mathcal{Y}_\psi(\mathbb{R}^d)\, \mathcal{Y}_M(\mathbb{R}^d),\; \mathcal{Y}_M(\mathbb{R}^d)\, \mathcal{Y}_\psi(\mathbb{R}^d) \subseteq \mathcal{Y}_M(\mathbb{R}^d) + \mathcal{Y}_G(\mathbb{R}^d)$$

and that we have the asserted rule for the symbols. This is an immediate application of 2.2.1. Theorem 24 to decompositions of the Mellin operators $\mathcal{M} = \mathcal{M}^0 + \mathcal{M}^1$, $\mathcal{M}^i$ satisfying the admissibility conditions in 2.2.1. $\square$

2.2.3. The Algebra $\mathcal{G}(\Omega \times \mathbb{R}^d)$ and Ellipticity

In the following we shall extend the algebra $\mathcal{G}(\mathbb{R}^d)$ by admitting extra parameters y varying in an open set $\Omega \subseteq \mathbb{R}^m$. This gives rise to a class $\mathcal{G}(\Omega \times \mathbb{R}^d)$. We shall study some effects that occur later on in an analogous form in the boundary symbolic calculus of operators in higher dimensions. In this sense $\mathcal{G}(\Omega \times \mathbb{R}^d)$ is a model for references in Chapter 3.

Denote by $\mathcal{G}_G(\Omega \times \mathbb{R}^d)$ the set of all families $op_G(b)(y,\eta)$ with
$$b(y,\eta,t,s) \in C^\infty(\Omega \times \mathbb{R}^d, \mathcal{S}_{\underline{\Delta}} \hat{\otimes}_\pi \mathcal{S}_{\underline{\Delta}})$$
and
$$op_G(b)(y,\lambda\eta) = \varkappa_\lambda op_G(b)(y,\eta)\, \varkappa_\lambda^{-1}$$
for $\lambda \geq 1$, $\eta \in \mathbb{R}^d$, $|\eta| \geq c$ with a constant $c > 0$ depending on b.

Next consider the space $C^\infty(\Omega, \mathcal{M}_Z^{(k)}[\eta]) =: C^\infty(\Omega, \mathcal{M}_Z)^{(k)}[\eta]$ of all homogeneous polynomials of order k in η with coefficients in $C^\infty(\Omega, \mathcal{M}_Z)$, $Z \in \underline{Z}$. Since $C^\infty(\Omega, \mathcal{M}_Z) = C^\infty(\Omega) \hat{\otimes}_\pi \mathcal{M}_Z$, for $Z = Z^1 + Z^2$, we obviously have
$$C^\infty(\Omega, \mathcal{M}_Z)^{(k)}[\eta] = C^\infty(\Omega, \mathcal{M}_{Z^1})^{(k)}[\eta] + C^\infty(\Omega, \mathcal{M}_{Z^2})^{(k)}[\eta] .$$
Thus, if we are given sequences $Z^j \in \underline{Z}$, $h_j \in C^\infty(\Omega, \mathcal{M}_{Z^j})^{(j)}[\eta]$,
$$(\gamma_{j,i})_{j \in \mathbb{Z}_+}, \quad i = 1,2, \quad 0 \leq \gamma_{j,i} \leq j,$$
$$\gamma_{j,i} \longrightarrow \infty , \quad j - \gamma_{j,i} \longrightarrow \infty , \quad i = 1,2,$$
$$Z^j = Z^{j,o} + Z^{j,1}, \quad \Gamma_{\frac{1}{2}} \cap T^{-\gamma_{j,i}} Z^{j,i} = \emptyset,$$

$Z^o = Z^{j,o}$, we can proceed in the same manner as in 2.1.5. Proposition 19 and form convergent sums
$$m(y,\eta) = \sum_{j=o}^{\infty} \omega_{|\eta|,j}(t)\, op_M^{-j}(h_j)(y,\eta)\, \omega_{|\eta|,j}(t)$$
uniformly in $(y,\eta) \in K \subset\subset \Omega \times \mathbb{R}^d$. Denote by $\mathcal{G}_M(\Omega \times \mathbb{R}^d)$ the class of all these operators.

Let $\mathcal{G}_\psi(\Omega \times \mathbb{R}^d)$ be the class of all families of operators $op_\psi(a)(y,\eta)$ where $a(y,\xi)$ is an element in $S^o(\Omega \times \mathbb{R}^d)$, $a(y,\lambda\xi) = a(y,\xi)$ for $\lambda \geq 1$,

$|\xi| \geq$ const., $\xi = (\eta, \tau)$. Set

$$\mathcal{G}(\Omega \times \mathbb{R}^d) = \mathcal{G}_\psi(\Omega \times \mathbb{R}^d) + \mathcal{G}_M(\Omega \times \mathbb{R}^d) + \mathcal{G}_G(\Omega \times \mathbb{R}^d).$$

For $\alpha(y,\eta) = \mathrm{op}_\psi(a)(y,\eta) + \mathcal{M}(y,\eta) + g(y,\eta) \in \mathcal{G}(\Omega \times \mathbb{R}^d)$ we have the interior symbol $\sigma_\psi(\alpha)(y,\xi)$ defined as the homogeneous extension of $a|_{|\xi| \geq \mathrm{const}}$ to $\mathbb{R}^{d+1} \setminus 0$ and the sequence of conormal symbols

$$\sigma_M^{-j}(\alpha)(y,\eta) \in C^\infty(\Omega, \mathfrak{M}_{\underline{Z}} + \mathfrak{M}^\psi)^{(j)}[\eta] \ , \ j \in \mathbb{Z}_+$$

by applying the definitions of 2.2.1.

Then there is an analogue of 2.2.1. Theorem 24 with the corresponding composition rule for symbols .

Note that for every fixed (y_o, η_o) we get an algebra

$$\mathcal{G}_{(y_o,\eta_o)} := \{\alpha(y_o,\eta_o) : \alpha \in \mathcal{G}(\Omega \times \mathbb{R}^d)\}$$

and $\mathcal{G}_{(y_o,\eta_o)} = \mathcal{G}_{(y_1,\eta_1)}$ for all $y_o, y_1 \in \Omega$, $\eta_o, \eta_1 \in \mathbb{R}^d$. Hence it is justified to denote this algebra by $\mathcal{G}$ with the subalgebras $\mathcal{G}_M + \mathcal{G}_G$ and $\mathcal{G}_G$.

We have

$$\mathcal{G}(\Omega \times \mathbb{R}^d) \subset C^\infty(\Omega \times \mathbb{R}^d, \mathcal{L}(\mathcal{S}_{\underline{\Delta}}, \mathcal{S}_{\underline{\Delta}})) \ .$$

For the closure $\bar{\alpha}$ of $\alpha \in \mathcal{G}(\Omega \times \mathbb{R}^d)$ in $L^2(\mathbb{R}_+)$ we get

$$\bar{\alpha} \in C^\infty(\Omega \times \mathbb{R}^d, \mathcal{L}(L^2(\mathbb{R}_+), L^2(\mathbb{R}_+)))$$

In the concept of ellipticity it is adequate also to deal with systems, i.e. with the algebra $\mathcal{G}(\Omega \times \mathbb{R}^d) \times \mathrm{hom}(\mathbb{C}^p, \mathbb{C}^p)$.

An $\alpha \in \mathcal{G}(\Omega \times \mathbb{R}^d) \times \mathrm{hom}\,(\mathbb{C}^p, \mathbb{C}^p)$ is called elliptic if $\sigma_\psi(\alpha)(y,\xi)$, $y \in \Omega$, $\xi \neq 0$, and $\sigma_M^0(\alpha)(y,\eta,z)$ are families of invertible pxp matrices (the latter including the limits as $|\mathrm{Im}\, z| \longrightarrow \infty$). The definition also applies pointwise, so incidentally we will talk about ellipticity at a fixed (y_o, η_o).

1. <u>Definition</u>. An operator family $b \in \mathcal{G}(\Omega \times \mathbb{R}^d) \otimes \mathrm{hom}\,(\mathbb{C}^p, \mathbb{C}^p)$ is called a parametrix of $\alpha \in \mathcal{G}(\Omega \times \mathbb{R}^d) \times \mathrm{hom}\,(\mathbb{C}^p, \mathbb{C}^p)$ if

$$\alpha b - 1, \ b\alpha - 1 \in \mathcal{G}_G(\Omega \times \mathbb{R}^d) \times \mathrm{hom}(\mathbb{C}^p, \mathbb{C}^p) \ .$$

An analogous definition holds pointwise, i.e. for fixed (y_o, η_o).

The following result is an analogue of 1.5.3. Theorem 4 for the algebra $\mathcal{G}(\Omega \times \mathbb{R}^d) \times \hom(\mathbb{C}^p, \mathbb{C}^p)$.

2. Theorem. Let $\mathfrak{a}(y, \eta) \in \mathcal{G}(\Omega \times \mathbb{R}^d) \times \hom(\mathbb{C}^p, \mathbb{C}^p)$, $(y_o, \eta_o) \in \Omega \times \mathbb{R}^d$ be fixed. The following conditions are equivalent

(i) $\mathfrak{a}(y_o, \eta_o)$ is elliptic,

(ii) $\bar{\mathfrak{a}}(y_o, \eta_o)$ is Fredholm.

If $\mathfrak{a}(y_o, \eta_o)$ is elliptic, there exists some $\mathfrak{b} \in \mathcal{G}(\Omega \times \mathbb{R}^d) \times \hom(\mathbb{C}^p, \mathbb{C}^p)$ such that $\mathfrak{b}(y_o, \eta_o)$ is a parametrix of $\mathfrak{a}(y_o, \eta_o)$ and there are finite-dimensional spaces $\mathcal{N}_i(y_o, \eta_o) \subset \mathcal{S}_{\underline{\Delta}} \otimes \mathbb{C}^p$, $i = 1,2$, with

$$\operatorname{im} \bar{\mathfrak{a}}(y_o, \eta_o) \oplus \mathcal{N}_1(y_o, \eta_o) = L^2(\mathbb{R}_+, \mathbb{C}^p), \quad \mathcal{N}_2(y_o, \eta_o) = \ker \bar{\mathfrak{a}}(y_o, \eta_o).$$

Thus $\operatorname{ind} \mathfrak{a}(y_o, \eta_o) = \operatorname{ind} \bar{\mathfrak{a}}(y_o, \eta_o)$. Moreover, $\mathfrak{a}u = f \in \mathcal{S}_{\underline{\Delta}} \otimes \mathbb{C}^p$, $u \in L^2(\mathbb{R}_+) \otimes \mathbb{C}^p$ implies $u \in \mathcal{S}_{\underline{\Delta}} \otimes \mathbb{C}^p$.

If $\mathfrak{a} \in \mathcal{G}(\Omega \times \mathbb{R}^d) \times \hom(\mathbb{C}^p, \mathbb{C}^p)$ is elliptic for all (y, η), then for every open $\Omega' \subset \Omega$ with $\bar{\Omega}'$ compact there is a parametrix $\mathfrak{b}' \in \mathcal{G}(\Omega' \times \mathbb{R}^d) \otimes \hom(\mathbb{C}^p, \mathbb{C}^p)$ of $\mathfrak{a}' := \mathfrak{a}|_{\Omega' \times \mathbb{R}^d}$.

Proof: The equivalence (i)$\Longleftrightarrow$(ii) is known, cf. [R2], since $\mathcal{G}$ is a subalgebra of the norm-closed algebra there. So we have to construct a parametrix of $\mathfrak{a}$ in our class. We may do this at once for families, since the ellipticity of $\mathfrak{a}(y_o, \eta_o)$ implies the ellipticity of $\mathfrak{a}(y, \eta)$ for all (y, η) in an open neighbourhood of (y_o, η_o). Then we have to give the arguments that there exists a parametrix which smoothly depends on the parameters with homogeneity in η for large $|\eta|$ and that the conormal symbols are polynomials in η of the corresponding degrees. First remark that the principal conormal symbol $\sigma_M^o(\mathfrak{a})(y,z)$ is holomorphic in $z \in \mathbb{C} \setminus \Pi$ for some $\Pi \in \underline{Z}$, $\Gamma_{\frac{1}{2}} \cap \Pi = \emptyset$, and all $y \in \Omega$. Let

$$A_{x_1, x_2} = \{ z : x_1 < \operatorname{Re} z < x_2 \} \cap \{ \mathbb{C} \setminus \Pi \},$$

$x_1, x_2 \in \mathbb{R}$. Then A_{x_1, x_2} only contains a finite number of zeros $z_j(y)$ of $\sigma_M^o(\mathfrak{a})(y,z)$ with $\operatorname{dist}(z_j, \Pi) > \varepsilon$ for all $x_1, x_2 \in \mathbb{R}$, $\varepsilon > 0$, $y \in \Omega'$.

$\overline{\Omega'} \subset\subset \Omega$. Thus $(\sigma_M^0(\mathfrak{a}))^{-1}(y,z), y \in \Omega'$, is holomorphic in $\mathbb{C}\setminus Z$ for some $Z \in \underline{Z}$, $\Gamma_{\frac{1}{2}} \cap Z = \emptyset$ and we have

$$h_o(y,z) := (\sigma_M^0(\mathfrak{a}))^{-1}(y,z) - \sigma_M^0(\mathrm{op}(a^{-1}))(y,z) \in C^\infty(\Omega', \mathfrak{M}_Z).$$

Let $a^{(-1)} \in S^0(\Omega \times \mathbb{R}^d)$, $a^{(-1)}a = 1$ for $|\xi| \geq$ const., and set

$$\mathfrak{b}^0(y,\eta) := \mathrm{op}_\psi(a^{(-1)})(y,\eta) + \omega(t\,\xi(|\eta|))\mathrm{op}_M(h_o)(y)\omega(t\,\xi(|\eta|)).$$

In view of the symbolic calculus in the algebra $\mathfrak{G}(\Omega \times \mathbb{R}^d)$ we know that

$$\sigma_\psi(\mathfrak{b}\mathfrak{a} - 1) = 0,\ \sigma_M^0(\mathfrak{b}^0\mathfrak{a} - 1) = 0.$$

Thus $\mathfrak{b}^0\mathfrak{a} = 1 + \mathfrak{c}$ with some $\mathfrak{c} \in \mathfrak{G}_{M+G}(\Omega' \times \mathbb{R}^d) \otimes \mathrm{hom}(\mathbb{C}^p, \mathbb{C}^p)$ of conormal order ≤ -1. Thus we may proceed in a standard way. We consider the formal Neumann series $(1 + \mathfrak{c})^{-1} = \sum_{j=0}^{\infty} (-1)^j \mathfrak{c}^{-j}$, which does not necessarily converge. But we know that $(-1)^j \mathfrak{c}^j$ is of conormal order $\leq -j$, and $h_k(y,\eta,z) := \sigma_M^{-k}(\sum_{j=0}^{k} (-1)^j \mathfrak{c}^j)(y,\eta,z)$ is a sequence in $C^\infty(\Omega', \mathfrak{M}_{\underline{Z}})^{(k)}[\eta]$. Thus, as above we find a Mellin family $\mathfrak{m}(y,\eta)$ with $h_k(y,\eta,z) = \sigma_M^{-k}(\mathfrak{m})(y,\eta,z)$, $k \in \mathbb{Z}_+$, with C^∞ dependence on y,η. Hence $\mathfrak{m}\mathfrak{b}^0\mathfrak{a} = 1 + \mathfrak{g}$ with some $\mathfrak{g} \in \mathfrak{G}_G(\Omega' \times \mathbb{R}^d) \otimes \mathrm{hom}(\mathbb{C}^p, \mathbb{C}^p)$. Thus we get a parametrix in the form $\mathfrak{b} = \mathfrak{m}\mathfrak{b}^0$ over $\Omega' \times \mathbb{R}^d$. The proof of (1) is analogous to that of the corresponding assertion in 1.5.3. Theorem 4, where 2.2.1. Proposition 5 is to be used.

Finally $\mathfrak{a}u = f \in \mathcal{S}_\Delta \otimes \mathbb{C}^p$ implies $\mathfrak{b}\mathfrak{a}u = (1 + \mathfrak{g})u \in \mathcal{S}_\Delta \otimes \mathbb{C}^p$ at the given (y,η). Since $\mathfrak{g} : L^2(\mathbb{R}_+) \otimes \mathbb{C}^p \longrightarrow \mathcal{S}_\Delta \otimes \mathbb{C}^p$, we get as desired $u \in \mathcal{S}_\Delta \otimes \mathbb{C}^p$. □

We also have a generalization of 1.5.3. Proposition 5 with the same proof, except for some slight modification, namely

3. Theorem. Let $\mathfrak{a} \in \mathfrak{G}(\Omega \times \mathbb{R}^d) \otimes \mathrm{hom}(\mathbb{C}^p, \mathbb{C}^p)$, and $\bar{\mathfrak{a}}(y,\eta)$ be invertible for all $y \in \Omega, \eta \in \mathbb{R}^d$. Then, for every open $\Omega' \subset \Omega$, $\overline{\Omega'} \subset\subset \Omega$, there exists a $\mathfrak{b}' \in \mathfrak{G}(\Omega' \times \mathbb{R}^d) \otimes \mathrm{hom}(\mathbb{C}^p, \mathbb{C}^p)$ with $\bar{\mathfrak{b}}'(y,\eta) = (\bar{\mathfrak{a}}'(y,\eta))^{-1}$, $\mathfrak{a}' := \mathfrak{a}|_{\Omega' \times \mathbb{R}^d}$.

As in 1.5.3. an algebra is defined which contains $\mathfrak{G}$ in the left upper corners and also trace and potential operators, which are defined with $\mathcal{S}_\Delta$ instead of $\mathcal{S}_\gamma$. Again there is a reduction to left upper corners and we get from Theorem 3 as a corollary an analogue of 1.5.3. Theorem 1.

2.3. Boundary symbolic calculus

2.3.1. The Mellin transform of Sobolev spaces

Recall that the space $\mathcal{H}^r(\mathbb{R}_+)$, $r \in \mathbb{Z}_+$, was defined in 1.2.1. by $\mathcal{H}^r(\mathbb{R}_+) = \{u \in \mathcal{D}'(\mathbb{R}_+) : \vartheta^l u \in L^2(\mathbb{R}_+),\ 0 \leq l \leq r\}$. It can be characterized in the Mellin image by

$$\frac{1}{2\pi i} \int_{\Gamma_{\frac{1}{2}}} (1+|z|^2)^r \, |\tilde{u}(z)|^2 dz < \infty .$$

This formula makes sense for each $r \in \mathbb{R}$. It will be taken as a definition for $\mathcal{H}^r(\mathbb{R}_+)$, $r \in \mathbb{R}$. In $\mathcal{H}^r(\mathbb{R}_+)$ we introduce the scalar product $(u,v)_{\mathcal{H}^r(\mathbb{R}_+)} = \frac{1}{2\pi i} \int (1+|z|^2)^r \, \tilde{u}(z)\overline{\tilde{v}(z)} dz$ (integral over $\Gamma_{\frac{1}{2}}$; then $\mathcal{H}^r(\mathbb{R}_+)$ is a Hilbert space, $\mathcal{H}^0 = L^2$). By extending the L^2 - scalar product we get a canonical identification between $\mathcal{H}^{-r}(\mathbb{R}_+)$ and $(\mathcal{H}^r(\mathbb{R}_+))'$, the dual space.

Let us introduce the weighted spaces

$$\mathcal{H}^{s,\lambda}(\mathbb{R}_+) := t^{\lambda}\, \mathcal{H}^s(\mathbb{R}_+), \quad \lambda \in \mathbb{R},$$

cf. the notation $L^{2,\lambda}$ of 1.3.2. in the case $s=0$. The Mellin image of $\mathcal{H}^{s,\lambda}$ is defined on $\Gamma_{\frac{1}{2}-\lambda}$ and equals the shift of $M\mathcal{H}^s$ from $\Gamma_{\frac{1}{2}}$ to $\Gamma_{\frac{1}{2}-\lambda}$.

Denote by $\mathcal{H}^{s,\lambda}_{(0)}$ the set of all $u \in \mathcal{H}^{s,\lambda}$ with bounded support. Then we obviously have

$$\mathcal{H}^{s,\lambda}_{(0)} \hookrightarrow \mathcal{H}^{s',\lambda'}_{(0)} \quad \text{for} \quad s \geq s',\ \lambda \geq \lambda'.$$

If $E \subseteq \mathcal{D}'(\mathbb{R}_+)$ is a class of distributions, we denote by $E[a,b]$ the set of all elements of E supported by $[a,b]$, $0 \leq a \leq b \leq \infty$. In particular we consider $\mathcal{H}^{s,\gamma}[0,d]$, $H^s[c,\infty]$, $c,d > 0$, with the Hilbert space structures induced by $\mathcal{H}^{s,\lambda}(\mathbb{R}_+)$ and $H^s(\mathbb{R}_+)$, respectively.

1. Definition. Set for $0 < c < d < \infty$

$$\mathcal{K}^{s,\gamma}(\mathbb{R}_+) := \mathcal{H}^{s,\gamma}[0,d] + H^s[c,\infty],$$

$s, \gamma \in \mathbb{R}$, endowed with the norm of the sum.

In view of

$$\mathcal{H}^{s,\gamma}[c,d] = H^s[c,d] = \mathcal{H}^{s,\gamma}[0,d] \cap H^s[c,\infty]$$

we get an equivalent definition for another choice of c,d, in other words $\mathcal{K}^{s,\gamma}(\mathbb{R}_+)$ is independent of c,d. Remark that $C_0^\infty(\mathbb{R}_+)$ is dense in $\mathcal{K}^{s,\gamma}(\mathbb{R}_+)$ for all $s,\gamma \in \mathbb{R}$. Moreover, $\varphi\mathcal{K}^{s,\gamma}(\mathbb{R}_+) \subset \mathcal{K}^{s,\gamma}(\mathbb{R}_+)$ for every $\varphi \in C_0^\infty(\mathbb{R}_+)$. This follows easily from the corresponding property for $\mathcal{H}^{s,\gamma}(\mathbb{R}_+)$.

Since $\mathcal{H}^{s,\gamma}[0,d]$, $H^s[c,\infty]$ are Hilbert spaces, we also can introduce a Hilbert space structure in the sum.

It will be fixed then once and for all. So $\mathcal{K}^{s,\gamma}(\mathbb{R}_+)$ is a Hilbert space.

From $H^s[c,\infty] \hookrightarrow H^{s'}[c,\infty]$ for $s \geq s'$, $\mathcal{H}^{s,\gamma}[0,d] \hookrightarrow \mathcal{H}^{s',\gamma'}[0,d]$ for $s \geq s'$, $\gamma \geq \gamma'$, we obtain

$$\mathcal{K}^{s,\gamma} \hookrightarrow \mathcal{K}^{s',\gamma'} \quad \text{for } s \geq s', \; \gamma \geq \gamma'.$$

Moreover, all inclusions are dense embeddings and $\varphi\mathcal{K}^{s,\gamma} \hookrightarrow \varphi\mathcal{K}^{s',\gamma'}$ is compact for $s > s'$, $\gamma > \gamma'$, $\varphi \in C_0^\infty(\overline{\mathbb{R}}_+)$.

Furthermore, we have

$$(\mathcal{K}^{s,\gamma}(\mathbb{R}_+))' = \mathcal{K}^{-s,-\gamma}(\mathbb{R}_+)$$

for all $s,\gamma \in \mathbb{R}$.

In the following we shall define the action of ψDO's on $\mathbb{R}_+$ in spaces with different weights. Set

$$l_\pm(\delta,\tau) = \delta \pm i\tau ,$$

$\delta \in \mathbb{R}_+$ a parameter, and let $op(l_\pm)(\delta)$ be the usual ψDO action on $\mathbb{R}$. For $\delta = 1$ we simply write $op(l_\pm)$.

Remember that

$$op(l_+^s)(\delta) : H_0^r(\overline{\mathbb{R}}_+) \xrightarrow[\cong]{} H_0^{r-\mathrm{Re}\, s}(\mathbb{R}_+)$$

for all $r \in \mathbb{R}$, $s \in \mathbb{C}$, $\delta \in \mathbb{R}_+$, and $op(l_+^{-s}) = op(l_+^s)^{-1}$. Here

$$H_0^r(\overline{\mathbb{R}}_+) = \{u \in H^r(\mathbb{R}) : \operatorname{supp} u \subseteq \overline{\mathbb{R}}_+\}.$$

As usually $L^2(\mathbb{R}_+)$ and $H^0_o(\overline{\mathbb{R}}_+)$ will be canonically identified, i.e.

$$\mathrm{op}(l^s_+)(\delta) : L^2(\mathbb{R}_+) \xrightarrow[\cong]{} H^{-s}_o(\overline{\mathbb{R}}_+). \qquad (1)$$

The notation op_ψ will be used as usually $\mathrm{op}_\psi = r^+ \mathrm{op}\, e^+$, r^+ the restriction operator to $\mathbb{R}_+$, e^+ the operator of extension by zero.

Moreover, recall the following asymptotic behaviour of the Γ function (cf. T 1, Section 4.4).

2. Lemma. For every fixed $p \in \mathbb{C}$ and $\delta_1, \delta_2 \in \mathbb{R}$, $c \in \mathbb{R}_+$, there are constants $c_1, c_2 > 0$ such that

$$c_1 |\mathrm{Im}\, z|^{\mathrm{Re}\, p} \leq \left| \frac{\Gamma(z+p)}{\Gamma(z)} \right| \leq c_2 \, |\mathrm{Im}\, z|^{\mathrm{Re}\, p}$$

for $\delta_1 \leq \mathrm{Re}\, z \leq \delta_2$, $|\mathrm{Im}\, z| \geq c$, where c is so large that there are no poles of $\Gamma(z+p)$ in $\delta_1 \leq \mathrm{Re}\, z \leq \delta_2$, $|\mathrm{Im}\, z| \geq c$.

Define the binomial coefficients $\binom{z}{k}$, $z \in \mathbb{C}$ by $(1+x)^z = \sum_{k=0}^{\infty} \binom{z}{k} x^k$.
Then we have the identity

$$\binom{z}{k} = \frac{(-1)^k \Gamma(-z+k)}{\Gamma(-z)\, k!}. \qquad (2)$$

Now introduce the functions

$$m_+^{\alpha-k}(z) := \frac{\Gamma(1-z)}{\Gamma(1-z-\alpha+k)}, \qquad (3)$$

$\alpha \in \mathbb{C}$, $k \in \mathbb{Z}_+$.

3. Proposition. Let $\alpha \in \mathbb{C}$, $r = \mathrm{Re}\, \alpha$. Then

$$\omega\, \mathrm{op}^o_\psi(l^\alpha_+)(\delta) := \omega\, t^{-\alpha} \sum_{k=0}^{\infty} (t\,\delta)^k \binom{\alpha}{k} \mathrm{op}_M(m_+^{\alpha-k}) \qquad (4)$$

converges in $\mathcal{L}(L^2(\mathbb{R}_+), \mathcal{H}^{-r,-r}(\mathbb{R}_+))$, and we have

$$\omega \mathrm{op}^o_\psi(l^\alpha_+)(\delta)\, u = \omega\, r^+ \mathrm{op}(l^\alpha_+)(\delta) e^+ u$$

for all $u \in C^\infty_o(\mathbb{R}_+)$.

Proof: Let $u \in C_0^\infty(\mathbb{R}_+)$. Assume for the moment $r<0$. Then

$$r^+ op(l_+^\alpha)(\delta)u(t) = (2\pi)^{-1} r^+ \int_0^\infty \Big\{\int_{-\infty}^\infty e^{i(t-s)\tau}(\delta+i\tau)^\alpha \, d\tau\Big\} u(s)ds.$$

In view of

$$\int_{-\infty}^\infty e^{it\tau}(\delta+i\tau)^\alpha \, d\tau = \begin{cases} \frac{2\pi}{\Gamma(-\alpha)} \, t^{-\alpha-1} e^{-\delta t} & \text{for } t>0, \\ 0 & \text{for } t \leq 0 \end{cases}$$

we obtain

$$r^+ op(l_+^\alpha)(\delta)e^+u(t) = \frac{1}{\Gamma(-\alpha)} \int_0^t (t-s)^{-\alpha-1} \, e^{-\delta(t-s)} \, u(s) \, ds.$$

Inserting the expansion $\quad e^{-\delta(t-s)} = \sum_{k=0}^\infty \frac{(-1)^k}{k!} \delta^k (t-s)^k$

which converges uniformly as $|t-s|$ is bounded, we get

$$r^+ op(l_+^\alpha)(\delta)e^+u(t) = \frac{1}{\Gamma(-\alpha)} \sum_{k=0}^\infty \frac{(-1)^k}{k!} \delta^k \int_0^t (t-s)^{k-\alpha-1} \, u(s) \, ds.$$

Now

$$\int_0^t (t-s)^{k-\alpha-1} u(s)ds = \int_1^\infty (t-\tfrac{t}{r})^{k-\alpha-1} \, t \, r^{-2} u(\tfrac{t}{r}) \, dr$$

$$= t^{k-\alpha} \int_1^\infty (s-1)^{k-\alpha-1} \, s^{-(k-\alpha)} \, u\left(\frac{t}{s}\right) \frac{ds}{s}.$$

Thus

$$\int_0^t (t-s)^{k-\alpha-1} \, u(s)ds = t^{k-\alpha} \int_0^\infty \varphi^{\alpha-k}(s) u\left(\frac{t}{s}\right) \frac{ds}{s}$$

with

$$\varphi^{\alpha-k}(s) := \begin{cases} (s-1)^{k-\alpha-1} s^{-(k-\alpha)} & \text{for } s>1, \\ 0 & \text{for } s \leq 1. \end{cases}$$

Hence

$$\int_0^t (t-s)^{k-\alpha-1} \, u(s)ds = t^{k-\alpha} \, op_M\left(f_+^{\alpha-k}\right) u,$$

where $f_+^{\alpha-k}(z) = \int_1^\infty s^{z-1-k+\alpha}(s-1)^{k-\alpha-1} \, ds.$

Since

$$\int_1^\infty s^{-\nu}(s-1)^{\mu-1}\,ds = B(\nu-\mu,\mu),\quad B(x,y) = \frac{\Gamma(x)\Gamma(y)}{\Gamma(x+y)}$$

(cf. [G 4], pp. 298,964) we have with $\mu = k-\alpha$, $\nu = -z+1+k-\alpha$

$$f_+^{\alpha-k}(z) = \frac{\Gamma(1-z)\Gamma(k-\alpha)}{\Gamma(1-z-\alpha+k)} .$$

Using (2) for $z=\alpha$ we obtain

$$r^+ op(1_+^\alpha)(\delta)u(t) = t^{-\alpha}\sum_{k=0}^\infty (t\delta)^k \binom{\alpha}{k} op_M(m_+^{\alpha-k})\, u(t).$$

Next we want to show the convergence of (4) in the operator norm. From Lemma 2 we get

$$\| op_M(m_+^{\alpha-k})\|_{\mathcal{L}(L^2(\mathbb{R}_+),\, \mathcal{H}^{-r}(\mathbb{R}_+))} \leqq c$$

for all k with c independent of k. For each cut-off function ω we have

$$\| \omega t^{-\alpha+k}\|_{\mathcal{L}(\mathcal{H}^{-r}(\mathbb{R}_+),\, \mathcal{H}^{-r,-r}(\mathbb{R}_+))} \leqq c ,$$

where c is also independent of k.

Because of $\sum_{k=0}^\infty \binom{\alpha}{k} \delta^k = (1+\delta)^\alpha$ we thus obtain

$$\| \omega\, op_\psi^0(1_+^\alpha)\|_{\mathcal{L}(L^2(\mathbb{R}_+),\, \mathcal{H}^{-r,-r}(\mathbb{R}_+))}$$

$$\leqq \sum_{k=0}^\infty \| \omega t^{-\alpha+k}\, \delta^k \binom{\alpha}{k}\, op_M(m_+^{\alpha-k})\|_{\mathcal{L}(L^2(\mathbb{R}_+),\, \mathcal{H}^{r,-r}(\mathbb{R}_+))}$$

$$\leqq c(1+\delta)^\alpha .$$

For arbitrary $\alpha\in\mathbb{C}$ we can do the same, since the restriction Re $\alpha < 0$ disappears after reformulating the right hand sides with $\binom{\alpha}{k}$ and using analytic extension with respect to α . □

4. <u>Definition.</u> The function

$$\sigma_M^{\alpha-k}(1_+^\alpha)(\delta,z) = \delta^k \binom{\alpha}{k}\, m_+^{\alpha-k}(z)$$

is called the conormal symbol of $1_+^\alpha(\delta,z)$ belonging to the conormal order $\alpha - k$.

Observe that

$$\Gamma(1-z) = (-1)^p z(z+1)\cdot\ldots\cdot(z+p-1)\,\Gamma(1-z-p)$$

for $p \in \mathbb{Z}_+$, $p \geq 1$, and hence, for $\alpha = N \in \mathbb{Z}_+$,

$$m_+^{N-k}(z) = \begin{cases} (-1)^{N-k}\, z(z+1)\cdot\ \ldots\ \cdot(z+(N-k)+1) & \text{for } 0 \leq k \leq N-1 \\ 1 & \text{for } k = N, \end{cases} \tag{5}$$

$$\sigma_M^{N-k}(1_+^N)(\delta,z) = 0 \quad \text{for } k > N.$$

On the other hand, for $N \in \mathbb{Z}_-, N \leq -1$,

we obtain

$$m_+^{N-k}(z) = \prod_{j=1}^{-N+k} (j-z)^{-1}$$

and thus

$$\sigma_M^{N-k}(1_+^N)(\delta,z) = \delta^k \binom{N}{k} \prod_{j=1}^{-N+k} (j-z)^{-1}.$$

These expressions coincide with those in Section 1.4.1.

We need a generalization of 1.3.1. Proposition 8 for the case of Mellin symbols h that are meromorphic and not necessarily rapidly decreasing as $|\mathrm{Im}\, z| \to \infty$. Assume that $h(z)$ behaves like $\Gamma(1-z)(\Gamma(1-z+\alpha+k))^{-1}$. For each fixed $\lambda \leq 0$ we have a decomposition $h = h_1 + h_2$, where h_1 is meromorphic and rapidly decreasing as in 1.3.1. and h_2 is holomorphic in the strip $\frac{1}{2} < \mathrm{Re}\, z < \frac{1}{2} - \lambda$ and has the same growth magnitude as h. Then

$$\omega\, t^\lambda \mathrm{op}_M(T^{-\lambda} h_2) t^{-\lambda} \omega u - \omega \mathrm{op}_M(h_2)\, \omega u = 0$$

for all $u \in \mathcal{H}^{s,\lambda}(\mathbb{R}_+)$. Indeed, for $u \in C_0^\infty(\mathbb{R}_+)$ this is obvious, since $\tilde{u}(z)$ is rapidly decreasing for $|\mathrm{Im}\, z| \to \infty$, but then it follows in general, since $C_0^\infty(\mathbb{R}_+)$ is dense in $\mathcal{H}^{s,\lambda}(\mathbb{R}_+)$.

Thus

$$\omega\left\{t^\lambda\ \mathrm{op}_M(T^{-\lambda} h)\ t^{-\lambda} - \mathrm{op}_M(h)\right\}\omega$$
$$= \omega\left\{t^\lambda\ \mathrm{op}_M(T^{-\lambda} h_1) t^{-\lambda} - \mathrm{op}_M(h_1)\right\}\omega$$

can be calculated by the same arguments as in 1.3.1. Proposition 8.

5. Proposition. $\omega \mathrm{op}_{\psi}^{o}(1_{+}^{\alpha})(\delta)\omega \big|_{C_0^\infty(\mathbb{R}_+)}$ extends by continuity to a continuous operator

$$\omega \mathrm{op}_{\psi}^{\gamma}(1_{+}^{\alpha})(\delta)\omega \ : \mathcal{H}^{s,\gamma}(\mathbb{R}_+) \longrightarrow \mathcal{H}^{s-r,\gamma-r}(\mathbb{R}_+)$$

for every $\gamma \geqq 0$, $s \in \mathbb{R}, \alpha \in \mathbb{C}$, $r = \mathrm{Re}\,\alpha$.

Proof: The series on the right in Proposition 3 obviously converges in the operator norm of $\mathcal{L}(\mathcal{H}^s(\mathbb{R}_+), \mathcal{H}^{s-r,-r}(\mathbb{R}_+))$. Since $C_0^\infty(\mathbb{R}_+)$ is dense in $\mathcal{H}^s(\mathbb{R}_+)$, we thus obtain the assertion for $\gamma = 0$. From

$$\omega t^{\gamma} \mathrm{op}_M(T^{-\gamma} m_{+}^{\alpha-k}) t^{-\gamma}\omega = \omega \mathrm{op}_M(m_{+}^{\alpha-k})\omega ,$$

for $\gamma \geqq 0$ then the result follows in general. $\square$

6. Remark. The series

$$\omega \mathrm{op}_{\psi}^{\gamma}(1_{+}^{\alpha})(\delta)\omega := \omega t^{-\alpha} \sum_{k=0}^{\infty} (t\delta)^k \binom{\alpha}{k} t^{\gamma} \mathrm{op}_M(T^{-\gamma} m_{+}^{\alpha-k}) \quad t^{-\gamma}\omega$$

converges in $\mathcal{L}(\mathcal{H}^{s,\gamma}(\mathbb{R}_+), \mathcal{H}^{s-r,\gamma-r}(\mathbb{R}_+))$ for all $s, \gamma \in \mathbb{R}, -\gamma \notin \frac{1}{2} + \mathbb{Z}_+$.

Let $\omega, \omega_1, \omega_2$ be cut-off functions, $\omega_1\omega = \omega_1, \omega_2\omega_1 = \omega_2$. Then

$$\mathrm{op}_{\psi}^{\gamma}(1_{+}^{\alpha})(\delta) := \omega \mathrm{op}_{\psi}^{\gamma}(1_{+}^{\alpha})(\delta)\,\omega_1 + (1-\omega_2)\mathrm{op}_{\psi}(1_{+}^{\alpha})(\delta)(1-\omega_1)$$

is a ψDO on $\mathbb{R}_+$ for which

$$\mathrm{op}_{\psi}^{\gamma}(1_{+}^{\alpha})(\delta) - \mathrm{op}_{\psi}(1_{+}^{\alpha})(\delta) \in C^\infty(\mathbb{R}_+ \times \mathbb{R}_+)$$

for every δ. Moreover, the definition of $\mathrm{op}_{\psi}^{\gamma}(1_{+}^{\alpha})(\delta)$ is unique modulo a Green operator in classes with weights (cf. Section 2.3.3 below).

7. Remark. $\mathrm{op}_{\psi}^{\gamma}(1_{+}^{\alpha})(\delta)$ induces a continuous operator

$$\mathrm{op}_{\psi}^{\gamma}(1_{+}^{\alpha})(\delta) \ : \mathcal{K}^{s,\gamma}(\mathbb{R}_+) \rightarrow \mathcal{K}^{s-r,\gamma-r}(\mathbb{R}_+)$$

for all $s, \gamma \in \mathbb{R}$, $-\gamma \notin \frac{1}{2} + \mathbb{Z}_+$.

Next we want to establish a connection between the spaces $H_0^s(\overline{\mathbb{R}}_+)$, $H^s(\mathbb{R}_+) = r^+H^s(\mathbb{R})$, and $\mathcal{K}^{s,s}(\mathbb{R}_+)$. First remember that $u \in \mathcal{H}^{s,s}(\mathbb{R}_+)$ supp u bounded, is equivalent to

(i) $\tilde{u}(z) \in \mathcal{A}(\{\mathrm{Re}\, z > \frac{1}{2} - s\})$

(ii) $\frac{1}{2\pi i} \int_{\Gamma_\beta} (1+|z|^2)^s \, |\tilde{u}(z)|^2 \, dz \leq c a^\delta$

for all $\beta = \frac{1}{2} - s + \delta$, $\delta > 0$ with constants c, a depending on u.

This is an immediate generalization of 1.1. Theorem 5. It follows that the limit of the integrals in (ii) for $\delta \to 0$ is also finite.

An analogous characterization can be given for $u \in \mathcal{H}^{s,s}(\mathbb{R}_+) + \mathcal{S}_\varphi(\mathbb{R}_+)$ with bounded support, $\varphi \in \underline{\varphi}$, $\pi_{\mathbb{C}}(\varphi) \cap \Gamma_{\frac{1}{2}-s} = \emptyset$, where we assume, instead of (i), that $\tilde{u}$ is meromorphic in $\operatorname{Re} z > \frac{1}{2} - s$ with the corresponding poles, and in (ii) $\pi_{\mathbb{C}}(\varphi) \cap \Gamma_\beta = \emptyset$. Note that for $s < 0$ the latter restrictions are not relevant and for $s > 0$, only a finite number of poles has to be counted.

Let $s \in \mathbb{R}$ and

$$\varkappa(s) := \max \left\{ j \in \mathbb{Z}_+ : j < |s| - \tfrac{1}{2} \right\}.$$

Denote by $\mathcal{D}^s$, $s \leq 0$, the linear span of the derivatives of orders $\leq \varkappa(s)$ of the Dirac-measure at the origin and $\mathcal{T}^s$, $s \geq 0$, the linear span of the functions $\{ t^j \omega(t) : j = 0, \dots, \varkappa(s) \}$.

8. <u>Theorem.</u> Let $s \in \mathbb{R}$, then

$$H_0^s(\overline{\mathbb{R}}_+) = \begin{cases} \mathcal{K}^{s,s}(\mathbb{R}_+) & \text{for } s \geq 0, \\ \mathcal{K}^{s,s}(\mathbb{R}_+) + \mathcal{D}^s & \text{for } s \leq 0,\ s \neq \frac{1}{2} \bmod \mathbb{Z} \end{cases}$$

$$H^s(\mathbb{R}_+) = \begin{cases} \mathcal{K}^{s,s}(\mathbb{R}_+) + \mathcal{T}^s & \text{for } s \geq 0,\ s \neq \frac{1}{2} \bmod \mathbb{Z} \\ \mathcal{K}^{s,s}(\mathbb{R}_+) & \text{for } s \leq 0. \end{cases}$$

The proof will be given in several steps. First show the following.

9. <u>Proposition.</u> We have for $s \geq 0$

$$\omega H_0^s(\overline{\mathbb{R}}_+) = \omega \mathcal{H}^{s,s}(\mathbb{R}_+),$$

with equivalence of the topologies, ω a cut-off function.

Proof: First we show that there is a continuous embedding $\omega H_0^s(\overline{\mathbb{R}}_+) \hookrightarrow \omega \mathcal{H}^{s,s}(\mathbb{R}_+)$. In view of the isomorphism (1) for every $u \in H_0^s(\mathbb{R}_+)$ there is a unique $v \in L^2(\mathbb{R}_+)$ with $u = op(1_+^{-s})v$. Let supp u be bounded, $\omega u = u$, and choose a sequence $\{v_j\}$ in $C_0^\infty(\mathbb{R}_+)$ with $v_j \longrightarrow v$ in $L^2(\mathbb{R}_+)$. Then $u_j = op(1_+^{-s})\, v_j \longrightarrow u$ in $H_0^s(\overline{\mathbb{R}}_+)$. From Proposition 3 we know that

$$\omega\, r^+ op(1_+^{-s})\, v_j = \omega\, op_\psi^o(1_+^{-s})\, v_j$$

for all j and that the right-hand side converges in $\mathcal{H}^{s,s}(\mathbb{R}_+)$ to some limit u'. Thus $u' = r^+u$ and hence $\omega H_0^s(\overline{\mathbb{R}}_+)|_{\mathbb{R}_+} \subseteqq \omega\, \mathcal{H}^{s,s}(\mathbb{R}_+)$ for all $s \in \mathbb{R}$. In particular for $s \geq 0$ we get

$$\omega\, op_\psi(1_+^{-s})\, v = \omega\, op_\psi^o(1_+^{-s})\, v \tag{6}$$

for all $v \in L^2(\mathbb{R}_+)$. Now let $u_j \in H_0^s(\overline{\mathbb{R}}_+)$, $u_j \longrightarrow u$ in $H_0^s(\overline{\mathbb{R}}_+)$. Then $v_j = op(1_+^s)\, u_j \longrightarrow v$ in $L^2(\mathbb{R}_+)$ and $\omega op^o(1_+^{-s})\, v_j \longrightarrow u'$ in $\mathcal{H}^{s,s}(\mathbb{R}_+)$. From (6) applied to v_j it follows $u' = u$, $u_j \longrightarrow u$ in $\mathcal{H}^{s,s}(\mathbb{R}_+)$.

For the reverse direction we apply Proposition 5 and obtain the continuity

$$\omega_1 op_\psi^s(1_+^s)\omega \quad : \mathcal{H}^{s,s}(\mathbb{R}_+) \longrightarrow L^2(\mathbb{R}_+),$$

$\omega\omega_1 = \omega$. For $u \in C_0^\infty(\mathbb{R}_+)$ we have

$$\begin{aligned}
\omega u &= \omega\, r^+ op(1_+^{-s}) op(1_+^s)\, e^+\omega u \\
&= \omega r^+ op(1_+^{-s})\, e^+\, (1-\omega_1)\, r^+ op(1_+^s)\, e^+\omega u \\
&+ \omega r^+ op(1_+^{-s})\, e^+\omega_1\, op(1_+^s)\, e^+\omega u \quad .
\end{aligned}$$

By Proposition 5 it follows

$$\begin{aligned}
&\omega\, r^+ op(1_+^{-s})\, e^+\omega_1\, op(1_+^s)\, e^+\omega u \\
&= \omega r^+ op(1_+^{-s})\, e^+\omega_1\, op_\psi^s(1_+^s)\, \omega u \ .
\end{aligned}$$

Now choose a sequence $\{v_j\}$ of C_0^∞ functions , $\omega v_j \longrightarrow u$, $u \in \mathcal{H}^{s,s}(\mathbb{R}_+)$. Then

$$f_j := \omega_1 op_\psi^s(1_+^s)\, \omega v_j \longrightarrow \omega_1\, op_\psi^s(1_+^s)\, \omega u$$

in $L^2(\mathbb{R}_+)$. Thus

$$u_j' := \omega r^+ \operatorname{op}(l_+^{-s})\, e^+ f_j \longrightarrow u' \in H_0^s(\overline{\mathbb{R}}_+).$$

Moreover,

$$u_j'' := \omega r^+ \operatorname{op}(l_+^{-s})\, e^+ (1-\omega_1) r^+ \operatorname{op}(l_+^{s})\, e^+ \omega v_j$$

converges in $C^\infty(\overline{\mathbb{R}}_+)$. Thus $u_j' + u_j''$ converges in $H_0^s(\overline{\mathbb{R}}_+)$ with the limit ωu.□

As a corollary we obtain

$$H_0^s(\overline{\mathbb{R}}_+) = \mathcal{K}^{s,s}(\mathbb{R}_+) \text{ for } s \gtrless 0, \tag{7}$$

$$H^s(\mathbb{R}_+) = \mathcal{K}^{s,s}(\mathbb{R}_+) \text{ for } s \lessgtr 0, \tag{8}$$

where (8) follows from (7) by duality. This is just a part of Theorem 8.

Next we shall consider a Mellin expansion of $\operatorname{op}_\psi(l_-^{-\alpha})(\delta)$ for $\alpha \in \mathbb{C}$, $l_-(\delta,\tau) = \delta - i\tau$. The mapping properties of $\omega \operatorname{op}_\psi(l_-^{-\alpha})(1-\omega_1)$, $\omega = \omega_1\omega$, show that

$$\omega \operatorname{op}_\psi(l_-^{-\alpha}) - \omega \operatorname{op}_\psi(l_-^{-\alpha})\,\omega_1$$

$$= \omega \operatorname{op}_\psi(l_-^{-\alpha})(1-\omega_1) \in \operatorname{op}_G(\mathcal{S}_{\varphi_0} \otimes_\pi \mathcal{S}_0)$$

for all $\alpha \in \mathbb{C}$. The Mellin expansion of $\operatorname{op}_\psi(l_-^{-\alpha})$ can be evaluated modulo Green operators along the lines of the proof of 1.4.1. Theorem 5. A more complete information about the Green operators may be obtained by repeating the arguments in the proof of Proposition 3.

Set

$$m_-^{\alpha-k}(z) := \frac{\Gamma(z)}{\Gamma(z-\alpha+k)} .$$

10. Proposition. Let $\alpha \in \mathbb{C}$, $r = \operatorname{Re}\alpha$. Then

$$\operatorname{op}_\psi^r(l_-^{\alpha})(\delta)\omega := \sum_{k=0}^{\infty} \delta^k \binom{\alpha}{k} \operatorname{op}_M(m_-^{\alpha-k})\, t^{k-\alpha}\omega \tag{9}$$

converges in $\mathcal{L}(\mathcal{H}^{s,r}(\mathbb{R}_+), \mathcal{H}^{s-r}(\mathbb{R}_+))$, $s \in \mathbb{R}$, and we have

$$\operatorname{op}_\psi^r(l_-^{\alpha})(\delta)\,\omega u = \operatorname{op}_\psi(l_-^{\alpha})(\delta)\,\omega u$$

for all $u \in C_0^\infty(\mathbb{R}_+)$.

Proof: First assume $r < 0$. Then, for $u \in C_0^\infty(\mathbb{R}_+)$ it follows

$$
\begin{aligned}
&\mathrm{op}_\psi(1_-^\alpha)(\delta)u(t) \\
&= \frac{1}{\Gamma(-\alpha)} \int_t^\infty (t-s)^{-\alpha-1} e^{-\delta(t-s)} u(s)ds \\
&= \frac{1}{\Gamma(-\alpha)} \sum_{k=0}^\infty \frac{(-1)^k}{k!} \delta^k \mathrm{op}_M(f_-^{\alpha-k}) s^{k-\alpha} u(s) \\
&= \sum_{k=0}^\infty \binom{\alpha}{k} \delta^k \mathrm{op}_M(m_-^{\alpha-k}) s^{k-\alpha} u(s).
\end{aligned}
$$

Here $f_-^{\alpha-k}(z) := \Gamma(-\alpha+k)\,\Gamma(z)\{\Gamma(z-\alpha+k)\}^{-1}$.

Now we can proceed in an analogous manner as in the proof of Proposition 3. We have

$$\| \omega t^{k-\alpha} \|_{\mathcal{L}(\mathcal{H}^{s,r}(\mathbb{R}_+),\, \mathcal{H}^s(\mathbb{R}_+))} \leq c,$$

$$\| \mathrm{op}_M(m_-^{\alpha-k}) \|_{\mathcal{L}(\mathcal{H}^s(\mathbb{R}_+),\, \mathcal{H}^{s-r}(\mathbb{R}_+))} \leq c$$

for all k with c independent of k.

Thus

$$\| \mathrm{op}_\psi(1_-^\alpha)(\delta)u(t) \|_{\mathcal{L}(\mathcal{H}^{s,r}(\mathbb{R}_+),\, \mathcal{H}^{s-r}(\mathbb{R}_+))} \leq c(1+\delta)^\alpha.$$

By analytic extension of

$$\mathrm{op}_\psi(1_-^\alpha)(\delta)u = \sum_{k=0}^\infty \binom{\alpha}{k} \delta^k \mathrm{op}_M(m_-^{\alpha-k})u$$

we then get the same result for all $\alpha \in \mathbb{C}$. □

11. Proposition. $\omega \mathrm{op}_\psi^r(1_-^\alpha)(\delta)\omega \big|_{C_0^\infty(\mathbb{R}_+)}$ extends by continuity to a continuous operator

$$\omega \mathrm{op}_\psi^{r+\gamma}(1_-^\alpha)(\delta)\omega : \mathcal{H}^{s,\gamma+r}(\mathbb{R}_+) \longrightarrow \mathcal{H}^{s-r,\gamma}(\mathbb{R}_+)$$

for every $\gamma \leq 0$, $s \in \mathbb{R}$, $\alpha \in \mathbb{C}$, $r = \mathrm{Re}\,\alpha$.

Proof: This follows by applying

$$\omega t^{\gamma} op_M(T^{-\gamma} m_-^{k-\alpha}) t^{-\gamma}\omega = \omega op_M(m_-^{k-\alpha})\omega \quad ,$$

$\gamma \leqq 0$, and the following

12. Remark. The series

$$\omega op_\psi^{\gamma}(1_-^{\alpha})(\delta)\omega \; := \; \omega \sum_{k=0}^{\infty} \delta^k \binom{\alpha}{k} t^{\gamma-r} \; op_M\left(T^{-\gamma+r} m_-^{k-\alpha}\right) t^{-\gamma+r-\alpha+k}\omega$$

converges in $\mathcal{L}(\mathcal{H}^{s,\gamma}(\mathbb{R}_+), \mathcal{H}^{s-r,\gamma-r}(\mathbb{R}_+))$ for all γ with $\gamma - r \leqq 0$ and $\gamma - r > 0$, $\gamma - r \neq \frac{1}{2} \mod \mathbb{Z}$.

Recall the notations

$$\mathfrak{z}_{\gamma,\lambda} = \mathfrak{z} \cap \left\{\tfrac{1}{2} + \gamma < \operatorname{Re} z < \tfrac{1}{2} + \lambda\right\}, \; \mathfrak{z} \in \underline{\mathfrak{z}},$$

cf. 1.3.1., and $\mathfrak{z}_0 = \left\{(j,0)\right\}_{j \in \mathbb{Z}}$.

13. Proposition. Let $\alpha \in \mathbb{C}$, $r = \operatorname{Re} \alpha$, $\lambda \leqq 0$, $\lambda \neq \frac{1}{2} \mod \mathbb{Z}$. Then, for cut-off functions ω, ω_1 with $\omega = \omega_1 \omega$ we have

$$\omega op_\psi^{r}(1_-^{\alpha})\, \omega_1 u = \omega op_\psi^{r+\lambda}(1_-^{\alpha})\, \omega_1 u + op_G(b)u$$

for all $u \in C_0^\infty(\mathbb{R}_+)$ with some $b \in \mathcal{S}_{\mathfrak{r}} \otimes_\pi \mathcal{S}_{\sigma}$ with $\sigma = \left(T^{\alpha}\mathfrak{z}_0\right)_{r,r+\lambda}$, $\mathfrak{r} = (\mathfrak{z}_0)_{\lambda,0}$.

Proof: For $u \in C_0^\infty(\mathbb{R}_+)$ we have

$$\omega op_M(m_-^{\alpha-k})t^{k-\alpha}\omega u - \omega t^{-\lambda} op_M\left(T^{\lambda} m_-^{\alpha-k}\right) t^{k-\alpha+\lambda}\omega u$$

$$= \frac{1}{2\pi i}\,\omega \int_{\Sigma_\lambda} t^{-z} m_-^{\alpha-k}(z)\; \widetilde{\omega u}(z+k-\alpha)\,dz, \tag{10}$$

where $\Sigma_\lambda = \Gamma_{\frac{1}{2}} \cup \Gamma_{\frac{1}{2}+\lambda}$ with orientation on $\Gamma_{\frac{1}{2}}$ from $\operatorname{Im} z = -\infty$ to $\operatorname{Im} z = +\infty$ and the reverse on $\Gamma_{\frac{1}{2}+\lambda}$.

The Γ function is holomorphic for $\operatorname{Re} z > 0$ and has simple poles at the points $z = -j$, $j \in \mathbb{Z}_+$, with the residues $(-1)^j j!$. In particular, the integral makes sense for $\lambda \neq \frac{1}{2} \mod \mathbb{Z}$. It may be replaced by a sum of integrals over small circles around $-j$, $\frac{1}{2}+\lambda < -j \leqq 0$.

Thus (10) equals

$$\omega \sum_{\frac{1}{2}+\lambda<-j\leq 0} t^j \frac{(-1)^j}{j!} \frac{1}{\Gamma(-j-\alpha+k)} \widetilde{\omega u}\,(-j+k-\alpha).$$

Since $\widetilde{\omega u}\,(-j+k-\alpha) = \int t^{-j+k-\alpha-1}\,\omega(t)u(t)dt$, (10) is an integral operator with kernel

$$\sum_{\frac{1}{2}+\lambda<-j\leq 0} \frac{(-1)^j}{j!} \frac{1}{\Gamma(-j-\alpha+k)} \omega(t)t^j s^{-j+k-\alpha-1}\omega(s) = b_k(t,s).$$

In view of the estimates $\left|\binom{\alpha}{k}\left(\Gamma(-j-\alpha+k)\right)^{-1}\right| \leq (k!)^{-1}\,c(\alpha,j)k^j$ the series

$$b(t,s) = \sum_{k=0}^{\infty} \binom{\alpha}{k} b_k\,(t,s)$$

converges in $\mathcal{S}_{\mathcal{U}} \hat{\otimes}_{\mathcal{J}} \mathcal{S}_q$.

14. Remark. $op_G(b)$ for $\alpha = \lambda = -s$, $s \geq 0$, extends to a continuous operator $op_G(b) : L^2(\mathbb{R}_+) \longrightarrow \mathcal{S}_{P_0}(\mathbb{R}_+)$.

15. Proposition. We have for $s \geq 0$, $s \neq \frac{1}{2} \bmod \mathbb{Z}$ $\omega H^s(\mathbb{R}_+) = \omega\{\mathcal{H}^{s,s}(\mathbb{R}_+) + \mathcal{T}^s\}$.

Proof: Let us first show the continuous inclusion

$$\omega H^s(\mathbb{R}_+) \hookrightarrow \omega\{\mathcal{H}^{s,s}(\mathbb{R}_+) \oplus \mathcal{T}^s\},$$

$s \geq 0$, $s \neq \frac{1}{2} \bmod \mathbb{Z}$. Let $u \in H^s(\mathbb{R}_+)$, $\omega u = u$. Then, in virtue of the isomorphism $op_\psi(1_-^{-s}) : L^2(\mathbb{R}_+) \longrightarrow H^s(\mathbb{R}_+)$ we find a $v \in L^2(\mathbb{R}_+)$ such that $\omega u = \omega op_\psi(1_-^{-s})\,v$. Let ω_1 be another cut-off function with $\omega = \omega\omega_1$. Choose a sequence $v_j \in C_0^\infty(\mathbb{R}_+)$ with $v = \lim v_j$ in $L^2(\mathbb{R}_+)$. Then, in view of the Propositions 10 and 13 we obtain

$$\omega\, op_\psi(1_-^{-s})\,\omega_1 v_j = \omega op_\psi^0(1_-^{-s})\,\omega_1 v_j + \omega\, op_G(b)\,\omega_1 v_j$$

for all j. Remark 14 shows that $\omega op_G(b)\,\omega_1 v_j$ converges in $\mathcal{S}_{P_0}(\mathbb{R}_+)$. Moreover, $\omega op_\psi^0(1_-^{-s})\,\omega_1 v_j$ converges in $\mathcal{H}^{s,s}(\mathbb{R}_+)$. It is obvious that $\omega op_\psi(1_-^{-s})(1-\omega_1)v_j$ converges in $\mathcal{S}_{P_0}(\mathbb{R}_+)$. Thus we get the desired continuous inclusion. For the reverse it suffices to check $\omega\mathcal{H}^{s,s}(\mathbb{R}_+) \subseteq \omega H^s(\mathbb{R}_+)$. Let $u \in \mathcal{H}^{s,s}(\mathbb{R}_+)$, $u = \omega u$, ω_1 as mentioned. Choose a sequence $u_j \in C_0^\infty(\mathbb{R}_+)$, $u_j \longrightarrow u$ in $\mathcal{H}^{s,s}(\mathbb{R}_+)$. Then, by Proposition 10 we have $op_\psi(1_-^s)\,\omega_1 u_j = op_\psi^s(1_-^s)\,\omega_1 u_j = v_j$ and $v_j \longrightarrow v$ in $L^2(\mathbb{R}_+)$ for some v. By a standard property of $op_\psi(1_-^s)$ we know that

$$\omega u_j = \omega\, op_\psi(1_-^{-s})\, op_\psi(1_-^s)\, \omega_1 u_j$$
$$= \omega\, op_\psi(1_-^{-s})\, v_j$$

and it is clear that $op_\psi(1_-^{-s})\, v_j$ converges in $H^s(\mathbb{R}_+)$. This gives us the second inclusion. $\square$

2.3.2. Spaces with asymptotics

This section considers spaces with asymptotics (discrete, continuous) based on the $\mathcal{K}^{s,\gamma}$ spaces. The asymptotic types here play a similar role as those in Chapter 1. The following calculus is more precise in the sense of spaces with weights γ and finite smoothness orders s. We also admit finite parts of asymptotics on the right of $\{\mathrm{Re}\, z = \frac{1}{2} - \gamma\}$. This makes it necessary to introduce some cutting operations over singularity types.

1. Definition. Denote by $\underline{\mathcal{P}}(\gamma)$, $\gamma \in \mathbb{R}$, the set of all elements $\mathcal{P} \in \underline{\mathfrak{z}}$ for which

$$\pi_{\mathbb{C}} \mathcal{P} \subset \{\mathrm{Re}\, z < \tfrac{1}{2} - \gamma\}.$$

Set $\underline{\underline{\mathcal{P}}} = \bigcup_\gamma \underline{\mathcal{P}}(\gamma)$. For each $\delta \in \mathbb{R}$ we define the operations $l(\delta)$, $\bar{l}(\delta)$, $r(\delta)$, $\bar{r}(\delta)$ on $\underline{\mathfrak{z}}$ by

$$l(\delta)\mathfrak{z} = \{(p_j, m_j) \in \mathfrak{z} : \mathrm{Re}\, p_j < \tfrac{1}{2} - \delta\},$$

$$\bar{l}(\delta)\mathfrak{z} = \{(p_j, m_j) \in \mathfrak{z} : \mathrm{Re}\, p_j \leq \tfrac{1}{2} - \delta\},$$

$$r(\delta)\mathfrak{z} = \{(p_j, m_j) \in \mathfrak{z} : \mathrm{Re}\, p_j > \tfrac{1}{2} - \delta\},$$

$$\bar{r}(\delta)\mathfrak{z} = \{(p_j, m_j) \in \mathfrak{z} : \mathrm{Re}\, p_j \geq \tfrac{1}{2} - \delta\}.$$

Obviously, $\mathfrak{z} = l(\delta)\mathfrak{z} + \bar{r}(\delta)\mathfrak{z} = \bar{l}(\delta)\mathfrak{z} + r(\delta)\mathfrak{z}$ for each $\delta \in \mathbb{R}$ and $\mathfrak{z} \in \underline{\mathfrak{z}}$.

For $\mathcal{P} = \{(p_j, m_j)\} \in \underline{\underline{\mathcal{P}}}$ we have the translation $T^\lambda \mathcal{P} = \{(p_j + \lambda, m_j)\} \in \underline{\underline{\mathcal{P}}}$, $\lambda \in \mathbb{C}$, which defines a bijection

$$T^{\lambda} : \underline{\mathcal{P}}(\gamma) \longrightarrow \underline{\mathcal{P}}(\gamma-\lambda).$$

In particular $T^{\gamma}\underline{\mathcal{P}}(\gamma) = \underline{\mathcal{P}}(0) = \underline{\mathcal{P}}$.

For every $\mathcal{P} \in \underline{\mathcal{P}}(\gamma)$ we define the space

$$\mathcal{S}_{\mathcal{P}}(\mathbb{R}_+) := t^{\lambda}\, \mathcal{S}_{T^{\lambda}\mathcal{P}}(\mathbb{R}_+) \qquad (1)$$

for $\lambda \leqq \gamma$. The right-hand side is well-defined, cf. Chapter 1, and $\mathcal{S}_{\mathcal{P}}(\mathbb{R}_+)$ is equipped with the Fréchet space structure induced by the bijection. It can easily be proved that it is independent of the choice of λ.

We have $\mathcal{S}_{\mathcal{P}}(\mathbb{R}_+) \subset L^{2,\gamma}(\mathbb{R}_+)$, $\mathcal{P} \in \underline{\mathcal{P}}(\gamma)$, the space is nuclear, and the properties established in Chapter 1 for $\gamma = 0$ are valid in general in an analogous form. Remember that the Mellin transform on $L^{2,\gamma}$ is defined by $T^{\gamma} M t^{-\gamma}$. It will be denoted by M again if the weight is clear, and $Mu = \tilde{u}$.

$\mathcal{S}_{\mathcal{P}}(\mathbb{R}_+)$ many be written as projective limit

$$\mathcal{S}_{\mathcal{P}}(\mathbb{R}_+) = \varprojlim_{k \in \mathbb{Z}_+} \mathcal{S}^{(k)}_{\mathcal{P}}(\mathbb{R}_+),$$

where $\mathcal{S}^{(k)}_{\mathcal{P}}(\mathbb{R}_+)$ is defined as the space of all $u \in L^{2,\gamma}(\mathbb{R}_+)$ for which

(i) $Mu(z)$ is meromorphic in $\left\{\frac{1}{2} - \gamma - k < \mathrm{Re}\, z < \frac{1}{2} - \gamma + k\right\}$ with the poles corresponding to $\mathcal{P}$,

(ii) $\frac{1}{2\pi i} \int_{\Gamma_\rho} |\chi(z)\, (1 + |z|)^k\, Mu(z)|^2\, dz < \infty$

for all ρ, $\frac{1}{2} - \gamma - k < \rho < \frac{1}{2} - \gamma + k$, and every $\mathcal{P}$-excision function χ.

Remember that a $\mathcal{P}$-excision function is of module $\leqq 1$ and vanishes in a neighbourhood of $\pi_{\mathbb{C}}\mathcal{P}$.

Let $\mathfrak{q} = \{(q_1, m_1), \ldots, (q_N, m_N)\} \in \underline{\underline{\mathcal{P}}}$ be finite and $\omega(t)$ a fixed cut-off function. Then

$$\mathcal{E}(\mathfrak{q}) = \left\{ \sum_{j=1}^{N} \sum_{k=0}^{m_j} \zeta_{jk} t^{-q_j} \log^k t\, \omega(t) : \zeta_{jk} \in \mathbb{C} \right\}$$

is a finite-dimensional space. The dimension is

$$\dim \mathfrak{q} = \sum_{j=1}^{N} (m_j + 1). \qquad (2)$$

The latter notation will be used below for finite singularity types, i.e. when $\pi_{\mathbb{C}}\mathcal{O}$ is a finite set.

For every $\mathcal{P} \in \underline{\underline{\mathcal{P}}}$, $\gamma \in \mathbb{R}$, we have an analogue of the remainder operator $r_{\mathcal{P}}^{\gamma}$ of Section 1.2.1

$$r_{\mathcal{P}}^{\gamma} : \mathcal{S}_{\mathcal{P}} \longrightarrow \mathcal{S}_{\ell(\gamma)\mathcal{P}} ,$$

which is a projection with kernel $\mathcal{E}(\bar{r}(\gamma)\mathcal{P})$.

2. <u>Definition.</u> Let $\mathcal{P} \in \underline{\mathcal{P}}(\gamma)$ and denote by $\mathcal{H}^s(\mathbb{R}_+)$ the subspace of all $u \in \mathcal{H}^{s,\gamma}(\mathbb{R}_+) = t^{\gamma}\mathcal{H}^s(\mathbb{R}_+)$ for which

(i) $\tilde{u}(z) \in \mathcal{A}(\mathbb{C} \setminus \pi_{\mathbb{C}}\mathcal{P})$ is meromorphic with poles at p_j of order $m_j + 1$ for all j

(ii)

$$\frac{1}{2\pi i} \int_{\Gamma_\rho} (1+|z|^2) \quad | \quad \chi(z)\tilde{u}(z) \; |^2 \, dz \quad < \infty$$

for all $\rho \in \mathbb{R}$ and every $\mathcal{P}$-excision function χ.

The spaces $\mathcal{H}^s_{\mathcal{P}}$ admit a similar calculus as the $\mathcal{S}_{\mathcal{P}}$ spaces. The main difference is that we lose the nuclearity. The other obvious details are left to the reader, in particular that the spaces are Fréchet and

$$\mathcal{S}_{\mathcal{P}}(\mathbb{R}_+) = \varprojlim_{s \in \mathbb{R}} \mathcal{H}^s_{\mathcal{P}}(\mathbb{R}_+), \; \mathcal{P} \in \underline{\underline{\mathcal{P}}}.$$

Note that

$$\mathcal{H}^s_{\mathcal{P}}(\mathbb{R}_+) = \mathcal{H}^s_0(\mathbb{R}_+) + \mathcal{S}_{\mathcal{P}}(\mathbb{R}_+), \tag{3}$$

where O indicates the empty singularity type.

The remainder operator $r_{\mathcal{P}}^{\gamma}$ has a natural extension

$$r_{\mathcal{P}}^{\gamma} : \mathcal{H}^s_{\mathcal{P}}(\mathbb{R}_+) \longrightarrow \mathcal{H}^s_{1(\gamma)\mathcal{P}}(\mathbb{R}_+)$$

with $\ker r_{\mathcal{P}}^{\gamma} = \mathcal{E}(\bar{r}(\gamma)\mathcal{P})$. The topology of $\mathcal{H}^s_{\mathcal{P}}$ is that of the sum

$$\mathcal{H}^s_{\mathcal{P}} = \mathcal{H}^s_{1(\gamma)\mathcal{P}} + \mathcal{E}(\bar{r}(\gamma)\mathcal{P}).$$

We also will have to use the spaces $\mathcal{H}^{s,\gamma} + \mathcal{S}_{\mathcal{P}}$, $\mathcal{P} \in \underline{\underline{\mathcal{P}}}$, where, of course,

$$\mathcal{H}^{s,\gamma} + \mathcal{S}_{\mathfrak{p}} = \mathcal{H}^{s,\gamma} + \mathcal{E}(\bar{r}(\gamma)\mathfrak{p}) \qquad (4)$$

and $\mathcal{E}(\bar{r}(\gamma)\mathfrak{p})$ disappears for $\mathfrak{p} \in \underline{\mathfrak{p}}(\gamma)$. The equality (4) shows that the topology of sums is often very simple. Namely, whenever we have two Fréchet spaces E_1, E_2, and E_2 may be written as a direct sum $E_2 = E_2' + E_2''$ and $\dim E_2'' < \infty$, and when we have a **continuous embedding** $E_2' \hookrightarrow E_1$, then

$$E_1 + E_2 = E_1 + E_2''.$$

In other words, E_2' is absorbed and the sum appears as a finite-dimensional modification of E_1.

Analogous notations shall be used with respect to the $\mathcal{K}^{s,\gamma}$ spaces. In particular we have

$$\mathcal{K}^s_{\mathfrak{p}}(\mathbb{R}_+) = \mathcal{K}^s_0(\mathbb{R}_+) + \mathcal{S}_{\mathfrak{p}}(\mathbb{R}_+), \ \mathfrak{p} \in \underline{\underline{\mathfrak{p}}},$$

where

$$\mathcal{K}^s_0(\mathbb{R}_+) = \left\{ \omega u + (1-\omega)v \ : \ u \in \mathcal{H}^s_0(\mathbb{R}_+), v \in H^s(\mathbb{R}_+) \right\}.$$

Then

$$\mathcal{K}^s_{\mathfrak{p}}(\mathbb{R}_+) = \mathcal{K}^s_{1(\gamma)\mathfrak{p}}(\mathbb{R}_+) + \mathcal{E}(\bar{r}(\gamma)\mathfrak{p}),$$

$$\mathcal{K}^{s,\gamma}_{\mathfrak{p}}(\mathbb{R}_+) := \mathcal{K}^{s,\gamma}(\mathbb{R}_+) + \mathcal{S}_{\mathfrak{p}}(\mathbb{R}_+) = \mathcal{K}^{s,\gamma}(\mathbb{R}_+) + \mathcal{E}(\bar{r}(\gamma)\mathfrak{p}). \qquad (10)$$

It is clear that

$$\mathcal{K}^{s,\gamma}_{\mathfrak{p}}(\mathbb{R}_+) = g^{\gamma-\lambda}\mathcal{K}^{s,\lambda}_{T^{-\lambda}\mathfrak{p}}, \qquad (6)$$

$\lambda \in \mathbb{R}$. Thus, these spaces could be transformed to some fixed weight. For notational convenience we also write

$$\mathcal{K}^{s,\infty}_{\mathfrak{p}}(\mathbb{R}_+) =: \mathcal{K}^s_{\mathfrak{p}}(\mathbb{R}_+).$$

Remark that $\mathcal{K}^s_{\mathfrak{p}}(\mathbb{R}_+)$ can be written as a projective limit of Hilbert spaces $\mathcal{K}^s(\mathbb{R}_+)^{(j)}$, $j \in \mathbb{Z}_+$. The choice of such a scale is not canonical and follows from corresponding scales for $\mathcal{K}^s_0(\mathbb{R}_+)$ and $\mathcal{S}_{\mathfrak{p}}(\mathbb{R}_+)$. For references below the scalar products and norms belonging to $\mathcal{K}^s(\mathbb{R}_+)^{(j)}$ will be denoted by

$$(\ldots)_{j,\mathcal{K}^s_\varphi(\mathbb{R}_+)} \quad \text{and} \quad \|\cdot\|_{j,\mathcal{K}^s_\varphi(\mathbb{R}_+)}, \tag{7}$$

respectively.

The notation κ^γ_φ will also be used for the $\mathcal{K}$ - spaces,

$$\kappa^\gamma_\varphi : \mathcal{K}^s_\varphi \to \mathcal{K}^s_{l(\gamma)\varphi}, \quad \mathcal{K}^{s,\gamma} + \mathcal{S}_\varphi \to \mathcal{K}^{s,\gamma}.$$

The description in the Mellin image will always be given for ωu, whereas the influence of $M(1-\omega)u$ will be neglected at all. This is necessary here, since $M(1-\omega)u$ is holomorphic only in $\operatorname{Re} z < \frac{1}{2}$ (e.g. when $u \in L^2$) but $M\omega u$ is of interest in $\operatorname{Re} z < \frac{1}{2} - \gamma$, where γ may be negative, and then we would have an overlapping strip.

Let us also introduce the following small modification of our spaces (5) that only contain an asymptotic information in the open half space $\operatorname{Re} z > \frac{1}{2} - \gamma$. First observe that

$$\mathcal{K}^{s,\gamma} + \mathcal{S}_\varphi \hookrightarrow \mathcal{K}^{s',\gamma'} + \mathcal{S}_\varphi$$

for $s \geqq s'$, $\gamma \geqq \gamma'$ with continuous embedding.
Then we can define the projective limit

$$\mathcal{K}^{s,\gamma-0} + \mathcal{S}_\varphi := \varprojlim_{\gamma'<\gamma} (\mathcal{K}^{s,\gamma'} + \mathcal{S}_\varphi).$$

Note that

$$\mathcal{K}^{s,\gamma-0} + \mathcal{S}_\varphi = \Big\{ \varprojlim_{\gamma'<\gamma} \mathcal{K}^{s,\gamma'} \Big\} + \mathcal{S}_\varphi .$$

One may ask whether the spaces with discrete singularity types are invariant under certain natural operations such as multiplication by functions in $\mathcal{S}(\overline{\mathbb{R}}_+)$ or coordinate diffeomorphisms. In this connection it is natural to introduce the condition that

$$\sum_{j=0}^{\infty} T^{-j}\varphi \subseteq \varphi . \tag{8}$$

Observe that the Taylor singularity $\varphi_0 = \{(-j,0)\}_{j \in \mathbb{Z}_+}$ is of that type.

3. <u>Proposition</u>. The operator of multiplication by $\varphi \in \mathcal{S}(\overline{\mathbb{R}}_+)$ $(= \mathcal{S}_{\varphi_0}(\overline{\mathbb{R}}_+))$ induces a continuous operator

$$\varphi : \mathcal{K}^{s,\gamma}_{\mathfrak{p}}(\mathbb{R}_+) \longrightarrow \mathcal{K}^{s,\gamma}_{\mathfrak{q}}(\mathbb{R}_+),\ s,\gamma \in \mathbb{R},\ \mathfrak{p} \in \underline{\underline{\mathfrak{p}}}$$

$\mathfrak{q} = \sum_{j=0}^{\infty} T^{-j}\mathfrak{p}$ and the same without γ. More generally, for $\varphi \in \mathcal{S}_{\mathfrak{r}}(\mathbb{R}_+)$ and $\pi_{\mathbb{C}}\mathfrak{r} \subseteq \{\mathrm{Re}\ z < 0\}$ we have an analogous result with $\mathfrak{q} = \mathfrak{p}\mathfrak{r}$ (cf. 1.2.2. (22) and 1.2.3. Lemma 1).

Proof: For $\varphi \in \mathcal{S}(\mathbb{R}_+)$ we first have $\varphi : \mathcal{K}^{s,\gamma} \longrightarrow \mathcal{K}^{s,\gamma}$, since $\mathcal{K}^{s,\gamma} = g^{\gamma-s}\mathcal{K}^{s,s}$ and $\mathcal{K}^{s,s} = H^s_0$ for $s > -\frac{1}{2}$, $= H^s$ for $s < \frac{1}{2}$, cf. 2.3.1. Theorem 8 (remember that $g^{\rho}(t)$ equals t^{ρ} close to $t = 0$). Moreover, $\varphi : \mathcal{S}_{\mathfrak{p}} \longrightarrow \mathcal{S}_{\mathfrak{q}}$ is obvious, cf. 1.2.3. Lemma 1. Then the result follows by taking the sum, cf. (5). A similar consideration applies for the case without γ. The simple arguments for the general case are left to the reader. $\square$

Now let $\varkappa_0 : \mathbb{R} \longrightarrow \mathbb{R}$ be a diffeomorphism that induces a diffeomorphism $\varkappa : \mathbb{R}_+ \longrightarrow \mathbb{R}_+$ (such that in particular $\varkappa_0(0) = 0$) and let $(\varkappa^*\varphi)(r) = \varphi(\varkappa(r))$. Moreover, let $\varkappa(r) = r$ for $r \geq c$ with some constant $c > 0$.

4. Proposition. Let $\varkappa$ be as mentioned, then $\varkappa^*$ induces continuous operators

$$\varkappa^* : \mathcal{K}^{s,\gamma}_{\mathfrak{p}} \longrightarrow \mathcal{K}^{s,\gamma}_{\mathfrak{q}}, \tag{9}$$

$\mathfrak{q} = \sum_{j=0}^{\infty} T^{-j}\mathfrak{p}$, $s,\gamma \in \mathbb{R}, \mathfrak{p} \in \underline{\underline{\mathfrak{p}}}$, and the same without γ. Moreover, $\varkappa^*$ induces a continuous operator

$$\varkappa^* : \mathcal{S}_{\mathfrak{p}} \longrightarrow \mathcal{S}_{\mathfrak{q}}. \tag{10}$$

Proof. Let $t = \varkappa(r)$ and $\lambda \in \mathbb{C}$. Then the singular function $t^{\lambda}\,\omega(t)$ is represented in new coordinates by

$$\varkappa(r)^{\lambda}\ \omega(\varkappa(r)) = \Big(\sum_{j=0}^{N-1} c_j\ r^{\lambda+j} + r^{N+\lambda}\varphi_N(r)\Big)\,\omega_1(r)\ , \tag{11}$$

where $N \in \mathbb{Z}_+$ is arbitrarily given, $\omega_1(r) = \omega(\varkappa(r))$, $\varphi_N \in C^{\infty}(\overline{\mathbb{R}}_+)$. This follows from the Taylor expansion of $\varkappa$ at $r = 0$

$$\varkappa(r) = a_1 r + a_2 r^2 + \quad + a_{N-1}r^{n-1} + a_N(r)r^N$$

$$= a_1 r(1 + \frac{a_2}{a_1} r + \dots + \frac{a_N}{a_1} r^N)$$

where $a_1 \neq 0$, $a_N \in C^\infty(\overline{\mathbb{R}}_+)$ and

$$\varkappa(r)^\lambda = a_1^\lambda r^\lambda (1 + \frac{a_2}{a_1} r + \dots + \frac{a_N}{a_1} r^N)^\lambda$$

$$= a_1^\lambda r^\lambda \left(1 + b_1 r + \dots + b_N(r) r^N\right),$$

$b_N(r) \in C^\infty(\overline{\mathbb{R}}_+)$. Differentiation of (11) with respect to λ yields the transformation of singular functions with logarithmic terms. Then (10) follows from the invariance of $\mathcal{S}_0(\mathbb{R}_+) \cong \{u \in \mathcal{S}(\mathbb{R}) : \operatorname{supp} u \subseteq \overline{\mathbb{R}}_+\}$ and the behaviour of the singular functions under $\varkappa^*$. In view of 2.3.1. Theorem 8 and the invariance of the Sobolev spaces we get $\varkappa^* : \mathcal{K}^{s,s} \longrightarrow \mathcal{K}^{s,s}$ for all $s \in \mathbb{R}$. As proved above $\varkappa^* g^{\gamma-s}(r) = r^{\gamma-s} + o(r^{\gamma-s})$ for r near 0. Then the isomorphism

$$g^{\gamma-s} : \mathcal{K}^{s,\gamma} \longrightarrow \mathcal{K}^{s,s}$$

shows that also $\mathcal{K}^{s,\gamma}$ is invariant under $\varkappa^*$. Taking the sum, (9) is proved. □

5. Definition. Denote by $\underline{\Delta}(\gamma)$, $\gamma \in \mathbb{R}$, the set of all elements $\Delta \in \underline{Z}$ for which $\Delta \subset \{\operatorname{Re} z < \frac{1}{2} - \gamma\}$. Set $\underline{\underline{\Delta}} = \bigcup_\gamma \underline{\Delta}(\gamma)$.
Then we have the notion of spaces $\mathcal{S}_\Delta$ with continuous asymptotics, namely

$$\mathcal{S}_\Delta(\mathbb{R}_+) := t^\lambda \mathcal{S}_{T^\lambda \Delta}(\mathbb{R}_+)$$

$\Delta \in \underline{\Delta}(\gamma)$, $\lambda \leq \gamma$. The right-hand side is then well-defined and it can be proved again that it is independent of λ.

This leads to the spaces

$$\mathcal{H}^s_\Delta(\mathbb{R}_+) := \mathcal{H}^s_0(\mathbb{R}_+) + \mathcal{S}_\Delta(\mathbb{R}_+)$$

and

$$\mathcal{K}^s_\Delta(\mathbb{R}_+) := \mathcal{K}^s_0(\mathbb{R}_+) + \mathcal{S}_\Delta(\mathbb{R}_+)$$

$\Delta \in \underline{\underline{\Delta}}$. In case of continuous singularity types we also admit spaces of the sort

$$\mathcal{K}^{s,\varrho}_{\Delta}(\mathbb{R}_+) + \mathcal{S}_{\varphi}(\mathbb{R}_+) := \mathcal{K}^{s,\varrho}(\mathbb{R}_+) + \mathcal{S}_{\Delta}(\mathbb{R}_+) + \mathcal{S}_{\varphi}(\mathbb{R}_+) .$$

In this situation we keep in mind a fixed weight γ and assume $\varphi \in \bar{F}(\gamma)\underline{\underline{\varphi}}$, moreover $\varrho \geq \gamma$ and $\Delta \in \underline{\Delta}(\gamma)$. The reason is that in our calculus we have fixed reference weights, here γ, where on the right of $\Gamma_{\frac{1}{2}-\gamma}$ in the complex plane we only admit a discrete finite singularity type φ. Continuous parts occur only on the left. Moreover, we also have the spaces

$$\mathcal{K}^{s,\varrho-0}(\mathbb{R}_+) + \mathcal{S}_{\varphi}(\mathbb{R}_+) = \underset{\varrho'<\varrho}{\underleftarrow{\lim}} \left(\mathcal{K}^{s,\varrho'}_{\Delta}(\mathbb{R}_+) + \mathcal{S}_{\varphi}(\mathbb{R}_+)\right).$$

2.3.3. Green operators

1. Definition. Let $\gamma \in \mathbb{R}$ and $\varphi \in \bar{F}(\gamma)\underline{\underline{\varphi}}$. An operator

$$\mathcal{G} \in \bigcap_{s \in \mathbb{R}} \mathcal{L}(\mathcal{K}^{s,\gamma} + \mathcal{S}_{\varphi}, \mathcal{K}^{-\infty,-\infty}) \tag{1}$$

is called a left Green operator of discrete singularity type if it induces continuous operators

$$\mathcal{G} : \mathcal{K}^{s,\gamma} + \mathcal{S}_{\varphi} \longrightarrow \mathcal{S}_{\psi}, \ s \in \mathbb{R} . \tag{2}$$

for some $\psi \in \underline{\underline{\varphi}}$ depending on $\mathcal{G}$.

Remark that such operators have a kernel in $\mathcal{S}_{\psi} \otimes_{\pi} (\mathcal{K}^{\infty,-\gamma} + \mathcal{S}_{\varphi})$.

We shall often have to deal with couples $P=(\gamma,\varphi)$, $\gamma \in \mathbb{R}$, $\varphi \in \bar{F}(\gamma)\underline{\underline{\varphi}}$. Denote by $\underline{P}$ the set of all those P. Below we often write $\mathcal{K}^s_P(\mathbb{R}_+)$ instead of $\mathcal{K}^{s,\gamma} + \mathcal{S}_{\varphi}$.

Fix another couple $Q = (\lambda,\eta) \in \underline{P}$ and denote by (1) $- \mathcal{L}_G(Q,P)$ the subspace of all $\mathcal{G}$ of Definition 1 for which $\eta = \bar{F}(\lambda)\psi$. In other words we have fixed that part of the resulting singularity type which lies on the right of $\Gamma_{\frac{1}{2}-\lambda}$ (including this line itself) whereas the part on the left remains variable.

Thus

$$(1)-\mathcal{L}_G(Q,P) = \bigcap_{s\in\mathbb{R}} \bigcup_{\substack{\mathfrak{v}\in\underline{\mathfrak{p}} \\ \bar{\tau}(\lambda)\mathfrak{v}=\mathfrak{q}}} \mathcal{L}\left(\mathcal{K}^{s,\gamma}+\mathcal{S}_{\mathfrak{p}}, \mathcal{S}_{\mathfrak{v}}\right).$$

Write also

$$(1)-\mathcal{L}_G(\lambda,\gamma) := (1)-\mathcal{L}_G\left((\lambda,0),(\gamma,0)\right).$$

Remember that we have isomorphisms

$$\mathcal{K}^{s,\gamma}+\mathcal{S}_{\mathfrak{p}} \cong \mathcal{K}^{s,\gamma}+\mathcal{E}(\mathfrak{p}),$$

$$\mathcal{K}^{s,\lambda}+\mathcal{S}_{\mathfrak{q}} \cong \mathcal{K}^{s,\lambda}+\mathcal{E}(\mathfrak{q}),$$

where the sums on the right are direct. Thus every $\mathfrak{g}\in(1)-\mathcal{L}_G(Q,P)$ can be represented by a matrix

$$\begin{pmatrix} \mathfrak{g}_0 & \mathfrak{c} \\ \mathfrak{d} & \mathfrak{u} \end{pmatrix} : \begin{matrix} \mathcal{K}^{s,\gamma} \\ \oplus \\ \mathcal{E}(\mathfrak{p}) \end{matrix} \longrightarrow \begin{matrix} \mathcal{K}^{s,\lambda} \\ \oplus \\ \mathcal{E}(\mathfrak{q}) \end{matrix} \qquad (3)$$

Here

$$\mathfrak{g}_0 : \mathcal{K}^{s,\gamma} \to \mathcal{S}_{\mathfrak{w}},$$

and

$$\mathfrak{d} : \mathcal{K}^{s,\gamma} \to \mathcal{E}(\mathfrak{q}),$$

$$\mathfrak{c} : \mathcal{E}(\mathfrak{p}) \to \mathcal{S}_{\mathfrak{w}}, \quad \mathfrak{u} : \mathcal{E}(\mathfrak{p}) \to \mathcal{E}(\mathfrak{q}),$$

where $\mathfrak{w}\in\underline{\mathfrak{p}}(\lambda)$ depends on $\mathfrak{g}$.

Thus $\mathfrak{g}_0$ has a kernel in $\mathcal{S}_{\mathfrak{w}}\otimes_\pi\mathcal{K}^{\infty,-\gamma}$, i.e. $\mathfrak{g}_0 = \mathrm{op}_G(g_0)$, where $\mathrm{op}_G(g_0)$ is understood as an extension of $\int g_0(t,t')u(t')dt'$ to a pairing in t' between $\mathcal{K}^{\infty,-\gamma}$ and $\mathcal{K}^{-\infty,\gamma}$. Similarly $\mathfrak{d}$ has a kernel in $\mathcal{E}(\mathfrak{q})\otimes\mathcal{K}^{\infty,-\gamma}$.

The projections $\mathfrak{v}_{\mathfrak{q}}^{\lambda}$, $\mathfrak{v}_{\mathfrak{p}}^{\gamma}$ are fixed once and for all. So it is justified to identify g_0 and $\mathfrak{v}_{\mathfrak{q}}^{\lambda}\,\mathfrak{g}\,\mathfrak{v}_{\mathfrak{p}}^{\gamma}$.

The latter operator is called the regular part and $\mathcal{G}_1 = \mathcal{G} - \varkappa_{q}^{\lambda}\, \mathcal{G}\, \varkappa_{\varphi}^{\gamma}$ the singular part of $\mathcal{G}$.

Let us now consider the matrix (3) as a sum

$$\mathcal{G} = \mathcal{G}' + \mathcal{G}'' , \quad \mathcal{G}' = \begin{pmatrix} \mathcal{G}_0 & 0 \\ \mathcal{T} & 0 \end{pmatrix}, \quad \mathcal{G}'' = \begin{pmatrix} 0 & \mathcal{K} \\ 0 & \mathcal{Q} \end{pmatrix},$$

where $\mathcal{G}''$ is determined by the operator

$$\mathcal{B} = \begin{pmatrix} \mathcal{K} \\ \mathcal{Q} \end{pmatrix} : \mathcal{E}(\varphi) \longrightarrow \mathcal{K}^{s,\lambda} + \mathcal{E}(q).$$

Then $\mathcal{B}$ can be written as a composition $\mathcal{B} = \mathcal{R} \circ \mathcal{A}$, where

$$\mathcal{A} : \mathcal{E}(\varphi) \longrightarrow \mathbb{C}^N, \quad N = \dim(\varphi) = \dim \mathcal{E}(\varphi),$$

$$\mathcal{R} : \mathbb{C}^N \longrightarrow \mathcal{K}^{s,\lambda} + \mathcal{E}(q)$$

are linear operators, and $\mathcal{A}$ is realized in the following way.

Let $\varphi = \{(p_1, m_1), \ldots, (p_l, m_l)\}$, and set

$h_{jk} = (z-p_j)^k$, $0 \leq k \leq m_j$, $j = 1, \ldots, l$.

Denote by $M\mathcal{E}(\varphi)$ the finite-dimensional space of meromorphic functions, obtained by applying the weighted Mellin transform $u \longrightarrow T^{\rho} M t^{-\rho} u$ to $u \in \mathcal{E}(\varphi)$, $\rho \in \mathbb{R}$ chosen in such a way that $\operatorname{Re}(p_j+\rho) < \frac{1}{2}$ for $j=1, \ldots, l$.

Let $C_j = \{|z-p_j| = \varepsilon\}$ with $\varepsilon > 0$ so small that $p_\nu \notin \{|z-p_j| \leq \varepsilon\}$ for all $\nu \neq j$, and define

$$\mathcal{A} = \{\mathcal{A}_{jk}\}_{0 \leq k \leq m_j,\ 1 \leq j \leq l}$$

by

$$\mathcal{A}_{jk} u := \frac{1}{2\pi i} \int_{C_j} h_{jk}(z) Mu(z)\, dz.$$

Then $\mathcal{A}$ is obviously as desired. Note that $\mathcal{A}$ is well defined also on $\mathcal{S}_{\varphi}(\mathbb{R}_+)$ or on every space $\mathcal{S}_{\tilde{\varphi}}$ where $\varphi \subseteq \tilde{\varphi}$.

2. <u>Example.</u> Let $\varphi = \{(-j, 0)\}_{j=0,\ldots,l-1}$. Then

$$\gamma_{jo} u = \left(\frac{d}{dt}\right)^j u(t)\Big|_{t=0}, \quad j=0,\ldots, l-1,$$

is a trace operator of positive type on $\mathcal{S}_{P_0}(\mathbb{R}_+)$ in the sense of boundary symbols in Boutet de Monvel's algebra. Thus the above γ extends the notion of such trace operators for the general conormal asymptotics.

3. Definition. An operator $\mathcal{G} \in (l)-\mathcal{L}_G(Q,P)$ is called a Green operator of discrete singularity type if $\mathcal{G}_0$ has a kernel in $\mathcal{S}_{\mathcal{R}} \hat{\otimes}_\pi J\mathcal{S}_{\mathcal{R}_0}$ and $\mathcal{T}$ in $\mathcal{E}(\mathcal{O}) \otimes J\mathcal{S}_{\mathcal{R}_1}$ for certain $\mathcal{R}_0, \mathcal{R}_1 \in \underline{\mathcal{P}}(-\gamma)$. Denote by $\mathcal{L}_G(Q,P)$ the class of all Green operators with given $Q=(\lambda,\mathcal{O})$, $P=(\gamma,\mathcal{P})$ ($\mathcal{R}, \mathcal{R}_0, \mathcal{R}_1$ vary with $\mathcal{G}$). We also write

$$\mathcal{L}_G(\lambda,\gamma) = \mathcal{L}_G\big((\lambda,0),\ (\gamma,0)\big).$$

It can easily be verified that this definition is correct, i.e. independent of the representation of $\mathcal{G}$ as a matrix (3).

It is obvious that

$$\mathcal{G}_1 \in (1) - \mathcal{L}_G(Q,\tilde{Q}),\ \mathcal{G}_2 \in (1) - \mathcal{L}_G(\tilde{Q},P)$$

implies $\mathcal{G}_1\mathcal{G}_2 \in (1)-\mathcal{L}_G(Q,P)$, and the same without (1).

All essential results hold both for left Green and the proper Green operators. So, from time to time we give the formulations only for one case, where the other one is then left to the reader.

It is convenient also to use Green operators that have finite smoothness both with respect to the interior and conormal orders.

To this end we first observe that every $\mathcal{G} \in (1)-\mathcal{L}_G(Q,P)$ induces continuous operators

$$\mathcal{G} : \mathcal{K}^{s,\gamma} + \mathcal{S}_{\mathcal{P}} \longrightarrow \mathcal{K}^{s-(\gamma-\lambda)+k,\lambda+k-0} + \mathcal{S}_{\mathcal{R}} \tag{4}$$

$$t^j(1-\omega)\mathcal{G} : \mathcal{K}^{s,\gamma} + \mathcal{S}_{\mathcal{P}} \longrightarrow H^{s-(\gamma-\lambda)+k}(\mathbb{R}_+), \tag{5}$$

for all $k \in \mathbb{R}_+$, $j \in \mathbb{Z}_+$, where $\mathcal{R} \setminus \mathcal{O}$ depends on $\mathcal{G}$.

Denote by

$$(1)-\mathcal{L}_G(Q,P)^{(-k)} \tag{6}$$

the set of all $\mathcal{G}$ as in (1) for which (4), (5) are continuous for all $j \in \mathbb{Z}_+$.

Then

$$(1)-\mathcal{L}_G(Q,P)^{(-k-1)} \subset (1)-\mathcal{L}_G(Q,P)^{(-k)} \tag{7}$$

and

$$(1)-\mathcal{L}_G(Q,P) = \bigcap_{k \in \mathbb{Z}_+} (1)-\mathcal{L}_G(Q,P)^{(-k)}. \tag{8}$$

Moreover,

$$\mathcal{G}_1 \in (1)-\mathcal{L}_G(Q,Q)^{(-k)}, \ \mathcal{G}_2 \in (1)-\mathcal{L}_G(Q,P)^{(-k)}$$

implies $\mathcal{G}_1\mathcal{G}_2 \in (1)-\mathcal{L}_G(Q,P)^{(-k)}$.

$\mathcal{L}_G(Q,P)^{(-k)}$ denotes the subclass of all $\mathcal{G} \in (1)-\mathcal{L}_G(Q,P)^{(-k)}$ for which $\mathcal{G}$ has a kernel in $\mathcal{E}(\mathfrak{q}) \otimes (\mathcal{K}^{\infty,-\gamma+k} + \mathcal{S}_{\mathcal{P}_1})$ for some $\mathcal{P}_1 \in \underline{\mathcal{P}}(-\gamma)$ and

$$\mathcal{G}_0^* : \mathcal{K}^{s,-\lambda} \to \mathcal{K}^{s,-\gamma}$$

induces continuous operators

$$\mathcal{G}_0^* : \mathcal{K}^{s,-\lambda} \to \mathcal{K}^{s-(\gamma-\lambda)+k,-\gamma+k-0} + \mathcal{S}_{\mathcal{P}_0},$$

$$t^j(1-\omega)\mathcal{G}_0^* : \mathcal{K}^{s,-\lambda} \to H^{s-(\gamma-\lambda)+k}(\mathbb{R}_+),$$

for all $j \in \mathbb{Z}_+$, $\mathcal{P}_0 \in \underline{\mathcal{P}}(-\gamma)$ depending on $\mathcal{G}$.

4. <u>Remark.</u> The notion of left Green operators also suggests the notion of a right Green operator. For instance, we can set

$$(r)-\mathcal{L}_G(-\gamma,-\lambda)^{(-k)} := \{\mathcal{G}^* : \mathcal{G} \in (1)-\mathcal{L}_G(\lambda,\gamma)^{(-k)}\},$$

$\gamma, \lambda \in \mathbb{R}$. The extension of this definition in the sense of classes

$$(r)-\mathcal{L}_G(P',Q'),$$

$P' = (-\gamma, \mathcal{P}')$, $Q' = (-\lambda, \mathfrak{q}')$, is obvious and left to the reader.

Now let us briefly discuss the composition of differential operators and Green operators. Let D be a differential operator of order μ with coefficients in $C^{\infty}(\mathbb{R}_+)$ that are constant for $t \geq$ const. Then D induces continuous operators

$$D : \mathcal{S}_{\varkappa} \longrightarrow \mathcal{S}_{\ell} \;,$$

for every $\varkappa \in \underline{\underline{\varphi}}$, $\ell = \sum_{j=0}^{\infty} T^{-j}(T^{\mu}\varkappa)$, and

$$D : \mathcal{K}^{s,\gamma}(\mathbb{R}_+) \longrightarrow \mathcal{K}^{s-\mu,\gamma-\mu}(\mathbb{R}_+),$$

$s, \gamma \in \mathbb{R}$. Then we also get continuous operators

$$D : \mathcal{K}^{s,\gamma}(\mathbb{R}_+) + \mathcal{S}_{\varkappa}(\mathbb{R}_+) \longrightarrow \mathcal{K}^{s-\mu,\gamma-\mu}(\mathbb{R}_+) + \mathcal{S}_{\ell}(\mathbb{R}_+).$$

From this we easily obtain the following.

5. Proposition. Let D be a differential operator as mentioned, $\mu = \mathrm{ord}\, D$, then

$$D\left((1)-\mathcal{L}_G(Q,P)^{(-k)}\right) \subseteq (1)-\mathcal{L}_G(\tilde{Q},P)^{(-k)}, P = (\gamma,\varphi), Q = (\lambda,\eta), \tilde{Q} = (\lambda-\mu,\tilde{\eta})$$

$$\tilde{\eta} := \bar{r}(\lambda-\mu)\sum_{j=0}^{\infty} T^{-j}(T^{\mu}\eta), \text{ and}$$

$$\left((1)-\mathcal{L}_G(Q,P)^{(-k)}\right) D \subseteq (1)-\mathcal{L}_G(Q,\tilde{P})^{(-k)}$$

when $\bar{r}(\gamma)\varphi \subseteq \bar{r}(\gamma) \sum_{j=0}^{\infty} T^{-j}\varphi$, $\tilde{P} = (\gamma+\mu, T^{-\mu}\varphi)$. Analogous relations hold for the classes without (1).

Now let us introduce the notion of flatness of an operator.

6. Definition. An operator $\alpha \in \bigcap_s \mathcal{L}(\mathcal{K}^{s,\gamma}, \mathcal{K}^{s-\mu,\lambda})$ is called flat of order k at t=0 with respect to the weights γ, λ, if α induces continuous operators

$$\alpha : \mathcal{K}^{s,\gamma} \longrightarrow \mathcal{K}^{s-\mu,\lambda+k} \;, \tag{9}$$

$$\alpha^{*} : \mathcal{K}^{s,-\lambda} \longrightarrow \mathcal{K}^{s-\mu,-\gamma+k} \tag{10}$$

for all $s \in \mathbb{R}$. Here α^{*} denotes the formal adjoint of α.

Denote by $\mathring{\mathcal{L}}^{\mu}(\lambda,\gamma)^{(-k)}$ the space of all operators that are flat of order k at t=0.

Incidentally we also will employ the class (1)- $\mathring{\mathcal{L}}^{\mu}(\lambda,\gamma)^{(-k)}$ of those

$\mathfrak{a}$ for which only (9) is satisfied.

7. <u>Remark.</u> From the definition it follows immediately

$$\overset{\circ}{\mathcal{L}}{}^{\gamma-\lambda-k}(\lambda,\gamma)^{(-k)} \subset \mathcal{L}_G(\lambda,\gamma)^{(-k)} .$$

Denote by $L^{\mu}_{cl,(0,1)}(\mathbb{R}_+)$ the space of all classical ψDO's $\mathfrak{a}$ over $\mathbb{R}_+$ with

$$\mathfrak{a}u = \int e^{i(t-t')\tau} a(\tau)u(t')dt'đ\tau + \mathfrak{g}u$$

for all $u \in C_0^\infty(\mathbb{R}_+)$, $\operatorname{supp} u \subset (1,\infty)$, with some $a(\tau) \in S^{\mu}_{cl}(\mathbb{R})$ and a Green operator with kernel in $\mathcal{S}(\overline{\mathbb{R}}_+) \otimes_\pi \mathcal{S}(\overline{\mathbb{R}}_+)$. This space has a natural Fréchet topology (cf. [R 6] in an analogous situation).

Note that the formal adjoint $\mathfrak{a}^*$ of $\mathfrak{a} \in L^{\mu}_{cl,(0,1)}(\mathbb{R}_+)$ also belongs to $L^{\mu}_{cl,(0,1)}(\mathbb{R}_+)$.

The space $\mathcal{L}(\mathcal{K}^{s,\gamma},\mathcal{K}^{s-\mu,\lambda})$ will always be considered in the topology of the norm convergence of operators. Then $\bigcap_s \mathcal{L}(\mathcal{K}^{s,\gamma},\mathcal{K}^{s-\mu,\lambda})$ is a Fréchet space.

Denote by $\overset{\circ}{\mathcal{L}}{}^{\mu}_{\psi}(\lambda,\gamma)^{(-k)}$ the set of all $\mathfrak{a} \in L^{\mu}_{cl,(0,1)}(\mathbb{R}_+)$ which are flat of order k at t=0 with respect to γ,λ. Consider $\overset{\circ}{\mathcal{L}}{}^{\mu}_{\psi}(\lambda,\gamma)^{(-k)}$ in the weakest locally convex topology for which the embeddings both to

$$L^{\mu}_{cl,(0,1)}(\mathbb{R}_+) \cap \bigcap_{s\in\mathbb{R}} \mathcal{L}(\mathcal{K}^{s,\gamma},\mathcal{K}^{s-\mu,\lambda+k})$$

and

$$L^{\mu}_{cl,(0,1)}(\mathbb{R}_+) \cap \bigcap_{s\in\mathbb{R}} \mathcal{L}(\mathcal{K}^{s,-\lambda},\mathcal{K}^{s-\mu,-\gamma+k})$$

are continuous (the second one induced by the formal adjoint). Set

$$\overset{\circ}{\mathcal{L}}{}^{\mu}_{\psi} = \bigcap_{k\in\mathbb{Z}_+} \overset{\circ}{\mathcal{L}}{}^{\mu}_{\psi}(\lambda,\gamma)^{(-k)} . \tag{11}$$

The notation on the left of (11) indicates independence of γ,λ. The independence of the intersection of λ is obvious from (9). Moreover, (10) shows that $\mathfrak{a} \in \overset{\circ}{\mathcal{L}}{}^{\mu}_{\psi}$ satisfies $\omega t^{-k}\mathfrak{a}^* \in \overset{\circ}{\mathcal{L}}{}^{\mu}_{\psi}$ for all k. In other words $\mathfrak{a} \in \overset{\circ}{\mathcal{L}}{}^{\mu}_{\psi}$ implies $\omega t^{-k'}\mathfrak{a}\, t^{-k}\omega \in \overset{\circ}{\mathcal{L}}{}^{\mu}_{\psi}$ for all $k,k' \in \mathbb{Z}_+$, and hence we get independence of the weights.

Next we want to discuss Green operators of continuous singularity types.

8. Definition. Let $\gamma \in \mathbb{R}$, $\varphi \in \bar{\Gamma}(\gamma)_{\underline{\underline{\varphi}}}$. An operator

$$\mathcal{G} \in \bigcap_{s \in \mathbb{R}} \mathcal{L}(\mathcal{K}^{s,\gamma} + \mathcal{S}_\varphi, \mathcal{K}^{-\infty,-\infty})$$

is called a left Green operator of continuous singularity type if it induces continuous operators

$$\mathcal{G} : \mathcal{K}^{s,\gamma} + \mathcal{S}_\varphi \longrightarrow \mathcal{S}_\Delta + \mathcal{S}_\sigma, \quad s \in \mathbb{R},$$

for some $\Delta \in \underline{\Delta}(\lambda)$, $\sigma \in \bar{\Gamma}(\lambda)_{\underline{\underline{\varphi}}}$.

Fix $P = (\gamma, \varphi)$, $Q = (\lambda, \sigma)$, and denote by $(1)\text{-}\mathcal{G}_G(Q,P)$ the set of all left Green operators with the given P,Q. The continuous singularity type Δ on the left of $\Gamma_{\frac{1}{2}-\lambda}$ is variable and depends on the concrete operator. The operators in $(1)\text{-}\mathcal{G}_G(Q,P)$ admit representations as matrices of the form (3).

9. Definition. An operator $\mathcal{G} \in (1)\text{-}\mathcal{G}_G(Q,P)$ is called a Green operator of continuous singularity type if $\mathcal{G}_0$ has a kernel in $\mathcal{S}_\Xi \otimes_\Gamma J\mathcal{S}_{\Pi_0}$ and ϑ in $\mathcal{E}(\sigma) \otimes J\mathcal{S}_{\Pi_1}$ for certain $\Xi \in \underline{\Delta}(\lambda)$, $\Pi_0, \Pi_1 \in \underline{\Delta}(-\gamma)$. Denote by $\mathcal{G}_G(Q,P)$ the class of all Green operators with given P,Q.

Again we have the notion of regular and singular parts of (left) Green operators. Then every $\mathcal{G} \in (1)\text{-}\mathcal{G}_G(Q,P)$ can be written as $\mathcal{G} = \mathcal{G}_1 + \mathcal{G}_0$, where $\mathcal{G}_0$ belongs to $(1)\text{-}\mathcal{G}_G(\lambda,\gamma)$ (i.e. σ, φ are empty in the latter class) and $\mathcal{G}_1$ is singular in the sense of the class with discrete asymptotics. In other words, the difference between $(1)\text{-}\mathcal{G}_G(Q,P)$ and $(1)\text{-}\mathcal{G}_G(\lambda,\gamma)$ consists of finite-dimensional Green operators in $(1)\text{-}\mathcal{L}_G(Q,P)$.

Let $(1)\text{-}\mathcal{G}_G(Q,P)^{(-k)}$ be the set of all $\mathcal{G}$ as in (1) that induce continuous operators

$$\mathcal{G} : \mathcal{K}^{s,\gamma} + \mathcal{S}_\varphi \longrightarrow \mathcal{K}^{s-(\gamma-\lambda)+k,\lambda+k-0} + \mathcal{S}_\Delta + \mathcal{S}_\sigma, \qquad (12)$$

$$t^j(1-\omega)\mathcal{G} : \mathcal{K}^{s,\gamma} + \mathcal{S}_{\varphi} \longrightarrow H^{s-(\gamma-\lambda)+k}(\mathbb{R}_+), \tag{13}$$

for all $s \in \mathbb{R}$, and $j \in \mathbb{Z}_+$, $\Delta \in \underline{\Delta}(\lambda)$ depending on $\mathcal{G}$. Moreover, let $\mathcal{G}_G(Q,P)^{(-k)}$ be the subset for which $\mathcal{G}$ has a kernel in

$\mathcal{E}(\sigma)\hat{\otimes}(\mathcal{K}^{\infty,-\gamma+k} + \mathcal{S}_{\Pi_1})$

and

$$\mathcal{G}_0^* : \mathcal{K}^{s,-\lambda} \longrightarrow \mathcal{K}^{s-(\gamma-\lambda)+k,-\gamma+k-0} + \mathcal{S}_{\Pi_1},$$

$$t^j(1-\omega)\,\mathcal{G}_0^* : \mathcal{K}^{s,-\lambda} \longrightarrow H^{s-(\gamma-\lambda)+k}(\mathbb{R}_+),$$

for all $s \in \mathbb{R}$, $j \in \mathbb{Z}_+$, $\Pi_0, \Pi_1 \in \underline{\Delta}(-\gamma)$ depending on $\mathcal{G}$.

Then we have relations analogous to (7), (8), as well as to the behaviour under compositions.

Remark that there is an immediate analogue of Proposition 5 for the classes with continuous asymptotics.

The following proposition is a more precise version of the general result that if $1 + K : H \longrightarrow H$ is bijective, H a Hilbert space, K compact, $(1 + K)^{-1}$ has the form $1 + K_1$ with K_1 compact. Particular cases have already been discussed in the previous chapter.

10. Proposition. Let $\mathcal{G} \in (1)\text{-}\mathcal{G}_G(P,P)^{(-k)}$, $P = (\gamma, \varphi)$ and assume that

$$1 + \mathcal{G} : \mathcal{K}^{s,\gamma} + \mathcal{S}_{\varphi} \longrightarrow \mathcal{K}^{s,\gamma} + \mathcal{S}_{\varphi}$$

is invertible. Then $(1 + \mathcal{G})^{-1} = 1 + \mathcal{G}_1$ with some $\mathcal{G}_1 \in (1)\text{-}\mathcal{G}_G(P,P)^{(-k)}$. The analogous statement holds for the classes $\mathcal{G}_G(\dots)^{(-k)}$ and $(1)\text{-}\mathcal{L}_G(\dots)^{(-k)}$, $\mathcal{L}_G(\dots)^{(-k)}$, as well as for the projective limits over all k.

Proof: The inverse of $1 + \mathcal{G}$ is obviously of the form $1 + \mathcal{G}_1$ with some $\mathcal{G}_1 \in \mathcal{L}(\mathcal{K}^{s,\gamma} + \mathcal{S}_{\varphi}, \mathcal{K}^{s,\gamma} + \mathcal{S}_{\varphi})$, and $\mathcal{G}_1 = -\mathcal{G} - \mathcal{G}\mathcal{G}_1$. From (12) for $\varphi = \sigma$, $\gamma = \lambda$, it easily follows that

$$\mathcal{G}_1 : \mathcal{K}^{s,\gamma} + \mathcal{S}_{\varphi} \longrightarrow \mathcal{K}^{s-(\gamma-\lambda)+k} + \mathcal{S}_{\Delta} + \mathcal{S}_{\varphi}$$

$$t^j(1-\omega)\,g_1 : \mathcal{K}^{s,\gamma}+\mathcal{S}_{\mathcal{P}} \to H^{s-(\gamma-\lambda)+k}(\mathbb{R}_+)$$

for some $\Delta \in \underline{\Delta}(\gamma)$. This is obviously true of all s and hence we have proved the first assertion. The result without (ℓ) follows by a simple reduction to the case with $\mathcal{P} = \emptyset$, which is possible, since the extra terms associated with $\mathcal{P}$ are of finite dimension, cf.(3). The proof for the $\mathcal{L}_G$-classes is then the same. □

The definition of Green operators induces natural locally convex topologies in the corresponding classes. First consider $(1)-\mathcal{L}_G(Q,P)^{(-k)}$ defined by the mapping properties. Every fixed $\mathfrak{r}\in\underline{\underline{\mathcal{P}}}$ gives rise to a subclass $(1)-\mathcal{L}_G(Q,P)^{(-k)}(\mathfrak{m})$ where $\mathfrak{m}+q=\mathfrak{r}$, $\mathfrak{m}\in\underline{\mathcal{P}}(\lambda)$ and linear operators

$$(1)-\mathcal{L}_G(Q,P)^{(-k)}(\mathfrak{m}) \to \bigcap_s \mathcal{L}(\mathcal{K}^{s,\gamma}+\mathcal{S}_{\mathcal{P}},\ \mathcal{K}^{s-(\gamma-\lambda)+k,\lambda+k-\sigma}+\mathcal{S}_{\mathfrak{r}}),$$

given by (4), (5). Consider $(1)-\mathcal{L}_G(Q,P)^{(-k)}(\mathfrak{m})$ in the weakest locally convex topology under which these mappings are continuous. Then we get on this space a system of norms under which it is Fréchet. Now consider $(1)-\mathcal{L}_G(Q,P)(\mathfrak{m}) = \varprojlim_k (1)-\mathcal{L}_G(Q,P)^{(-k)}(\mathfrak{m})$ in the topology of the projective limit.

inally define

$$(1)-\mathcal{L}_G(Q,P)^{(-k)} = \varinjlim (1)-\mathcal{L}_G(Q,P)^{(-k)}(\mathfrak{m}),$$

where the inductive limit is taken over all $\mathfrak{m}\in\underline{\mathcal{P}}(\lambda)$, and in the same way

$$(1)-\mathcal{L}_G(Q,P) = \varinjlim (1)-\mathcal{L}_G(Q,P)(\mathfrak{m}).$$

Now we can proceed in a similar way for the class without (1). First we have to introduce the subspace $\mathcal{L}_G(Q,P)^{(-k)}(\mathfrak{m},\mathfrak{n},\mathfrak{n}_1)$ of $(1)-\mathcal{L}_G(Q,P)^{(-k)}(\mathfrak{m})$, $\mathfrak{n}\in\underline{\mathcal{P}}(-\gamma)$, for which

$$g_0^* : \mathcal{K}^{s,-\lambda} \to \mathcal{K}^{s-(\gamma-\lambda)+k,-\gamma+k-0}+\mathcal{S}_{\mathfrak{n}} \tag{14}$$

$$t^j(1-\omega)\, g_0^* \quad : \mathcal{K}^{s,-\lambda} \longrightarrow H^{s-(\gamma-\lambda)-k}(\mathbb{R}_+) \tag{15}$$

are continuous for all $s \in \mathbb{R}$, $j \in \mathbb{Z}_+$, and for which ϑ has a kernel in $\mathcal{E}(\Omega)\hat{\otimes}(\mathcal{K}^{\infty,-\gamma+k} + J\mathcal{S}_{\mathcal{O}_1})$ for the given $\mathcal{O}_1 \in \underline{\mathcal{P}}(-\gamma)$. This leads to a Fréchet topology in this subspace (corresponding to the convergence of the operators (4), (5) and (14), (15) and of the ϑ parts), which is stronger than that induced by (1)-$\mathcal{L}_G(Q,P)^{(-k)}(\mathcal{M}\mathcal{O})$.

Then we obtain topologies in

$$\mathcal{L}_G(Q,P)^{(-k)} = \varinjlim_{\mathcal{M}\mathcal{O},\mathcal{O},\mathcal{O}_1} \mathcal{L}_G(Q,P)^{(-k)}(\mathcal{M}\mathcal{O},\mathcal{O},\mathcal{O}_1)$$

$$\mathcal{L}_G(Q,P) = \varprojlim \mathcal{L}_G(Q,P)^{(-k)},$$

and so on. For the $\mathcal{G}_G$-classes we can do the same. The obvious details are left to the reader.

2.3.4. Mellin operators

Let us now extend the calculus of Mellin operators of Section 1.3 with respect to arbitrary conormal orders.

Remember that $\underline{\mathfrak{z}}$ denotes the set of all countable subsets $\mathfrak{z}$ of $\mathbb{C} \times \mathbb{Z}_+$ such that

$$\pi_{\mathbb{C}}\mathfrak{z} \cap \left\{ \tfrac{1}{2} - \rho < \operatorname{Re} z < \tfrac{1}{2} + \rho \right\}$$

is finite for every $\rho \in \mathbb{R}_+$.

1. Definition. Let $\mathfrak{z} = \{(r_j, n_j)\} \in \underline{\mathfrak{z}}$, $\nu \in \mathbb{R}$, and denote by $\mathcal{M}_{\mathfrak{z}}^{\nu}$ the space of all $h(z) \in \mathcal{A}(\mathbb{C} \setminus \pi_{\mathbb{C}}\mathfrak{z})$ that are meromorphic with poles at the points r_j of multiplicities n_j+1, and

$$\sup_{-\rho < \operatorname{Re} z < \rho} \left| (1+|z|)^{-\nu} \chi(z) h(z) \right| \leq c(\rho)$$

for all $\rho \in \mathbb{R}$ and every $\mathfrak{z}$-excision function χ, $c(\rho) > 0$ a constant. The space $\mathcal{M}_{\mathfrak{z}}^{\nu}$ is Fréchet in a natural semi-norm system induced by the definition. We have

$$\mathcal{M}_{\mathfrak{Z}} = \mathcal{M}_{\mathfrak{Z}}^{-\infty} = \varprojlim_{\nu \in \mathbb{R}} \mathcal{M}_{\mathfrak{Z}}^{\nu}$$

where $\mathcal{M}_{\mathfrak{Z}}$ was defined in Section 1.3.1. By analogous considerations as for the proof of theorems of Borel type we have

$$\mathcal{M}_{\mathfrak{Z}}^{\nu} = \mathcal{M}_{\mathfrak{Z}} + \mathcal{M}_{O}^{\nu}.$$

where O indicates the empty singularity type.

Set

$$\underline{\mathfrak{Z}}[\gamma] = \{ \mathfrak{Z} \in \underline{\mathfrak{Z}} : \pi_{\mathbb{C}} \mathfrak{Z} \cap \Gamma_{\frac{1}{2}-\gamma} = \emptyset \}$$

and

$$\mathcal{M}_{\underline{\mathfrak{Z}}[\gamma]}^{\nu} = \varinjlim_{\mathfrak{Z} \in \underline{\mathfrak{Z}}[\gamma]} \mathcal{M}_{\mathfrak{Z}}^{\nu} .$$

Now let $h \in \mathcal{M}_{\underline{\mathfrak{Z}}[\gamma]}^{\mu}$, $\mu, \gamma \in \mathbb{R}$, and

$$op_M^{\gamma}(h) := t^{\gamma} \; op_M(T^{-\gamma}h)t^{-\gamma} .$$

Then the operator

$$\omega t^{-\mu} op_M^{\gamma}(h) \omega : \mathcal{H}^{s,\gamma}(\mathbb{R}_+) \longrightarrow \mathcal{H}^{s-\mu,\gamma-\mu}(\mathbb{R}_+) \tag{1}$$

is continuous, $s \in \mathbb{R}$, and induces continuous operators

$$\mathcal{H}_{\varphi}^{s}(\mathbb{R}_+) \longrightarrow \mathcal{H}_{\sigma}^{s-\mu}(\mathbb{R}_+)$$

for every $\varphi \in \underline{\varphi}(\gamma)$ with some $\sigma \in \underline{\varphi}(\gamma-\mu)$. It is clear that the $\mathcal{H}$-spaces might be replaced by the $\mathcal{K}$-spaces.

The formal adjoint of (1) is given by

$$\omega \, op_M^{-\gamma}(h^*) t^{-\mu} \omega : \mathcal{H}^{\mu-s,-\gamma+\mu}(\mathbb{R}_+) \longrightarrow \mathcal{H}^{-s,-\gamma}(\mathbb{R}_+),$$

$$h^*(z) = \bar{h}(1-\bar{z}) \in \mathcal{M}_{\underline{\mathfrak{Z}}[\gamma]}^{\mu} .$$

We are interested mainly in the case $\mu = -\infty$, but a number of assertions holds for arbitrary μ. So we shall formulate a part of the calculus in general and below apply it in the special case.

2. Proposition. Let $k \in \mathbb{Z}_+$, $k \geq 1$, and $\gamma - k \leq \lambda \leq \gamma$, $h \in \mathfrak{M}^{\mu}_{\underline{\mathfrak{Z}}[\lambda]}$. Then

$$\omega t^{-\mu+k} \operatorname{op}_M^{\lambda}(h)\omega \in \mathcal{L}_G(\gamma-\mu,\gamma)^{(-k)} + \overset{\circ}{\mathcal{L}}{}^{\mu}(\gamma-\mu,\gamma)^{(-k)} .$$

Proof: h may be written in the form $h = h_o + h_1$, $h_o \in \mathfrak{M}_{\underline{\mathfrak{Z}}[\lambda]}$, $h_1 \in \mathfrak{M}^{\mu}_{0}$. Then it suffices to show that

$$\omega t^{-\mu+k} \operatorname{op}_M^{\lambda}(h_o)\omega \in \mathcal{L}_G(\gamma-\mu,\gamma)^{(-k)},$$

$$\omega t^{-\mu+k} \operatorname{op}_M^{\lambda}(h_1)\omega \in \overset{\circ}{\mathcal{L}}{}^{\mu}(\gamma-\mu,\gamma)^{(-k)}.$$

Choose a small $\varepsilon > 0$ such that $h \in \mathfrak{M}^{\mu}_{\underline{\mathfrak{Z}}[\gamma-\varepsilon]}$. Then we know by the results of Section 1.3.1. that

$$\omega t^{-\mu+k} \operatorname{op}_M^{\lambda}(h_o)\omega - \omega t^{-\mu+k} \operatorname{op}_M^{\gamma-\varepsilon}(h_o)\omega \in \mathcal{L}_G(\gamma-\mu,\gamma).$$

Moreover,

$$\omega t^{-\mu+k} \operatorname{op}_M^{\gamma-\varepsilon}(h_o)\omega : \mathcal{K}^{s,\gamma} \to \mathcal{K}^{\infty,\gamma-\mu+k-\varepsilon}$$

is continuous. Thus $\omega t^{-\mu+k} \operatorname{op}_M^{\lambda}(h_o)\omega \in (1)-\mathcal{L}_G(\gamma-\mu,\gamma)^{(-k)}$.

For the adjoint operator we may conclude in the same way. Thus the assertion holds without 1.

Now we have

$$\omega t^{-\mu+k} \operatorname{op}_M^{\lambda}(h_1)\omega = \omega t^{-\mu+\gamma+k} \operatorname{op}_M^{\gamma}(h_1)\omega$$

and hence

$$\omega t^{-\mu+k} \operatorname{op}_M^{\lambda}(h_1)\omega \; : \mathcal{K}^{s,\gamma} \to \mathcal{K}^{s-\mu,\gamma-\mu+k}.$$

By the same argument follows

$$(\omega t^{-\mu+k} \operatorname{op}_M^{\lambda}(h_1)\omega)^* : \mathcal{K}^{s,-(\gamma-\mu)} \to \mathcal{K}^{s-\mu,-\gamma+k}. \quad \square$$

3. Proposition. Let $k \in \mathbb{Z}_+$, $k \geq 1$, $\gamma - k \leq \lambda, \delta \leq \gamma$, $h \in \mathfrak{M}^{\mu}_{\underline{\mathfrak{Z}}[\lambda] \cap \underline{\mathfrak{Z}}[\delta]}$. Then

$$\omega_1 t^{-\mu+k} \operatorname{op}_M^{\lambda}(h)\omega_2 - \tilde{\omega}_1 t^{-\mu+k} \operatorname{op}_M^{\delta}(h)\tilde{\omega}_2$$
$$\in \mathcal{L}_G(\gamma-\mu,\gamma)^{(-k)} + \overset{\circ}{\mathcal{L}}{}^{\mu}(\gamma-\mu,\gamma)^{(-k)}$$

for arbitrary cut-off functions $\omega_i, \tilde{\omega}_i$, $i=1,2$.

The flat term vanishes when $\omega_1=\tilde{\omega}_1, \omega_2=\tilde{\omega}_2$.

4. Definition. An operator

$$\mathfrak{a} \in \bigcap_{s\in\mathbb{R}} \mathcal{L}(\mathcal{K}^{s,\gamma}, \mathcal{K}^{s-\mu,\gamma-\mu}) \tag{2}$$

is said to have a (left-) Mellin expansion near t=0 with the conormal symbols

$$\sigma_M^{\mu-j}(\mathfrak{a}) = a_j \in \mathfrak{M}^{\mu}_{\mathfrak{Z}_j}\ , \ j \in \mathbb{Z}_+,$$

if for certain $\gamma_j \in \mathbb{R}$, $\gamma - j \leq \gamma_j \leq \gamma$, $j \in \mathbb{Z}_+$, and every $N \in \mathbb{Z}_+$

$$\omega_1\{\mathfrak{a} - t^{-\mu} \sum_{j=0}^{N-1} t^j op_M^{\gamma_j}(a_j)\}\omega_2 \tag{3}$$

$$\in (1)-\mathcal{L}_G(\gamma-\mu,\gamma)^{(-N)} + (1)\text{-}\mathring{\mathcal{L}}^{\mu}(\gamma-\mu,\gamma)^{(-N)}$$

ω_1, ω_2 arbitrary cut-off functions.

We talk about a Mellin expansion if (3) holds without (1) on the right.

Note that the definition implies $\gamma_0=\gamma$ and hence $a_0 \in \mathfrak{M}^{\mu}_{\underline{\mathfrak{Z}}[\gamma]}$.

5. Remark. Let $\mathfrak{a}$ have a left-Mellin expansion. Then the conormal symbols $\sigma_M^{\mu-j}(\mathfrak{a})$ are uniquely determined, $j \in \mathbb{Z}_+$. They are also unique for $j=0,\dots,N-1$ if the expansion only holds with fixed N. This is a consequence of 1.3.1. Theorem 11.

6. Proposition. Let $\mathfrak{a}$ be an operator satisfying Definition 4. Then $\mathfrak{a}$ induces continuous operators

$$\mathfrak{a} : \mathcal{K}^s_{\mathcal{P}}(\mathbb{R}_+) \to \mathcal{K}^{s-\mu}_{\mathcal{Q}}(\mathbb{R}_+) \tag{4}$$

for every $s \in \mathbb{R}$, $\mathcal{P} \in \underline{\mathcal{P}}(\gamma)$ with some $\mathcal{Q} \in \underline{\mathcal{P}}(\gamma-\mu)$ depending on $\mathcal{P}$ and $\mathfrak{a}$.

If $\mathfrak{a}$ is an operator satisfying Definition 4 we can define an extension to $\mathcal{K}^{s,\gamma} + \mathcal{S}_{\mathcal{P}}$, $\mathcal{P} \in \bar{F}(\gamma)\underline{\underline{\mathcal{P}}}$ by $\mathfrak{a}\mathfrak{A}^{\gamma}_{\mathcal{P}}$. Then $\mathfrak{a}$ defines continuous operators

$$\mathfrak{a} : \mathcal{K}^{s,\gamma} + \mathcal{S}_{\mathcal{P}} \to \mathcal{K}^{s-\mu,\gamma-\mu}$$

for arbitrary $\mathcal{P} \in \underline{\underline{\mathcal{P}}}$ and also continuous operators in the sense of (4).

7. Definition. Let $\gamma, \mu \in \mathbb{R}$, $\varphi \in \bar{\Gamma}(\gamma)\underline{\underline{\mathfrak{p}}}$, $\mathfrak{q} \in \bar{\Gamma}(\gamma-\mu)\underline{\underline{\mathfrak{p}}}$, $P = (\varphi, \gamma)$, $Q = (\mathfrak{q}, \gamma-\mu)$, and denote by $\mathcal{L}_{M+G}(Q,P)^{(-k)}$ the space of all

$$\mathfrak{a} \in \bigcap_{s \in \mathbb{R}} \mathcal{L}(\mathcal{K}^{s,\gamma} + \mathcal{S}_{\varphi}, \mathcal{K}^{\infty,\gamma-\mu} + \mathcal{S}_{\mathfrak{q}}) \tag{5}$$

for which

$$\mathfrak{a} - \omega_1 \Big\{ \sum_{j=0}^{k-1} t^{-\mu+j} \mathrm{op}_M^{\gamma_j}(a_j) \Big\} \omega_2 \in \mathcal{L}_G(Q,P)^{(-k)} \tag{6}$$

for certain $a_j \in \mathfrak{M}_{\mathfrak{z}_j}$, $\gamma - j \leq \gamma_j \leq \gamma$, ω_1, ω_2 arbitrary cut-off functions. If $\mathfrak{a}$ satisfies (6) with $(1)\text{-}\mathcal{L}_G$ instead, we get by definition the class $(1)\text{-}\mathcal{L}_{M+G}(Q,P)^{(-k)}$.

The subclasses for which $\varphi = \mathfrak{q} = \emptyset$ will be denoted by $\mathcal{L}_{M+G}(\gamma-\mu,\gamma)^{(-k)}$ and $(1)\text{-}\mathcal{L}_{M+G}(\gamma-\mu,\gamma)^{(-k)}$, respectively.

For $\mathfrak{a} \in (1)\text{-}\mathcal{L}_{M+G}(Q,P)^{(-k)}$ we write $\sigma_M^{\mu-j}(\mathfrak{a}) = \sigma_M^{\mu-j}(\mathfrak{a}_0)$, $j=0,\dots,k-1$.

8. Remark. $\mathfrak{a} \in \mathcal{L}_{M+G}(\gamma-\mu,\gamma)^{(-k)}$ implies $\mathfrak{a}^* \in \mathcal{L}_{M+G}(-\gamma,-\gamma+\mu)^{(-k)}$ and

$$\sigma_M^{\mu-j}(\mathfrak{a}^*)(z) = \sigma_M^{\mu-j}(\mathfrak{a})^*(z+\mu-j), \quad j=0,\dots,k-1.$$

Here $h^*(z) = \bar{h}(1-\bar{z})$.

The following calculus will be formulated for the $\mathcal{L}_{M+G}$-classes. For $(1)\text{-}\mathcal{L}_{M+G}$, which is the larger class, we can easily do the same. The details may be dropped here.

By definition we have

$$\mathcal{L}_{M+G}(Q,P)^{(-k)} = \mathcal{L}_{M+G}(\gamma-\mu,\gamma)^{(-k)} + \mathcal{L}_G(Q,P)^{(-k)},$$

where the elements $\mathfrak{a}_0 \in \mathcal{L}_{M+G}(\gamma-\mu,\gamma)^{(-k)}$ act on $\mathcal{K}^{s,\gamma} + \mathcal{S}_{\varphi}$ by $\mathfrak{a} := \mathfrak{a}_0 \pi_{\varphi}^{\gamma}$.

From this we easily obtain the following.

For every $\mathfrak{a} \in \mathcal{L}_{M+G}(Q,P)^{(-k)}$ we can recover the conormal symbols

$$\sigma_M^{\mu}(\mathfrak{a})(z), \dots, \sigma_M^{\mu-(k-1)}(\mathfrak{a})(z)$$

by the standard procedure, cf. Remark 5. This yields a well-defined linear mapping

$$\{\sigma_M^{\mu-j}\} : \mathcal{L}_{M+G}(Q,P)^{(-k)} \to \mathfrak{M}_{\underline{z}[\gamma]} \times \mathop{\times}_{j=1}^{k-1} \mathfrak{M}_{\underline{z}} .$$

By definition

$$0 \to \mathcal{L}_G(Q,P)^{(-k)} \to \mathcal{L}_{M+G}(Q,P)^{(-k)} \xrightarrow{\{\sigma_M^{\mu-j}\}} \tag{7}$$

$$\mathfrak{M}_{\underline{z}[\gamma]} \times \mathop{\times}_{j=1}^{k-1} \mathfrak{M}_{\underline{z}} \to 0$$

is exact.

9. <u>Theorem.</u> The sequence

$$0 \to \mathcal{L}_G(Q,P) \to \mathcal{L}_{M+G}(Q,P) \xrightarrow{\{\sigma_M^{\mu-j}\}}$$

$$\mathfrak{M}_{\underline{z}[\gamma]} \times \mathop{\times}_{j=1}^{\infty} \mathfrak{M}_{\underline{z}} \to 0$$

is exact, $Q = (\gamma-\mu,\varphi)$, $P = (\gamma,\psi)$.

This can be proved by a similar construction as the corresponding result in 1.3.1. Corollary 14.

10. <u>Theorem.</u> Let $\gamma,\mu,\nu \in \mathbb{R}$, $P = (\gamma,\psi)$, $Q = (\varphi,\gamma-\mu)$, $R = (\sigma,\gamma-\mu-\nu)$. Then

$$\mathfrak{a} \in \mathcal{L}_{M+G}(R,Q)^{(-k)},\ \mathfrak{b} \in \mathcal{L}_{M+G}(Q,P)^{(-k)}$$

implies $\mathfrak{a}\mathfrak{b} \in \mathcal{L}_{M+G}(R,Q)^{(-k)}$ and

$$\sigma_M^{\mu+\nu-j}(\mathfrak{a}\mathfrak{b})(z) = \sum_{j=j_1+j_2} \sigma_M^{\nu-j_1}(\mathfrak{a})(z+\mu-j_2)\,\sigma_M^{\mu-j_2}(\mathfrak{b})(z),$$

$j=0,\ldots,k-1$. The Green operators form a two-sided ideal under this composition.

Now we want to introduce in $\mathcal{L}_{M+G}(Q,P)$ a natural locally convex topology. It suffices to consider $\mathcal{L}_{M+G}(Q,P)^{(-k)}$ and then to set

$$\mathcal{L}_{M+G}(Q,P) = \varprojlim_k \mathcal{L}_{M+G}(Q,P)^{(-k)}.$$

Let k be fixed and consider a fixed tuple

$\mathfrak{z} = \{z_0, z_1, \ldots, z_{k-1}\} \in \underline{z}[\gamma] \times \underline{z}^{k-1}$. Denote by

$\mathcal{L}_{M+G}(\ldots)_{\mathfrak{V}}^{(-k)}$ the subspace of those operators $\mathfrak{a}$ in $\mathcal{L}_{M+G}(\ldots)^{(-k)}$ for which $\sigma_M^{\mu-l}(\mathfrak{a}) \in \mathfrak{M}_{\mathfrak{z}_l}$, $l=0,\ldots,k-1$. Then the exact sequence

$$0 \to \mathcal{L}_G(Q,P)^{(-k)} \to \mathcal{L}_{M+G}(Q,P)^{(-k)} \xrightarrow{\{\sigma_M^{\mu-j}\}} \mathop{\times}_{j=0}^{k-1} \mathfrak{M}_{\mathfrak{z}_j} \to 0$$

splits. Thus

$$\mathcal{L}_{M+G}(Q,P)_{\mathfrak{V}}^{(-k)} \cong \mathcal{L}_G(Q,P)^{(-k)} \oplus \sum_{j=0}^{k-1} \mathfrak{M}_{\mathfrak{z}_j}.$$

In the previous section we have defined a topology in $\mathcal{L}_G(\ldots)^{(-k)}$. The spaces $\mathfrak{M}_{\mathfrak{z}_j}$ also have a given topology. Then $\mathcal{L}_{M+G}(\ldots)^{(-k)}$ can be equipped with the topology of the sum. Finally we can define

$$\mathcal{L}_{M+G}(\ldots)^{(-k)} = \varinjlim_{\mathfrak{V}} \mathcal{L}_{M+G}(\ldots)_{\mathfrak{V}}^{(-k)}.$$

Next let D be a differential operator as in 2.3.3. Proposition 5, $\nu = \text{ord } D$.

11. <u>Proposition</u>. We have

$$D\left((1)-\mathcal{L}_{M+G}(Q,P)^{(-k)}\right) \subseteq (1)-\mathcal{L}_{M+G}(\tilde{Q},P)^{(-k)},$$

$$P=(\gamma,\varphi),\ Q=(\gamma-\mu,\mathfrak{q}),\ \tilde{Q}=(\gamma-\mu-\nu,\tilde{\mathfrak{q}}),$$

$$\tilde{\mathfrak{q}} = \bar{F}(\gamma-\mu-\nu) \sum_{j=0}^{\infty} T^{-j}(T^{\mu}\mathfrak{q}),\ \text{and}$$

$$\left((1)-\mathcal{L}_G(Q,P)^{(-k)}\right) D \subseteq (1)-\mathcal{L}_G(Q,\tilde{P})^{(-k)}$$

if $\bar{F}(\gamma)\varphi \subseteq \bar{F}(\gamma) \sum_{j=0}^{\infty} T^{-j}\varphi$, $\tilde{P} = (\gamma+\nu, T^{-\nu}\varphi)$. Analogous relations hold for the classes without (1).

For the continuous singularity types we can do the same. First we can easily formulate an extension of the Definition 1 to continuous singularity types and thus get the classes $\mathfrak{M}_Z^{\nu}$, $Z \in \underline{Z}$. Then $\mathfrak{M}_Z^{\nu} = \mathfrak{M}_0^{\nu} + \mathfrak{M}_Z$, $\mathfrak{M}_Z = \mathfrak{M}_Z^{-\infty}$, cf. Section 2.2.2.

Moreover, set

$$\underline{Z}[\gamma] = \{Z \in \underline{Z} : Z \cap \Gamma_{\frac{1}{2}-\gamma} = \emptyset\},$$

$$\mathfrak{M}^{\nu}_{\underline{Z}[\gamma]} = \lim_{\overrightarrow{Z \in \underline{Z}[\gamma]}} \mathfrak{M}^{\nu}_{Z} .$$

Every $h \in \mathfrak{M}^{\nu}_{\underline{Z}[\gamma]}$ induces continuous operators

$$\omega t^{-\mu} \mathrm{op}_M^{\gamma}(h)\omega : \mathcal{H}^{s,\gamma}(\mathbb{R}_+) \longrightarrow \mathcal{H}^{s-\nu,\gamma-\mu}(\mathbb{R}_+)$$

with continuous restrictions

$$\mathcal{H}^{s}_{\Delta}(\mathbb{R}_+) \longrightarrow \mathcal{H}^{s-\nu}_{\Sigma}(\mathbb{R}_+)$$

for every $\Delta \in \underline{\Delta}(\gamma)$ and some resulting $\Sigma \in \underline{\Delta}(\gamma-\mu)$. Here the $\mathcal{H}$-spaces also can be replaced by the $\mathcal{K}$-spaces.

In the framework of the continuous asymptotics we also consider expressions like

$$t^{-\mu} \sum_{i=1}^{p_j} t^j \mathrm{op}_M^{\gamma_{ij}}(a_{ij}),$$

$a_{ij} \in \mathfrak{M}^{\nu}_{\underline{Z}[\gamma_{ij}]}$, $\gamma - j \leqq \gamma_{ij} \leqq \gamma$, cf. Section 2.2.2.

12. Definition. An operator (2) is said to have a Mellin expansion near t=0 with the conormal symbols

$$\sigma_M^{\mu-j}(\mathfrak{a}) = a_j \in \mathfrak{M}^{\mu}_{Z_j}, \quad Z_j \in \underline{Z},$$

$j \in \mathbb{Z}_+$, if for certain decompositions $a_j = \sum_{i=1}^{p_j} a_{ij}$, $a_{ij} \in \mathfrak{M}^{\mu}_{Z_{ij}}$ and numbers γ_{ij} with $\gamma - j \leqq \gamma_{ij} \leqq \gamma$, $Z_{ij} \in \underline{Z}[\gamma_{ij}]$, for every k

$$\omega_1\{\mathfrak{a} - t^{-\mu} \sum_{j=0}^{k-1} \sum_{i=1}^{p_j} t^j \mathrm{op}_M^{\gamma_{ij}}(a_{ij})\} \omega_2$$
$$\in \mathcal{C}_G(\gamma-\mu,\gamma)^{(-k)} + \mathring{\mathcal{L}}(\gamma-\mu,\gamma)^{(-k)},$$

ω_1, ω_2 arbitrary cut-off functions.

13. Proposition. Let $\mathfrak{a}$ be an operator satisfying Definition 12. Then $\mathfrak{a}$ induces continuous operators

$$\mathfrak{a} : \mathcal{K}^{s}_{\Delta}(\mathbb{R}_+) \longrightarrow \mathcal{K}^{s-\nu}_{\Sigma}(\mathbb{R}_+)$$

for every $s \in \mathbb{R}$, $\Delta \in \underline{\Delta}(\gamma)$ with some $\Sigma \in \underline{\Delta}(\gamma-\mu)$ depending on Δ and $\mathfrak{a}$.

14. Definition. Denote by $\mathcal{Y}_{M+G}(Q,P)^{(-k)}$, $P=(\gamma,\varphi)$, $\varphi \in \bar{F}(\gamma)\underline{\underline{\varphi}}$, $Q=(\gamma-\mu,\mathfrak{q})$, $\mathfrak{q} \in \bar{F}(\gamma-\mu)\underline{\underline{\varphi}}$, the space of all

$$\mathfrak{a} \in \bigcap_s \mathcal{L}(\mathcal{K}^{s,\gamma}+\mathcal{S}_\varphi, \mathcal{K}^{\infty,\gamma-\mu}+\mathcal{S}_\mathfrak{q})$$

for which

$$\mathfrak{a}-\omega_1 \sum_{j=0}^{k-1} \sum_{i=1}^{p_j} t^{-\mu+j} \operatorname{op}_M^{\gamma_{ij}}(a_{ij})\omega_2 \in \mathcal{Y}_G(Q,P)^{(-k)}$$

for certain $a_{ij} \in \mathfrak{M}_{Z_{ij}}$ and $\gamma-j \leq \gamma_{ij} \leq \gamma$, ω_1,ω_2 arbitrary cut-off functions.

Define in the same way the class (l)-$\mathcal{Y}_{M+G}(\ldots)^{(-k)}$ with (l)-$\mathcal{Y}_G(\ldots)^{(-k)}$ instead.

For brevity we formulate the calculus with continuous singularity types in the proper classes.

The left classes are then completely analogous.

Set

$$\mathcal{Y}_{M+G}(Q,P) = \bigcap_k \mathcal{Y}_{M+G}(Q,P)^{(-k)}.$$

We have by definition

$$\mathcal{Y}_{M+G}(Q,P)^{(-k)} = \mathcal{Y}_{M+G}(\gamma-\mu,\gamma)^{(-k)} + \mathcal{Y}_G(Q,P)^{(-k)},$$

where the elements $\mathfrak{a}_0 \in \mathcal{Y}_{M+G}(\gamma-\mu,\gamma)^{(-k)}$ act on $\mathcal{K}^{s,\gamma}+\mathcal{S}_\varphi$ by $\mathfrak{a}_0 \mathfrak{u}_\varphi^\gamma$.

If we replace $\mathcal{L}$ by $\mathcal{Y}$, the assertion of Theorem 10 remains valid. Moreover, we have an immediate analogue of Proposition 11.

As for operators with discrete asymptotics for every $\mathfrak{a} \in \mathcal{Y}_{M+G}(Q,P)^{(-k)}$ we can recover the conormal symbols, and then

$$0 \longrightarrow \mathcal{Y}_G(Q,P)^{(-k)} \longrightarrow \mathcal{Y}_{M+G}(Q,P)^{(-k)} \xrightarrow{\{\sigma_M^{\mu-j}\}} \qquad (8)$$

$$\mathfrak{M}_{\underline{Z}[\gamma]} \times \mathop{\times}_{j=1}^{k-1} \mathfrak{M}_{\underline{Z}} \longrightarrow 0$$

is exact.

For $\mathcal{Y}_{M+G}$ we have an immediate analogue of Theorem 9. The precise formulation is obvious and left to the reader.

In $\mathcal{G}_{M+G}(Q,P)$ we have a natural locally convex topology that can be defined by a similar procedure as in the discrete case.

Now let us make two remarks that are valid in the corresponding versions both for the discrete and the continuous asymptotics.

15. Remark. The exact sequence (8) for $k \leq \infty$ shows that the data $\gamma, \mu \in \mathbb{R}$ and $\{a_j\} \in \mathfrak{M}_{\underline{Z}[\gamma]} \times \overset{k-1}{\underset{1}{\times}} \mathfrak{M}_{\underline{Z}}$ are completely independent, with the only exception that $a_0 \in \mathfrak{M}_{\underline{Z}[\gamma]}$. Thus, for every sequence of conormal symbols we can generate a corresponding operator in $\mathcal{G}_{M+G}(\gamma-\mu,\gamma)$, provided $a_0 \in \mathfrak{M}_{\underline{Z}[\gamma]}$. Thus, if at the same time $a_0 \in \mathfrak{M}_{\underline{Z}[\delta]}$, we also get an operator in $\mathcal{G}_{M+G}(\delta-\nu,\delta)$ for every $\nu \in \mathbb{R}$. For the discrete case every δ fits except for a discrete subset of $\mathbb{R}$ for every individual a_0.

16. Remark. The calculus of the classes $\mathcal{L}_{M+G}(\gamma-\mu,\gamma)$ and $\mathcal{G}_{M+G}(\gamma-\mu,\gamma)$ can be generalized to the case of conormal symbol spaces $\mathfrak{M}^{\mu}_{\underline{Z}}$ and $\mathfrak{M}^{\mu}_{\underline{\gamma}}$, respectively. The class of Green operators then has to be enlarged by the flat ones. We also could distinguish between the shifts of conormal and Sobolev space order, but in our applications in the next section we have equality. We shall deal in fact with appropriate subspaces of $\mathfrak{M}^{\mu}$. The cone algebra of Chapter 4 is another generalization of the concept of operators with higher order Mellin symbols.

Note that there also exists a Mellin ψDO calculus on $\mathbb{R}_+$ where the Fourier transform is replaced by the Mellin transform, but without any specific asymptotic information. A series of papers of Lewis Parenti [L 2], [L 3],... is devoted to this question. There also exists a higher-dimensional analogue in the sense of a Mellin transform with respect to several t-variables, cf. Schulze [S 4]. This calculus represents an operator convention which is different from that with the Fourier transform. Nevertheless the class of Mellin operators in this sense over $(\mathbb{R}_+)^k$ coincides with the class of standard ψDO's.

2.3.5. Pseudo-differential operators of arbitrary orders

This section considers the Mellin expansions of pseudo-differential operators on the half axis of arbitrary orders generalizing 1.4.1. First assume $a(\tau) \in S^{\mu}_{cl}(\mathbb{R})$. Below we then pass to symbols that also may depend on t.

As in Section 1.4.1. we start from the asymptotic expansion

$$a(\tau) \sim \sum_{k=0}^{\infty} (i\tau)^{\mu-k}\, a_k^{\pm} \quad \text{for } \tau \longrightarrow \pm\infty ,$$

$|\arg(i\tau)| \leq \frac{\pi}{2}$.

Then, for each $\delta > 0$ and $l_+(\delta,\tau) = \delta + i\tau$

$$(i\tau)^{\mu-k} = (l_+(\delta,\tau) - \delta)^{\mu-k} = l_+^{\mu-k}(\delta,\tau)\ (1 - \delta\, l_+^{-1}(\delta,\tau))^{\mu-k} =$$

$$= l_+^{\mu-k}(\delta,\tau) \sum_{p=0}^{\infty} \gamma_{kp}(\delta)\ l_+^{-p}(\delta,\tau)$$

where $\gamma_{kp}(\delta) = \binom{\mu-k}{p}(-1)^p\, \delta^p$. This yields

$$a(\tau) \sim \sum_{j=0}^{\infty} \left\{ A_j^+(\delta)\ \theta^+(\tau) + A_j^-(\delta)\ \theta^-(\tau) \right\} l_+^{\mu-j}(\delta,\tau) \qquad (1)$$

as $\tau \longrightarrow \pm\infty$, where

$$A_j^{\pm}(\delta) = \sum_{k+p=j} a_k^{\pm}\ \gamma_{kp}(\delta).$$

Repeating the conclusions in the proof of 1.4.1. Proposition 3 we get (with the notations introduced in the beginning of 2.3.3.)

1. Proposition. For each $N \in \mathbb{Z}_+$, $N > \mu + 1$ and $\delta > 0$ there exists $g_{N,\delta}$ such that for each λ, $0 \leq \lambda \leq N - \mu - 1$, $\lambda \neq \frac{1}{2} \bmod \mathbb{Z}$,

$$g_{N,\delta} : L^2(\mathbb{R}_+) \longrightarrow \mathcal{K}^{\lambda}_{P_0(\lambda)}(\mathbb{R}_+),$$

$$g^*_{N,\delta} : L^2(\mathbb{R}_+) \longrightarrow \mathcal{K}^{\lambda}_{P_0(\lambda)}(\mathbb{R}_+),$$

where $P_0(\lambda) = (\lambda, r(\lambda)\varphi_0)$, such that for all $u \in C_0^{\infty}(\mathbb{R}_+)$

$$\omega\, op_{\psi}(a)u = \omega \sum_{j=0}^{N-1} \{A_j^+(\delta) op_M(g^+) + A_j^-(\delta) op_M(g^-)\}\, op_{\psi}(l_+^{\mu-j}(\delta,\tau))u + g_{N,\delta}u. \qquad (2)$$

Proof. Set

$$b_{N,\delta}(t) = \int e^{it\tau} a_{N,\delta}(\tau)\, d\tau ,$$

where

$$a_{N,\delta}(\tau) = a(\tau) - \sum_{j=0}^{N-1} \left\{ A_j^+(\delta)\theta^+(\tau) + A_j^-(\delta)\theta^-(\tau) \right\} l_+^{\mu-j}(\delta,\tau).$$

Since $|a_{N,\delta}(\tau)| \leq c\langle\tau\rangle^{\mu-N}$ we have $b_{N,\delta} \in H^{N-\mu-1}(\mathbb{R})$.

In view of $op_\psi(\theta^\pm) = op_M(g^\pm)$ (2) follows, where $g_{N,\delta}$ is the operator with kernel $\omega b_{N,\delta}(t-s)$. Then, for λ under consideration

$$\| g_{N,\delta} u \|_{\mathcal{K}^\lambda_{P_0(\lambda)}} = \left\| \omega \int_0^\infty b_{N,\delta}(t-s)u(s)ds \right\|_{\mathcal{K}^\lambda_{P_0(\lambda)}}$$

$$\leq c \| u \|_{L^2} .$$

The mapping property for $g^*_{N,\delta}$ follows in the same way, since $g^*_{N,\delta}$ has $\overline{b_{N,\delta}(s-t)}\omega$ as kernel. □

Note that the mapping properties of $g_{N,\delta}$ imply that $g_{N,\delta} \in \mathcal{L}_G(\gamma,0)^{(\gamma-N+\mu+1)}$ for $\gamma \neq \frac{1}{2}$ mod $\mathbb{Z}$, $0 \leq \gamma < N-\mu-1$.

2. Definition. The function

$$\sigma_M^{\mu-1}(a)(z) := \left\{ a_l^+ g^+(z+\mu) + a_l^- g^-(z+\mu) \right\} \frac{\Gamma(1-z)}{\Gamma(1-z+l-\mu)}, \qquad (3)$$

$l \in \mathbb{Z}_+$, is called the conormal symbol of $a \in S^\mu_{cl}(\mathbb{R})$ belonging to the conormal order $\mu-l$.

Note that $\sigma_M^{\mu-l}(a) \in \mathfrak{M}^{\mu-l}_{\mathfrak{z}_{l,\mu}}$, where $\mathfrak{z}_{l,\mu} = \{ (1-j-\mu,0), (1+j,0) \}_{j \in \mathbb{Z}_+}$, (cf. 2.3.4. Definition 1). For $-\mu \notin \mathbb{Z}_+$ we have $\mathfrak{z}_{l,\mu} = \{ (1-j-\mu,0), (1+j,0) \}_{j \in \mathbb{Z}_+}$.

In case $a_l^+ = a_l^-$ for $l = 0,1,\dots,L-1$, we have $\sigma_M^{\mu-l}(a) \in \mathfrak{M}^{\mu-l}_{\mathfrak{z}_0^+}$, where $\mathfrak{z}_0^+ = \{ (1+j,0) \}_{j \in \mathbb{Z}_+}$ for $l < L$. If, in addition, $\mu \in \mathbb{Z}_+$, $\mu \geq L$, $\mathfrak{z}_0^+$ can be replaced by O, since $\frac{\Gamma(1-z)}{\Gamma(1-z+l-\mu)}$ is a polynomial then. If $\mu \in \mathbb{Z}$ and $a_l^+ \neq a_l^-$ for some l, the conormal symbol $\sigma_M^{\mu-l}(a)$ has poles of second order at the points $z = 1,2,\dots, l-\mu$.

Our next objective is to derive the Mellin expansion for higher order ψDO's on $\mathbb{R}_+$. First we consider the case when the weight γ of the desired domain of definition satisfies

$$\gamma > \max\left(-\tfrac{1}{2}, \mu\right). \tag{4}$$

Then we know from 2.2.1. that there are continuous extensions of $\mathrm{op}_\psi(l_+^{\mu-j})$, $j \in \mathbb{Z}_+$,

$$\mathrm{op}_\psi(l_+^{\mu-j}) : \mathcal{K}^{s,\gamma} \longrightarrow \mathcal{K}^{s-\mu,\gamma-\mu}$$

for all $s \in \mathbb{R}$. The Mellin operators $\omega\,\mathrm{op}_M(g^{\pm})$ define continuous mappings

$$\omega\,\mathrm{op}_M(g^{\pm}) : \mathcal{K}^{s-\mu,0} \longrightarrow \mathcal{K}^{s-\mu,0},$$

since $|g^{\pm}(z)| \leqq$ const., $\mathrm{Re}\, z = \frac{1}{2}$. Provided that

$$\gamma - \mu \neq \tfrac{1}{2} \mod \mathbb{Z}, \tag{5}$$

the restriction to $\mathcal{K}^{s-\mu,\gamma-\mu}$ gives a continuous operator

$$\omega\,\mathrm{op}_M(g^{\pm}) : \mathcal{K}^{s-\mu,\gamma-\mu} \longrightarrow \mathcal{K}^{s-\mu}_{P_0}(\gamma-\mu).$$

Then the right-hand side of the representation of $\mathrm{op}_\psi(a)$ in Proposition 1 induces a continuous mapping

$$\mathrm{op}_\psi(a) : \mathcal{K}^{s,\gamma} \longrightarrow \mathcal{K}^{s-\mu}_{P_0}(\gamma-\mu) \tag{6}$$

for each $s \in \mathbb{R}$ and γ satisfying (4) and (5). Note that (6) is uniquely determined by its restriction to the dense subspace $C_0^\infty(\mathbb{R}_+) \subset \mathcal{K}^{s,\gamma}$.

Inserting now the Mellin expansion of $\mathrm{op}_\psi(l_+^{\mu-j}(\delta,\tau))$ in (2) we derive a Mellin expansion for (6). Consider $\omega\,\mathrm{op}_\psi(a)\omega$ and use that

$$\mathrm{op}_\psi(l_+^{\mu-j}(\delta,\tau))\omega = t^{j-\mu}\sum_{k=0}^{\infty} (t\delta)^k\, \mathrm{op}_M^{\gamma}(h_{k,j-\mu}(z))\omega, \tag{7}$$

where $h_{k,j-\mu}(z) = \binom{\mu-j}{k} \frac{\Gamma(1-z)}{\Gamma(1-z+j+k-\mu)}$ (cf.1.4.1.)

and the right-hand side converges in $\mathcal{L}(\mathcal{K}^{s,\gamma}, \mathcal{K}^{s-\mu,\gamma-\mu})$, is the Mellin expansion of $op_\psi(1_+^{\mu-j}(\delta,\tau)\omega$. We get from (2) a finite sum of terms of the form

$$\sum_{k=0}^{\infty} A_j^{\pm}(\delta)\delta^k \omega\, op_M(g^{\pm})t^{\gamma+j+k-\mu}\, op_M(T^{\gamma}h_{k,j-\mu})t^{-\gamma}\omega .$$

For given $l \in \mathbb{Z}_+$ all terms with $j+k = l$ contribute to the Mellin term with conormal order $-(\gamma-\mu+l)$. Since the left-hand side is independent of δ, we may simplify the calculation of the Mellin symbols letting $\delta \rightarrow 0$, i.e. we take into account only the absolute terms of the polynomials in δ. We have

$$A_j^{\pm}(0) = a_j^{\pm}$$

and only $k=0$ occurs , i.e. $j=l$. In view of

$$\sigma_M^{-\gamma+\mu-l}(\omega\, op_M(g^{\pm})t^{\gamma+l-\mu}) = T^{\mu-\gamma}g^{\pm}$$

we get

$$\sigma_M^{\mu-l}(op_\psi(a))(z) = \left\{a_l^{+}g^{+}(z+\mu) + a_l^{-}g^{-}(z+\mu)\right\}\frac{\Gamma(1-z)}{\Gamma(1-z+l-\mu)} = \sigma_M^{\mu-l}(a)(z).$$

It remains to find the singularity type of the Green remainder terms. Clearly

$$t^{j-\mu}\sum_{k=N}^{\infty} (t\delta)^k op_M^{\gamma}(h_{k,j-\mu})\omega \in \mathcal{L}(\mathcal{K}^{s,\gamma}, \mathcal{K}^{s-\mu+N,\gamma-\mu+j+N})$$

and

$$\omega\, op_M(g^{\pm}) \in \mathcal{L}(\mathcal{K}^{s-\mu+N,\gamma-\mu+j+N}, \mathcal{K}^{s-\mu+N}_{P_o(\lambda)})$$

for each $\lambda \leqq \gamma-\mu+j+N$, $\lambda \neq \frac{1}{2} \bmod \mathbb{Z}$, therefore

$$\omega\, op_M(g^{\pm})t^{j-\mu}\sum_{k=N}^{\infty} (t\delta)^k op_M^{\gamma}(h_{k,j-\mu})\omega \in \mathcal{L}(\mathcal{K}^{s,\gamma}, \mathcal{K}^{s-\mu+N}_{P_o(\lambda)}),$$

and setting

$$\mathcal{g}_1 = \omega \sum_{j=0}^{N-1} \left\{ A_j^+(\delta) op_M(g^+) + A_j^-(\delta) op_M(g^-) \right\} t^{j-\mu} \sum_{k=N}^{\infty} (t\delta)^k op_M^\gamma(h_{k,j-\mu})\omega$$

we get $\mathcal{g}_1 \in \mathcal{L}(\mathcal{K}^{s,\gamma}, \mathcal{K}^{s-\mu+N}_{P_0(\lambda)})$.

Consider now

$$\mathcal{g}_2 = \omega \sum_{j=0}^{N-1} \left\{ A_j^+(\delta) op_M(g^+) + A_j^-(\delta) op_M(g^-) \right\} t^{j-\mu} \sum_{k=0}^{N-1} (t\delta)^k op_M^\gamma(h_{k,j-\mu})\omega -$$

$$- \omega t^{-\mu} \sum_{l=0}^{N-1} t^l op_M^\gamma(\sigma_M^{\mu-l}(a))\omega ,$$

where by definition of $\sigma_M^{\mu-l}(a)$ all Mellin symbols of order $> \mu - N$ vanish. Commuting on the right $op_M(g^\pm)$ and $t^{j-\mu+k+\gamma}$ we get in view of 1.3.1. Proposition 8 Green operators with image in $\mathcal{S}_{r(j+k-\mu+\gamma)\gamma_0}$. The arising Mellin operators of conormal order $> \mu - N$ disappear and those of order $\leq \mu - N$ have image in $\mathcal{K}^{s-\mu+N, N+\gamma-\mu}$.

Finally, we estimate the singularity type of the Green remainder terms in the second argument. For $\mathcal{g}_{N,\delta}$ we proved that it is again γ_0. For the other contributions we use that for $\lambda' \geq 0$, $\lambda' \neq \frac{1}{2} \bmod \mathbb{Z}$ and $u \in C_0^\infty(\mathbb{R}_+)$ such that $\widetilde{\omega u}(j) = 0, j = 1,2,\dots, [\frac{1}{2}+\lambda']$

$$t^{-\lambda'} op_M(h_{k,j-\mu})\omega u = op_M(T^{-\lambda'} h_{k,j-\mu})\, t^{-\lambda'} \omega u. \tag{8}$$

Note that for $\varphi \in C_0^\infty(\overline{\mathbb{R}}_+)$, $\varphi \geq 0$ there are no zeros of $\tilde{\varphi}(z)$ at $z \in \mathbb{R}_+$. Set

$$\mathcal{g}'_{\lambda'} u(t) = \sum_{k=0}^{[\frac{1}{2}+\lambda']} \frac{\omega(t)}{\tilde{\omega}(k)} \int_0^\infty s^{k-1} \omega(s) u(s) ds. \tag{9}$$

Then $\widetilde{\omega u}(z) - \widetilde{\mathcal{g}'_{\lambda'} u}(z)$ vanishes at $j = 1,2,\dots, [\frac{1}{2}+\lambda']$ and (8) implies for arbitrary $u \in C_0^\infty(\overline{\mathbb{R}}_+)$

$$t^{-\lambda'} op_M(h_{k,j-\mu})\omega u = t^{-\lambda'} op_M(h_{k,j-\mu})(\omega u - \mathcal{g}'_\lambda u + \mathcal{g}'_\lambda u) =$$

$$op_M(T^{-\lambda'} h_{k,j-\mu}) t^{-\lambda'}(\omega u - \mathcal{g}'_{\lambda'} u) + \mathcal{g}''_{\lambda'} u ,$$

where $\mathcal{G}''_{\lambda'}$ is a Green operator of the form

$$\mathcal{G}''_{\lambda'}u(t) = \sum_{k=0}^{[\frac{1}{2}+\lambda']} f_k(t) \int_0^\infty s^{k-1}\omega(s)u(s)\,ds, \quad f_k \in \mathcal{S}_{T}{}^{\lambda'}{}_{\varphi_0} .$$

It follows that for $\mathcal{G} = \mathcal{G}_1 + \mathcal{G}_2$

$$\mathcal{G}(1-\mathcal{G}'_{\lambda'}) \in \mathcal{L}(\mathcal{K}^{s,-\lambda'}, \mathcal{K}^{s-\mu+N}_{P_0(\lambda)}) .$$

In view of (9) this means that the kernel $g(t,s)$ of $\mathcal{G}$ has the form

$$g(t,s) = \sum_{k=1}^{[\frac{1}{2}+\lambda']} g_k(t)s^{k-1}\omega(s) + g_1(t,s). \tag{10}$$

$g_k \in \mathcal{S}_{\varphi_0}$, and

$$op_G(g_1) \in \mathcal{L}(\mathcal{K}^{s,-\lambda'}, \mathcal{K}^{s-\mu+N}_{P_0(\lambda)}) .$$

Since λ' can be chosen arbitrarily large, we obtain that $\mathcal{G}$ has asymptotics in the second argument.

Thus we have proved the following

3. <u>Theorem.</u> Let $a \in S^{\mu}_{cl}(\mathbb{R})$, $\gamma \in \mathbb{R}$ such that (4) and (5) are satisfied. Consider $op_{\psi}(a)$ as in (6). Then, for each $\lambda \leqq \gamma - \mu + N$, $\lambda \neq \frac{1}{2} \mod \mathbb{Z}$,

$$\omega(op_{\psi}(a) - t^{-\mu}\sum_{l=0}^{N-1} t^{l} op_M^{\gamma}(\sigma_M^{\mu-l}(a))\omega \in \mathcal{L}(\mathcal{K}^{s,\gamma}, \mathcal{K}^{s-\mu+N}_{P_0(\lambda)})$$

and the kernel of this operator has the form (10).

In the notation of 2.3.3. we have then that the difference in Theorem 3 belongs to $\mathcal{L}_G(Q,\gamma)^{(-N)}$, $Q = (\gamma-\mu, r(\gamma-\mu)\varphi_0)$. Clearly the choice of the cut-off function localizing the action near the origin is not essential.

For $\mu > 0$ we are also interested in those γ which satisfy $\gamma > -\frac{1}{2}$, possibly violating (4). Then the action of $op_{\psi}(1_+^{\mu-j})$ on $\mathcal{K}^{s,\gamma}$ can give δ-derivatives at 0 which are not reflected immediately by the Mellin transform (the Mellin transform of distributions which have compact support in $\overline{\mathbb{R}}_+$ can be defined by duality). For this reason the representation of $op_\psi(a)$ in Proposition 1 has to be modified (cf.1.4.1. Proposition 3').

1'. <u>Proposition.</u> For each $N \in \mathbb{Z}_+, N > \mu + 1, \delta > 0$ there exists $\mathcal{G}'_{N,\delta}$ such that for each $\lambda \geq 0$, $\lambda \leq N-\mu-1$, $\lambda \neq \frac{1}{2} \bmod \mathbb{Z}$,

$$\mathcal{G}'_{N,\delta} : L^2(\mathbb{R}_+) \longrightarrow \mathcal{K}^\lambda_{P_0}(\lambda),$$

$$(\mathcal{G}'_{N,\delta})^*: L^2(\mathbb{R}_+) \longrightarrow \mathcal{K}^\lambda_{P_0}(\lambda),$$

such that for all $u \in C_0^\infty(\mathbb{R}_+)$

$$op_\psi(a)u = \sum_{j=0}^{N-1} op_\psi(1_-^{\mu-j}(\delta,\tau)) \left\{ B_j^+(\delta) op_M(g^+) + B_j^-(\delta) op_M(g^-) \right\} u + \mathcal{G}'_{N,\delta} u\,, \tag{11}$$

where $B_j^\pm(\delta)$ are uniquely determined from

$$a(\tau) \sim \sum_{j=0}^{\infty} \left\{ B_j^+(\delta)\, \theta^+(\tau) + B_j^-(\delta)\, \theta^-(\tau) \right\} 1_-^{\mu-j}(\delta,\tau),\ |\tau| \to \infty.$$

The proof is similar to that of Proposition 1 and therefore dropped.

Assume now that γ satisfies

$$-\frac{1}{2} < \gamma < \mu + \frac{1}{2},\ \gamma \neq \frac{1}{2} \bmod \mathbb{Z}. \tag{12}$$

Then the Mellin operators $op_M(g^\pm)\omega$ define continuous mappings

$$op_M(g^\pm)\omega \ : \mathcal{K}^{s,\gamma} \longrightarrow \mathcal{K}^s_{P_0}(\gamma)$$

for each $s \in \mathbb{R}$, and $op_\psi(1_-^{\mu-j})$ have continuous extensions

$$op_\psi(1_-^{\mu-j}) \ : \mathcal{K}^s_{P_0}(\gamma) \longrightarrow \mathcal{K}^{s-\mu+j,\gamma-\mu}$$

(cf. 2.1.1.). Then Proposition 1' gives that $op_\psi(a)$ induces a continuous mapping

$$op_\psi(a) \ : \mathcal{K}^{s,\gamma} \longrightarrow \mathcal{K}^{s-\mu,\gamma-\mu} \tag{13}$$

for each $s \in \mathbb{R}$.

Repeating the arguments of the proof of Theorem 3 with simple modifications we obtain

4. <u>Theorem.</u> Let $a \in S^{\mu}_{cl}(\mathbb{R})$ and $\gamma \in \mathbb{R}$ such that (12) is satisfied. Consider $op_{\psi}(a)$ as in (13). Then, for each $\lambda \leq \gamma-\mu+N$, $\lambda \neq \frac{1}{2} \mod \mathbb{Z}$,

$$\omega(op_{\psi}(a) - t^{-\mu}\sum_{l=0}^{N-1} t^{l} op_M(\sigma_M^{\mu-l}(a))\omega \in \mathcal{L}(\mathcal{K}^{s,\gamma}, \mathcal{K}^{s-\mu+N}_{P_0(\lambda)})$$

and the kernel of this operator has the form (10).

From Theorem 4 it is clear that we have also an action of $op_{\psi}(a)$ as in (13) for γ such that $-\frac{1}{2}<\gamma<\mu+\frac{1}{2}$, γ possibly a half-integer. In fact, the Mellin operators in the expansion in Theorem 4 have this property and the Green difference, too.

5. <u>Remark.</u> A symbol $a(\tau) \in S^{\mu}_{cl}(\mathbb{R})$ is said to satisfy the transmission property if $\mu \in \mathbb{Z}$ and

$$a_j^{+} = a_j^{-} \quad \text{for all } j \in \mathbb{Z}_{+}. \tag{14}$$

Then the Mellin symbols are given by

$$\sigma_M^{\mu-l}(a)(z) = a_l \frac{\Gamma(1-z)}{\Gamma(1-z+l-\mu)}$$

$(a_l = a_l^{+} = a_l^{-})$.

For a with the transmission property Theorem 3 remains valid even without the assumption (5).

It remains the discussion of the case $\gamma<-\frac{1}{2}$. The first observation is that even operators with smooth kernels do not give, in general, a well-defined action on $\mathcal{K}^{s,\gamma}$ in this case. However, a finite-dimensional modification makes this possible. In fact, let $b(t,s) \in C^{\infty}(\overline{\mathbb{R}}_{+} \times \overline{\mathbb{R}}_{+})$, $b \equiv 0$ for $s >$ const. and set for $N \in \mathbb{Z}_{+}$

$$b_1^{(N)}(t,s) = \sum_{k=0}^{N-1} \frac{s^k}{k!} \partial_s^k\, b(t,0)\, \omega(s)$$

and $b_2^{(N)}(t,s) = b(t,s) - b_1^{(N)}(t,s)$. Then $op_G(b_1^{(N)})$ is finite-dimensional and

$$op_G(b_2^{(N)}) : \mathcal{K}^{s,\gamma} \longrightarrow C^{\infty}(\overline{\mathbb{R}}_{+})$$

is continuous for all $\gamma > -N - \frac{1}{2}$.

Note that similar arguments were used already in the proof of Theorem 3. It will be proved below that an arbitrary ψDO on $\mathbb{R}_+$ admits an extension to spaces with negative weights after modification by a finite-dimensional operator.

6. Theorem. Let $a \in S^{\mu}_{cl}(\mathbb{R})$ and $\gamma < -\frac{1}{2}$, $\gamma \neq \frac{1}{2} \mod \mathbb{Z}$. For $\mu \geq 0$ let $\lambda \in \mathbb{R}$, $-\frac{1}{2} < \lambda < \mu + \frac{1}{2}$, $\lambda \neq \frac{1}{2} \mod \mathbb{Z}$, $N > 2 - \gamma$. Denote by $op_G(b_N)$ the Green operator in Theorem 4 and put

$$b^N_{N,1}(t,s) = \sum_{k=0}^{N-1} \frac{s^k}{k!} \partial^k_s\, b(t,0)\, \omega(s).$$

Moreover, define

$$g_N(t,s) = \omega(t) t^{-\mu} \sum_{l=0}^{N-1} \sum_{j,k=1}^{[\frac{1}{2}-\gamma]} c_{jkl}\, t^{l-j} s^{k-1} \omega(s)$$

where

$$c_{jkl} = \frac{(-1)^{k-1}}{(k-1)!} \left\{ a^+_l \frac{g^+(\mu)}{\Gamma(1+l-k-\mu)} + a^-_l \frac{g^-(\mu)}{\Gamma(1+l-k-\mu)} \right\} \frac{\tilde{\omega}(j)}{\tilde{\omega}(k)}.$$

Then the operator

$$\omega\, op_\psi(a)\, \omega : \mathcal{K}^{s,\lambda} \longrightarrow \mathcal{K}^{s-\mu,\lambda-\mu}$$

(cf. (13)) modified by the finite-dimensional operator

$$op_G(b^N_{N,1}) + op_G(g_N) : \mathcal{K}^{s,\lambda} \longrightarrow \mathcal{K}^{s-\mu,\lambda-\mu}$$

admits for $N > \frac{1}{2} - \gamma$ an extension as

$$\omega\, op_\psi(a)\omega - op_G(b^N_{N,1}) - op_G(g_N) : \mathcal{K}^{s,\gamma} \longrightarrow \mathcal{K}^{s-\mu,\gamma-\mu}. \tag{15}$$

For $\mu \leq 0$ let $\lambda \in \mathbb{R}$, $\lambda > -\frac{1}{2}$, $\lambda - \mu \neq \frac{1}{2} \mod \mathbb{Z}$. Then

$$op_\psi(a) : \mathcal{K}^{s,\lambda} \longrightarrow \mathcal{K}^{s-\mu}_{P_o}(\lambda-\mu)$$

with a similar modification yields for $N > \frac{1}{2} - \gamma$ a continuous extension

$$\omega op_\psi(a)\omega - op_G(b^N_{N,1}) - op_G(g_N) : \mathcal{K}^{s,\gamma} \longrightarrow \mathcal{K}^{s-\mu,\gamma-\mu} \tag{16}$$

if $\gamma-\mu<\frac{1}{2}$ and

$$\omega \mathrm{op}_{\psi}(a)\omega - \mathrm{op}_G(b^N_{N,1}) - \mathrm{op}_G(g_N) : \mathcal{K}^{s,\gamma} \longrightarrow \mathcal{K}^{s-\mu}_{P_0}(\gamma-\mu) \tag{17}$$

if $\gamma-\mu>\frac{1}{2}$, $\gamma-\mu \neq \frac{1}{2} \bmod \mathbb{Z}$.

If we denote by $\mathrm{op}^{\gamma}_{\psi}(a)$ the action defined by (15), (16) or (17), then for each $\lambda \leq \gamma-\mu+N'$, $\lambda \neq \frac{1}{2} \bmod \mathbb{Z}$

$$\omega(\mathrm{op}^{\gamma}_{\psi}(a) - t^{-\mu}\sum_{l=0}^{N'-1} t^l \,\mathrm{op}^{\gamma}_M(\sigma^{\mu-l}_M(a)))\omega \in \mathcal{L}(\mathcal{K}^{s,\gamma}, \mathcal{K}^{s-\mu+N'}_{P_0}(\lambda)),$$

where $N' < N - \max(\frac{1}{2}-\gamma+\mu, \frac{1}{2}-\gamma)$ and the kernel of this operator has the form: for each $\lambda' > \gamma$, $\lambda' \neq \frac{1}{2} \bmod \mathbb{Z}$

$$g(t,s) = \sum_{[\frac{1}{2}+\gamma]<k\leq[\frac{1}{2}+\lambda']} g_k(t)\, s^{k-1}\omega(s) + g_1(t,s),$$

$\mathrm{op}_G(g_1) \in \mathcal{L}(\mathcal{K}^{s,-\lambda'}, \mathcal{K}^{s-\mu+N'}_{P_0}(\lambda))$, $g_k \in \mathcal{S}_{T^{\mu+[-\frac{1}{2}-\gamma]}P_0}$

Proof. Consider the case $\mu \geq 0$. In view of Theorem 4

$$\omega\, \mathrm{op}_{\psi}(a)\omega = \omega\Big\{t^{-\mu}\sum_{l=0}^{N-1} t^{\ell}\mathrm{op}^{\lambda}_M(\sigma^{\mu-l}_M(a))\Big\}\omega + \mathrm{op}_G(b_N)$$

and

$$b^{(N)}_{N,1}(t,s) = \sum_{k=0}^{N-1} \frac{s^k}{k!}\,\partial^k_s\, b_N(t,0)\,\omega(s).$$

Then $\mathrm{op}_G(b^{(N)}_{N,2}) = \mathrm{op}_G(b_N) - \mathrm{op}_G(b^{(N)}_{N,1})$ extends to an operator in $\mathcal{L}(\mathcal{K}^{s,\gamma}, \mathcal{K}^{s-\mu,\gamma-\mu})$ for $N>1-\gamma$. It remains to consider the Mellin operators. Splitting $\widetilde{\omega u}(z)$ into a finite sum and a holomorphic function with zeros at the positive integers $<\frac{1}{2}-\gamma$

$$\widetilde{\omega u}(z) = \sum_{k=1}^{[\frac{1}{2}-\gamma]} \tilde{\omega}(z)\,\frac{1}{\tilde{\omega}(k)}\,\widetilde{\omega u}(k) + \tilde{u}_1(z)$$

we can shift the integration contour

$$\omega\, \mathrm{op}^{\lambda}_M(\sigma^{\mu-l}_M(a))(u_1)(t) = \omega(t)\int_{\Gamma_{\frac{1}{2}-\lambda}} t^{-z}\,\sigma^{\mu-l}_M(a)(z)\,\tilde{u}_1(z)\frac{dz}{2\pi i} =$$

$$= \omega(t) \int_{\Gamma_{\frac{1}{2}-\gamma}} t^{-z} \sigma_M^{\mu-1}(a)(z) \tilde{u}_1(z) \frac{dz}{2\pi i}$$

since the zeros of $\tilde{u}_1$ cancel the relevant poles of $\sigma_M^{\mu-1}(a)$. The integral over $\Gamma_{\frac{1}{2}-\gamma}$ defines an operator in $\mathcal{L}(\mathcal{K}^{s,\gamma}, \mathcal{K}^{s-\mu,\gamma})$. The finite sum gives by the Residue Theorem

$$\omega \, op_M^{\lambda}\big(\sigma_M^{\mu-1}(a)\big)\Big(\sum_{k=1}^{[\frac{1}{2}-\gamma]} \frac{\tilde{\omega u}(k)}{\tilde{\omega}(k)} \, \omega\Big)(t) = \omega \, op_M^{\gamma}\big(\sigma_M^{\mu-1}(a)\big)\Big(\sum_{k=1}^{[\frac{1}{2}-\gamma]} \frac{\tilde{\omega u}(k)}{\tilde{\omega}(k)} \, \omega\Big)(t)$$

$$= \omega(t) \sum_{k=1}^{[\frac{1}{2}-\gamma]} \frac{\tilde{\omega u}(k)}{\tilde{\omega}(k)} \sum_{j=1}^{[\frac{1}{2}-\gamma]} t^{-j} \, \tilde{\omega}(j) \operatorname*{Res}_{z=j} \sigma_M^{\mu-1}(a)(z) .$$

This proves (15). In the same way (16) and (17) are proved. Then the mapping properties of the Green remainder operators of the Mellin expansion are proved by the same arguments as in the proof of Theorem 3. □

7. Remark. The modification of the pseudo-differential action by a finite-dimensional operator for obtaining an extension to negative weights is necessary also for ψDO's with the transmission property. In this special case it is well-known that without modification we cannot expect a continuous action even on $H^s(\mathbb{R}_+)$, $s < -\frac{1}{2}$. However, $op_\psi(1_-^\alpha)$ defines a continuous action

$$op_\psi(1_-^\alpha) : H^s(\mathbb{R}_+) \longrightarrow H^{s-\alpha}(\mathbb{R}_+)$$

for all $s \in \mathbb{R}$.

For given $\mu \in \mathbb{R}$ we call a weight γ admissible (with respect to μ) if

$$\gamma - \mu \neq \tfrac{1}{2} \bmod \mathbb{Z}_+ \text{ and } -\gamma \neq \tfrac{1}{2} \bmod \mathbb{Z}_+ .$$

In particular, for $\gamma > \max(-\frac{1}{2}, \mu + \frac{1}{2})$ there remains only the first condition and for $\gamma < \min(-\frac{1}{2}, \mu + \frac{1}{2})$ only the second one.

Above we defined for each fixed $N \in \mathbb{Z}_+$ (non-canonical) mappings

$$op_\psi^\gamma : S_{cl}^\mu(\mathbb{R}) \longrightarrow \mathcal{L}(\mathcal{K}^{s,\gamma}, \mathcal{K}_{P_0(\gamma-\mu)}^{s-\mu}), \; \gamma > \max(-\tfrac{1}{2}, \mu + \tfrac{1}{2})$$

and

$$\mathrm{op}_\psi^\gamma : S^\mu_{cl}(\mathbb{R}) \longrightarrow \mathcal{L}(\mathcal{K}^{s,\gamma}, \mathcal{K}^{s-\mu,\gamma-\mu}),\ \gamma < \max(-\tfrac{1}{2}, \mu + \tfrac{1}{2})$$

understood as the extension from C_0^∞ of a suitable finite-dimensional modification of the usual ψDO action. Then the evaluation of the symbol in $S^\mu_{cl}(\mathbb{R})$ of $\mathrm{op}_\psi^\gamma(a)$ gives $a(\tau)$, i.e.

$$\sigma_\psi \circ \mathrm{op}_\psi^\gamma = \mathrm{id}.$$

Now we want to extend Theorem 3,4,6 to symbols $a(t,\tau)$. Since we are interested in the behaviour for small t we shall usually assume that $a(t,\tau)$ is independent of t for $|t| \geq$ const. Set

$$S^\mu_{cl}(\overline{\mathbb{R}}_+ \times \mathbb{R}) = \{a(t,\tau) = \tilde{a}(t,\tau)|_{\overline{\mathbb{R}}_+} : \tilde{a} \in S^\mu_{cl}(\mathbb{R}\times\mathbb{R})\}.$$

It can easily be proved that $a \in S^\mu_{cl}(\overline{\mathbb{R}}_+ \times \mathbb{R})$ is equivalent to

$$a(t,\tau) \in S^\mu_{cl}(\mathbb{R}_+ \times \mathbb{R}) \cap C^\infty(\overline{\mathbb{R}}_+ \times \mathbb{R})$$

and the symbol estimates hold uniformly in t near $t=0$.

8. <u>Definition.</u> Let $a(t,\tau) \in S^\mu_{cl}(\overline{\mathbb{R}}_+ \times \mathbb{R})$ and $a(t,\tau) \sim \sum_{j=0}^{\infty} t^j a^{[j]}(\tau)$ be the Taylor expansion for $t \to 0$, $a^{[j]}(\tau) = \frac{1}{j!}\left(\frac{\partial^j}{\partial t^j} a\right)(0,\tau) \in S^\mu_{cl}$. Then

$$\sigma_M^{\mu-l}(a)(z) := \sum_{j+k=l} \sigma_M^{\mu-k}\left(a^{[j]}\right)(z)$$

is called the conormal symbol of a belonging to the conormal order $\mu - l$.

The sequence

$$\sigma_M(a) = \{\sigma_M^{\mu-j}(a)\}_{j\in\mathbb{Z}_+} \in \times_j \mathfrak{M}^{\mu-j}_{\mathfrak{z}_0(\mu)}$$

where $\mathfrak{z}_0(\mu) = \{(-j-\mu,0),\ (1+j,0)\}_{j\in\mathbb{Z}_+}$, is called the complete conormal symbol of $a \in S^\mu_{cl}(\overline{\mathbb{R}}_+ \times \mathbb{R})$.

Consider the decomposition

$$a(t,\tau) = \sum_{j=0}^{N-1} t^j a^{[j]}(\tau) + a_N(t,\tau).$$

Then

$$b_N(t,\tau) = t^{-N} a_N(t,\tau) \in S^{\mu}_{cl}(\bar{\mathbb{R}}_+ \times \mathbb{R}).$$

Now we can apply Theorems 3,4 and 6 to $op_\psi(a^{[j]})$. It remains to consider $op_\psi(a_N)$. For each $s \in \mathbb{R}$ $op_\psi(b_N)$ has a continuous extension in $\mathcal{L}(\mathcal{K}^{s,0}, \mathcal{K}^{s-\mu,0})$, therefore

$$op_\psi(a_N) : \mathcal{K}^{s,0} \longrightarrow \mathcal{K}^{s-\mu,N}$$

and for the formal adjoint

$$op_\psi(a_N)^* : \mathcal{K}^{-s,0} \longrightarrow \mathcal{K}^{-s-\mu}_{P_0(N)}.$$

This proves that for $\gamma > -\frac{1}{2}$, $\gamma - \mu \neq \frac{1}{2} \bmod \mathbb{Z}$

$$op_\psi(a_N) \in \mathcal{L}_G(P_0(\gamma-\mu),\gamma)^{(-k)} + \mathcal{L}_\psi(\gamma-\mu,\gamma)^{(-k)}$$

for $k \leq N-\mu+\gamma$. Choosing N sufficiently large we obtain

3'. <u>Theorem.</u> Let $a \in S^{\mu}_{cl}(\bar{\mathbb{R}}_+ \times \mathbb{R})$ be independent of t for $t \geq$ const. Then, for γ, μ as in Theorem 3 and N sufficiently large

$$op_\psi(a) = \sum_{j=0}^{N-1} t^j \, op_\psi(a^{[j]}) + op_\psi(a_N) \tag{18}$$

defines an operator in $\mathcal{L}(\mathcal{K}^{s,\gamma}, \mathcal{K}^{s-\mu}_{P_0(\gamma-\mu)})$.

Then, for each $\lambda \leq \gamma-\mu+N'$, $\lambda \neq \frac{1}{2} \bmod \mathbb{Z}$,

$$\omega\Big(op_\psi(a) - t^{-\mu}\sum_{l=0}^{N'-1} t^l \, op_M^\gamma(\sigma_M^{\mu-l}(a))\Big)\omega \in \mathcal{L}(\mathcal{K}^{s,\gamma}, \mathcal{K}^{s-\mu}_{P_0(\lambda)}),$$

where $\sigma_M^{\mu-l}(a)$ is defined by Definition 8. Moreover, for sufficiently large N

$$\omega\Big(op_\psi(a) - op_\psi(a_N) - t^{-\mu}\sum_{l=0}^{N'-1} t^l \, op_M^\gamma\big(\sigma_M^{\mu-l}(a)\big)\Big)\omega \in \mathcal{L}(\mathcal{K}^{s,\gamma}, \mathcal{K}^{s-\mu+N'}_{P_0(\lambda)})$$

and the kernel has the form (10), hence

$$\omega\Big(op_\psi(a) - t^{-\mu}\sum_{l=0}^{N'-1} t^l \, op_M^\gamma\big(\sigma_M^{\mu-l}(a)\big)\Big)\omega \in \mathring{\mathcal{L}}^\mu_\psi(\gamma-\mu,\gamma)^{(-k)} + \mathcal{L}_G(P_0(\gamma-\mu),\gamma)^{(-k)}$$

for $k \leq N-\mu+\gamma$, $k \leq N'$.

This is an immediate consequence of the properties of $op_\psi(a_N)$ and Theorem 3 applied to $op_\psi(a^{[j]})$. Denote the operation described in Theorem 3' by $op_\psi^\gamma(a)$. Then, for each admissible weight γ, $\gamma > \max(-\frac{1}{2}, \mu)$, we have a mapping

$$op_\psi^\gamma : S^\mu_{cl,st}(\overline{\mathbb{R}}_+ \times \mathbb{R}) \longrightarrow \mathcal{L}(\mathcal{K}^{s,\gamma}, \mathcal{K}^{s-\mu}_{P_0}(\gamma-\mu)).$$

$S^\mu_{cl,st}$ denotes the subspace of S^μ_{cl} of symbols independent of t for $t \geq$ const, which has the property that the symbol of $op_\psi^\gamma(a)$ is equal to $a(t,\tau)$ modulo $S^{-\infty}_{(o)}(\mathbb{R}_+ \times \mathbb{R})$ ($S^{-\infty}_{(o)}$ denotes the subspace of $S^{-\infty}$ of all symbols vanishing for sufficiently large t).

For γ satisfying $-\frac{1}{2} < \gamma < \mu + \frac{1}{2}$ we get from Theorem 4

<u>Theorem 4'.</u> Let $a \in S^\mu_{cl,st}(\overline{\mathbb{R}}_+ \times \mathbb{R})$. For $\gamma \in \mathbb{R}$, $-\frac{1}{2} < \gamma < \mu + \frac{1}{2}$ and N sufficiently large by (18) and Theorem 4 we get $op_\psi(a) \in \mathcal{L}(\mathcal{K}^{s,\gamma}, \mathcal{K}^{s-\mu,\gamma-\mu})$. For $op_\psi(a) - op_\psi(a_N)$ the conclusions of Theorem 4 are valid, therefore

$$\omega(op_\psi(a) - t^{-\mu}\sum_{l=0}^{N'-1} t^l \, op_M^\gamma(\sigma_M^{\mu-l}(a)))\omega \in \mathring{\mathcal{L}}^\mu_\psi(\gamma-\mu,\gamma)^{(-k)} + \mathcal{L}_G(\gamma-\mu,\gamma)^{(-k)}$$

for $k \leqq N-\mu+\gamma$, $k \leqq N'$.

Finally, the extension to admissible weights $\gamma < -\frac{1}{2}$ is possible after a finite-dimensional modification.

<u>Theorem 6'.</u> Let $a \in S^\mu_{cl}(\overline{\mathbb{R}}_+ \times \mathbb{R})$ be independent of t for $t \geq$ const., $\gamma < -\frac{1}{2}$, $\gamma \neq \frac{1}{2}$ mod $\mathbb{Z}$. For N sufficiently large there exists

$$g(t,s) = \sum_{k=0}^{[-\frac{1}{2}-\gamma]} g_k(t)s^k$$

$g_k \in \mathcal{S}_{T^{\mu+[-\frac{1}{2}-\gamma]}}\mathcal{P}$. depending on $a_l^{[j]\pm}$, $j+l \leqq N$ such that $op_\psi(a) - op_G(g)$ extends to an operator in $\mathcal{L}(\mathcal{K}^{s,\gamma}, \mathcal{K}^{s-\mu}_{P_0}(\gamma-\mu))$ for $\gamma-\mu > \frac{1}{2}$, $\gamma-\mu \neq \frac{1}{2}$ mod $\mathbb{Z}$ and in $\mathcal{L}(\mathcal{K}^{s,\gamma}, \mathcal{K}^{s-\mu,\gamma-\mu})$ for $\gamma-\mu < \frac{1}{2}$. Moreover, to $op_\psi(a) - op_\psi(a_N)$ the conclusions of Theorem 6 apply.

In view of Theorem 4' and 6' we have for all admissible weights $\gamma < \mu + \frac{1}{2}$ an operator convention

$$\mathrm{op}_{\psi}^{\gamma} : S^{\mu}_{cl,st}(\overline{\mathbb{R}}_+ \times \mathbb{R}) \longrightarrow \mathcal{L}(\mathcal{K}^{s,\gamma}, \mathcal{K}^{s-\mu,\gamma-\mu})$$

which is a right inverse of the symbol mapping σ_{ψ} in $S^{\mu}_{cl}/S^{-\infty}_{(o)}$.

The Mellin expansions imply immediately that $\mathrm{op}_{\psi}^{\gamma}(a)$ maps subspaces with discrete or continuous asymptotics into subspaces with the same type of asymptotics.

7. Theorem. Let $a \in S^{\mu}_{cl}(\overline{\mathbb{R}}_+ \times \mathbb{R})$ be independent of t for $t \gtrsim$ const. and γ an admissible weight, $\mathrm{op}_{\psi}^{\gamma}(a)$ a realization according to Theorem 3', 4' or 6'. Then the restriction to $\mathcal{K}^s_P$, $P = (\delta, \wp) \in \underline{P}$, $\delta > \gamma$ gives for each $\varrho \leq \delta - \mu$, $\varrho \neq \frac{1}{2}$ mod $\mathbb{Z}$ a continuous operator

$$\mathrm{op}_{\psi}(a) : \mathcal{K}^s_P \longrightarrow \mathcal{K}^{s-\mu}_Q ,$$

where $Q = (\varrho, q) \in \underline{P}$, $q = \bigcup_{l \in \mathbb{Z}_+} r(\varrho) \; (T^{\mu-l}\wp + \wp_0)$.

For $\Delta_1 \in \underline{\Delta}$, $\Delta_1 \subset \left\{\mathrm{Re}\ z < \frac{1}{2} - \gamma\right\}$ and ϱ as above we get a continuous operator

$$\mathrm{op}_{\psi}(a) : \mathcal{K}^{s,\delta} + \mathcal{S}_{\Delta_1} \longrightarrow \mathcal{K}^{s-\mu,\varrho} + \mathcal{S}_{\Delta_2} ,$$

where $\Delta_2 = \bigcup_{l \in \mathbb{Z}_+} T^{\mu-l}\Delta_1 \cup \pi_{\mathbb{C}} \wp_0$.

Proof. The stated mapping properties are clear for the Green remainder operators in the Mellin expansion. For the Mellin operators they were proved in 2.3.4. $\square$

Next we shall establish the behaviour of the conormal symbols $\sigma_M^{\mu-j}(a)$ of a under compositions of the operators.

Let $\mathcal{M}$ be the space of all sequences of functions $\{h_j(z)\}_{j \in \mathbb{Z}_+}$, where $h_j(z)$ is holomorphic in z for $c_1 < \mathrm{Re}\ z < c_2$, $|\mathrm{Im}\ z| > d$, for all $c_1, c_2 \in \mathbb{R}$ with certain $d = d(c_1, c_2, h_j) > 0$. Then $\mathcal{M}$ is an algebra with the composition $\{h_j(z)\} \#_M \{g_k(z)\} = \{f_l(z)\}$, where $f_l(z) := \sum_{j+k=l} h_j(z-k) g_k(z)$. Note that $\mathcal{M}$ is closed under translation $T^{\gamma}h := \{h_j(z+\gamma)\}$, $\gamma \in \mathbb{C}$.

Moreover let $a \in S^{\mu}_{cl}(\overline{\mathbb{R}}_+ \times \mathbb{R})$, $b \in S^{\nu}_{cl}(\overline{\mathbb{R}}_+ \times \mathbb{R})$ and denote by $a \#_\psi b$ a symbol in $S^{\mu+\nu}_{cl}(\overline{\mathbb{R}}_+ \times \mathbb{R})$ with

$$a \#_\psi b \sim \sum_{k=0}^{\infty} \frac{1}{k!} \left(\partial_\tau^k a\right) D_t^k b \ .$$

It is clear that $a \#_\psi b$ always exists for every a, b and is uniquely determined modulo $S^{-\infty}(\overline{\mathbb{R}}_+ \times \mathbb{R})$.

8. <u>Theorem.</u> For $a \in S^{\mu}_{cl}(\overline{\mathbb{R}}_+ \times \mathbb{R})$, $b \in S^{\nu}_{cl}(\overline{\mathbb{R}}_+ \times \mathbb{R})$ we have

$$\sigma_M(a \#_\psi b) = (T^{\nu}\sigma_M(a)) \#_M \sigma_M(b) \bmod \overset{\infty}{\underset{j=0}{\mathsf{X}}} \underline{\mathfrak{M}}_{\mathfrak{z}}.$$

<u>Proof</u>: First we show the assertion for symbols a and b that are independent of t. The coefficients $a_j^{\pm}$, $b_k^{\pm}$ obviously satisfy the identities

$$(ab)_l^{\pm} = \sum_{j+k=l} a_j^{\pm} b_k^{\pm}. \tag{19}$$

Now

$$\sigma_M^{\mu-j}(a)(z+\nu-k) = \left\{a_j^+ g^+(z+\nu+\mu) + a_j^- g^-(z+\nu+\mu)\right\} m_+^{\mu-j}(z+\nu-k),$$

$$\sigma_M^{\nu-k}(b)(z) = \left\{b_k^+ g^+(z+\nu) + b_k^- g^-(z+\nu)\right\} m_+^{\nu-k}(z).$$

Using

$$\frac{\Gamma(1-z)}{\Gamma(1-z+l-(\mu+\nu))} = \frac{\Gamma(1-z-\nu+k)}{\Gamma(1-z+k+j-(\mu+\nu))} \cdot \frac{\Gamma(1-z)}{\Gamma(1-z+k-\nu)}$$

we obtain

$$\sum_{j+k=l} \sigma_M^{\mu-j}(a)(z+\nu-k)\sigma_M^{\nu-k}(b)(z)$$

$$= \left\{ \sum_{j+k=l} \left\{a_j^+ g^+(z+\mu+\nu) + a_j^- g^-(z+\mu+\nu)\right\} \left\{b_k^+ g^+(z+\nu) + b_k^- g^-(z+\nu)\right\}\right\} m_+^{\mu+\nu-l}(z).$$

Now we use that

$$(g^{\pm})^2(z) - g^{\pm}(z),\ g^+(z+\alpha)g^-(z+\beta) \in \underline{\mathfrak{M}}_{\mathfrak{z}}$$

for all $\alpha, \beta \in \mathbb{C}$. Then, in view of $g^+ + g^- = 1$ we get

$$g^+(z+\mu+\nu)g^+(z+\nu)-g^+(z+\mu+\nu)g^+(z+\mu+\nu)$$

$$= g^+(z+\mu+\nu)\left\{g^+(z+\nu)+g^-(z+\nu)\right\} -g^+(z+\mu+\nu)g^-(z+\nu)-(g^+)^2(z+\mu+\nu)\cdot$$

$$= g^+(z+\mu+\nu)-(g^+)^2(z+\mu+\nu)-g^+(z+\mu+\nu)g^-(z+\nu) \in \mathfrak{M}_{\underline{z}}.$$

In the same way we obtain

$$g^-(z+\mu+\nu)g^-(z+\nu) = (g^-)^2(z+\mu+\nu) \bmod \mathfrak{M}_{\underline{z}} .$$

This yields

$$\sum_{j+k=1} \sigma_M^{\mu-j}(a)(z+\nu-k)\sigma_M^{\nu-k}(b)(z)$$

$$= \left\{ \sum_{j+k=1} a_j^+ b_k^+ g^+(z+\mu+\nu)-a_j^- b_k^- g^-(z+\mu+\nu) \right\} m_+^{\mu+\nu-\ell}(z)$$

mod $\mathfrak{M}_{\underline{z}}$. Thus the desired identity follows from(19). It remains to discuss the case of t-depending symbols. It obviously suffices to consider symbols of the form

$$a(t,\tau) = t^j a^{[j]}(\tau), \; b(t,\tau) = t^k b^{[k]}(\tau).$$

Since the $\#_M$ composition is associative, we may restrict ourselves to the cases j,k = 0,1, where j=1 is trivial, if we have the result for j=0, k=0;1.

Moreover, for k=1 we may assume $b^{[k]} = 1$. Since j=k=0 was already considered, it remains

$$a(t,\tau) = a(\tau), \; b(t,\tau) = t.$$

By definition we have

$$\sigma_M^{-k}(t) = \begin{cases} 0 & \text{for } k \neq 1, \\ 1 & \text{for } k=1, \end{cases}$$

and $a \#_\psi t = t\,a + D_\tau a$. Now

$$\sigma_M^{\mu-j}(ta)(z) = \sigma_M^{\mu-j+1}(a)(z)$$

$$= \left\{ a_{j-1}^{+} g^{+}(z+\mu) + a_{j-1}^{-} g^{-}(z+\mu) \right\} \frac{\Gamma(1-z)}{\Gamma(1-(z+\mu-j+1))}$$

and

$$\sigma_M^{\mu-j}(D_\tau a)(z) = \left\{ (D_\tau a)_j^{+} g^{+}(z+\mu) + (D_\tau a)_j^{-} g^{-}(z+\mu) \right\} \frac{\Gamma(1-z)}{\Gamma(1-(z+\mu-j))} .$$

From $a(\tau) \sim \sum a_j^{\pm} (i\tau)^{\mu-j}$ it follows $(D_\tau a)_j^{\pm} = (\mu-j+1) a_{j-1}^{\pm}$.

Thus

$$\sigma_M^{\mu-j}(D_\tau a)(z) = \left\{ a_{j-1}^{+} g^{+}(z+\mu) + a_{j-1}^{-} g^{-}(z+\mu) \right\} \frac{(\mu-j+1)\, \Gamma(1-z)}{\Gamma(1-(z+\mu-j))} .$$

We have to show

$$\sigma_M^{\mu-l}(a \#_\psi t)(z) = \sum_{j+k=l} \sigma_M^{\mu-j}(a)(z-k)\sigma_M^{-k}(t)(z) = \sigma_M^{\mu-l+1}(a)(z-1),$$

i.e.

$$\sigma_M^{\mu-j}(ta)(z) + \sigma_M^{\mu-j}(D_\tau a)(z) = \sigma_M^{\mu-j+1}(a)(z-1). \qquad (20)$$

It suffices to check that the coefficients of the $g^{\pm}$ coincide. The coefficient of $g^{\pm}$ on the left of (20) is

$$a_{j-1}^{\pm} \left\{ \frac{\Gamma(1-z)}{\Gamma(1-(z+\mu-j+1))} + \frac{(\mu-j+1)\Gamma(1-z)}{\Gamma(1-(z+\mu-j))} \right\}$$

and that on the right

$$a_{j-1}^{\pm} \frac{\Gamma(1-(z-1))}{\Gamma(1-(z-1+\mu-j+1))} = a_{j-1}^{\pm} \frac{(1-z)\, \Gamma(1-z)}{\Gamma(1-(z+\mu-j))} .$$

Since $\Gamma(1-(z+\mu-j)) = -(z+\mu-j)\,\Gamma(1-(z+\mu-j+1))$, we get

$$\frac{1}{\Gamma(1-(z+\mu-j+1))} + \frac{(\mu-j+1)}{\Gamma(1-(z+\mu-j))} = \frac{-(z+\mu-j) + (\mu-j+1)}{\Gamma(1-(z+\mu-j))} = \frac{1-z}{\Gamma(1-(z+\mu-j))}$$

and hence the equality of the coefficients. $\square$

In general, for $a \in S^{\mu}_{cl,st}(\overline{\mathbb{R}}_+ \times \mathbb{R})$, $b \in S^{\nu}_{cl,st}(\overline{\mathbb{R}}_+ \times \mathbb{R})$, we cannot expect that the image of $op_\psi^{\gamma}(b)$ fits with the domain of $op_\psi^{\gamma-\nu}(a)$ because of the additional finite-dimensional space in the image of $op_\psi^{\gamma}(b)$.

Since in our calculus Green operators (on $\mathbb{R}_+$) are negligible, we take into account also finite-dimensional modifications such that

$$op_\psi^\gamma(b) : \mathcal{K}^{s,\gamma} \longrightarrow \mathcal{K}^{s-\nu,\gamma-\nu},$$

or, the other way round, enlarging the domain of definition

$$op_\psi^{\gamma-\nu}(a) : \mathcal{K}^{s-\nu}_{P_0}(\gamma-\nu) \longrightarrow \mathcal{K}^{s-\mu-\nu}_{P_0}(\gamma-\mu-\nu)$$

($\gamma-\nu \neq \frac{1}{2} \bmod \mathbb{Z}$, $\gamma-\mu-\nu \neq \frac{1}{2} \bmod \mathbb{Z}$) where the action is defined by means of a (non-canonical) projection $\mathcal{K}^{s-\nu}_{P_0}(\gamma-\nu) \longrightarrow \mathcal{K}^{s-\nu,\gamma-\nu}$.
In the latter case, we get, of course, a realization different from that obtained by restriction from suitable $\mathcal{K}^{s-\nu,\delta}$. All these modifications do not affect the Mellin and interior symbols. Then we obtain

9. Corollary. Let $a \in S^\mu_{cl,st}(\overline{\mathbb{R}}_+ \times \mathbb{R})$, $b \in S^\nu_{cl,st}(\overline{\mathbb{R}}_+ \times \mathbb{R})$.
Then

$$op_\psi^{\gamma-\nu}(a) op_\psi^\gamma(b) - op_\psi^\gamma(a \#_\psi b) \in \mathring{\mathcal{L}}_\psi^{\mu+\nu}(\gamma-\mu-\nu,\gamma) + \mathcal{L}_{M+G}(\gamma-\mu-\nu,\gamma).$$

2.3.6. The algebras $\mathcal{L}^\infty$ and $\mathcal{G}^\infty$.

Now we extend the classes $\mathcal{L}_{M+G}$ and $\mathcal{G}_{M+G}$ by pseudo-differential operators. For the resulting classes $\mathcal{L}$ and $\mathcal{G}$ we derive a precise symbolic calculus.

1. Definition. Let $\mu, \gamma \in \mathbb{R}$, $k \in \mathbb{Z}_+$ be given. Denote by $\mathcal{L}^\mu(\gamma-\mu,\gamma)^{(-k)}$ the subspace of all $\mathfrak{a} \in \bigcap_{s \in \mathbb{R}} \mathcal{L}(\mathcal{K}^{s,\gamma}(\mathbb{R}_+), \mathcal{K}^{s-\mu,\gamma-\mu}(\mathbb{R}_+))$ such that there exists $a \in S^\mu_{cl,st}(\overline{\mathbb{R}}_+ \times \mathbb{R})$ and an operator convention op_ψ^γ which gives operators in $\bigcap_s \mathcal{L}(\mathcal{K}^{s,\gamma}(\mathbb{R}_+), \mathcal{K}^{s-\mu,\gamma-\mu}(\mathbb{R}_+))$ such that

$$\mathfrak{a} - op_\psi^\gamma(a) \in \mathcal{L}_{M+G}(\gamma-\mu,\gamma)^{(-k)}$$

is satisfied.

In 2.3.5. it was proved that different operator conventions amount to a difference of the resulting operators in $\mathcal{L}^{(-k)}_{M+G}$. Therefore one can take any choice of op^{γ}_{ψ} in order to verify Definition 1. Since neither the Mellin nor the Green operator contribute to the complete interior symbol, we have a well-defined mapping

$$\sigma_{\psi} : \mathcal{L}^{\mu}(\gamma-\mu,\gamma)^{(-k)} \longrightarrow S^{\mu}_{cl,st}/S^{-\infty}_{(o)}$$

the kernel of which is precisely $\mathcal{L}_{M+G}(\gamma-\mu,\gamma)^{(-k)}$. Therefore the sequence

$$0 \longrightarrow \mathcal{L}_{M+G}(\gamma-\mu,\gamma)^{(-k)} \hookrightarrow \mathcal{L}^{\mu}(\gamma-\mu,\gamma)^{(-k)} \xrightarrow{\sigma_{\psi}} S^{\mu}_{cl,st}/S^{-\infty}_{(o)} \longrightarrow 0 \tag{1}$$

is exact.

Now we introduce a locally convex topology in $\mathcal{L}^{\mu}(\gamma-\mu,\gamma)^{(-k)}$ in the following way. In 2.3.4. a topology of $\mathcal{L}_{M+G}(\gamma-\mu,\gamma)^{(-k)}$ was described. For each fixed $\nu\in\mathbb{R}$ denote by $S^{\nu}_{cl,(o)}$ the subspace of $S^{\nu}_{cl,st}(\overline{\mathbb{R}}_+\times\mathbb{R})$ of symbols $a(t,\tau)$ vanishing for t larger than a constant depending on a. Consider $S^{\mu}_{cl,st}$ in the topology of the inductive limit over increasing constants c, where the symbols are independent of t for $t\geq c$. For each $N\in\mathbb{Z}_+$ we get a natural topology on $S^{\mu}_{cl,st}/S^{\mu-N}_{cl,(o)}$. Then

$$S^{\mu}_{cl,st}/S^{-\infty}_{(o)} = \bigcap_{N} S^{\mu}_{cl,st}/S^{\mu-N}_{cl,(o)}$$

and $S^{\mu}_{cl,st}/S^{-\infty}_{(o)}$ is endowed with the projective limit topology taken over increasing N.

The sequence (1) does not split. Therefore, we proceed as in the case of ordinary ψDOs. Fix $N\in\mathbb{Z}_+$ such that $N>k$. Set

$$\mathcal{L}^{\mu-N}_{(\gamma-\mu,\gamma)}{}^{(-k)} = \bigcap_{\substack{s\in\mathbb{R},\, s'\geq 0\\ \varepsilon>0}} \mathcal{L}(\mathcal{K}^{s,\gamma}(\mathbb{R}_+),(1+t)^{s'}\mathcal{K}^{s-\mu+N,\gamma-\mu+k-\varepsilon}(\mathbb{R}_+))$$

with the projective limit topology.

Then we have a well-defined symbol map

$$\sigma^{(N)}_{\psi} : \mathcal{L}^{\mu}_{(\gamma-\mu,\gamma)}{}^{(-k)} + \mathcal{L}^{\mu-N}_{(\gamma-\mu,\gamma)}{}^{(-k)} \longrightarrow S^{\mu}_{cl,st}/S^{\mu-N}_{cl,(o)},$$

since $\sigma_{\psi}(\mathcal{L}^{\mu}_{(\gamma-\mu,\gamma)}{}^{(-k)} \cap \mathcal{L}^{\mu-N}_{(\gamma-\mu,\gamma)}{}^{(-k)}) \subset S^{\mu-N}_{cl,(o)}$, and the sequence

$$0 \to \mathcal{L}_{M+G}^{(\gamma-\mu,\gamma)(-k)} + \mathcal{L}_{(\gamma-\mu,\gamma)}^{\mu-N(-k)} \hookrightarrow \mathcal{L}_{(\gamma-\mu,\gamma)}^{\mu\ (-k)} + \mathcal{L}_{(\gamma-\mu,\gamma)}^{\mu-N(-k)} \xrightarrow{\sigma_\psi^{(N)}} S_{cl,st}^{\mu} / S_{cl,(o)}^{\mu-N} \longrightarrow 0$$

is exact. Moreover, it splits because we can choose a continuous linear section

$$S_{cl,st}^{\mu} / S_{cl,(o)}^{\mu-N} \longrightarrow S_{cl,st}^{\mu}$$

mapping an equivalence class into a well-defined representative. The composition of this mapping with an op_ψ^γ convention is the desired right inverse of $\sigma_\psi^{(N)}$. Therefore

$$\mathcal{L}_{(\gamma-\mu,\gamma)}^{\mu\ (-k)} + \mathcal{L}_{(\gamma-\mu,\gamma)}^{\mu-N(-k)} \cong (\mathcal{L}_{M+G}^{(\gamma-\mu,\gamma)(-k)} + \mathcal{L}_{(\gamma-\mu,\gamma)}^{\mu-N(-k)}) \oplus (S_{cl,st}^{\mu} / S_{cl,(o)}^{\mu-N}),$$

which induces a topology on $\mathcal{L}_{(\gamma-\mu,\gamma)}^{\mu\ (-k)} + \mathcal{L}_{(\gamma-\mu,\gamma)}^{\mu-N(-k)}$.

Then

$$\mathcal{L}_{(\gamma-\mu,\gamma)}^{\mu\ (-k)} = \varprojlim_{N} (\mathcal{L}_{(\gamma-\mu,\gamma)}^{\mu\ (-k)} + \mathcal{L}_{(\gamma-\mu,\gamma)}^{\mu-N(-k)})$$

gives the topology on $\mathcal{L}_{(\gamma-\mu,\gamma)}^{\mu\ (-k)}$.

Combining the results of 2.3.5. and 2.3.4. we get also conormal symbols of operators in $\mathcal{L}^\mu(\gamma-\mu,\gamma)^{(-k)}$. In fact, in 2.3.5. we constructed a mapping

$$\sigma_M^{(-k)} : S_{cl}^{\mu} / S^{-\infty} \longrightarrow \mathop{\times}_{j=0}^{k-1} \mathfrak{M}_{\underline{z}}^{\mu}$$

which takes into account only the coefficients of the asymptotics for $\tau \longrightarrow \pm\infty$ of the Taylor expansion of the symbol $a(t,\tau)$ at $t=0$. Therefore we can omit the assumptions about the behaviour for large t. Exactness of (1) yields that each $\mathfrak{a} \in \mathcal{L}^\mu(\gamma-\mu,\gamma)^{(-k)}$ has the form

$$\mathfrak{a} = op_\psi^\gamma(a) + \mathfrak{m}$$

where $a = \sigma_\psi(\mathfrak{a})$, $\mathfrak{m} \in \mathcal{L}_{M+G}(\gamma-\mu,\gamma)^{(-k)}$ and op_ψ^γ is a fixed operator convention.

We set

$$\sigma_M^{(-k)}(\mathfrak{a}) = \sigma_M^{(-k)}(a) + \sigma_M^{(-k)}(\mathfrak{m})$$

where $\sigma_M^{(-k)}(\mathfrak{m})$ denotes the sequence of the conormal symbols of $\mathfrak{m}$. Clearly, this definition does not depend on the choice of the op_ψ^γ, which may alter $\mathfrak{m}$ only by a Green operator. Therefore, we get a well-defined mapping

$$\sigma_M^{(-k)} \;:\; \mathcal{L}^\mu(\gamma-\mu,\gamma)^{(-k)} \longrightarrow \mathop{X}_{j=0}^{k-1} \mathfrak{M}_{\underline{z}}^\mu .$$

Put $\sigma^{(-k)} = (\sigma_\psi, \sigma_M^{(-k)})$ and denote by $\mathcal{S}^{\mu,(-k),\gamma}$ its image in $(S_{cl,st}^\mu / S_{(o)}^{-\infty}) \mathop{X}_{j=0}^{k-1} \mathfrak{M}_{\underline{z}}^\mu$. An intrinsic characterization of $\mathcal{S}^{\mu,(-k),\gamma}$ in $(S_{cl,st}^\mu / S_{(o)}^{-\infty}) \mathop{X}_{j=0}^{k-1} \mathfrak{M}_{\underline{z}}^\mu$ is given in the following

1. <u>Lemma.</u> $(a,m) \in (S_{cl,st}^\mu / S_{(o)}^{-\infty}) \mathop{X}_{j=0}^{k-1} \mathfrak{M}_{\underline{z}}^\mu$ belongs to $\mathcal{S}^{\mu,(-k),\gamma}$ iff

$$\sigma_M^{(-k)}(a) - m \in \mathfrak{M}_{\underline{z}[\gamma]} \mathop{X}_{j=1}^{k-1} \mathfrak{M}_{\underline{z}} . \tag{2}$$

<u>Proof.</u> Given $\mathfrak{a} \in \mathcal{L}^\mu(\gamma-\mu,\gamma)^{(-k)}$. Then (2) is satisfied since it contains on the left the sequence of conormal symbols of an operator in $\mathcal{L}_{M+G}(\gamma-\mu,\gamma)^{(-k)}$. Conversely, given (a,m) satisfying (2) we get $\mathfrak{a} \in \mathcal{L}^\mu(\gamma-\mu,\gamma)^{(-k)}$ with the given symbol by setting $\mathfrak{a} = op_\psi^\gamma(a) + op_M(m-\sigma_M^{(-k)}(a))$, where op_M denotes the assignment of an operator in $\mathcal{L}_{\dot{M}+G}(\gamma-\mu,\gamma)^{(-k)}$ to a symbol in $\mathfrak{M}_{\underline{z}[\gamma]} \mathop{X}_{j=1}^{k-1} \mathfrak{M}_{\underline{z}}$. □

In particular, the conormal symbols occurring in $\mathcal{S}^{\mu,(-k),\gamma}$ belong to $\mathfrak{M}_{\underline{z}} \oplus V_k(\mu)$, where $V_k(\mu)$ denotes the finite-dimensional space spanned by the functions

$$g^{\pm}(z+\mu)\,\frac{\Gamma(1-z)}{\Gamma(1-z+l-\mu)} \quad , \quad l=0,1,\dots,k-1.$$

Therefore, we have a unique topology induced by $\mathfrak{M}_{\underline{z}}$.

The conormal symbols together with the interior symbol yield a control of operators in $\mathcal{L}^\mu$ modulo Green operators. In fact, from 2.3.4. we get

2. Theorem. The sequence

$$0 \longrightarrow \mathcal{L}_G(\gamma-\mu,\gamma)^{(-k)} \hookrightarrow \mathcal{L}^\mu(\gamma-\mu,\gamma)^{(-k)} \xrightarrow{\sigma^{(-k)}} \mathcal{T}^{\mu,(-k),\gamma} \longrightarrow 0$$

is exact.

Remark that it does not split.

Passing to the projective limit over increasing k we get the operator class $\mathcal{L}^\mu(\gamma-\mu,\gamma)$ and the symbol class $\mathcal{T}^{\mu,\gamma}$. Then the sequences

$$0 \longrightarrow \mathcal{L}_{M+G}(\gamma-\mu,\gamma) \hookrightarrow \mathcal{L}^\mu(\gamma-\mu,\gamma) \xrightarrow{\sigma_\psi} S^\mu_{cl,st}/S^{-\infty}_{(o)} \longrightarrow 0$$

and

$$0 \longrightarrow \mathcal{L}_G(\gamma-\mu,\gamma) \hookrightarrow \mathcal{L}^\mu(\gamma-\mu,\gamma) \xrightarrow{\sigma} \mathcal{T}^{\mu,\gamma} \longrightarrow 0$$

are exact. Below we shall formulate the results for the classes with $k \leqq \infty$ ($k=\infty$ understood as projective limit).

Let $P = (\gamma,p) \in \underline{P}$, $Q = (\gamma-\mu,q) \in \underline{P}$. A minor modification of the above approach will give the class $\mathcal{L}^\mu(Q,P)^{(-k)} \subset \bigcap_{s\in\mathbb{R}} \mathcal{L}(\mathcal{K}^s_P(\mathbb{R}_+), \mathcal{K}^{s-\mu}_Q(\mathbb{R}_+))$. Here we set $\mathcal{K}^s_P(\mathbb{R}_+) = \mathcal{K}^{s,\gamma}(\mathbb{R}_+) + \mathcal{S}_p$. The op^γ_ψ convention has to be replaced by an (arbitrary) convention giving operators in $\bigcap_{s\in\mathbb{R}} \mathcal{L}(\mathcal{K}^s_P(\mathbb{R}_+), \mathcal{K}^{s-\mu}_Q(\mathbb{R}_+))$, and $\mathcal{L}_{M+G}(Q,P)^{(-k)}$ replaces $\mathcal{L}_{M+G}(\gamma-\mu,\gamma)^{(-k)}$.

Then we have the exact sequences

$$0 \longrightarrow \mathcal{L}_{M+G}(Q,P)^{(-k)} \hookrightarrow \mathcal{L}^\mu(Q,P)^{(-k)} \xrightarrow{\sigma_\psi} S^\mu_{cl,st}/S^{-\infty}_{(o)} \longrightarrow 0$$

and

$$0 \longrightarrow \mathcal{L}_G(Q,P)^{(-k)} \hookrightarrow \mathcal{L}^\mu(Q,P)^{(-k)} \xrightarrow{\sigma} \mathcal{T}^{\mu,(-k),\gamma} \longrightarrow 0 .$$

Note that neither the interior nor the conormal symbols are affected by the choice of P and Q.

Now we are ready to investigate compositions and adjoints of operators in $\mathcal{L}^\infty = \bigcup_\mu \mathcal{L}^\mu$. We consider the composition first. In the previous sections we proved that $\mathcal{L}_{M+G}$ is closed under compositions (2.3.4.

Theorem 10) and that the composition of ψDOs is again a ψDO modulo $\mathcal{L}_{M+G}$(2.3.5. Corollary). Therefore we get immediately

3. Theorem. Let $\gamma,\mu,\nu\in\mathbb{R}$, $P = (\gamma,\mathcal{P})$, $Q = (\gamma-\mu,\mathcal{Q})$, $R = (\gamma-\mu-\nu,\mathcal{W})\in\underline{P}$. Then $\mathfrak{a}\in\mathcal{L}^{\mu}(R,Q)^{(-k)}$ and $\mathfrak{b}\in\mathcal{L}^{\mu}(Q,P)^{(-k)}$ implies $\mathfrak{a}\mathfrak{b}\in\mathcal{L}^{\mu+\nu}(Q,P)^{(-k)}$,

$$\sigma_{\psi}(\mathfrak{a}\mathfrak{b}) = \sigma_{\psi}(\mathfrak{a}) \# \sigma_{\psi}(\mathfrak{b}),$$

and for the conormal symbols the rule in 2.3.4. Theorem 10 holds.

In particular, $\mathcal{L}^{(-k)}_{M+G}$ and $\mathcal{L}^{(-k)}_{G}$ are two-sided ideals, since these subspaces are precisely $\ker\,\sigma_{\psi}^{(-k)}$ and $\ker\,\sigma^{(-k)}$, respectively.

When discussing adjoints it is reasonable to assume the operators acting between non-perturbed weighted spaces. Otherwise new types of spaces would occur.

4. Proposition. Let $\gamma,\mu\in\mathbb{R}$, $P = (\gamma,0)$, $Q = (\gamma-\mu,0)$ and $\mathfrak{a}\in\mathcal{L}^{\mu}(Q,P)^{(-k)}$. Then the formal adjoint $\mathfrak{a}^{*}$ belongs to $\mathcal{L}^{\mu}(P^{*},Q^{*})^{(-k)}$, where $P^{*}= (-\gamma,0)$, $Q^{*}= (\mu-\gamma,0)$.

Proof. For Mellin and (two-sided) Green operators the assertion follows from 1.3.1. and the definition of Green operators. For the ψDO it was proved in 2.3.5. $\square$

The restriction of an operator $\mathfrak{a}\in\mathcal{L}^{\mu}(Q,P)^{(-k)}$ to a subspace with asymptotics has an image in another space with asymptotics. More precisely, let $\mathfrak{a}\in\mathcal{L}^{\mu}(Q,P)^{(-k)}$. For each $P_1 = (\lambda,\mathcal{P}_1)\in\underline{P}$, $\gamma\leqq\lambda<\gamma+k$, $r(\gamma)\,\mathcal{P}_1\leqq\mathcal{P}$ and each $\varepsilon>0$ there exists a $Q_1 = (\lambda-\mu-\varepsilon,\mathcal{Q}_1)$, $r(\gamma-\mu)\,\mathcal{Q}_1\leqq\mathcal{Q}$ such that the restriction of $\mathfrak{a}$ to $\mathcal{K}^{\bullet}_{P_1}(\mathbb{R}_+)$ defines a continuous operator

$$\mathfrak{a} : \mathcal{K}^{s}_{P_1}(\mathbb{R}_+) \longrightarrow \mathcal{K}^{s-\mu}_{Q_1}(\mathbb{R}_+).$$

For Green operators this property was discussed in 2.3.3. and for Mellin operators and ψDOs in 2.3.4. and 2.3.5.

Replacing in Definition 1 $\mathcal{L}_{M+G}(\gamma-\mu,\gamma)^{(-k)}$ by $\mathcal{G}_{M+G}(\gamma-\mu,\gamma)^{(-k)}$ we define the class $\mathcal{G}^{\mu}(\gamma-\mu,\gamma)^{(-k)}$. We get an exact sequence

$$0 \longrightarrow \mathcal{G}_{M+G}(\gamma-\mu,\gamma)^{(-k)} \hookrightarrow \mathcal{G}^{\mu}(\gamma-\mu,\gamma)^{(-k)} \xrightarrow{\sigma_\psi} S^{\mu}_{cl,st}/S^{-\infty}_{(o)} \longrightarrow 0$$

and the same for $\mathcal{G}^{\mu}(Q,P)^{(-k)}$, $P,Q \in \underline{P}$. Again the topology on $\mathcal{G}_{M+G}(Q,P)^{(-k)}$ induces in a natural way a topology on $\mathcal{G}^{\mu}(Q,P)^{(-k)}$. Symbols of operators in $\mathcal{G}^{\mu}(Q,P)^{(-k)}$ consist of interior symbols and a sequence of conormal symbols. Their compatibility is expressed as in Lemma 1 with the classes of Mellin symbols with continuous asymptotics. Also the sequence

$$0 \longrightarrow \mathcal{G}_{G}(Q,P)^{(-k)} \hookrightarrow \mathcal{G}^{\mu}(Q,P)^{(-k)} \xrightarrow{\sigma^{(-k)}} \mathcal{T}(\mathcal{G})^{\mu,(-k),\gamma} \longrightarrow 0$$

is exact, where $\mathcal{T}(\mathcal{G})^{\mu,(-k),\gamma} \subset (S^{\mu}_{cl,st}/S^{-\infty}_{(o)}) \times \mathfrak{M}_{\underline{\Lambda}[\gamma]} \mathop{\times}\limits_{j=1}^{k=1} \mathfrak{M}_{\underline{\Lambda}}$ is characterized by the mentioned compatibility condition. Finally, for the class $\mathcal{G}^{\mu}(Q,P)^{(-k)}$ we have analogues of Theorem 3 and Proposition 4.

Our next objective is to establish the concept of ellipticity for operators in $\mathcal{L}^{\mu}(Q,P)^{(-k)}$ and $\mathcal{G}^{\mu}(Q,P)^{(-k)}$. It is very similar to that in Section 1.5.3. Since here we take into account also t – depending interior symbols, an additional condition at ∞ appears.

Recall that for $\alpha \in \mathcal{G}^{\mu}(Q,P)^{(-k)}$ there is a principal interior symbol $\sigma^{\mu}_{\psi}(\alpha)(t,\tau)$, $\tau \neq 0, t \geqq 0$ (homogeneous in τ of order μ), a principal conormal symbol $\sigma^{\mu}_{M}(\alpha)(z), z \in \Gamma_{\frac{1}{2}-\gamma}$, $P = (\gamma,\mathcal{P})$, and a well-defined (complete) symbol at $t=\infty$ $\sigma_{\psi}(\alpha)(\infty,\tau)$, $\tau \in \mathbb{R}$.

5. <u>Definition.</u> An operator $\alpha \in \mathcal{G}^{\mu}(Q,P)^{(-k)}$, $P = (\gamma,\mathcal{P}), Q = (\gamma-\mu,\mathcal{Q})$ is called <u>elliptic</u> if

(i) $\sigma^{\mu}_{\psi}(\alpha)(t,\tau) \neq 0$ for all $\tau \neq 0$, $t \geqq 0$,

(ii) $\sigma^{\mu}_{M}(\alpha)(z) \neq 0$ for all $z \in \Gamma_{\frac{1}{2}-\gamma}$,

(iii) $\sigma_{\psi}(\alpha)(\infty,\tau) \neq 0$ for all $\tau \in \mathbb{R}$.

Note that ellipticity depends on γ but not on φ, $\mathfrak{q}$ and k.

6. Definition. An operator $\mathfrak{b} \in \mathcal{C}^{-\mu}(P,Q)^{(-k)}$ is called a parametrix of $\mathfrak{a} \in \mathcal{C}^{\mu}(Q,P)^{(-k)}$ (modulo $\mathcal{C}_G^{(-k)}$) if

$$\mathfrak{a}\mathfrak{b} - 1 \in \mathcal{C}_G(Q,Q)^{(-k)}, \ \mathfrak{b}\mathfrak{a} - 1 \in \mathcal{C}_G(P,P)^{(-k)} .$$

For $\mathfrak{a} \in \mathcal{L}^{\mu}(Q,P)^{(-k)}$ we talk about a parametrix $\mathfrak{b} \in \mathcal{L}^{-\mu}(P,Q)^{(-k)}$ (modulo $\mathcal{L}_G^{(-k)}$) if

$$\mathfrak{a}\mathfrak{b} - 1 \in \mathcal{L}_G(Q,Q)^{(-k)}, \ \mathfrak{b}\mathfrak{a} - 1 \in \mathcal{L}_G(P,P)^{(-k)}.$$

Note that if $\mathfrak{b} \in \mathcal{C}^{-\mu}(P,Q)^{(-k-1)}$ is a parametrix of $\mathfrak{a} \in \mathcal{C}^{\mu}(Q,P)^{(-k-1)}$ mod $\mathcal{C}_G^{(-k-1)}$, then $\mathfrak{b}' \in \mathcal{C}^{-\mu}(P,Q)^{(-k)}$ is a parametrix of $\mathfrak{a}' \in \mathcal{C}^{\mu}(Q,P)^{(-k)}$ mod $\mathcal{C}_G^{(-k)}$. Here the prime indicates the induced operators in the corresponding larger operator classes. A similar remark holds for the $\mathcal{L}$ - classes.

7. Proposition. Let $h \in \mathcal{T}(\mathcal{C})^{\mu,(-k),\gamma}$ be elliptic. Then there exists $g \in \mathcal{T}(\mathcal{C})^{-\mu,(-k),\gamma}$ which is inverse to h with respect to the composition law mentioned in Theorem 3.

Proof. The assertion is well-known for the interior symbol. The leading Mellin symbol exists, is holomorphic in an appropriate subset of $\mathbb{C}$. Successively, the lower order conormal symbols are determined. Moreover, the compatibility condition of Lemma 1 is satisfied, which is proved in the same way as in 1.3.2. Theorem 4. $\square$

8. Theorem. For $\mathfrak{a} \in \mathcal{C}^{\mu}(Q,P)^{(-k)}$ $(\mathcal{L}^{\mu}(Q,P)^{(-k)})$ the following conditions are equivalent

(i) $\mathfrak{a}$ is elliptic,

(ii) $\mathfrak{a}$ possesses a parametrix modulo $\mathcal{C}_G(Q,P)^{(-k)}$ $(\mathcal{L}_G(Q,P)^{(-k)})$,

(iii) $\mathfrak{a} \in \bigcap_s \mathcal{L}(\mathcal{K}_P^s(\mathbb{R}_+), \mathcal{K}_Q^{s-\mu}(\mathbb{R}_+))$ is Fredholm.

Proof. (i) $\Rightarrow$ (ii): A parametrix is given by any operator in the class with the inverse symbol (with respect to the composition rule mentioned in Theorem 3). (ii) $\Rightarrow$(iii) since $\mathcal{G}_G(Q,P)^{(-k)}$ consists of compact operators in $\mathcal{L}(\mathcal{K}_P^s(\mathbb{R}_+),\mathcal{K}_Q^{s-\mu}(\mathbb{R}_+))$. (iii) $\Rightarrow$(i) follows in an analogous way as the proof of 1.5.3. Theorem 4. In fact, the injectivity of the principal interior symbol and the symbol at $t=\infty$ follows from the Fredholm property in the usual way. For the principal conormal symbol the procedure from [R 2] applies.

Provided the conditions of Theorem 8 are satisfied, we have the following regularity result.

9. Proposition. Let $\mathfrak{a}$ be elliptic and $u\in\mathcal{K}_P^s(\mathbb{R}_+)$ any solution of the equation $\mathfrak{a}u=f$. If the right-hand side $f\in\mathcal{K}_Q^{s-\mu}(\mathbb{R}_+)$ belongs to a space with asymptotics, then u has asymptotics, too. More precisely, let $\mathfrak{a}\in\mathcal{C}_{\mathcal{G}}^{\mu}(Q,P)^{(-k)}$ and $f\in\mathcal{K}^{s,\lambda}(\mathbb{R}_+)+\mathcal{S}_{\Delta}(\mathbb{R}_+)+\mathcal{S}_{q}(\mathbb{R}_+)$, $\lambda<\gamma-\mu+k$, $\Delta\in\underline{\Delta}(\gamma-\mu)$. Then each $u\in\mathcal{K}^{s',\gamma}(\mathbb{R}_+)+\mathcal{S}_{p}(\mathbb{R}_+)$ satisfying $\mathfrak{a}u=f$ belongs to $\mathcal{K}^{s+\mu,\lambda+\mu}(\mathbb{R}_+)+\mathcal{S}_{\Sigma}(\mathbb{R}_+)+\mathcal{S}_{p}(\mathbb{R}_+)$ for some $\Sigma\in\underline{\Delta}(\gamma)$ depending on Δ and $\mathfrak{a}$. For each norm $\|.\|$ on $\mathcal{K}^{s+\mu,\lambda+\mu}(\mathbb{R}_+)+\mathcal{S}_{\Sigma}(\mathbb{R}_+)+\mathcal{S}_{p}(\mathbb{R}_+)$ there is a norm $\|\ \|'$ on $\mathcal{K}^{s,\lambda}(\mathbb{R}_+)+\mathcal{S}_{\Delta}(\mathbb{R}_+)+\mathcal{S}_{q}(\mathbb{R}_+)$ such that

$$\|u\|\leqq c\,(\|f\|'+\|u\|'')$$

where $\|.\|''$ denotes the norm on $\mathcal{K}^{s',\gamma}(\mathbb{R}_+)+\mathcal{S}_{p}(\mathbb{R}_+)$.

This follows from the property of a (left) parametrix $\mathfrak{b}$ of $\mathfrak{a}$ that $1-\mathfrak{b}\mathfrak{a}\in\mathcal{L}_G$ and the mapping properties of $\mathfrak{b}$ and Green operators. Fo elliptic $\mathfrak{a}\in\mathcal{L}^{\mu}(Q,P)^{(-k)}$ we have in addition the analogous result for spaces with discrete asymptotics.

The generalization to systems of operators is immediate. Then all symbols take values in appropriate spaces of matrices and ellipticity means bijectivity of the corresponding linear mappings, which is equivalent to the condition that the determinants are not vanishing. We do not go into the details. We shall fix some of the notation to be used below.

We shall extend the concept of Green operators by trace and potential operators (cf. 1.5.3.).

10. Definition. Let $k \in \mathbb{Z}_+$, $P = (\gamma, \mathfrak{p})$, $Q = (\lambda, \eta)$. Denote by $\mathcal{L}_G(Q)^{(-k)}(\mathbb{R}_+, \partial\mathbb{R}_+)$ the space of all operators $\mathfrak{a} \in \bigcap_s \mathcal{L}(\mathbb{C}, \mathcal{K}^s_Q(\mathbb{R}_+))$ such that there exists $R = (\lambda+k, \mathfrak{r}) \in \underline{P}$, $R = R(\mathfrak{a})$ and $\mathfrak{a}$ induces continuous operators

$$(1+t)^{s'}\mathfrak{a} \;:\; \mathbb{C} \longrightarrow \mathcal{K}^s_{R_1}(\mathbb{R}_+)$$

for each s, s' and each $R_1 = (\lambda+k-\varepsilon, r(\lambda+k-\varepsilon)\mathfrak{r})$, $\varepsilon > 0$. Denote by $\mathcal{L}_G(P)^{(-k)}(\partial\mathbb{R}_+, \mathbb{R}_+)$ the space of all operators $\mathfrak{a} \in \bigcap_s \mathcal{L}(\mathcal{K}^s_P(\mathbb{R}_+), \mathbb{C})$ such that their restriction to $\mathcal{K}^{s,\gamma}(\mathbb{R}_+)$ is the formal adjoint of an operator in $\mathcal{L}_G((-\gamma, 0))^{(-k)}(\mathbb{R}_+, \partial\mathbb{R}_+)$. We also use the notation $\mathcal{L}_G(Q,P)^{(-k)}(\mathbb{R}_+, \mathbb{R}_+)$ for the class introduced in 2.3.3.

Let $M, M', N, N' \in \mathbb{Z}_+$ be fixed. An operator

$$\mathfrak{a} : \begin{array}{c} \mathcal{K}^s_P(\mathbb{R}_+) \otimes \mathbb{C}^M \\ \oplus \\ \mathbb{C}^{M'} \end{array} \longrightarrow \begin{array}{c} \mathcal{K}^{s-\mu}_Q(\mathbb{R}_+) \otimes \mathbb{C}^N \\ \oplus \\ \mathbb{C}^{N'} \end{array}$$

is said to belong to the class $\mathcal{L}^{\mu}(Q,P)^{(-k)}(N,N',M,M')$ if each component belongs to $\mathcal{L}^{\mu}$. This implies that all components not acting from $\mathbb{R}_+$ to $\mathbb{R}_+$ are Green operators in the above classes. The remaining components belong to $\mathcal{L}^{\mu}(Q,P)^{(-k)}$.

It is clear from Theorem 3 and Proposition 4 that compositions and adjoints of such operators again belong to the class provided they are defined. In fact, these properties can be checked component-wise.

In the same way the $\mathcal{C}^{\mu}$ classes with continuous asymptotics are defined and have the same property. Also the symbolic calculus remains valid with the same rules where, of course, all symbols are matrix valued.

Note that only the block in the left upper corner $\mathfrak{h}$ in the representation

$$\mathfrak{a} = \begin{pmatrix} \mathfrak{r} & \mathfrak{k} \\ \mathfrak{t} & \mathfrak{q} \end{pmatrix} : \begin{matrix} \mathcal{K}_P^s(\mathbb{R}_+) \otimes \mathbb{C}^M \\ \oplus \\ \mathbb{C}^{M'} \end{matrix} \longrightarrow \begin{matrix} \mathcal{K}_Q^{s-\mu}(\mathbb{R}_+) \otimes \mathbb{C}^N \\ \oplus \\ \mathbb{C}^{N'} \end{matrix}$$

contributes to the symbols. Since the remaining operators are all finite-dimensional they cannot affect the Fredholm property of $\mathfrak{a}$. Then Theorem 8 and Proposition 9 remain valid with obvious modifications. Clearly,

$$\operatorname{ind} \mathfrak{a} = \operatorname{ind} \mathfrak{r} + M' - N'.$$

For each fixed elliptic $\mathfrak{r}$ one can choose M', N' such that $\operatorname{ind} \mathfrak{a} = 0$ Modifying then the Green operators we can obtain a bijective operator $\mathfrak{a}'$, $\mathfrak{a} - \mathfrak{a}' \in \mathcal{L}_G$. This is possible in view of the regularity result in Proposition 9. In particular, given a bijective operator $\mathfrak{a} \in \mathcal{G}^\mu(Q,P)^{(-k)}(N,N',M,M')$ we find a bijective parametrix $\mathfrak{a}^{(-1)} \in \mathcal{G}^{-\mu}(P,Q)^{(-k)}(M,M',N,N')$. Then the inverse $\mathfrak{a}^{-1}$ is equal to $\mathfrak{a}^{(-1)}$ modulo $\mathcal{G}_G(P,Q)^{(-k)}(M,M',N,N')$ and belongs, therefore, to the class $\mathcal{G}^{-\mu}(P,Q)^{(-k)}(M,M',N,N')$.

11. <u>Corollary</u>. Let $\mathfrak{a} \in \mathcal{G}^\mu(Q,P)^{(-k)}(N,N',M,M')$ be bijective. Then $\mathfrak{a}^{-1} \in \mathcal{G}^{-\mu}(P,Q)^{(-k)}(M,M',N,N')$. The same is true for the $\mathcal{L}$ -classes.

Note that Corollary 11 is true also for continuous or smooth families of operators with values in $\mathcal{G}^\mu(Q,P)^{(-k)}$ or $\mathcal{L}^\mu(Q,P)^{(-k)}$.

2.4. <u>Notes</u>

The calculus of the preceding Sections 2.3.1.-2.3.5. is the higher order analogue of that in Chapter 1. It contains some essentially new elements and motivations, which are already rather complex. So another notes section should make it more transparent for the reader.

A basic idea for dealing with boundary value problems is the calculus transversal to the boundary which depends on the tangent variables and covariables as parameters. As soon as they are fixed they can be omitted. Then the discussion concerns operators on $\mathbb{R}_+ \ni t$, where $t = 0$ refers to the boundary.

Now a starting question may be to establish a calculus of ψDO's on $\mathbb{R}_+$

with classical symbols $a(t,\tau) \in S_{cl}(\overline{\mathbb{R}}_+ \times \mathbb{R})$, $\mu \in \mathbb{R}$. What can we say about a "natural action" in adequate analogues of Sobolev spaces? Are the standard Sobolev spaces the "true ones", what about the algebra properties and the symbolic structure of operators,...? In Chapter 1 we have studied zero order operators and the action in $L^2(\mathbb{R}_+)$. The action was introduced by $r^+op(a)e^+$, which contains the elements e^+, r^+ and a singularity from a jump of extended distributions over $t = 0$. We have developed the idea of the Mellin expansion of the operator. This gave then another independent definition of the action. The intuitive meaning of using the Mellin transform may be to avoid any knowledge on distributions and symbols over $\mathbb{R}_-$. In other words, the Mellin operator convention emphasizes the role of constructions over the open $\mathbb{R}_+$ close to $t = 0$. It should be more natural than the convention with extensions and restrictions. On the other hand, it is a priori by no means obvious how to organize the details.

In Chapter 1 we could start with the action $r^+op(a)e^+$ and then obtained the Mellin expansion by direct calculations. In the higher order case the action itself is unclear, and we have to invent at the same time the "Sobolev spaces" with order numbers that are shifted by the operations. A result of the finished theory is that we have to take the $\mathcal{K}^{s,\gamma}$ spaces which contain the interior order s and the conormal order γ (which is the weight).

A crucial observation is 2.3.1. Theorem 8 which gives a link to the standard Sobolev spaces. At the same time it becomes clear that the Sobolev space smoothness in $H^s(\mathbb{R}_+)$, $H^s_o(\overline{\mathbb{R}}_+)$ is a hidden weight which also could be chosen independently. Moreover, we see that for $H^s(\mathbb{R}_+)$ on the right of the weight line $\mathrm{Re}\, z = 1/2 - s$ for $s > 0, s \neq 1/2 \bmod \mathbb{Z}$, we have extra poles in the Mellin image which come from $\mathcal{J}^s$ in the corresponding strip. Thus the $\mathcal{K}^{s,\gamma}(\mathbb{R}_+)$ spaces should also be enriched in general by extra singular terms with finitely many poles p_j on the right of $\mathrm{Re}\, z = 1/2 - \gamma$ on the Mellin image side. In the preimage there are poles at $t = 0$ of increasing orders with increasing real parts of p_j. This explains a part of the troubles with operator actions close to $t = 0$. For instance, an operator A with kernel in $C_0^\infty(\overline{\mathbb{R}}_+ \times \overline{\mathbb{R}}_+)$ does not act on $t^{-p}\omega(t)$ for $\operatorname{Re} p$ sufficiently large. On the other hand, it is obvious that for every real γ_0 there is a finite-dimensional operator A_0 with kernel in $C_0^\infty(\overline{\mathbb{R}}_+ \times \overline{\mathbb{R}}_+)$ such that $A-A_0$ acts canonically on $t^{-p}\omega(t)$ for all p with $\operatorname{Re} p < 1/2 - \gamma_0$. Now one could hope that this idea applies in general. This was realized in 2.3.5., namely if A is a ψDO first defined over $C_0^\infty(\mathbb{R}_+)$, there is a finite-dimensional A_0 in the Green class such that $A-A_0$ extends by continuity to $\mathcal{K}^{s,\gamma} + \{\text{singular terms}\}$. We also have a natural weight restriction with respect to γ. These results follow from the Mellin expansion of A, which is again obtained by a direct computation. The meromorphic Mellin symbols in the Mellin operators admit shifts of the integration line which lead to changes by Green operators. Without this change the action of A would be limited to sufficiently large γ_0. It is well-known in the standard calculus of ψDO's with the transmission property (cf. [B1], [R1], [G6]) that for the action in $H^s(\mathbb{R}_+)$ there is a typical restriction, namely $s > -1/2$. This means for the $\mathcal{K}^{s,\gamma}$ spaces $\gamma > -1/2$. Thus, in our calculus we get the explanation that this is a weight condition, whereas the "interior" s remains untouched. More precisely s is then shifted by ord A for all real s. Another novelty for higher order ψDO's on $\mathbb{R}_+$ is that we employ "positive class" Green operators as analogues of the corresponding ones in Boutet de Monvel's algebra. These operators have a regular part which is analogous as in Chapter 1 (here also with respect to finite conormal orders; this is a minor modification) and a singular part which applies specifically to the extra singular terms with respect to the weight line and generates other singular terms.

Finally we have again the proper Mellin operators with meromorphic symbols with strong decrease for $|\operatorname{Im} z| \to \infty$. All the operators arise under natural constructions if we first start with ψDO's. These operators (ψDO + Mellin + Green) generate a minimal algebra which is closed under compositions and parametrix construction (under the ellipticity condition). In case of invertibility also the inverse belongs to the algebra.

The operations correspond to a beautiful symbolic structure. Besides the ψDO symbols, which behave as usually, we have the sequence of Mellin terms of decreasing conormal orders and the Mellin translation product under compositions. This is essential also for the calculation of parametrices by a step by step procedure (or indirectly by a formal Neumann series argument). The ellipticity gives rise to Fredholm operators. In the boundary symbolic calculus we have to generate isomorphisms (in order to substitute the Shapiro-Lopatinskij condition). They are attained by adding extra matrix elements which are finite-dimensional operators of trace and potential type, and scalars as right lower corners. These matrices also form an algebra and inverses can be carried out in it. This extends a well-known observation of Boutet de Monvel [B1] that an invertible boundary symbol with the transmission property has its inverse in the same class. Here we have an analogous result for invertible boundary symbols that are more general in various respects, for instance without the transmission property.

The trace objects also may contain singular parts, similarly as the Green operators, and compositions of trace and potential operators belong to the class of Green operators. The trace operators allow to define a natural generalization of the traces $(\partial/\partial t)^k H^s(\mathbb{R}_+)|_{t=0}$ for $s - k > 1/2$. The generalization concerns functionals on $\mathcal{K}^{s,\gamma}(\mathbb{R}_+) + \{\text{singular terms}\}$, acting on the singular terms of the form $ct^{-p}\log^j t\,\omega(t)$ by carrying out an integration in the Mellin image around the pole p (after multiplying by $(z-p)^j$). In this form we also could express the traces on $H^s(\mathbb{R}_+)$; in this case the poles are the negative integers in the corresponding weight strip, and $j = 0$.

As in the zero order case there also is a variant of the calculus with continuous asymptotics, which is again motivated by the desire to understand branchings of the asymptotics in higher dimensions, cf. [R7], [S6] .

Let us now come to some more comment on the idea of continuous conormal asymptotics, which is presented in the Sections 2.1.1.-2.2.3. In the typical applications the half axis theory depends on parameters. Then, in particular the Mellin symbols of the operators are parameter depending. They are then C^{∞} families of meromorphic functions with varying poles and changing multiplicities. It is well-known that this behaviour can be very capricious, the poles may form chaotic clouds in the complex plane.

In general it is not possible to find a reasonable parameter-dependent numeration. Since the operators on the half axis are part of the symbolic calculus, we also have to apply differentiations with respect to the parameters (in particular in Leibniz products).

This makes it necessary to rely on differentiable terms that preserve the structure under differentiations. The idea is now to interprete the varying poles (including multiplicities and coefficients), or alternatively the asymptotic data, in terms of C^{∞} families of analytic functionals, at least locally close to a fixed parameter value y_o.

In view of a standard continuity of poles under small perturbations there is a neighbourhood U of y_o such that the poles for $y \in U$ are clustered around those for y_o. This behaviour may be approximateded by sets Λ_j which are just the carriers of C^{∞} families of analytic functionals that are point-wise discrete. Intuitively it is quite natural also to talk about exponents in the asymptotics that are "continuously" smeared over Λ_j with certain densities. This is by definition just the case if we admit arbitrary analytic functionals on Λ_j. In describing functional analytic properties of the operators in adequate locally convex topologies this point of view is needed anyway. In other words, a consequent theory with continuous asymptotics contains the corresponding distribution spaces, the Mellin symbol spaces and Green operators. Now the operations in the calculus, in particular the Mellin translation products or sums of operators generate a pattern of continuous asymptotics where the patches Λ_j are not separated of each other but form large "nose shaped" sets of carriers of infinite length in real direction.

This again causes a discussion on such distribution spaces, Mellin symbols and Green operators.
Finally the whole calculus with discrete asymptotics has an analogue with continuous asymptotics. In particular we have the elliptic regularity in the sense of the continuous asymptotics. Since in particular the continuous asymptotics contains the discrete one as special case, we also get the elliptic regularity in these spaces for the operators of Chapter 1. This shows a large variety of interesting distribution spaces which are respected by an elliptic theory in the mentioned sense.

This observation is also of principal value for interpreting the sort of spaces in problems with singular or degenerate symbols. The calculus in higher dimensions, which will be developed below, also has the continuous analogue although we then mainly study the discrete case. The continuous case will be commented in terms of supplementary remarks.
The main reason for this style of exposition is that otherwise we would need much more formalism. On the other hand, this program in higher dimensions is worth being developed separately including the interesting question of branching of asymptotics of solutions for boundary problems without the transmission property that was discussed in [R 7], [S 6]. More information on the functional analysis for the continuous asymptotics may be found in [S 5], [S 9].

For the material as a whole the authors do not know any model in the

literature though some particular results exist. For instance, the Mellin characterization of Sobolev spaces on $\mathbb{R}_+$ for integer s may be found in Eskin [E4]. In Chapter 4 we also shall see that the existence of a Mellin expansion is linked to results of Plamenevskij [P4], [P7] on meromorphic Mellin symbols belonging to ψDO's on $\mathbb{R}^n \setminus \{0\}$.
Let us further mention a series of papers of Lewis, Parenti (see, for instance [L2], [L3]) with a calculus of Mellin ψDO's on $\mathbb{R}_+$. These papers have a certain intersection with our approach, but we are mainly interested in the Mellin expression of standard ψDO action whereas Lewis, Parenti treat the operators as analogues of ψDO's where the Fourier transform is replaced by the Mellin transform.

The boundary symbolic calculus for ψDO's without the transmission property can be considered the "Boutet de Monvel completion" of Višik, Eskin's work, here on $\mathbb{R}_+$. In the following section we shall do the same in higher dimensions and also for matrix-valued operators. It is known that in the latter case the range of the Wiener-Hopf technique is rather limited. Another difficulty with this method is that the factorization indices of parameter-dependent symbols depend on the parameters. This is typical of applications in higher dimensions. Thus an alternative approach that avoids the factorizations was necessary anyway, and the alternative is just the idea of the Mellin symbolic calculus.

3. Boundary value problems

3.1. Function spaces

We start from the description of function spaces in $\mathbb{R}^n_+$. The operator classes turn out to define continuous operators between the corresponding spaces. Moreover, negligible operators (of a fixed order) are defined by their mapping properties.

In 3.1.1. the Hilbert space scales $\mathcal{K}^{s,\gamma}(\mathbb{R}^n_+)$ are introduced. They give a framework for the Fredholm theory. Narrower classes of spaces with asymptotics are studied in 3.1.2. They allow in the final calculus the description of the asymptotics of solutions.

We use the notation $x = (x',t)$, $x' = (x_1,\dots,x_{n-1})$, $t = x_n$ for the standard coordinates in $\mathbb{R}^n, n \geq 2$, $\mathbb{R}^n_+ = \{(x',t) \in \mathbb{R}^n,\ t > 0\}$.

3.1.1. The spaces $\mathcal{K}^{s,\gamma}(\mathbb{R}^n_+)$

The spaces $\mathcal{K}^{s,\gamma}(\mathbb{R}^n_+)$ are the higher-dimensional generalization of the spaces $\mathcal{K}^{s,\gamma}(\mathbb{R}_+)$ which were studied in section 2.3.1.

First we give a non-standard description of the standard Sobolev spaces $H^s(\mathbb{R}^n)$, which will motivate the definitions below.

1. Lemma. For $u \in C_0^\infty(\mathbb{R}^n)$, (x',t) coordinates as above, set

$$(\varkappa^{-1}(\xi')F'u)(\xi',t) = \langle\xi'\rangle^{-\frac{1}{2}} \int_{\mathbb{R}^{n-1}} e^{-ix'\xi'}\, u(x',\langle\xi'\rangle^{-1}t)\, dx'$$

$\langle\xi'\rangle = (1+|\xi'|^2)^{1/2}$, F' is the partial Fourier transform w.r.t. x' and $\varkappa^{-1}(\xi') = \varkappa^{-1}_{\langle\xi'\rangle}$ is given by

$$\varkappa^{-1}(\xi')v(t) = \langle\xi'\rangle^{-\frac{1}{2}} v(\langle\xi'\rangle^{-1}t),\quad v \in C_0^\infty(\mathbb{R}_+).$$

Then

$$\| u\|^2_{H^s(\mathbb{R}^n)} := \int_{\mathbb{R}^n} (1+|\xi|^2)^s\, |Fu(\xi)|^2\, d\xi$$

(F is the Fourier transform in $\mathbb{R}^n$)

is equal to

$$\int_{\mathbb{R}^{n-1}} \langle\xi'\rangle^{2s} \| \varkappa^{-1}(\xi')F'u(\xi',.)\|^2_{H^s(\mathbb{R})}\, d\xi'.$$

Proof. We have

$$\| \varkappa^{-1}(\xi')F'u(\xi',.)\|^2_{H^s(\mathbb{R})} = \int_{\mathbb{R}} (1+|\tau|^2)^s |Fu(\xi',\langle\xi'\rangle\tau)|^2 \langle\xi'\rangle\, d\tau$$

$$= \int_{\mathbb{R}} (1+\langle\xi'\rangle^{-2}\tau^2)^s |Fu(\xi',\tau)|^2\, d\tau$$

and

$$\int_{\mathbb{R}^{n-1}} \langle \xi' \rangle^{2s} \| \varkappa^{-1}(\xi') F' u(\xi', .)\|^2_{H^s(\mathbb{R})} d\xi' =$$

$$= \int_{\mathbb{R}^{n-1}} \int_{\mathbb{R}} (\langle \xi' \rangle^2 + \tau^2)^s |Fu(\xi',\tau)|^2 d\tau d\xi'$$

$$= \int_{\mathbb{R}^n} (1+|\xi|^2)^s |Fu(\xi)|^2 d\xi . \quad \square$$

The norm in Lemma 1 treats in a special way the distinguished t coordinate which will later be the normal direction to the boundary.

Later on we shall often use the following point of view. Given a normed space $\mathcal{K}$ of functions (distributions) on $\mathbb{R}_+$ where $\varkappa_\lambda$ acts continuously (in general not norm preserving). Then we introduce a ξ'-depending norm

$$\| v\|_{\mathcal{K}(\xi')} := \|\varkappa^{-1}(\xi')\, v\|_{\mathcal{K}} .$$

Clearly, all the norms for different $\xi' \in \mathbb{R}^{n-1}$ are equivalent, but in general not uniformly for arbitrary large ξ'. Then the result of Lemma 1 can be reformulated

$$\| u\|^2_{H^s(\mathbb{R}^n)} = \int_{\mathbb{R}^{n-1}} \langle \xi' \rangle^{2s} \| F'u(\xi',.)\|^2_{H^s(\mathbb{R})(\xi')} d\xi' .$$

We shall use, in particular, the spaces $\mathcal{K}^{s,\gamma}(\mathbb{R}_+)$. For an arbitrary cut-off function ω we have $\mathcal{K}^{s,\gamma}(\mathbb{R}_+) = \omega\mathcal{K}^{s,\gamma}(\mathbb{R}_+) + (1-\omega)H^s_0(\overline{\mathbb{R}}_+)$ and

$$\| v\|_{\mathcal{K}^{s,\gamma}(\mathbb{R}_+)} \sim \| \omega v\|_{\mathcal{K}^{s,\gamma}(\mathbb{R}_+)} + \| (1-\omega) v\|_{H^s_0(\overline{\mathbb{R}}_+)}$$

$\sim$ means equivalence of norms, i.e. there is a constant c s.th. $\frac{1}{c}\|u\| \leq \|u\|_1 \leq c\|u\|$ for all u.

For the ξ'-depending norms we get

$$\| (1-\omega) v\|^2_{H^s_0(\overline{\mathbb{R}}_+)(\xi')} = \| \varkappa^{-1}(\xi') e^+ (1-\omega) v\|^2_{H^s(\mathbb{R})} =$$

$$= \int (1+ \langle \xi' \rangle^{-2}\tau^2)^s |F(e^+(1-\omega)v(\tau)|^2 d\tau$$

(cf. the proof of Lemma 1). Therefore

$$\langle\xi'\rangle^{-s}\|(1-\omega)v\|_{H_0^s(\overline{\mathbb{R}}_+)} \leqq \|(1-\omega)v\|_{H_0^s(\overline{\mathbb{R}}_+)(\xi')} \leqq \|(1-\omega)v\|_{H_0^s(\overline{\mathbb{R}}_+)} \quad \text{for } s \geqq 0 ,$$

(1)

$$\|(1-\omega)v\|_{H_0^s(\overline{\mathbb{R}}_+)} \leqq \|(1-\omega)v\|_{H_0^s(\overline{\mathbb{R}}_+)(\xi')} \leqq \langle\xi'\rangle^{-s}\|(1-\omega)v\|_{H_0^s(\overline{\mathbb{R}}_+)} \quad \text{for } s \leqq 0 .$$

The ξ'-depending norm of ωv in $\mathcal{K}^{s,\gamma}(\mathbb{R}_+)$ can be calculated explicitly.

$$M(\varkappa^{-1}(\xi')(\omega v))(z) = \int_0^\infty t^z \langle\xi'\rangle^{-1/2}(\omega v)(\langle\xi'\rangle^{-1}t)\frac{dt}{t}$$

$$= \langle\xi'\rangle^{-\frac{1}{2}+z}\int_0^\infty t^z(\omega v)(t)\frac{dt}{t} = \langle\xi'\rangle^{-\frac{1}{2}+z} M(\omega v)(z),$$

hence

$$\|\omega v\|^2_{\mathcal{K}^{s,\gamma}(\mathbb{R}_+)(\xi')} = \|\varkappa^{-1}(\xi')(\omega v)\|^2_{\mathcal{K}^{s,\gamma}(\mathbb{R}_+)}$$

$$= \int_{\Gamma_{\frac{1}{2}-\gamma}} \langle z\rangle^{2s}\left|\langle\xi'\rangle^{-\frac{1}{2}+z} M(\omega v)(z)\right|^2 đz$$

$$= \langle\xi'\rangle^{-2\gamma}\|\omega v\|^2_{\mathcal{K}^{s,\gamma}(\mathbb{R}_+)},$$

$$\|\omega v\|_{\mathcal{K}^{s,\gamma}(\mathbb{R}_+)(\xi')} = \langle\xi'\rangle^{-\gamma}\|\omega v\|_{\mathcal{K}^{s,\gamma}(\mathbb{R}_+)} . \tag{2}$$

As a conclusion we see that $\varkappa_\lambda$ acts continuously on $\mathcal{K}^{s,\gamma}(\mathbb{R}_+)$. However, the norms of functions which are not supported near 0 depend on ξ' in a complicated way, in contrast to the very simple rule for the ξ'-dependence of the norms of functions supported near 0.

2. <u>Definition.</u> Denote by $\mathcal{K}^{s,\gamma}(\mathbb{R}^n_+)$, $s,\gamma\in\mathbb{R}$, the completion of $C_0^\infty(\mathbb{R}^n_+)$ with respect to the norm

$$\|u\|_{\mathcal{K}^{s,\gamma}(\mathbb{R}^n_+)} := \left\{\int_{\mathbb{R}^{n-1}} \langle\xi'\rangle^{2s}\|F'u(\xi',\cdot)\|^2_{\mathcal{K}^{s,\gamma}(\mathbb{R}_+)(\xi')} đ\xi'\right\}^{\frac{1}{2}}.$$

Obviously, each scalar product $(\cdot,\cdot)_{\mathcal{K}^{s,\gamma}(\mathbb{R}_+)}$ on $\mathcal{K}^{s,\gamma}(\mathbb{R}_+)$ turns $\mathcal{K}^{s,\gamma}(\mathbb{R}^n_+)$ into a Hilbert space with the scalar product

$$(u,v)_{\mathcal{K}^{s,\gamma}(\mathbb{R}^n_+)} = \int_{\mathbb{R}^{n-1}} \langle\xi'\rangle^{2s} (F'u(\xi',.),\ F'v(\xi',.))_{\mathcal{K}^{s,\gamma}(\mathbb{R}_+)(\xi')}\, d\xi'$$

where $(.,.)_{\mathcal{K}^{s,\gamma}(\mathbb{R}_+)(\xi')}$ denotes the ξ'-depending scalar product

$$(w_1,w_2)_{\mathcal{K}^{s,\gamma}(\mathbb{R}_+)(\xi')} = (\varkappa^{-1}(\xi')w_1,\ \varkappa^{-1}(\xi')w_2)_{\mathcal{K}^{s,\gamma}(\mathbb{R}_+)}.$$

It is not hard to see that $\mathcal{K}^{s,\gamma}(\mathbb{R}^n_+)$ consists of all $u \in \mathcal{S}'(\mathbb{R}^n_+)$ such that for almost all $\xi' \in \mathbb{R}^{n-1}$ $F'u(\xi',.)$ exists in $\mathcal{K}^{s,\gamma}(\mathbb{R}_+)$ and the norm $\|u\|_{\mathcal{K}^{s,\gamma}(\mathbb{R}^n_+)}$ is finite.

From the characterization of $H^s(\mathbb{R}_+)$ and $H^s_0(\overline{\mathbb{R}}_+)$ in the Mellin image (cf. 2.3.1. Theorem 8, 2.3.2. Proposition 15) we get immediately from Lemma 1

$$\begin{aligned} H^s_0(\overline{\mathbb{R}^n_+}) &= \mathcal{K}^{s,s}(\mathbb{R}^n_+) \quad \text{for} \quad s > -\tfrac{1}{2}, \\ H^s(\mathbb{R}^n_+) &= \mathcal{K}^{s,s}(\mathbb{R}^n_+) \quad \text{for} \quad s < \tfrac{1}{2}. \end{aligned} \tag{3}$$

Obviously, for $s \geq s'$, $\gamma \geq \gamma'$ we have the continuous embedding

$$\mathcal{K}^{s,\gamma}(\mathbb{R}^n_+) \hookrightarrow \mathcal{K}^{s',\gamma'}(\mathbb{R}^n_+).$$

Since $\varkappa^{-1}(\xi')$ preserves the $L^2(\mathbb{R}_+)$ scalar product we get in the usual way that the $L^2(\mathbb{R}^n_+)$ scalar product extends to a non-degenerate separately continuous sesqui-linear pairing

$$\mathcal{K}^{s,\gamma}(\mathbb{R}^n_+) \times \mathcal{K}^{-s,-\gamma}(\mathbb{R}^n_+) \longrightarrow \mathbb{C}$$

which identifies $(\mathcal{K}^{s,\gamma}(\mathbb{R}^n_+))'$ with $\mathcal{K}^{-s,-\gamma}(\mathbb{R}^n_+)$. In fact, we have for $u,v \in C_0^\infty(\mathbb{R}^n_+)$

$$(u,v)_{L^2(\mathbb{R}^n_+)} = \int_{\mathbb{R}^{n-1}} \int_{\mathbb{R}_+} F'u(\xi',t)\overline{F'v(\xi',t)}\,dt\, d\xi' =$$

$$= \int_{\mathbb{R}^{n-1}} \int_{\mathbb{R}_+} \langle\xi'\rangle^{s} \varkappa^{-1}(\xi')F'u(\xi',t)\overline{\langle\xi'\rangle^{-s} \varkappa^{-1}(\xi')F'v(\xi',t)}\,dt\, d\xi'$$

$$\leq \|u\|_{\mathcal{K}^{s,\gamma}(\mathbb{R}^n_+)} \|v\|_{\mathcal{K}^{-s,-\gamma}(\mathbb{R}^n_+)}.$$

Applying the estimates (1) and the equality (2) to $F'u(\xi',t)$ we get

$$\frac{1}{c}\langle\xi'\rangle^{-s}\|F'u(\xi',.)\|_{\mathcal{K}^{s,\gamma}(\mathbb{R}_+)} \leqq \|F'u(\xi',.)\|_{\mathcal{K}^{s,\gamma}(\mathbb{R}_+)(\xi')} \leqq c\|F'u(\xi',.)\|_{\mathcal{K}^{s,\gamma}(\mathbb{R}_+)}$$

for $s \geqq \gamma \geqq 0$,

$$\frac{1}{c}\langle\xi'\rangle^{-\gamma}\|F'u(\xi',.)\|_{\mathcal{K}^{s,\gamma}(\mathbb{R}_+)} \leqq \|F'u(\xi',.)\|_{\mathcal{K}^{s,\gamma}(\mathbb{R}_+)(\xi')} \leqq c\|F'u(\xi',.)\|_{\mathcal{K}^{s,\gamma}(\mathbb{R}_+)}$$

for $\gamma \geqq s \geqq 0$,

$$\frac{1}{c}\|F'u(\xi',.)\|_{\mathcal{K}^{s,\gamma}(\mathbb{R}_+)} \leqq \|F'u(\xi',.)\|_{\mathcal{K}^{s,\gamma}(\mathbb{R}_+)(\xi')} \leqq c\langle\xi'\rangle^{-s}\|F'u(\xi',.)\|_{\mathcal{K}^{s,\gamma}(\mathbb{R}_+)}$$

for $s \leqq \gamma \leqq 0$,

$$\frac{1}{c}\|F'u(\xi',.)\|_{\mathcal{K}^{s,\gamma}(\mathbb{R}_+)} \leqq \|F'u(\xi',.)\|_{\mathcal{K}^{s,\gamma}(\mathbb{R}_+)(\xi')} \leqq c\langle\xi'\rangle^{-\gamma}\|F'u(\xi',.)\|_{\mathcal{K}^{s,\gamma}(\mathbb{R}_+)}$$

for $\gamma \leqq s \leqq 0$,

$$\frac{1}{c}\langle\xi'\rangle^{-s}\|F'u(\xi',.)\|_{\mathcal{K}^{s,\gamma}(\mathbb{R}_+)} \leqq \|F'u(\xi',.)\|_{\mathcal{K}^{s,\gamma}(\mathbb{R}_+)(\xi')}$$

$$\leqq c\langle\xi'\rangle^{-\gamma}\|F'u(\xi',.)\|_{\mathcal{K}^{s,\gamma}(\mathbb{R}_+)}$$

for $s \geqq 0 \geqq \gamma$,

$$\frac{1}{c}\langle\xi'\rangle^{-\gamma}\|F'u(\xi',.)\|_{\mathcal{K}^{s,\gamma}(\mathbb{R}_+)} \leqq \|F'u(\xi',.)\|_{\mathcal{K}^{s,\gamma}(\mathbb{R}_+)(\xi')}$$

$$\leqq c\langle\xi'\rangle^{-s}\|F'u(\xi',.)\|_{\mathcal{K}^{s,\gamma}(\mathbb{R}_+)}$$

for $\gamma \geqq 0 \geqq s$.

We denote by $H^{s'}(\mathbb{R}^{n-1}, \mathcal{K}^{s,\gamma}(\mathbb{R}_+))$ the space of all u with finite norm

$$\left\{\int_{\mathbb{R}^{n-1}} \langle\xi'\rangle^{2s'}\|F'u(\xi',.)\|^2_{\mathcal{K}^{s,\gamma}(\mathbb{R}_+)}\, d\xi'\right\}^{1/2}$$

Then we get the continuous embeddings

$$\begin{aligned}
&\mathcal{K}^{s,\gamma}(\mathbb{R}^n_+) \hookrightarrow H^0(\mathbb{R}^{n-1}, \mathcal{K}^{s,\gamma}(\mathbb{R}_+)) \quad \text{for } s \geqq 0,\ s \geqq \gamma,\\
&\mathcal{K}^{s,\gamma}(\mathbb{R}^n_+) \hookrightarrow H^{s-\gamma}(\mathbb{R}^{n-1}, \mathcal{K}^{s,\gamma}(\mathbb{R}_+)) \text{ for } \gamma \geqq 0,\ \gamma \geqq s\quad ,\\
&\mathcal{K}^{s,\gamma}(\mathbb{R}^n_+) \hookrightarrow H^{s}(\mathbb{R}^{n-1}, \mathcal{K}^{s,\gamma}(\mathbb{R}_+)) \quad \text{for } s,\gamma \leqq 0
\end{aligned} \tag{4}$$

and

$$\begin{aligned}
&H^s(\mathbb{R}^{n-1}, \mathcal{K}^{s,\gamma}(\mathbb{R}_+)) \hookrightarrow \mathcal{K}^{s,\gamma}(\mathbb{R}^n_+) \quad \text{for } \gamma, s \geqq 0\ ,\\
&H^0(\mathbb{R}^{n-1}, \mathcal{K}^{s,\gamma}(\mathbb{R}_+)) \hookrightarrow \mathcal{K}^{s,\gamma}(\mathbb{R}^n_+) \quad \text{for } s \leqq 0,\ s \leqq \gamma,\\
&H^{s-\gamma}(\mathbb{R}^{n-1}, \mathcal{K}^{s,\gamma}(\mathbb{R}_+)) \hookrightarrow \mathcal{K}^{s,\gamma}(\mathbb{R}^n_+) \text{ for } \gamma \leqq 0,\ \gamma \leqq s.
\end{aligned} \tag{5}$$

In fact, in view of the above estimates we have

$$\int_{\mathbb{R}^{n-1}} \langle\xi'\rangle^{0} \|F'u(\xi',.)\|^2_{\mathcal{K}^{s,\gamma}(\mathbb{R}_+)} d\xi' \leq c \int_{\mathbb{R}^{n-1}} \langle\xi'\rangle^{s} \|F'u(\xi',.)\|^2_{\mathcal{K}^{s,\gamma}(\mathbb{R}_+)(\xi')} d\xi',$$

which proves the first embedding. The others follow in the same way. Using in addition the obvious continuous embedding $\mathcal{K}^{s,\gamma}(\mathbb{R}_+) \hookrightarrow \mathcal{K}^{s',\gamma}(\mathbb{R}_+)$ for $s \geq s'$ we have for $s' \geq 0$, $s' \geq \gamma$

$$\int_{\mathbb{R}^{n-1}} \langle\xi'\rangle^{s-s'} \|F'u(\xi',.)\|^2_{\mathcal{K}^{s',\gamma}(\mathbb{R}_+)} d\xi' \leq \int_{\mathbb{R}^{n-1}} \langle\xi'\rangle^{s} \|F'u(\xi',.)\|^2_{\mathcal{K}^{s',\gamma}(\mathbb{R}_+)(\xi')} d\xi'$$

$$\leq \int_{\mathbb{R}^{n-1}} \langle\xi'\rangle^{s} \|F'u(\xi',.)\|^2_{\mathcal{K}^{s,\gamma}(\mathbb{R}_+)(\xi')} d\xi'.$$

This implies the continuous embedding

$$\mathcal{K}^{s,\gamma}(\mathbb{R}^n_+) \hookrightarrow H^{s-s'}(\mathbb{R}^{n-1}, \mathcal{K}^{s',\gamma}(\mathbb{R}_+)) \quad \text{for } s \geq s' \geq 0,\ s' \geq \gamma. \tag{6}$$

In the same way we obtain the continuous embedding

$$\mathcal{K}^{s,\gamma}(\mathbb{R}^n_+) \hookrightarrow H^{s-\lambda}(\mathbb{R}^{n-1}, \mathcal{K}^{s,\lambda}(\mathbb{R}_+)) \quad \text{for } \gamma \geq \lambda \geq 0,\ \lambda \geq s, \tag{7}$$

and

$$H^{s-s'}(\mathbb{R}^{n-1}, \mathcal{K}^{s',\gamma}(\mathbb{R}_+)) \hookrightarrow \mathcal{K}^{s,\gamma}(\mathbb{R}^n_+) \quad \text{for } s \leq s' \leq 0,\ s' \leq \gamma,$$

$$H^{s-\lambda}(\mathbb{R}^{n-1}, \mathcal{K}^{s,\lambda}(\mathbb{R}_+)) \hookrightarrow \mathcal{K}^{s,\gamma}(\mathbb{R}^n_+) \quad \text{for } \gamma \leq \lambda \leq 0,\ \lambda \leq s. \tag{8}$$

In particular,

$$\mathcal{K}^{s,\gamma}(\mathbb{R}^n_+) \hookrightarrow H^{s}(\mathbb{R}^{n-1}, \mathcal{K}^{0,\gamma}(\mathbb{R}_+)) \cap H^{0}(\mathbb{R}^{n-1}, \mathcal{K}^{s,\gamma}(\mathbb{R}_+)) \text{ for } s \geq 0, \gamma \leq 0, \tag{9}$$

$$\mathcal{K}^{s,\gamma}(\mathbb{R}^n_+) \hookrightarrow H^{s}(\mathbb{R}^{n-1}, \mathcal{K}^{s,0}(\mathbb{R}_+)) \cap H^{s-\gamma}(\mathbb{R}^{n-1}, \mathcal{K}^{s,\gamma}(\mathbb{R}_+)) \text{ for } s \leq 0, \gamma \geq 0.$$

3. <u>Lemma.</u> Let $\varphi \in C_0^\infty(\mathbb{R}^n_+)$, $\psi \in C_0^\infty(\mathbb{R}^n_+)$. Then, for arbitrary $s, \gamma, \lambda \in \mathbb{R}$ the operators of multiplication define continuous mappings (denoted by the same letter)

$$\varphi : \mathcal{K}^{s,\gamma}(\mathbb{R}^n_+) \longrightarrow \mathcal{K}^{s,\gamma}(\mathbb{R}^n_+)$$

$$\psi : \mathcal{K}^{s,\gamma}(\mathbb{R}^n_+) \longrightarrow \mathcal{K}^{s,\lambda}(\mathbb{R}^n_+) .$$

<u>Proof.</u> In view of the series decomposition

$$\varphi(x',t) = \sum_j \varphi_j^1(x')\, \varphi_j^2\,(t) \tag{10}$$

which converges in $C^\infty(\overline{\mathbb{R}^n_+})$, $\varphi_j^1 \in C_o^\infty(\mathbb{R}^{n-1})$, $\varphi_j^2 \in C_o^\infty(\overline{\mathbb{R}}_+)$, we are reduced to the case of functions of x' and functions of t only, provided the norm of the multiplication operator in $\mathcal{K}^{s,\gamma}(\mathbb{R}^n_+)$ (in $\mathcal{L}(\mathcal{K}^{s,\gamma}(\mathbb{R}^n_+), \mathcal{K}^{s,\lambda}(\mathbb{R}^n_+))$ in the second case) is bounded by certain semi-norms of the function in $C^\infty(\overline{\mathbb{R}^n_+})$.

Let $\varphi \in C_o^\infty(\mathbb{R}^{n-1})$ and $u \in \mathcal{K}^{s,\gamma}(\mathbb{R}^n_+)$. Then

$$(\varkappa^{-1}(\xi')F')(\varphi u)(\xi',t) = \int \hat{\varphi}(\xi'-\eta')(\varkappa^{-1}(\xi')F')(\eta',t)d\eta'$$

and

$$\langle\xi'\rangle^s \|(\varkappa^{-1}(\xi')F')(\varphi u)\|_{\mathcal{K}^{s,\gamma}(\mathbb{R}_+)} \leqq$$

$$\int \frac{\langle\xi'\rangle^s}{\langle\eta'\rangle^s} |\hat{\varphi}(\xi'-\eta')|\, \langle\eta'\rangle^s \|\varkappa^{-1}_{\frac{\langle\xi'\rangle}{\langle\eta'\rangle}}(\varkappa^{-1}(\eta')F')u(\eta',t)\|_{\mathcal{K}^{s,\gamma}(\mathbb{R}_+)} d\eta' .$$

For each $N \in \mathbb{Z}_+$ we have a constant $c = c(N)$ s.th.

$$|\hat{\varphi}(\xi'-\eta')| \leqq c\langle\xi'-\eta'\rangle^{-N}$$

c is bounded by a finite number of semi-norms of φ.

With arbitrary ω we use the decomposition

$$\varkappa^{-1}(\eta')F'u = \omega(t)\varkappa^{-1}(\eta')F'u + (1-\omega(t))\,\varkappa^{-1}(\eta')F'u.$$

For each $\lambda \in \mathbb{R}_+$ we have

$$\|\varkappa_\lambda(\omega u)\|_{\mathcal{K}^{s,\gamma}} = \lambda^{-\gamma}\|\omega u\|_{\mathcal{K}^{s,\gamma}}$$

(cf. (2)) and by (1)

$$\|\varkappa_\lambda(1-\omega)u\|_{\mathcal{K}^{s,\gamma}} \leqq (1+\lambda^{-s})\|(1-\omega)u\|_{\mathcal{K}^{s,\gamma}}.$$

Using Peetre's inequality and choosing N sufficiently large we get

$$\int [\langle\xi'\rangle^s \|(\kappa^{-1}(\xi')F')(\varphi u)(\xi',.)\|_{\mathcal{K}^{s,\gamma}(\mathbb{R}_+)}]^2 d\xi' \leqq$$

$$c\int [\langle\eta'\rangle^s \|\kappa^{-1}(\eta')F'u(\eta',.)\|_{\mathcal{K}^{s,\gamma}(\mathbb{R}_+)}]^2 d\eta' ,$$

where c is bounded by a finite number of semi-norms of φ.

For $\varphi \in C_0^\infty(\mathbb{R}_+)$ we have

$$(\kappa^{-1}(\xi')F')(\varphi(t)u(x',t))(\xi',t) = \varphi(\langle\xi'\rangle^{-1}t)\,\kappa^{-1}(\xi')F'u(\xi',t)$$

and

$$\|F'(\varphi u)\|_{\mathcal{K}^{s,\gamma}(\mathbb{R}_+)(\xi')} \leqq c\|F'u\|_{\mathcal{K}^{s,\gamma}(\mathbb{R}_+)(\xi')} ,$$

where the constant c is independent of ξ' and c can be estimated by a finite number of semi-norms of φ. Therefore, the first assertion is proved.

In the case $\psi \in C_0^\infty(\mathbb{R}_+^n)$ we get in the expansion (10) that $\psi^2 \in C_0^\infty(\mathbb{R}_+)$. Then there is a constant c independent of ξ' such that

$$\|F'(\psi u)\|_{\mathcal{K}^{s,\lambda}(\mathbb{R}_+)(\xi')} \leqq c\,\|F'u\|_{\mathcal{K}^{s,\gamma}(\mathbb{R}_+)(\xi')},$$

and c is bounded by semi-norms of ψ.

Then the second assertion follows. $\square$

Let $g \in C^\infty(\overline{\mathbb{R}}_+)$, $g > 0$, $g(t) = t$ for $0 < t < 1$, $g \equiv 1$ for $t > 2$. Then we define for $\lambda \in \mathbb{R}$

$$Op_\partial(g^\lambda)u(x',t) = \int_{\mathbb{R}^{n-1}} e^{ix'\xi'} g^\lambda(\langle\xi'\rangle t)F'u(\xi',t)\,đ\xi'.$$

Since

$$F'(Op_\partial(g^\lambda)u)(\xi',t) = g^\lambda(\langle\xi'\rangle t)F'u(\xi',t)$$

and

$$\kappa^{-1}(\xi')\left[g^\lambda(\langle\xi'\rangle t)F'u(\xi',t)\right] = g^\lambda(t)(\kappa^{-1}(\xi')F')u(\xi',.)$$

we obtain

4. Lemma. $Op_{\mathfrak{d}}(g^{\lambda})$ defines an isomorphism

$$Op_{\mathfrak{d}}(g^{\lambda}) : \mathcal{K}^{s,\gamma}(\mathbb{R}^n_+) \longrightarrow \mathcal{K}^{s,\gamma+\lambda}(\mathbb{R}^n_+)$$

with inverse $Op_{\mathfrak{d}}(g^{-\lambda})$.

For $s > -\frac{1}{2}$ we get in particular

$$Op_{\mathfrak{d}}(g^{s-\gamma}) : \mathcal{K}^{s,\gamma}(\mathbb{R}^n_+) \underset{\cong}{\longrightarrow} H^s_0(\overline{\mathbb{R}^n_+}) \tag{11}$$

and for $s < \frac{1}{2}$

$$Op_{\mathfrak{d}}(g^{s-\gamma}) : \mathcal{K}^{s,\gamma}(\mathbb{R}^n_+) \underset{\cong}{\longrightarrow} H^s(\mathbb{R}^n_+), \tag{12}$$

where the identity $H^s_0(\overline{\mathbb{R}^n_+}) = \mathcal{K}^{s,s}(\mathbb{R}^n_+)$, $s > -\frac{1}{2}$, and $H^s(\mathbb{R}^n_+) = \mathcal{K}^{s,s}(\mathbb{R}^n_+)$, $s < \frac{1}{2}$, was used.

If $E(\mathbb{R}^n_+)$ is a space of distributions on $\mathbb{R}^n_+$, we denote by $E_{comp}(\mathbb{R}^n_+)$ the subspace of distributions with compact support.

From the well-known facts about compact and Hilbert-Schmidt embeddings of the Sobolev spaces we conclude

5. Proposition.

(i) For $s > s'$, $\gamma > \gamma'$ the embedding

$$\mathcal{K}^{s,\gamma}_{comp}(\mathbb{R}^n_+) \hookrightarrow \mathcal{K}^{s',\gamma'}_{comp}(\mathbb{R}^n_+) \tag{13}$$

is compact.

(ii) For $s > s' + \frac{n}{2}$, $\gamma > \gamma' + \frac{n}{2}$ the embedding (13) is Hilbert-Schmidt.

(iii) For $s > s' + n$, $\gamma > \gamma' + n$ the embedding (13) is trace class.

In particular, for

$$\mathcal{K}^{\infty,\infty}(\mathbb{R}^n_+) = \underset{\overleftarrow{s,\gamma}}{\lim} \mathcal{K}^{s,\gamma}(\mathbb{R}^n_+)$$

we get that $\mathcal{K}^{\infty,\infty}_{comp}(\mathbb{R}^n_+)$ is nuclear.

The definition of the $\mathcal{K}^{s,\gamma}$-spaces together with the estimates (1), (2) implies immediately

6. <u>Proposition.</u> Let $\varphi \in C_0^\infty(\overline{\mathbb{R}_+^n})$ and $\lambda \geq 0$, $k \in \mathbb{Z}_+$, $\alpha \in \mathbb{Z}_+^{n-1}$. Then the operators $\varphi t^{\pm\lambda}$, D_x^α, ϑ_t^k define continuous mappings

$$\varphi t^\lambda \ : \mathcal{K}^{s,\gamma}(\mathbb{R}_+^n) \longrightarrow \mathcal{K}^{s,\gamma+\lambda}(\mathbb{R}_+^n),$$

$$\varphi t^{-\lambda} \ : \mathcal{K}^{s,\gamma}(\mathbb{R}_+^n) \longrightarrow \mathcal{K}^{s-\lambda,\gamma-\lambda}(\mathbb{R}_+^n),$$

$$D_x^\alpha,\ \vartheta_t^k \ : \mathcal{K}^{s,\gamma}(\mathbb{R}_+^n) \longrightarrow \mathcal{K}^{s-|\alpha|-k,\gamma}(\mathbb{R}_+^n)$$

for each $s \in \mathbb{R}$, $\gamma \in \mathbb{R}$.

3.1.2. <u>Subspaces with asymptotics</u>

Now we introduce the higher-dimensional analogue of the $\mathcal{K}_{\mathcal{P}}^s(\mathbb{R}_+)$ spaces from 2.3.3.

1. <u>Definition.</u> Let $s,\gamma \in \mathbb{R}$ and $\mathcal{P} \in \underline{\mathcal{P}}(\gamma)$. We denote by $\mathcal{K}^s(\mathbb{R}_+^n)$ the subspace of all $u \in \mathcal{K}^{s,\gamma}(\mathbb{R}_+^n)$ such that for each Hilbert norm $\|\cdot\|_{j,\mathcal{K}_{\mathcal{P}}^s(\mathbb{R}_+)}$ on $\mathcal{K}_{\mathcal{P}}^s(\mathbb{R}_+)$ the norm

$$\|u\|_{j,\mathcal{K}_{\mathcal{P}}^s(\mathbb{R}_+^n)} = \left\{\int_{\mathbb{R}^{n-1}} \langle\xi'\rangle^{2s} \|\varkappa^{-1}(\xi')F'u(\xi',\cdot)\|^2_{j,\mathcal{K}_{\mathcal{P}}^s(\mathbb{R}_+)} \, d\xi'\right\}^{\frac{1}{2}}$$

exists and is finite.

Obviously, $\mathcal{K}_{\mathcal{P}}^s(\mathbb{R}_+^n)$ is the projective limit of the Hilbert subspaces of $\mathcal{K}^{s,\gamma}(\mathbb{R}_+^n)$ defined by the scalar product

$$\int_{\mathbb{R}^{n-1}} \langle\xi'\rangle^{2s} (\varkappa^{-1}(\xi')F'u(\xi',\cdot),\varkappa^{-1}(\xi')F'(\xi',\cdot))_{j,\mathcal{K}_{\mathcal{P}}^s(\mathbb{R}_+)} \, d\xi'.$$

where $(\cdot,\cdot)$ in the integral denotes the j-th scalar product on $\mathcal{K}_{\mathcal{P}}^s(\mathbb{R}_+)$.

2. <u>Example.</u> For $p \in \mathbb{C}$, $k \in \mathbb{Z}_+$ consider

$$u(x',t) = \int_{\mathbb{R}^{n-1}} e^{ix'\xi'} \langle\xi'\rangle^{\frac{1}{2}} \omega(\langle\xi'\rangle t)\ (\langle\xi'\rangle t)^{-p} \log^k(\langle\xi'\rangle t) \hat{v}(\xi')\, d\xi' \quad (1)$$

where $v \in H^s(\mathbb{R}^{n-1})$. Then

$$(\kappa^{-1}(\xi')F'u)(\xi',t) = \omega(t)t^{-p}\log^k t\ \hat{v}(\xi'),$$

and therefore $u \in \mathcal{K}^s_{\varphi}(\mathbb{R}^n_+)$, $\varphi = \{(p,k)\}$. Such functions will play the role of singular functions in higher dimensions.

Below we shall present a general calculus which contains operators like $H^s(\mathbb{R}^{n-1}) \ni v \longrightarrow u \in \mathcal{K}^s_{\varphi}(\mathbb{R}^n_+)$. They contain as a special case the classical potentials in $\mathbb{R}^n_+$ of a multiple layer on $\partial\mathbb{R}^n_+$.

For $k=0$ we have

$$\int_{\mathbb{R}^{n-1}} e^{ix'\xi'}\langle\xi'\rangle^{\frac{1}{2}}\omega(\langle\xi'\rangle t)(\langle\xi'\rangle t)^{-p}\ \hat{v}(\xi')\,d\xi' =$$

$$= t^{-p}\int_{\mathbb{R}^{n-1}} e^{ix'\xi'}\omega(\langle\xi'\rangle t)\langle\xi'\rangle^{\frac{1}{2}-p}\ \hat{v}(\xi')\,d\xi' .$$

Therefore $u(x',t)$ behaves like t^{-p} for $t \longrightarrow +0$.

It is easy to see that each $u \in \mathcal{K}^s_{\varphi}(\mathbb{R}^n_+)$ may be represented as a sum of a function in an arbitrary weighted space $\mathcal{K}^{s,\delta}(\mathbb{R}^n_+)$ and a finite number of singular functions as in Example 2.

3. Proposition. Let $\pi^{\delta}_{\varphi} : \mathcal{K}^s(\mathbb{R}_+) \longrightarrow \mathcal{K}^s_{1(\delta)\varphi}(\mathbb{R}_+)$ be the unique projection with $\ker \pi^{\delta}_{\varphi} = \mathcal{E}(\bar{r}(\delta)\varphi)$ (cf. 2.3.3.). Then

$$u \longrightarrow \int_{\mathbb{R}^{n-1}} e^{ix'\xi'}\,\kappa(\xi')\,\pi^{\delta}_{\varphi}\,\kappa^{-1}(\xi')F'u(\xi',.)\,d\xi' \qquad (2)$$

defines a projection operator

$$\mathcal{K}^s_{\varphi}(\mathbb{R}^n_+) \longrightarrow \mathcal{K}^{s,\delta}(\mathbb{R}^n_+)$$

which is denoted by $\mathrm{Op}_{\partial}(\pi^{\delta}_{\varphi})$. Its kernel consists of singular functions of the form

$$\int_{\mathbb{R}^{n-1}} e^{ix'\xi'}\,\kappa(\xi')(w(t)\hat{v}(\xi'))\,d\xi',\quad w \in \mathcal{E}(\bar{r}(\delta)\varphi),\ v \in H^s(\mathbb{R}^{n-1}).$$

Note that the operation $\mathrm{Op}_{\partial}(\pi^{\delta}_{\varphi})$ on $\mathcal{K}^s(\mathbb{R}^n_+)$ cannot be induced from $\sigma_{\wp}^{\delta}$ on $\mathcal{K}_{\wp}^{s}(\mathbb{R}_+)$ for fixed x', since $\sigma_{\wp}^{\delta}$ and $\varkappa^{-1}(\xi')$ do not commute.

Proof. Set $v(x',t) := \mathrm{Op}_{\vartheta}(\sigma_{\wp}^{\delta})u(x',t)$. Then

$$(\varkappa^{-1}(\xi')F'v)(\xi',t) = \sigma_{\wp}^{\delta}((\varkappa^{-1}(\xi')F'u(\xi',t))$$

belongs to $\mathcal{K}^{s,\delta}(\mathbb{R}_+)$ for each ξ' and

$$\|v\|^2_{\mathcal{K}^{s,\delta}(\mathbb{R}^n_+)} = \int_{\mathbb{R}^{n-1}} \langle\xi'\rangle^{2s} \|\sigma_{\wp}^{\delta}((\varkappa^{-1}(\xi')F'u)(\xi',\cdot))\|^2_{\mathcal{K}^{s,\delta}(\mathbb{R}_+)} d\xi' \leq$$

$$c\int_{\mathbb{R}^{n-1}} \langle\xi'\rangle^{2s} \|\varkappa^{-1}(\xi')F'u(\xi',\cdot)\|^2_{j,\mathcal{K}^{s}_{\wp}(\mathbb{R}_+)} d\xi' = c\|u\|^2_{j,\mathcal{K}^{s}_{\wp}(\mathbb{R}^n_+)}$$

for a suitable norm $\|.\|_{j,\mathcal{K}^{s}_{\wp}(\mathbb{R}_+)}$ on $\mathcal{K}^{s}_{\wp}(\mathbb{R}_+)$. The last assertion follows immediately if we split $(\varkappa^{-1}(\xi')F'u)(\xi',t)$ into $\sigma_{\wp}^{\delta}(\varkappa^{-1}(\xi')F'u)$ and the remainder, which gives exactly the mentioned singular functions. $\square$

Let us use the notation

$$\underline{P} = \{P = (\gamma,\wp),\ \gamma\in\mathbb{R},\ \wp\in\underline{\underline{\wp}} : \bar{r}(\gamma)\wp = \wp\}.$$

Each $P\in\underline{P}$ consists of a finite set of points $p_j\in\mathbb{C}$ together with multiplicities $m_j\in\mathbb{Z}_+$, $j = 1,\dots,N$, such that

$$\mathrm{Re}\ p_j \geq \frac{1}{2} - \gamma.$$

For given points p_j there is, of course, a minimal γ compatible with the given points. It will be convenient, however, to fix the weight γ.

Set

$$\dim P = \sum_{j=1}^{N} (m_j+1).$$

Define $\mathcal{K}^{s}_{P}(\mathbb{R}^n_+)$, $s\in\mathbb{R}$, $P = (\gamma,\wp)\in\underline{P}$ by

$$\mathcal{K}^{s}_{P}(\mathbb{R}^n_+) = \mathcal{K}^{s,\gamma}(\mathbb{R}^n_+) + \mathcal{K}^{s}_{\wp}(\mathbb{R}^n_+).$$

Set $P\leq P'$ if $\gamma'\leq\gamma$, $\bar{r}(\gamma')\wp\leq\wp'$, turning $\underline{P}$ into a semi-ordered set. Then $P\leq P'$, $s\geq s'$ implies

$$\mathcal{K}^{s}_{P}(\mathbb{R}^n_+) \hookrightarrow \mathcal{K}^{s'}_{P'}(\mathbb{R}^n_+).$$

Let $P = (\gamma, \varphi) \in \underline{P}$ and define for each $\varepsilon > 0$

$$P_\varepsilon = (\gamma-\varepsilon,\ r(\gamma-\varepsilon)\varphi\ ).$$

Then we define

$$\mathcal{K}^{s}_{P^+}(\mathbb{R}^n_+) = \varprojlim_{\varepsilon} \mathcal{K}^{s}_{P_\varepsilon}(\mathbb{R}^n_+).$$

With the projective limit topology this is a Fréchet space. Because of the inclusions

$$\mathcal{K}^{s}_{P_\varepsilon}(\mathbb{R}^n_+) \hookrightarrow \mathcal{K}^{s}_{P_\delta}(\mathbb{R}^n_+) \quad \text{for} \quad \varepsilon \le \delta$$

we may restrict ourselves to a countable sequence of ε-values.

The proof of Proposition 3 shows that

$$u \longrightarrow (\mathrm{Op}_{\gamma}(\sigma^{\delta}_{\varphi})(u),\ (v_j)^{\dim P}_{j=1})$$

defines an isomorphism

$$\mathcal{K}^{s}_{P}(\mathbb{R}^n_+) \to \mathcal{K}^{s,\delta}(\mathbb{R}^n_+) \bigoplus_{j=1}^{\dim P} H^{s}(\mathbb{R}^{n-1}). \tag{3}$$

4. Example. For $s > -\frac{1}{2}$, $s \neq \frac{1}{2} \bmod \mathbb{Z}$ we have

$$H^{s}(\mathbb{R}^n_+) = \mathcal{K}^{s}_{P_0(s)}(\mathbb{R}^n_+),$$

where $P_0(s) = (s, r(s)\varphi_0)$. In this case the above decomposition corresponds to the well-known fact that $u \in H^s(\mathbb{R}^n_+)$ can be represented as a sum of $u_0 \in H^s_0(\mathbb{R}^n_+)$ and $[s - \frac{1}{2}]$ potentials of densities on $\partial\mathbb{R}^n_+$. The densities are (up to a constant coefficient ψDO) the traces of the normal derivatives.

Each $\varphi \in \underline{\underline{\varphi}}$ gives rise to a decreasing sequence

$$P(j) = (j, r(j)\varphi),\ j \in \mathbb{Z}_+.$$

Then

$$\mathcal{K}^s_{\mathcal{P}}(\mathbb{R}^n_+) = \varprojlim_j \mathcal{K}^s_{P(j)}(\mathbb{R}^n_+) \qquad (4)$$

In general, the spaces with asymptotics are not stable under multiplication by $C_0^\infty(\overline{\mathbb{R}_+^n})$ functions. Even for $n=1$ the product will have the singularity type $\sum_{j=0}^{\infty} T^{-j}\mathcal{P}$, in general. However, if we assume the singularity type to satisfy

$$\mathcal{P} \gtrsim T^{-j}\mathcal{P} \text{ for all } j \in \mathbb{Z}_+ \qquad (5)$$

and for $P = (\gamma, \mathcal{P})$

$$P \gtrsim (\gamma, \bar{r}(\gamma)(T^{-j}\mathcal{P})) \text{ for all } j \in \mathbb{Z}_+ . \qquad (6)$$

5. <u>Lemma.</u> (i) (6) implies that $\mathcal{K}^s_P(\mathbb{R}^n_+)$ is a $C_0^\infty(\overline{\mathbb{R}^n_+})$ module.
(ii) (5) implies that $\mathcal{K}^s(\mathbb{R}^n_+)$ is a $C_0^\infty(\overline{\mathbb{R}^n_+})$ module.

<u>Proof.</u> Obviously (i) $\Rightarrow$ (ii) in view of (4). Since the weighted spaces $\mathcal{K}^{s,\gamma}(\mathbb{R}^n_+)$ are $C_0^\infty(\overline{\mathbb{R}^n_+})$ modules, it remains to show that the products with singular functions are again in the space.

Given $\varphi \in C_0^\infty(\overline{\mathbb{R}^n_+})$. We get from the Taylor expansion for each M

$$\varphi(x',t) = \sum_{k=0}^{M-1} t^k \omega(t)\varphi_k(x') + t^M \varphi_M(x',t),$$

$\varphi_k \in C_0^\infty(\mathbb{R}^{n-1})$, $\varphi_M \in C_0^\infty(\overline{\mathbb{R}^n_+})$. Choose M so large that

$$\omega(t)t^M \mathcal{K}^s_P(\mathbb{R}^n_+) \subset \mathcal{K}^{s,\gamma}(\mathbb{R}^n_+).$$

First we consider products of the form

$$t^k \omega(t) \int_{\mathbb{R}^{n-1}} e^{ix'\xi'} \varkappa(\xi')(w(t))\hat{v}(\xi')\, d\xi' ,$$

where we may assume that $\omega(t)\varkappa(\xi')(w(t)) = \varkappa(\xi')(w(t))$ for all $\xi' \in \mathbb{R}^{n-1}$. Then the above product is equal to

$$\int_{\mathbb{R}^{n-1}} e^{ix'\xi'} \varkappa(\xi')(t^k w(t))\langle\xi'\rangle^{-k}\hat{v}(\xi')d\xi',$$

which has the desired form, since $(F')^{-1}\langle\xi'\rangle^{-k}\hat{v}(\xi')\in H^{s}(\mathbb{R}^{n-1})$. It remains to consider products with $\varphi(x')\in C_0^{\infty}(\mathbb{R}^{n-1})$.

Consider

$$\varphi(x')\int e^{ix'\xi'}\langle\xi'\rangle^{\frac{1}{2}}(t\langle\xi'\rangle)^{-p}\omega(t\langle\xi'\rangle)\hat{v}(\xi')đ\xi' =$$

$$= t^{-p}\iint e^{i(x'-y')\xi'}\omega(t\langle\xi'\rangle)\varphi(x')w(y')dy'đ\xi' ,$$

where

$$w(y') = \int e^{iy'\xi'}\langle\xi'\rangle^{\frac{1}{2}-p}\hat{v}(\xi')đ\xi'$$

belongs to $H^{s-\frac{1}{2}+\mathrm{Re}\,p}(\mathbb{R}^{n-1})$. We get from Taylor's formula

$$\varphi(x') = \varphi(y') + \sum_{|\alpha|=1}(x'-y')^{\alpha}\varphi_{\alpha}(x',y'),$$

where $\varphi_{\alpha}\in C_0^{\infty}(\mathbb{R}^{n-1}\times\mathbb{R}^{n-1})$.

The main contribution yields $\varphi(y')$

$$\iint e^{i(x'-y')\xi'}\langle\xi'\rangle^{\frac{1}{2}}(t\langle\xi'\rangle)^{-p}\omega(t\langle\xi'\rangle)\langle\xi'\rangle^{-\frac{1}{2}+p}(\varphi w)(y')\,dy'đ\xi' ,$$

which has the desired form since

$$F^{-1}\langle\xi'\rangle^{-\frac{1}{2}+p}F(\varphi w)\in H^{s}(\mathbb{R}^{n-1}).$$

For $|\alpha| = 1$ we use partial integrations and obtain

$$t^{-p}\iint e^{i(x'-y')\xi'}(-D_{\xi'})^{\alpha}\omega(t\langle\xi'\rangle)\varphi_{\alpha}(x',y')w(y')dy'\,đ\xi'$$

$$= \iint e^{i(x'-y')\xi'}\langle\xi'\rangle^{\frac{1}{2}}(t\langle\xi'\rangle)^{-p}(t\langle\xi'\rangle)\ \omega'(t\langle\xi'\rangle)\langle\xi'\rangle^{\frac{1}{2}+p}\varphi_{\alpha}(x',y')w(y')dy'đ\xi' .$$

Since $\varphi_{\alpha}\in C_0^{\infty}(\mathbb{R}^{n-1}\times\mathbb{R}^{n-1})$, we have an absolutely converging series expansion

$$\varphi_{\alpha}(x',y') = \sum_{l=0}^{\infty} f_l(x')\,g_l(y').$$

For each l the resulting function belongs to $\mathcal{K}_0^{s+1}(\mathbb{R}_+^n)$ and the series converges absolutely.

In the same way we conclude for singular functions containing logarithmic terms. This case follows also by differentiation with respect to p .

In the same way one can prove that the multiplication defines a mapping

$$\mathcal{K}^{\infty}_{\mathfrak{p},comp} \times \mathcal{K}^{s}_{\mathfrak{q}} \longrightarrow \mathcal{K}^{s}_{\mathfrak{r}}$$

for suitable $\mathfrak{r}$ depending on $\mathfrak{p}$ and $\mathfrak{q}$. Moreover, it depends continuously on the first argument if it is considered as

$$\mathcal{K}^{\infty}_{\mathfrak{p},comp} \longrightarrow \mathcal{L}(\mathcal{K}^{s}_{\mathfrak{q}}, \mathcal{K}^{s}_{\mathfrak{r}}),$$

$\mathcal{L}(\mathcal{K}^{s}_{\mathfrak{q}}, \mathcal{K}^{s}_{\mathfrak{r}})$ with the topology of uniform convergence on all bounded sets.

For a compact $K \subset \overline{\mathbb{R}^n_+}$ denote by $\mathcal{K}^s_P(K)$ the subspace of $\mathcal{K}^s_P(\mathbb{R}^n_+)$ of all elements supported by K. It is closed with the induced topology.

6. Proposition. (i) For $s > s' + \frac{n}{2}$, $\gamma > \gamma' + \frac{n}{2}$, $P \leq Q$, $K \subset \overline{\mathbb{R}^n_+}$ compact, the embedding

$$\mathcal{K}^s_P(K) \hookrightarrow \mathcal{K}^{s'}_Q(K)$$

is Hilbert-Schmidt.

(ii) For $s > s'+n$, $\gamma > \gamma'+n$, $P \leq Q$ the above embedding is trace class.

This is an immediate consequence of 3.1.1. Proposition 5 and the isomorphism (3). In particular,

$$\mathcal{K}^{\infty}_{\mathfrak{p},comp}(\mathbb{R}^n_+) = \varprojlim_{s} \mathcal{K}^{s}_{\mathfrak{p},comp}(\mathbb{R}^n_+)$$

is nuclear.

Let $V \subset \mathbb{R}^{n-1}$ open and set $U = V \times \mathbb{R}_+ \subset \mathbb{R}^n_+$. Then we define $\mathcal{K}^s_{P,comp}(U)$ as the subspace of $\mathcal{K}^s_P(\mathbb{R}^n_+)$ of all $u(x',t)$ with compact support in U. With the topology induced by $\mathcal{K}^s_P(\mathbb{R}^n_+)$ the space $\mathcal{K}^s_{P,comp}(U)$ is complete and therefore a closed subspace of $\mathcal{K}^s_P(\mathbb{R}^n_+)$.

Define $\mathcal{K}^s_{P,loc}(U)$ by the condition that $u \in \mathcal{K}^s_{P,loc}(U)$ if for each $\varphi \in C^{\infty}_0(V)$ the distribution φu belongs to $\mathcal{K}^s_P(\mathbb{R}^n_+)$. $\mathcal{K}^s_{P,loc}(U)$ is considered in the projective limit topology being the weakest topology such that all mappings $\mathcal{K}^s_{P,loc}(U) \ni u \longrightarrow \varphi u \in \mathcal{K}^s_P(\mathbb{R}^n_+)$ are continuous. Obviously, it is sufficient to consider a sequence $\{\varphi_j\}$ of $C^\infty_0(V)$ functions with the property that $K_j = \{x' \in V : \varphi_j \geq 1\}$ are such that $\bigcup_j K_j = V$. Therefore, $\mathcal{K}^s_{P,loc}(U)$ is a countable projective limit of Hilbert spaces.

Besides the discrete asymptotics we have to deal also with continuous asymptotics. We get the higher-dimensional analogue of the $\mathcal{K}^s_\Delta(\mathbb{R}_+)$ spaces.

7. <u>Definition</u>. Let $s, \gamma \in \mathbb{R}$, $\Delta \in \underline{\Delta}(\gamma)$. Denote by $\mathcal{K}^s_\Delta(\mathbb{R}^n_+)$ the subspace of $\mathcal{K}^{s,\gamma}(\mathbb{R}^n_+)$ of all u such that for each Hilbert norm $\|.\|_{j,\mathcal{K}^s_\Delta(\mathbb{R}_+)}$ on $\mathcal{K}^s_\Delta(\mathbb{R}_+)$ the norm

$$\|u\|_{j,\mathcal{K}^s_\Delta(\mathbb{R}^n_+)} = \left\{ \int_{\mathbb{R}^{n-1}} \langle \xi' \rangle^{2s} \|\varkappa^{-1}(\xi')F'u(\xi',.)\|^2_{j,\mathcal{K}^s_\Delta(\mathbb{R}_+)} \, d\xi' \right\}^{\frac{1}{2}}$$

exists and is finite.

Obviously, $\mathcal{K}^s_\Delta(\mathbb{R}^n_+)$ is the projective limit of the corresponding Hilbert spaces.

In the same way we define $K^s_\Lambda(\mathbb{R}^n_+)$ for $\Lambda \in \underline{\Lambda}(\gamma)$.

The main difference to the spaces with discrete asymptotics is that, in general, there are not unique typical singular functions, cf. the corresponding observation on $\mathbb{R}_+$, which is the reason for the behaviour in $\mathbb{R}^n_+$. The only exception is the case if $\Lambda \cap \Gamma_{\frac{1}{2}-\delta} = \emptyset$. Then we have

$$\mathcal{K}^{s,\delta}(\mathbb{R}_+) + \mathcal{K}^s_\Delta(\mathbb{R}_+) = \mathcal{K}^{s,\delta}(\mathbb{R}_+) \oplus \mathcal{E}(\bar{r}(\delta)\Delta).$$

Let us denote by $\mathcal{E}^s(\bar{r}(\delta)\Delta)(\mathbb{R}^n_+)$ the space of all

$$u(x',t) = \int_{\mathbb{R}^{n-1}} e^{ix'\xi'} \varkappa(\xi')(w(t))\hat{v}(\xi')\,d\xi' \qquad (7)$$

where $w \in \mathcal{E}(\bar{r}(\delta)\Delta)$ i.e. $w(t) = \langle \zeta_p, \omega(t)t^{-p} \rangle$, ζ a functional carried by $\bar{r}(\delta)\Delta$ and $v \in H^s(\mathbb{R}^{n-1})$, we get

$$\mathcal{K}^{s,\delta}(\mathbb{R}^n_+) + \mathcal{K}^s_\Delta(\mathbb{R}^n_+) = \mathcal{K}^{s,\delta}(\mathbb{R}^n_+) \oplus \mathcal{E}^s(\bar{r}(\delta)\Delta)(\mathbb{R}^n_+).$$

$\mathcal{E}^s(\bar{r}(\delta)\Delta)(\mathbb{R}^n_+)$ is a Fréchet space with the system of Hilbert norms

$$p_j(u)^2 = q_j(\zeta)^2 + \|v\|^2_{H^s(\mathbb{R}^{n-1})},\ j \in \mathbb{Z}_+,$$

where ζ is the analytic functional carried by $\bar{r}(\delta)\Delta$ which generates w in (7), and q_j denotes a Hilbert norm on this space of functionals

Also the other results on spaces with continuous asymptotics on $\mathbb{R}_+$ have straightforward generalizations to $\mathbb{R}^n_+$. We mention, for example, the following

7. Proposition. Let $\Delta = \Lambda_1 + \Lambda_2$, $\Delta \in \Delta(\gamma)$, $\Lambda_j \in \underline{\Lambda}(\gamma)$. Then each $u \in \mathcal{K}^s_\Delta(\mathbb{R}^n_+)$ admits a (non-unique) representation

$$u = u_1 + u_2, \text{ where } u_1 \in \mathcal{K}^s_{\Lambda_j}(\mathbb{R}^n_+).$$

3.2. Boundary value problems without the transmission property

3.2.1. Green operators without boundary symbols

Classes of Green operators play the role of negligible operators in our theory. It is natural to take into account the filtration by the orders (reflecting the shift of the smoothness index and the conormal order) and also a filtration by the number of homogeneous components of the boundary symbol. The classes without symbols are defined by their mapping properties.

Let X be a C^∞ manifold with boundary ∂X. First we describe Green operators which map functions on X into functions on X. We start from smoothing Green operators

1. Definition. Let $k \in \mathbb{Z}_+$, $P = (\gamma, \mathcal{P}) \in \underline{P}$, $Q = (\lambda, q) \in \underline{P}$. Denote by (1) $\mathcal{W}^{-\infty}_G(Q,P)^{(-k)}(X,X)$ the subspace of $\bigcap_{s,s' \in \mathbb{R}} \mathcal{L}(\mathcal{K}^s_{P,\mathrm{comp}}(X), \mathcal{K}^{s'}_{Q,\mathrm{loc}}(X))$ of all operators $\mathcal{A}$ such that there exists $R = (\lambda+k, \mathcal{r}) \in \underline{P}$, $R = R(\mathcal{A})$ and $\mathcal{A}$ is continuous

$$\mathcal{A}: \mathcal{K}^s_{P,\mathrm{comp}}(X) \longrightarrow \mathcal{K}^{s'}_{R^+,\mathrm{loc}}(X) \text{ for all } s,s' \in \mathbb{R},$$

where $\mathcal{K}^{s'}_{R^+,\mathrm{loc}} = \varprojlim \mathcal{K}^{s'}_{R_\varepsilon,\mathrm{loc}}$, $R_\varepsilon = (\lambda+k-\varepsilon, r(\lambda+k-\varepsilon)\mathcal{r})$.

Elements of this space are called smoothing(1)-Green operators of conormal order $-k$ (with respect to the given weights) of discrete singularity type.

For compact X we can, of course, drop the subscripts comp and loc.

Set

$$(1)\text{-}\mathcal{O}_G^{-\infty}(Q,P)(X,X) = \bigcap_k (1)\text{-}\mathcal{O}_G^{-\infty}(Q,P)^{(-k)}(X,X).$$

Since $\mathcal{K}^{\infty}_{\mathfrak{r},\mathrm{loc}}(X) = \varprojlim_s \mathcal{K}^{s}_{\mathfrak{r},\mathrm{loc}}(X)$ is nuclear, each $\mathcal{A} \in (1)\text{-}\mathcal{O}_G^{-\infty}(Q,P)(X,X)$ has a kernel in $\mathcal{K}^{\infty}_{\mathfrak{r},\mathrm{loc}}(X) \otimes_\pi (\mathcal{K}^{-\infty}_{P,\mathrm{comp}}(X))'$, where

$$\mathcal{K}^{-\infty}_{P,\mathrm{comp}}(X) = \varinjlim_s \mathcal{K}^{s}_{P,\mathrm{comp}}(X).$$

The converse is also true.

Temporarily we will assume that the coordinate neighbourhoods intersecting the boundary are contractible and the local model is $\mathbb{R}^n_+$. Later it will be proved that this is not really necessary. Coordinate neighbourhoods in the interior are arbitrary and the local model are open sets in $\mathbb{R}^n$. Then, for a partition of unity $\{\varphi_j\}_{j\in J}$ consider $\mathcal{A}_{ij} := \varphi_i \mathcal{A} \varphi_j$ in local coordinates :

$$\mathcal{A}_{ij} : \mathcal{K}^{s}_{P,\mathrm{loc}}(\mathbb{R}^n_+) \longrightarrow \mathcal{K}^{s'}_{Q,\mathrm{comp}}(\mathbb{R}^n_+) \tag{1}$$

or $\mathcal{K}^{s}_{P,\mathrm{loc}}(\mathbb{R}^n_+) \longrightarrow H^{s'}_{\mathrm{comp}}(\Omega)$, $\Omega \subset \mathbb{R}^n$ open,

or $H^{s}_{\mathrm{loc}}(\Omega) \longrightarrow \mathcal{K}^{s'}_{Q,\mathrm{comp}}(\mathbb{R}^n_+)$, or $H^{s}_{\mathrm{loc}}(\Omega_1) \longrightarrow H^{s'}_{\mathrm{comp}}(\Omega_2)$,

$\Omega_1, \Omega_2 \subset \mathbb{R}^n$ open. Consider first the case if both i and j correspond to a boundary neighbourhood, i.e. $\mathcal{A}_{ij}$ acts as (1). In view of the isomorphisms

$$\mathcal{K}^{s}_{P}(\mathbb{R}^n_+) \cong \mathcal{K}^{s,\gamma}(\mathbb{R}^n_+) \oplus (H^{s}(\mathbb{R}^{n-1}))^{\dim P},$$

$$\mathcal{K}^{s'}_{Q}(\mathbb{R}^n_+) \cong \mathcal{K}^{s',\lambda}(\mathbb{R}^n_+) \oplus (H^{s'}(\mathbb{R}^{n-1}))^{\dim Q},$$

given by 3.1.2 (3) we may represent $\mathcal{A}_{ij}$ as 2×2 matrix of operators

$$\begin{pmatrix} \mathcal{A}_{ij}^{11} & \mathcal{A}_{ij}^{12} \\ \mathcal{A}_{ij}^{21} & \mathcal{A}_{ij}^{22} \end{pmatrix} : \begin{matrix} \mathcal{K}^{s,\gamma}(\mathbb{R}_+^n) \\ \oplus \\ (H^s(\mathbb{R}^{n-1}))^{\dim P} \end{matrix} \longrightarrow \begin{matrix} \mathcal{K}^{s',\lambda}(\mathbb{R}_+^n) \\ \oplus \\ (H^{s'}(\mathbb{R}^{n-1}))^{\dim Q} \end{matrix} \qquad (2)$$

For the formal adjoints $(\mathcal{A}_{ij}^{11})^*$ and $(\mathcal{A}_{ij}^{21})^*$ we obtain

$$(\mathcal{A}_{ij}^{11})^* : \mathcal{K}^{-s',-\lambda}(\mathbb{R}_+^n) \longrightarrow \mathcal{K}^{-s,-\gamma}(\mathbb{R}_+^n)$$

$$(\mathcal{A}_{ij}^{21})^* : (H^{-s'}(\mathbb{R}^{n-1}))^{\dim Q} \longrightarrow \mathcal{K}^{-s,-\gamma}(\mathbb{R}_+^n).$$

In a similar way, in the two cases if one of the neighbourhoods intersects the boundary we get representations

$$(\mathcal{A}_{ij}^{11}, \mathcal{A}_{ij}^{12}) : \begin{matrix} \mathcal{K}^{s,\gamma}(\mathbb{R}_+^n) \\ \oplus \\ (H^s(\mathbb{R}^{n-1}))^{\dim P} \end{matrix} \longrightarrow H^{s'}(\mathbb{R}^{n-1}) \qquad (3)$$

and

$$\begin{pmatrix} \mathcal{A}_{ij}^{11} \\ \mathcal{A}_{ij}^{21} \end{pmatrix} : H^s(\mathbb{R}^{n-1}) \longrightarrow \begin{matrix} \mathcal{K}^{s',\lambda}(\mathbb{R}_+^n) \\ \oplus \\ (H^{s'}(\mathbb{R}^{n-1}))^{\dim Q} \end{matrix} \qquad (4)$$

and the formal adjoint in the first case

$$((\mathcal{A}_{ij}^{11})^* : H^{-s'}(\mathbb{R}^{n-1}) \longrightarrow \mathcal{K}^{-s,-\gamma}(\mathbb{R}_+^n)$$

2. <u>Definition.</u> Denote by $\mathcal{W}_G^{-\infty}(Q,P)^{(-k)}(X,X)$ the subspace of (1)-$\mathcal{W}_G^{-\infty}(Q,P)^{(-k)}(X,X)$ of all $\mathcal{A}$ such that for a partition of unity the above formal adjoints define continuous operators: in the first case

$$(\mathcal{A}_{ij}^{11})^* : \mathcal{K}^{s,-\lambda}(\mathbb{R}_+^n) \longrightarrow \mathcal{K}^{s'}_{R_1^+}(\mathbb{R}_+^n)$$

$$(\mathcal{A}_{ij}^{21})^* : (H^s(\mathbb{R}^{n-1}))^{\dim Q} \longrightarrow \mathcal{K}^{s'}_{R_1^+}(\mathbb{R}_+^n)$$

for some $R_1 = R_1(\mathcal{A}) = (-\gamma+k, \vartheta_1) \in \underline{P}$, s,s' arbitrary, and in the second one

$$(\mathcal{A}_{ij}^{11})^* : H^s(\mathbb{R}^{n-1}) \longrightarrow \mathcal{K}^{s'}_{R_1^+}(\mathbb{R}_+^n),$$

R_1 as above.

In a similar way we define smoothing Green operators acting from ∂X to X and conversely.

3. Definition. Let $k \in \mathbb{Z}_+$, $P = (\gamma, \wp) \in \underline{P}$, $Q = (\lambda, \eta) \in \underline{P}$. Denote by $\mathcal{W}_G^{-\infty}(Q)^{(-k)}(X,\partial X)$ the subspace of $\bigcap_{s\in\mathbb{R}} \mathcal{L}(H^s_{comp}(\partial X), \mathcal{K}_{Q,loc}(X))$ of all operators $\mathcal{A}$ such that there exists $R = (\lambda+k, \kappa) \in \underline{P}$, $R = R(\mathcal{A})$ and $\mathcal{A}$ is continuous

$$\mathcal{A} : H^s_{comp}(\partial X) \longrightarrow \mathcal{K}^{s'}_{R^+,loc}(X).$$

Denote by (1)-$\mathcal{W}_G^{-\infty}(P)(\partial X, X)$ the space $\bigcap_{s\in\mathbb{R}} \mathcal{L}(\mathcal{K}^s_{P,comp}(X), C^\infty(\partial X))$. Let $\mathcal{W}_G^{-\infty}(P)^{(-k)}(\partial X, X)$ be the subspace of (1)-$\mathcal{W}_G^{-\infty}(P)(\partial X,X)$ of all $\mathcal{A}$ such that for a partition of unity the operators

$$\mathcal{A}^1_{ij} : \mathcal{K}^{s,\gamma}(\mathbb{R}^n_+) \longrightarrow C^\infty(\Omega),$$

$\Omega \subset \mathbb{R}^{n-1}$ open, have the property

$$(\mathcal{A}^1_{ij})^* : \mathcal{E}'(\Omega) \longrightarrow \mathcal{K}^s_{R_1^+}(\mathbb{R}^n_+)$$

for some $R_1 = R_1(\mathcal{A}) = (-\gamma+k, \kappa_1) \in \underline{P}$, s arbitrary.

Set

$$\mathcal{W}_G^{-\infty}(Q)(X,\partial X) = \bigcap_k \mathcal{W}_G^{-\infty}(Q)^{(-k)}(X,\partial X)$$

and similarly for $\mathcal{W}_G^{-\infty}(P)(\partial X, X)$.

Operators in $\mathcal{W}_G^{-\infty}(Q)(X,\partial X)((1)$-$\mathcal{W}_G^{-\infty}(P)(\partial X,X))$ are characterized by a kernel in $\mathcal{K}^\infty_{\kappa,loc}(X) \otimes_\pi C^\infty(\partial X)$ (in $C^\infty(\partial X) \otimes_\pi (\mathcal{K}^s_{P,comp}(X))'$). We shall also use the notation $\mathcal{W}_G^{-\infty}(\partial X,\partial X)$ for operators on ∂X with C^∞ kernel.

The composition of smoothing Green operators is again a smoothing Green operator provided it is defined. More precisely, we have

4. Proposition. Let $\varphi \in C_0^\infty(X)$, $\psi \in C_0^\infty(\partial X)$, $\mathcal{G} \in \mathcal{W}_G^{-\infty}(Q,P)^{(-k)}(X,X)$, $\mathcal{T} \in \mathcal{W}_G^{-\infty}(Q)^{(-k)}(\partial X, X)$, $\mathcal{K} \in \mathcal{W}_G^{-\infty}(P)^{(-k)}(X,\partial X)$, $\mathcal{C} \in \mathcal{W}_G^{-\infty}(\partial X,\partial X)$. Then

$$\mathcal{T}\varphi\mathcal{G} \in \mathcal{W}_G^{-\infty}(P)^{(-k)}(\partial X, X),$$

$$\mathcal{E}\psi\mathcal{T} \in \mathcal{W}_G^{-\infty}(Q)^{(-k)}(\partial X, X),$$

$$\mathcal{G}\varphi\mathcal{K} \in \mathcal{W}_G^{-\infty}(Q)^{(-k)}(X, \partial X),$$

$$\mathcal{K}\psi\mathcal{E} \in \mathcal{W}_G^{-\infty}(P)^{(-k)}(X, \partial X),$$

$$\mathcal{K}\psi\mathcal{T} \in \mathcal{W}_G^{-\infty}(P,Q)^{(-k)}(X, X).$$

The same is true without k. For $P = Q$ we have

$$\mathcal{T}\varphi\mathcal{K} \in \mathcal{W}_G^{-\infty}(\partial X, \partial X).$$

The statements for the (1)-classes follow immediately from the definitions. For the classes without (1) we use the following simple observation concerning adjoints.

For $P = (\gamma, 0)$, $Q = (\lambda, 0)$ the formal adjoints of smoothing Green operators are again in this class, where the role of X and ∂X is interchanged. We have

5. <u>Proposition.</u> Let $P = (\gamma, 0)$, $Q = (\lambda, 0)$ and $\mathcal{G}$, $\mathcal{T}$, $\mathcal{K}$, $\mathcal{E}$ as in Proposition 4. Then the formal adjoints belong to the classes:

$$\mathcal{G}^* \in \mathcal{W}_G^{-\infty}(P^*, Q^*)^{(-k)}(X, X),$$

$$\mathcal{T}^* \in \mathcal{W}_G^{-\infty}(Q^*)^{(-k)}(X, \partial X),$$

$$\mathcal{K}^* \in \mathcal{W}_G^{-\infty}(P^*)^{(-k)}(\partial X, X),$$

$$\mathcal{E}^* \in \mathcal{W}_G^{-\infty}(\partial X, \partial X),$$

where $P^* = (-\gamma, 0)$. The same is true without k.

Smoothing Green operators form the smallest possible class of negligible operators in the calculus. We describe now larger negligible classes which appear, for instance, if there are only finite sequences of symbols.

6. Definition. Let $\mu \in \mathbb{R}, k \in \mathbb{Z}_+, P = (\gamma, \wp) \in \underline{P}, Q = (\lambda, \sigma) \in \underline{P}$. Denote by $(1)\text{-}\mathcal{W}_G^{\mu,(o)}(Q,P)^{(-k)}(X,X)$ the subspace of $\bigcap_{s\in\mathbb{R}} \mathcal{L}(K^s_{P,comp}(X), K^{s-\mu}_{Q,loc}(X))$ of all operators $\mathcal{A}$ such that there exists $R = (\lambda+k, \mathcal{N}) \in \underline{P}$, $R = R(\mathcal{A})$ such that

$$\mathcal{A} : \mathcal{K}^s_{P,comp}(X) \longrightarrow K^{s-\mu}_{R^+,loc}(X)$$

and for arbitrary $\varphi, \psi \in C_0^\infty(X), \varphi\psi = 0$ we have $\psi\mathcal{A}\varphi \in (1)\text{-}\mathcal{W}_G^{-\infty}(Q,P)^{(-k)}(X,X)$. Elements of this space are called <u>(1)-Green operators of order μ without boundary symbols</u> acting on X (of discrete singularity type).

In contrast to the classes of order $-\infty$ we have no simple characterization by kernels.

Obviously

$$\bigcap_\mu (1)\text{-}\mathcal{W}_G^{\mu,(o)}(Q,P)^{(-k)}(X,X) = (1)\text{-}\mathcal{W}_G^{-\infty}(Q,P)^{(-k)}(X,X).$$

Note that the definition contains a certain pseudolocality condition.

With the notations used in Definition 2 we define the two-sided class.

7. Definition. Denote by $\mathcal{W}_G^{\mu,(o)}(Q,P)^{(-k)}(X,X)$ the subspace of $(1)\text{-}\mathcal{W}_G^{\mu,(o)}(Q,P)^{(-k)}(X,X)$ of all $\mathcal{A}$ such that the operators $\mathcal{A}^{11}_{ij}$ and $\mathcal{A}^{21}_{ij}$ have the property

$$(\mathcal{A}^{11}_{ij}) \quad : \mathcal{K}^{s,-\lambda}_{loc}(\Omega_i \times \mathbb{R}_+) \longrightarrow \mathcal{K}^{s-\mu}_{R^+_1,loc}(\Omega_j \times \mathbb{R}_+),$$

$$(\mathcal{A}^{21}_{ij}) \quad : H^s_{loc}(\Omega_i)^{\dim \sigma} \longrightarrow \mathcal{K}^{s-\mu}_{R^+_1,loc}(\Omega_j \times \mathbb{R}_+),$$

where $R_1 = R_1(\mathcal{A}) = (-\gamma+k, \mathcal{N}_1) \in \underline{P}$, s arbitrary.

Now it is clear by which modifications in the Definition 3 the classes $\mathcal{W}_G^{\mu,(o)}(Q)^{(-k)}(X,\partial X)$ and $(1)\text{-}\mathcal{W}_G^{\mu,(o)}(P)^{(-k)}(\partial X,X)$, $\mathcal{W}_G^{\mu,(o)}(P)^{(-k)}(\partial X,X)$ are defined.

Again we use the convention that in the intersection over all k the index k is dropped.

We denote by $\mathcal{B}_G^{\mu,(o)}(\partial X,\partial X)$ the space $\bigcap_s \mathcal{L}(H^s_{comp}(\partial X), H^{s-\mu}_{loc}(\partial X))$.

Then instead of Proposition 4 we get

8. <u>Proposition.</u> Let $\varphi \in C_0^\infty(X)$, $\psi \in C_0^\infty(\partial\Omega)$, $\mathcal{G} \in \mathcal{B}_G^{\mu_1,(o)}(Q,P)^{(-k)}(X,X)$, $\mathcal{T} \in \mathcal{B}_G^{\mu_2,(o)}(Q)^{(-k)}$, $\mathcal{K} \in \mathcal{B}_G^{\mu_3,(o)}(P)^{(-k)}(X,\partial X)$, $\mathcal{E} \in \mathcal{B}_G^{\mu_4,(o)}(\partial X,\partial X)$. Then

$$\mathcal{T}\varphi\mathcal{G} \in \mathcal{B}_G^{\mu_1+\mu_2,(o)}(P)^{(-k)}(\partial X,X),$$

$$\mathcal{E}\psi\mathcal{T} \in \mathcal{B}_G^{\mu_4+\mu_2,(o)}(Q)^{(-k)}(\partial X,X),$$

$$\mathcal{G}\varphi\mathcal{K} \in \mathcal{B}_G^{\mu_1+\mu_3,(o)}(Q)^{(-k)}(X,\partial X),$$

$$\mathcal{K}\psi\mathcal{E} \in \mathcal{B}_G^{\mu_3+\mu_4,(o)}(P)^{(-k)}(X,\partial X),$$

$$\mathcal{K}\psi\mathcal{T} \in \mathcal{B}_G^{\mu_2+\mu_3,(o)}(P,Q)^{(-k)}(X,X).$$

The same is true without k.

For $P = Q$ we have

$$\mathcal{T}\varphi\mathcal{K} \in \mathcal{B}_G^{\mu_2+\mu_3,(o)}(\partial X,\partial X).$$

A generalization of Proposition 5 provides

9. <u>Proposition.</u> Let $P = (\gamma,0)$, $Q = (\lambda,0)$ and $\mathcal{G}, \mathcal{T}, \mathcal{K}, \mathcal{E}$ as in Proposition 8. Then we get for the formal adjoints

$$\mathcal{G}^* \in \mathcal{B}_G^{\mu_1,(o)}(P^*,Q^*)^{(-k)}(X,X),$$

$$\mathcal{T}^* \in \mathcal{B}_G^{\mu_2,(o)}(Q^*)^{(-k)}(X,\partial X),$$

$$\mathcal{K}^* \in \mathcal{B}_G^{\mu_3,(o)}(P^*)^{(-k)}(\partial X,X),$$

$$\mathcal{E}^* \in \mathcal{B}_G^{\mu_4,(o)}(\partial X,\partial X).$$

The same is true without k.

10. <u>Example.</u> Let $X = \mathbb{R}^n_+$. According to 3.1.2. Proposition 3 singular functions in $\mathcal{K}^s_P(X)$, $P = (\delta,\varphi) \in \underline{P}$, which do not belong to $\mathcal{K}^{s,\delta}(X)$ are given by

$$v \longrightarrow \int_{\mathbb{R}^{n-1}} e^{ix'\xi'} \varkappa(\xi')(w(t))\, \hat{v}(\xi')\, d\xi'$$

with fixed $w \in \mathcal{E}(\mathfrak{p})$, $v \in H^s(\mathbb{R}^{n-1})$.

This mapping belongs to

$$\bigcap_{s \in \mathbb{R}} \mathcal{L}(H^s(\mathbb{R}^{n-1}), \mathcal{K}^s_{\mathfrak{p}}(\mathbb{R}^n_+))$$

and therefore to

$$\mathcal{W}^{0,(0)}_G(P)(\overline{\mathbb{R}^n_+}, \mathbb{R}^{n-1}).$$

Below we shall define narrower classes of Green operators possessing boundary symbols which also contain the above potential operators.

11. Example. Consider the operation of evaluation of a density v corresponding to a fixed singular term of $u \in \mathcal{K}^s_P(X)$, which is locally given by

$$v(x') = \int e^{ix'\xi'}\, r'(p_j,k)(\varkappa^{-1}(\xi')F'u(\xi',.))\, d\xi',$$

where $(p_j,m_j) \in P$, $0 \leq k \leq m_j-1$, and $r'(p_j,k)$ denotes the functional on $\mathcal{S}_{\mathfrak{p}}(\mathbb{R}_+)$ of evaluating the coefficient of $t^{-p_j}\log^k t$ for $t \to 0$. This defines a continuous mapping in

$$\bigcap_s \mathcal{L}(\mathcal{K}^s_{P,comp}(X), H^s_{loc}(\partial X)).$$

A standard argument from the theory of ψDO's shows the pseudolocality. Therefore the mapping

$$\mathcal{K}^s_{P,comp} \ni u \longrightarrow v \in H^s_{loc}$$

belongs to $\mathcal{W}^{0,(0)}_G(P)(\partial X, X)$.

Let us conclude by a few remarks on the way how Green operators enter our theory. Boundary value problems considered as mappings between function spaces act simultaneously on functions on X and on ∂X. Thus we have to take into account matrices of operators which obey, in the simplest case, the form

$$\mathcal{A} = \begin{pmatrix} \mathcal{A}_{11} & \mathcal{A}_{12} \\ \mathcal{A}_{21} & \mathcal{A}_{22} \end{pmatrix} : \begin{matrix} \mathcal{K}^{s}_{P,comp}(X) \\ \oplus \\ H^{s}_{comp}(\partial X) \end{matrix} \longrightarrow \begin{matrix} \mathcal{K}^{s-\mu}_{Q,loc}(X) \\ \oplus \\ H^{s-\mu}_{loc}(\partial X) \end{matrix} , \quad (5)$$

where $\mathcal{A}_{11} \in (1)-\mathcal{D}^{\mu,(o)}_{G}(Q,P)^{(-k)}(X,X)$, $\mathcal{A}_{12} \in (1)-\mathcal{D}^{\mu,(o)}_{G}(Q)^{(-k)}(X,\partial X)$, $\mathcal{A}_{21} \in \mathcal{D}^{\mu,(o)}_{G}(P)(\partial X,X)$, $\mathcal{A}_{22} \in \mathcal{D}^{\mu,(o)}_{G}(\partial X,\partial X)$ are in our case simply scalar operators. In practice, one gets also systems, but this case reduces often to more simpler operators, at least, as far as compositions, adjoints and the transition between different local representation are concerned. The solvability theory, of course, does not admit a simple reduction to pieces.

Denote the class of operators of the form (5) by $(1)-\mathcal{D}^{\mu,(o)}_{G}(Q,P)^{(-k)}(X,\partial X;X,\partial X)$. It is clear from the assumptions about the components that modulo $(1)-\mathcal{D}^{-\infty}_{G}(Q,P)^{(-k)}(X,\partial X;X,\partial X)$ we can restrict ourselves to the local situation. Then we have the isomorphisms

$$\mathcal{K}^{s}_{P}(\mathbb{R}^{n}_{+}) \cong \mathcal{K}^{s,\gamma}(\mathbb{R}^{n}_{+}) \bigoplus_{j=1}^{\dim P} H^{s}(\mathbb{R}^{n-1})$$

$$\mathcal{K}^{s-\mu}_{Q}(\mathbb{R}^{n}_{+}) \cong \mathcal{K}^{s-\mu,\lambda}(\mathbb{R}^{n}_{+}) \bigoplus_{j=1}^{\dim Q} H^{s-\mu}(\mathbb{R}^{n-1})$$

and $\mathcal{A}_{11}$ can be represented by a (dim Q+1)X(dim P+1) matrix of operators. The left upper corner

$$\mathcal{A}^{o}_{11} : \mathcal{K}^{s,\gamma}_{comp}(\mathbb{R}^{n}_{+}) \longrightarrow \mathcal{K}^{s-\mu,\lambda}_{loc}(\mathbb{R}^{n}_{+})$$

is called <u>regular part</u> of $\mathcal{A}_{11}$. The remaining parts are called <u>singular</u> (the same terminology was used also for Green operators on $\mathbb{R}_{+}$). The regular part possesses a formal adjoint

$$(\mathcal{A}^{o}_{11})^{*} : \mathcal{K}^{\mu-s,-\lambda}_{comp}(\mathbb{R}^{n}_{+}) \longrightarrow \mathcal{K}^{-s,-\gamma}_{loc}(\mathbb{R}^{n}_{+}).$$

We define classes of two-sided Green operators of order μ on $(X,\partial X)$, demanding that they can be localized modulo $\mathcal{D}^{-\infty}_{G}$ (the two-sided class), and in the local situation we refer to the above representation and repeat the definition of the class $\mathcal{B}_G$ on $\mathbb{R}_+$.

In particular, for $P = (\gamma,0)$, $Q = (\lambda,0)$ we have formal adjoints and the class of two-sided Green operators is closed under this operation. Moreover, it is an algebra filtered by the order.

We define now classes of Green operators with continuous singularity types. The special case $X = \mathbb{R}_+$ was considered in 2.3.3. The classes are defined by a simple modification of the definition of the classes $\mathcal{R}_G^{\mu,(o)}(\,.\,)^{(-k)}$: The quality of the mapping for weights λ' better than λ are described by means of the $\mathcal{K}_Q^s + \mathcal{K}^s_{\Lambda(\lambda')^+}$ spaces, $\Lambda \in \underline{\Lambda}$.

12. <u>Definition.</u> Let $k \in \mathbb{Z}_+$, $P = (\gamma,\varphi) \in \underline{P}$, $Q = (\lambda,\eta) \in \underline{P}$. Denote by (1)-$\mathcal{M}\mathcal{R}_G^{-\infty}(Q,P)^{(-k)}(X,X)$ the subspace of $\bigcap_{s\in\mathbb{R}} \mathcal{L}(\mathcal{K}^s_{P,comp}(X), \mathcal{K}^\infty_{Q,loc}(X))$ of all operators $\mathcal{A}$ such that there exists a $\Lambda \in \underline{\Lambda}(\lambda)$, $\Lambda = \Lambda(\mathcal{A})$ and $\mathcal{A}$ is continuous

$$\mathcal{A} : \mathcal{K}^s_{P,comp}(X) \longrightarrow \mathcal{K}^\infty_{\Lambda(\lambda+k)^+,loc}(X) + \mathcal{K}^\infty_{Q,loc}(X).$$

With the same modification the spaces

$$\mathcal{M}\mathcal{R}_G^{-\infty}(Q,P)^{(-k)}(X,X), \mathcal{M}\mathcal{R}_G^{-\infty}(Q)^{(-k)}(X,\partial X), \mathcal{M}\mathcal{R}_G^{-\infty}(P)^{(-k)}(\partial X,X)$$

and also the spaces with $\mu,(o)$ instead of $-\infty$ are defined.

Note that the continuous asymptotics appears only in $\{\mathrm{Re}\, z < \frac{1}{2} - \lambda\}$, z the variable of the Mellin image, and depends on the concrete operator.

The results about the $\mathcal{R}_G^{\mu,(o)}$ classes remain valid for the $\mathcal{M}\mathcal{R}_G^{\mu,(o)}$ classes. We have analogues of the Propositions 4,5,8 and 9, where everywhere in the formulation $\mathcal{R}$ has to be replaced by $\mathcal{M}\mathcal{R}$. Therefore, it is not necessary to repeat the corresponding formulations.

3.2.2. Green operators with boundary symbols

A considerable refinement of the theory of Green operators is achieved by restricting to those which possess (a finite number of) boundary symbols. Then we get, in particular, a precise description of the graded sequence associated with the filtration.

We start from the local situation. Later on we prove that the classes of operators are independent of the choice of coordinates, which yields immediately the global theory.

Now we introduce classes of Green boundary symbols. They contain as a special case the boundary symbols of the algebra with the transmission property. As usual, boundary symbols consist of families of operators of corresponding type on the half line, depending on the base space variable x' and the covariable ξ'. The definition of symbol classes includes a control of the growth of the norms for large ξ' with respect to ξ'-dependent norms on $\mathbb{R}_+$ (cf. 3.1.1.).

We begin with boundary symbols of operators which act between functions (distributions) on $\Omega\times\mathbb{R}_+$, $\Omega\subseteq\mathbb{R}^{n-1}$ open. They take values in the operator class $\mathcal{L}_G(Q,P)^{(-k)}$ introduced in 2.3.3.

Let $d'\in\mathbb{Z}_+$ and $\Omega\subseteq\mathbb{R}^{d'}$ be open.

1. Definition. Given $\mu\in\mathbb{R}, k\in\mathbb{Z}_+, P=(\gamma,\varphi)\in\underline{P}$, $Q=(\lambda,q)\in\underline{P}$. A family of operators $g(y,\xi)$, $y\in\Omega$, $\xi\in\mathbb{R}^d\setminus 0$, is said to belong to $\mathcal{R}^{(\mu)}((1)-\mathcal{L}_G(Q,P)^{(-k)}(\mathbb{R}_+,\mathbb{R}_+))(\Omega\times(\mathbb{R}^d\setminus 0))$ if it is a C^∞ family on $\Omega\times(\mathbb{R}^d\setminus 0)$ with values in $(1)-\mathcal{L}_G(Q,P)^{(-k)}(\mathbb{R}_+,\mathbb{R}_+)$ and satisfies the homogeneity condition

$$g(y,\lambda\xi) = \lambda^{\mu}\,\varkappa_\lambda g(y,\xi)\varkappa_\lambda^{-1} \quad \text{for each } \lambda>0 \tag{1}$$

and all $(y,\xi)\in\Omega\times(\mathbb{R}^d\setminus 0)$.

Here we denote by $\varkappa_\lambda$, $\lambda>0$, the family of operators on functions on $\mathbb{R}_+$ given by

$$\varkappa_\lambda v(t) = \lambda^{\frac{1}{2}} v(\lambda t), t \in \mathbb{R}_+ .$$

If no confusion is possible, we write shorter $\mathcal{R}^{(\mu)}((1)-\mathcal{L}_G(Q,P)^{(-k)})$. Elements of this space are called (1)-Green boundary symbols homogeneous of order μ (of conormal order $-k$).

Set

$$g_o(y,\xi) := \varkappa^{-1}_{|\xi|}\, g(y,\xi)\, \varkappa_{|\xi|} .$$

$g_o(y,\xi)$ is positively homogeneous in $|\xi|$ in the usual sense, because

$$g_o(y,\lambda\xi) = \varkappa^{-1}_{|\lambda\xi|} g(y,\lambda\xi)\, \varkappa_{|\lambda\xi|} =$$

$$= \varkappa^{-1}_{|\xi|}\, \varkappa^{-1}_{\lambda}\, g(y,\lambda\xi) \varkappa_\lambda \varkappa_{|\xi|} =$$

$$= \varkappa^{-1}_{|\xi|}\, \lambda^\mu\, g(y,\xi)\, \varkappa_{|\xi|} = \lambda^\mu g_o(y,\xi).$$

For fixed μ a boundary symbol $g \in \mathcal{R}^{(\mu)}((1)-\mathcal{L}_G(Q,P)^{(-k)})$ is uniquely determined by its values at $|\xi|$ = const. Each smooth family on $\Omega \times S^{d-1}$, $S^{d-1} = \{ \xi \in \mathbb{R}^d : |\xi| = 1 \}$ gives a boundary symbol by extension by homogeneity.

In the same way $\mathcal{R}^{(\mu)}(\mathcal{L}_G(Q,P))$ is defined.

We consider $\mathcal{R}^{(\mu)}((1)-\mathcal{L}_G(Q,P)^{(-k)})$ in the topology of $C^\infty(\Omega \times S^{d-1}, (1)-\mathcal{L}_G(Q,P)^{(-k)})$, which yields an (LF)-space. Since

$$C^\infty(\Omega \times S^{d-1}, (1)-\mathcal{L}_G(Q,P)^{(-k)}) = C^\infty(\Omega \times S^{d-1}) \otimes_\pi (1)-\mathcal{L}_G(Q,P)^{(-k)}$$

$$= C^\infty(\Omega) \otimes_\pi C^\infty(S^{d-1}, (1)-\mathcal{L}_G(Q,P)^{(-k)}),$$

each $g(y,\xi) \in \mathcal{R}^{(\mu)}((1)-\mathcal{L}_G(Q,P)^{(-k)})$ admits a representation

$$g(y,\xi) = \sum_{l=0}^{\infty} a_l(y)\, g_l(\xi),\quad a_l \in C^\infty(\Omega), \tag{2}$$

$g_l \in \mathcal{R}^{(\mu)}((1)-\mathcal{L}_G(Q,P)^{(-k)})$ is independent of y, and the series converges absolutely.

Note that

$$\mathcal{L}_G(Q,P) = \lim_{\overleftarrow{k}} \mathcal{L}_G(Q,P)^{(-k)}$$

induces

$$\mathcal{R}^{(\mu)}(\mathcal{L}_G(Q,P)) = \lim_{\overleftarrow{k}} \mathcal{R}^{(\mu)}(\mathcal{L}_G(Q,P)^{(-k)})$$

and the same for $(1)-\mathcal{L}_G$). Then Lemma 2 and the above decomposition remain valid for $\mathcal{R}^{(\mu)}(\mathcal{L}_G(Q,P))$.

Given $g(x',y',\xi') \in \mathcal{R}^{(\mu)}((1)-\mathcal{L}_G(Q,P)^{(-k)})(V\times V\times(\mathbb{R}^{n-1}\setminus 0))$, $V \subseteq \mathbb{R}^{n-1}$ open. We shall use g as amplitude function defining a ψDO (with operator-valued symbol) on $U := V\times\mathbb{R}_+$. Let $\chi(\xi')$ be an excision function, i.e. $\chi \in C^\infty(\mathbb{R}^{n-1})$, $\chi = 0$ near 0 and $\chi = 1$ outside a neighbourhood of 0.

Set

$$Op_\eta(\chi g)u(x') = \int_{\mathbb{R}^{n-1}} \int_V e^{i(x'-y')\xi'} \chi(\xi')\, g(x',y',\xi')u(y')dy'\,d\!\!{}^-\xi' \quad (3)$$

where $u \in C_0^\infty(U)$ is regarded as vector-valued function of y' (also the result is treated in this way).

Consider the simple case

$$g(x',y',\xi') = a(x')g_1(\xi')b(y')$$

$a,b \in C^\infty(\Omega)$ and g_1 independent of x',y'. Then

$$Op_\eta(\chi g) = a\, Op_\eta(\chi g_1)\, b$$

where now a,b denote the corresponding multiplication operators.

The mapping properties of $Op_\eta(\chi g)$ show that this operator belongs to $(1)-\mathcal{W}^{\mu,(0)}_G(Q,P)(U,U)$.

2. <u>Proposition.</u> Let $g(x',y',\xi') \in \mathcal{R}^{(\mu)}((1)-\mathcal{L}_G(Q,P)^{(-k)})(V\times V\times(\mathbb{R}^{n-1}\setminus 0))$. Then $Op_\eta(\chi g) \in (1)-\mathcal{W}^{\mu,(0)}_G(Q,P)^{(-k)}(U,U)$ and for another excision function χ_1: $Op_\eta((\chi-\chi_1)g) \in (1)-\mathcal{W}^{-\infty}_G(Q,P)^{(-k)}(U,U)$.

<u>Proof.</u> By the representation (2) of $g(x',y',\xi')$ we are lead to

compositions of multiplication operators and operators with x', y'- independent amplitude functions. Since the multiplication operator is continuous in $\mathcal{K}^s_{P,\mathrm{comp}}(U)$ and $\mathcal{K}^s_{P,\mathrm{loc}}(U)$ it remains to consider $\mathrm{Op}_\partial(\chi(\xi')g_1(\xi'))$. For $u \in \mathcal{K}^s_{P,\mathrm{comp}}(U)$ we have

$$F'(\mathrm{Op}_\partial(\chi g_1)u)(\xi') = \chi(\xi')g_1(\xi')F'u(\xi')$$

and therefore, for each fixed $\xi' \neq 0$, and suitable $R \in \underline{P}$, $R = (\lambda + k, \mathfrak{r})$, $\mathfrak{r}(\lambda)\mathfrak{r} = q$

$$g_1(\xi') : \mathcal{K}^s_P(\mathbb{R}_+) \longrightarrow \mathcal{K}^{s-\mu}_{R^+}(\mathbb{R}_+)$$

and

$$\|\chi(\xi')g_1(\xi')F'u(\xi')\|_{\mathcal{K}^{s-\mu}_{R^+}(\mathbb{R}_+)(\xi')} = \|\varkappa^{-1}(\xi')\,\chi(\xi')g_1(\xi')F'u(\xi')\|_{\mathcal{K}^{s-\mu}_{R^+}(\mathbb{R}_+)}$$

$$= \|\varkappa^{-1}(\xi')\,\chi(\xi')g_1(\xi')\,\varkappa(\xi')\,\varkappa^{-1}(\xi')F'u(\xi')\|_{\mathcal{K}^{s-\mu}_{R^+}(\mathbb{R}_+)}$$

$$\leq c\langle\xi'\rangle^\mu\|\varkappa^{-1}(\xi')F'u(\xi')\|_{\mathcal{K}^s_P(\mathbb{R}_+)}$$

$$= c\langle\xi'\rangle^\mu\,\|F'u(\xi')\|_{\mathcal{K}^s_P(\mathbb{R}_+)(\xi')}$$

where the estimate

$$\sup_{\xi' \in \mathbb{R}^{n-1}} \|\varkappa^{-1}(\xi')\,\chi(\xi')g_1(\xi')\,\varkappa(\xi')\|_{\mathcal{L}(\mathcal{K}^s_P(\mathbb{R}_+),\,\mathcal{K}^{s-\mu}_{R^+}(\mathbb{R}_+))} \leq c\langle\xi'\rangle^\mu \qquad (4)$$

following from the homogeneity assumption was used.

Then

$$\|\mathrm{Op}_\partial(\chi g_1)u\|^2_{\mathcal{K}^{s-\mu}_{R^+}(U)} = \int_{\mathbb{R}^{n-1}} \langle\xi'\rangle^{2(s-\mu)}\,\|\chi(\xi')g_1(\xi')F'u(\xi')\|^2_{\mathcal{K}^{s-\mu}_{R^+}(\mathbb{R}_+)(\xi')}\,d\xi'$$

$$\leq c^2 \int_{\mathbb{R}^{n-1}} \langle\xi'\rangle^{2s}\|F'u(\xi')\|^2_{\mathcal{K}^s_P(\mathbb{R}_+)(\xi')}\,d\xi'$$

$$= c^2\|u\|^2_{\mathcal{K}^s_P(U)}$$

which proves the asserted continuity for x', y'- independent amplitude functions. Since the sum in (2) converges absolutely and the operator norm is bounded by certain semi-norms of the amplitude function and of the multiplicator, respectively, also the series of operators converges in the operator norm. For $(\chi_1(\xi')-\chi(\xi'))g(x',y',\xi')$ we conclude in the same way. Instead of (4) we get an estimate by $c\langle\xi'\rangle^{-N}$ for each N, which proves the second assertion. In order to show the pseudo-locality let $\varphi,\psi\in C_0^\infty(V)$ with $\varphi\psi=0$. Then, in the integral (3) we introduce

$$e^{i(x',y')\xi'}=\frac{1}{|x'-y'|^{2N}}(-\Delta_{\xi'})^N(e^{i(x',y')\xi'}),$$

$N\in\mathbb{Z}_+$ arbitrary large. By partial integration we obtain an amplitude function $g_N(x',y',\xi')$ satisfying an estimate (4) with $\mu-N$ instead of μ. Since N is arbitrary we get the mapping properties of the class $(1)\text{-}\mathcal{W}_G^{-\infty}$. $\square$

<u>3. Remark.</u> In the proof we used only the estimates (4) and not the homogeneity itself. Moreover, we get continuity $\mathcal{K}_P^s(U)\longrightarrow\mathcal{K}_{R^+}^{s-\mu}(U)$ (i.e. without comp, loc) provided $g(x',y',\xi')$ vanishes for (x',y') outside a compact set in $\Omega\times\Omega$.

Replacing $(1)\text{-}\mathcal{L}_G(Q,P)^{(-k)}(\mathbb{R}_+,\mathbb{R}_+)$ in Definition 1 by one of the classes

$$\mathcal{L}_G(Q)^{(-k)}(\mathbb{R}_+,\partial\mathbb{R}_+),\ \mathcal{L}_G(P)^{(-k)}(\partial\mathbb{R}_+,\mathbb{R}_+),$$

we define $\mathcal{R}^{(\mu)}(\mathcal{L}_G(Q)^{(-k)}(\mathbb{R}_+,\partial\mathbb{R}_+))(\Omega\times(\mathbb{R}^d\setminus 0))$ and $\mathcal{R}^{(\mu)}(\mathcal{L}_G(P)^{(-k)}(\partial\mathbb{R}_+,\mathbb{R}_+))(\Omega\times(\mathbb{R}^d\setminus 0))$. The homogeneity is to be understood with the action of $\varkappa_\lambda$ as identity on $\mathbb{C}$. This implies for $g\in\mathcal{R}^{(\mu)}(\mathcal{L}_G(P)^{(-k)}(\partial\mathbb{R}_+,\mathbb{R}_+))$, for instance, that

$$g(y,\lambda\xi)=\lambda^\mu g(y,\xi)\varkappa_\lambda^{-1}.$$

Since $\mathcal{L}_G(\partial\mathbb{R}_+,\partial\mathbb{R}_+)$ are simply the complex numbers considered as linear operators $\mathbb{C}\longrightarrow\mathbb{C}$, the space $\mathcal{R}^{(\mu)}(\mathcal{L}_G(\partial\mathbb{R}_+,\partial\mathbb{R}_+))$ consists of all functions $g(y,\xi)\in C^\infty(\Omega\times(\mathbb{R}^d\setminus 0))$ that are positively homogeneous in ξ of order μ.

Clearly

$$D_y^{\alpha} D_{\xi}^{\beta} \mathcal{R}^{(\mu)}(\,.\,) \subset \mathcal{R}^{(\mu - |\beta|)}(\,.\,)$$

where the dot represents one of the classes of Green operators on the half line. In view of the results of 2.3.3. the composition of boundary symbols of Green type taken pointwise for fixed y and ξ (provided it is defined) is again a boundary symbol of Green type.

<u>4. Proposition.</u> Let $g(\mathcal{G}) \in \mathcal{R}^{(\mu_1)}(\mathcal{L}_G(Q,P)^{(-k)}(\mathbb{R}_+,\mathbb{R}_+))$, $g(\mathcal{T}) \in \mathcal{R}^{(\mu_2)}(\mathcal{L}_G(Q)^{(-k)}(\partial\mathbb{R}_+,\mathbb{R}_+))$, $g(\mathcal{K}) \in \mathcal{R}^{(\mu_3)}(\mathcal{L}_G(P)^{(-k)}(\mathbb{R}_+,\partial\mathbb{R}_+))$, $g(\mathcal{E}) \in \mathcal{R}^{(\mu_4)}(\mathcal{L}_G(\partial\mathbb{R}_+,\partial\mathbb{R}_+))$. Then

(i) $g(\mathcal{T})g(\mathcal{G}) \in \mathcal{R}^{(\mu_1+\mu_2)}(\mathcal{L}_G(P)^{(-k)}(\partial\mathbb{R}_+,\mathbb{R}_+))$,

(ii) $g(\mathcal{E})g(\mathcal{T}) \in \mathcal{R}^{(\mu_2+\mu_4)}(\mathcal{L}_G(Q)^{(-k)}(\partial\mathbb{R}_+,\mathbb{R}_+))$,

(iii) $g(\mathcal{G})g(\mathcal{K}) \in \mathcal{R}^{(\mu_1+\mu_3)}(\mathcal{L}_G(Q)^{(-k)}(\mathbb{R}_+,\partial\mathbb{R}_+))$,

(iv) $g(\mathcal{K})g(\mathcal{E}) \in \mathcal{R}^{(\mu_3+\mu_4)}(\mathcal{L}_G(P)^{(-k)}(\mathbb{R}_+,\partial\mathbb{R}_+))$,

(v) $g(\mathcal{K})\, g(\mathcal{E}) \in \mathcal{R}^{(\mu_2+\mu_3)}(\mathcal{L}_G(P,Q)^{(-k)}(\mathbb{R}_+,\mathbb{R}_+))$.

For $P = Q$

(vi) $g(\mathcal{T})\, g(\mathcal{K}) \in \mathcal{R}^{(\mu_2+\mu_3)}(\mathcal{L}_G(\partial\mathbb{R}_+,\partial\mathbb{R}_+))$.

<u>Proof.</u> We show (v). The other assertions may be proved in the same way. For fixed $y \in \Omega$, $\xi \in \mathbb{R}^d \setminus 0$ we have

$$g(\mathcal{K})(y,\xi)g(\mathcal{T})(y,\xi) \in \mathcal{L}_G(P,Q)^{(-k)}(\mathbb{R}_+,\mathbb{R}_+).$$

The smooth dependence on y and ξ is clear. The homogeneity follows from the homogeneity of the given boundary symbols:

$$g(\mathcal{K})(y,\lambda\xi)g(\mathcal{T})(y,\lambda\xi) = \lambda^{\mu_2+\mu_3} \varkappa_\lambda g(\mathcal{K})(y,\xi)g(\mathcal{T})(y,\xi)\, \varkappa_\lambda^{-1}. \quad \square$$

For $P = (\gamma,0)$, $Q = (\lambda,0)$ we have pointwise formal adjoints. With the notations of Proposition 4 we get

$$g(\mathcal{G})^* \in \mathcal{R}^{(\mu_1)}(\mathcal{L}_G(P^*,Q^*)^{(-k)}(\mathbb{R}_+,\mathbb{R}_+)),$$

$$g(\mathcal{T})^* \in \mathcal{R}^{(\mu_2)}(\mathcal{L}_G(Q^*)^{(-k)}(\mathbb{R}_+,\partial\mathbb{R}_+)),$$
$$g(\mathcal{K})^* \in \mathcal{R}^{(\mu_3)}(\mathcal{L}_G(P^*)^{(-k)}(\partial\mathbb{R}_+,\mathbb{R}_+)), \qquad (5)$$

where $P^* = (-\gamma, 0)$.

Let now $\Omega = V\times V$, $V \subset \mathbb{R}^{n-1}$ open, and

$$g(\mathcal{T})(x',y',\xi') \in \mathcal{R}^{(\mu)}((1)-\mathcal{L}_G(P)^{(-k)}(\partial\mathbb{R}_+,\mathbb{R}_+))(V\times V\times(\mathbb{R}^{n-1}\setminus 0)),$$
$$g(\mathcal{K})(x',y',\xi') \in \mathcal{R}^{(\mu)}(\mathcal{L}_G(Q)^{(-k)}(\mathbb{R}_+,\partial\mathbb{R}_+))(V\times V\times(\mathbb{R}^{n-1}\setminus 0)).$$

Then we get in an analogous manner as Proposition 2 the following

5. Proposition. Let χ be an excision function as above and $Op_\partial(\chi g(\mathcal{T}))$ and $Op_\partial(\chi g(\mathcal{K}))$ defined by (3). Then

$$Op_\partial(\chi g(\mathcal{T})) \in (1)-\mathcal{W}_G^{\mu,(0)}(P)^{(-k)}(V,U),$$
$$Op_\partial(\chi g(\mathcal{K})) \in \mathcal{W}_G^{\mu,(0)}(Q)^{(-k)}(U,V).$$

Clearly, $Op_\partial(\chi g(\mathcal{C}))$, $g(\mathcal{C})(x',y',\xi') \in \mathcal{R}^{(\mu)}(\mathcal{L}_G(\partial\mathbb{R}_+,\partial\mathbb{R}_+))(V\times V\times(\mathbb{R}^{n-1}\setminus 0))$, is a classical ψDO of order μ. Its continuity properties imply that

$$Op_\partial(\chi g(\mathcal{C})) \in \mathcal{W}_G^{\mu,(0)}(V,V).$$

We shall simplify the notation of the classes of boundary symbols of Green type by the following convention: The omitted singularity types, combination of $\mathbb{R}_+$ and $\partial\mathbb{R}_+$ and so on can be specified in an adequate way.

Below we shall use standard results about ψDO's with operator-valued symbol. The calculus is valid also in case the norms in the function spaces depend on ξ' in view of 3.1.1.(1),(2) (cf. [S8], ch. 4 for basic arguments on the calculus). In fact, the homogeneity assumption implies the estimate (4), which means exactly the desired estimate of the operator norm in the ξ'-depending norms.

In particular, we have

6. Proposition. Let $g(\xi') \in \mathcal{R}^{(\mu)}(\mathcal{L}_G^{(-k)})$ and $\varphi(x') \in C_0^\infty(V)$, χ an excision function. Then

$$\mathrm{op}_\partial(\chi g)\varphi - \sum_{|\alpha|<\ell} \frac{1}{\alpha!} D_{x'}^{\alpha}\varphi(x')\mathrm{op}_\partial(\chi(\xi')\partial_{\xi'}^{\alpha} g(\xi'))$$
$$\in \mathcal{W}_G^{\mu-1,(0)}(\ .\ )^{(-k)},$$

and the operator norms of the difference can be estimated by certain semi-norms of g.

As a consequence we obtain that each operator of the form $\mathrm{op}_\partial(\chi(\xi')g(x',y',\xi'))$, g with compact support in (x',y'), can be written as $\mathrm{op}_\partial(\chi(\xi')g_1(x',\xi'))$ modulo $\mathcal{W}_G^{\mu-1,(0)}(\ .\ )^{(-k)}$ for each $l \in \mathbb{Z}_+$.

Proposition 6 also implies

7. Proposition. Let $g_j(x',\xi') \in \mathcal{R}^{(\mu_j)}(\mathcal{L}_G^{(-k)})$ such that $g_1 g_2$ is defined and set for fixed $l \in \mathbb{Z}_+$ and $\varphi \in C_0^\infty(V)$

$$g(x',\xi') = \sum_{|\alpha|<\ell} \frac{1}{\alpha!} \partial_{\xi'}^{\alpha} g_1(x',\xi') D_{x'}^{\alpha}(\varphi(x')g_2(x',\xi')). \tag{6}$$

Then $\mathrm{op}_\partial(\chi g_1)\varphi\,\mathrm{op}_\partial(\chi g_2) - \mathrm{op}_\partial(\chi g) \in \mathcal{W}_G^{\mu_1+\mu_2-1,(0)}(\ .\ )^{(-k)}$.

Proof. We have the decompositions

$$g_j(x',\xi') = \sum_{m=0}^{\infty} \varphi_{jm}(x')\, g_{jm}(\xi'),\ j = 1,2,$$

where $\varphi_{jm} \in C_0^\infty(V)$. Then the assertion follows by applying Proposition 6 to $\mathrm{op}_\partial(\chi g_{1m})\varphi_{2m}$. •

Let $P = (\gamma,0)$, $Q = (\lambda,0)$ and $l \in \mathbb{Z}_+$ be fixed.

For

$$g(x',\xi') \in \mathcal{R}^{(\mu)}(\mathcal{L}_G(Q,P)^{(-k)}(\mathbb{R}_+,\mathbb{R}_+))(V\times(\mathbb{R}^{n-1}\setminus 0)),$$
$$a(x',\xi') \in \mathcal{R}^{(\mu)}(\mathcal{L}_G(P)^{(-k)}(\mathbb{R}_+,\mathbb{R}_+))(V\times(\mathbb{R}^{n-1}\setminus 0)),$$
$$k(x',\xi') \in \mathcal{R}^{(\mu)}(\mathcal{L}_G(Q)^{(-k)}(\mathbb{R}_+,\mathbb{R}_+))(V\times(\mathbb{R}^{n-1}\setminus 0))$$

we set

$$g^*(x',\xi') = \sum_{|\alpha|<\ell} \frac{1}{\alpha!} \partial_{\xi'}^{\alpha} D_{x'}^{\alpha} g(x',\xi')^*,$$

where $g(x',\xi')^*$ denotes the formal adjoint taken for fixed x',ξ'. f^* and k^* are defined in the same way and we have

$$g^*(x',\xi') \in \mathcal{R}^{(\mu)}(\mathcal{L}_G(P^*,Q^*)^{(-k)}(\mathbb{R}_+,\mathbb{R}_+))(V\times(\mathbb{R}^{n-1}\setminus 0)),$$

$$f^*(x',\xi') \in \mathcal{R}^{(\mu)}(\mathcal{L}_G(P^*)^{(-k)}(\mathbb{R}_+,\partial\mathbb{R}_+))(V\times(\mathbb{R}^{n-1}\setminus 0)),$$

$$k^*(x',\xi') \in \mathcal{R}^{(\mu)}(\mathcal{L}_G(Q^*)^{(-k)}(\partial\mathbb{R}_+,\mathbb{R}_+))(V\times(\mathbb{R}^{n-1}\setminus 0)). \tag{7}$$

8. <u>Proposition.</u> We have

$$Op_\partial(\chi g)^* - Op_\partial(\chi g^*) \in \mathcal{D}_G^{\mu-1,(o)}(P^*,Q^*)^{(-k)}(U,U),$$

$$Op_\partial(\chi f)^* - Op_\partial(\chi f^*) \in \mathcal{D}_G^{\mu-1,(o)}(P^*)^{(-k)}(U,V),$$

$$Op_\partial(\chi k)^* - Op_\partial(\chi k^*) \in \mathcal{D}_G^{\mu-1,(o)}(Q^*)^{(-k)}(V,U).$$

As a consequence we conclude that Proposition 2 and 5 are valid without (1).

9. <u>Definition.</u> Let $\mu\in\mathbb{R}, k,l\in\mathbb{Z}_+$. Set

$$\mathcal{R}^{\mu,(l)}(\mathcal{L}_G^{(-k)}) = \sum_{m=o}^{l-1} \mathcal{R}^{(\mu-m)}(\mathcal{L}_G^{(-k)}).$$

For $g^j = \sum_{m=o}^{l-1} g_m^j$, $j = 1,2$, such that $g^1 g^2$ is defined we set

$$g^1 \# g^2 = \sum_{|\alpha|+m+m'<l} \frac{1}{\alpha!} \partial_{\xi'}^{\alpha} g_m^1 D_{x'}^{\alpha} g_{m'}^2. \tag{8}$$

Let $P = (\gamma,0)$, $Q = (\lambda,0)$,

$$g \in \mathcal{R}^{\mu,(l)}(\mathcal{L}_G(Q,P)^{(-k)}(\mathbb{R}_+,\mathbb{R}_+))(V\times(\mathbb{R}^{n-1}\setminus 0)),$$

$$f \in \mathcal{R}^{\mu,(l)}(\mathcal{L}_G(P)^{(-k)}(\partial\mathbb{R}_+,\mathbb{R}_+))(V\times(\mathbb{R}^{n-1}\setminus 0)),$$

$$k \in \mathcal{R}^{\mu,(l)}(\mathcal{L}_G(Q)^{(-k)}(\mathbb{R}_+,\partial\mathbb{R}_+))(V\times(\mathbb{R}^{n-1}\setminus 0)).$$

Set

$$g^* = \sum_{|\alpha|+m<l} \frac{1}{\alpha!} \partial^{\alpha}_{\xi'} D^{\alpha}_{x'} g_m(x',\xi') \ , \tag{9}$$

f^* and k^* defined in the same way.

From Proposition 4 we get that $\#$ defines a mapping

$$\# : \mathcal{R}^{\mu_1,(l)}(\mathcal{L}_G^{(-k)}) \times \mathcal{R}^{\mu_2,(l)}(\mathcal{L}_G^{(-k)}) \longrightarrow \mathcal{R}^{\mu_1+\mu_2,(l)}(\mathcal{L}_G^{(-k)})$$

and from (7) that

$$* : \mathcal{R}^{\mu,(l)}(\mathcal{L}_G(Q,P)^{(-k)}(\mathbb{R}_+,\mathbb{R}_+)) \longrightarrow \mathcal{R}^{\mu,(l)}(\mathcal{L}_G(P^*,Q^*)^{(-k)}(\mathbb{R}_+,\mathbb{R}_+)),$$

$$* : \mathcal{R}^{\mu,(l)}(\mathcal{L}_G(P)^{(-k)}(\partial\mathbb{R}_+,\mathbb{R}_+)) \longrightarrow \mathcal{R}^{\mu,(l)}(\mathcal{L}_G(P^*)^{(-k)}(\mathbb{R}_+,\partial\mathbb{R}_+)),$$

$$* : \mathcal{R}^{\mu,(l)}(\mathcal{L}_G(Q)^{(-k)}(\mathbb{R}_+,\partial\mathbb{R}_+)) \longrightarrow \mathcal{R}^{\mu,(l)}(\mathcal{L}_G(Q^*)^{(-k)}(\partial\mathbb{R}_+,\mathbb{R}_+)).$$

10. <u>Definition.</u> Let $\mu \in \mathbb{R}, k,l \in \mathbb{Z}_+, P,Q \in \underline{P}$. We denote by $\mathcal{W}_G^{\mu,(l)}(Q,P)^{(-k)}(U,U)$ the space of all operators $\mathcal{G}$ such that there exists

$$g(x',\xi') \in \mathcal{R}^{\mu,(l)}(\mathcal{L}_G(Q,P)^{(-k)}(\mathbb{R}_+,\mathbb{R}_+))(V\times(\mathbb{R}^{n-1}\setminus 0))$$

with

$$\mathcal{G} - op_{\eta}(\chi g) \in \mathcal{W}_G^{\mu-l,(o)}(Q,P)^{(-k)}(U,U).$$

In the same way we introduce $\mathcal{W}_G^{\mu,(l)}(P)^{(-k)}(V,U)$, $\mathcal{W}_G^{\mu,(l)}(Q)^{(-k)}(U,V)$ and $\mathcal{W}_G^{\mu,(l)}(V,V)$.

Note that $\mathcal{W}_G^{\mu,(l)}(V,V)$ consists simply of ψDO's on V possessing l homogeneous components of the symbol.

11. <u>Theorem.</u> The symbol in $\mathcal{R}^{\mu,(l)}(\mathcal{L}_G(\ .\ )^{(-k)})$ of an operator in $\mathcal{W}_G^{\mu,(l)}(\ .\ )^{(-k)}$ is uniquely determined and the sequence

$$0 \longrightarrow \mathcal{W}_G^{\mu-l,(o)}(\ .\ )^{(-k)} \hookrightarrow \mathcal{W}_G^{\mu,(l)}(\ .\ )^{(-k)} \longrightarrow \mathcal{R}^{\mu,(l)}(\mathcal{L}_G(\ .\ )^{(-k)}) \longrightarrow 0$$

is exact.

This follows from general facts about ψDO's with operator-valued symbols. For convenience of the reader we indicate another proof which is based on the reduction to ordinary ψDO. First we show that compositions and adjoints are again in the class provided they are defined.

12. Theorem (i). Let $\mathcal{G}_j \in \mathcal{B}_G^{\mu_j,(1)}(\,.\,)^{(-k)}$ with symbols $g_j \in \mathcal{R}^{\mu_j,(1)}(\mathcal{L}_G(\,.\,)^{(-k)})$ such that $g_1 \# g_2$ is defined. Then, for each $\varphi \in C_0^\infty(V)$, $\mathcal{G} := \mathcal{G}_1 \varphi \mathcal{G}_2 \in \mathcal{B}_G^{\mu_1+\mu_2,(1)}(\,.\,)^{(-k)}$ and $\mathcal{G}$ has $g_1 \# (\varphi g_2)$ as a symbol.

(ii) Let $P = (\gamma, 0)$, $Q = (\lambda, 0)$ and $\mathcal{G} \in \mathcal{B}_G^{\mu,(1)}(Q,P)^{(-k)}(U,U)$, $\mathcal{T} \in \mathcal{B}_G^{\mu,(1)}(P)^{(-k)}(V,U)$, $\mathcal{K} \in \mathcal{B}_G^{\mu,(1)}(Q)^{(-k)}(U,V)$ with symbols g, t, k. Then symbols of $\mathcal{G}^*, \mathcal{T}^* \mathcal{K}^*$ are g^*, t^*, k^*.

Proof. (i) Apply Proposition 7 to the homogeneous components. (ii) Use Proposition 8 and 3.2.1. Proposition 9. $\square$

Proof of Theorem 11. In order to show that the symbol is unique we fix $x_0' \in V$ arbitrarily and $\varphi \in C_0^\infty(V)$, $\varphi \equiv 1$ near x_0'. Consider $\mathcal{T} \in \mathcal{B}_G^{\mu,(1)}(P)^{(-k)}(V,U)$. For each $k(\xi') \in \mathcal{R}^{(\nu)}(\mathcal{L}_G(P)^{(-k)}(\mathbb{R}_+, \partial\mathbb{R}_+))(\mathbb{R}^{n-1} \setminus 0)$ we have in view of Theorem 10

$$\mathcal{T}\varphi\, \mathrm{op}_\partial(\chi k) \in \mathcal{B}_G^{\mu+\nu,(1)}(V,V).$$

This is a usual ψDO with uniquely determined principal symbol

$$\sigma^{(\mu+\nu)}(\mathcal{T}\varphi\, \mathrm{op}_\partial(\chi k))(x', \xi') = \varphi(x')\, t(x', \xi') \circ k(\xi'),$$

where $t \in \mathcal{R}^{(\mu)}(\mathcal{L}_G(P)^{(-k)}(\partial\mathbb{R}_+, \mathbb{R}_+))(V \times (\mathbb{R}^{n-1} \setminus 0))$. Near x_0' we can omit φ and get that the leading homogeneous part is uniquely determined, since k was arbitrary. For $\mathcal{K} \in \mathcal{B}_G^{\mu,(1)}(Q)^{(-k)}(U,V)$ $\subset \mathcal{B}_G^{\mu,(1)}(Q')^{(-k-\lambda'+\lambda)}(U,V)$, $Q' = (\lambda', 0)$, we consider the adjoint (with respect to the weight λ') $\mathcal{T} = \mathcal{K}^* \in \mathcal{B}_G^{\mu,(1)}(Q'^*)^{(-k-\lambda'+\lambda)}(V,U)$. Then the above conclusion applies. Finally, for $\mathcal{G} \in \mathcal{B}_G^{\mu,(1)}(Q,P)^{(-k)}(U,U)$ we get that

$$\mathcal{G}\varphi\, \mathrm{op}_\partial(\chi k) \in \mathcal{B}_G^{\mu+\nu,(1)}(Q)^{(-k)}(U,V)$$

is a Green operator of potential type. Above we proved already that it has a uniquely determined principal symbol.

By induction we recover in this way all homogeneous components which give operators that do not belong to $\mathcal{W}_G^{\mu-1,(o)}(\,.\,)^{(-k)}$.

The surjectivity follows, since the symbol map is a left inverse of Op_{∂}.

The above conclusion shows also that an operator with vanishing symbol belongs to $\mathcal{W}_G^{\mu-1,(o)}(\,.\,)^{(-k)}$. □

13. Remark. It is possible to establish a calculus of operators having an infinite sequence of homogeneous boundary symbols. The essential additional point is the asymptotic summation of a sequence of homogeneous boundary symbols, which is possible, since on the subspace of $\mathcal{L}_G(\,.\,)^{(-k)}$ with fixed singularity type of the image and that of the adjoint the topology can be given by a countable set of norms.

The generalization of the local theory of Green operators with boundary symbols to manifolds is based on the proof of the invariance of the classes under changes of coordinates. If we assume the diffeomorphisms acting only on $V \subset \mathbb{R}^{n-1}$, the result follows from the theory of ψDO's with operator-valued symbol.

Let $\varkappa : V \longrightarrow V_1$ be a diffeomorphism. Denote by $\varkappa^*$ the pull-back of functions on V_1 and also on $U_1 = V_1 \times \mathbb{R}_+$

$$\varkappa^* u(x',t) = u(\varkappa(x'),t).$$

For an operator $\mathcal{G}$ we set $\varkappa_* \mathcal{G} = (\varkappa^*)^{-1} \mathcal{G}\, \varkappa^*$. Denote by x' the coordinate on V and by y' the coordinate on V_1.

14. Theorem. Let $\mathcal{G} \in \mathcal{W}_G^{\mu,(1)}(\,.\,)^{(-k)}(V)$ with symbol $g(x',\xi')$. Then $\varkappa_* \mathcal{G} \in \mathcal{W}_G^{\mu,(1)}(\,.\,)^{(-k)}(V_1)$ and it has the symbol

$$g_1(y',\eta') = \sum_{|\alpha|+m<1} \frac{1}{\alpha!}\, \partial_{\xi'}^{\alpha}\, g(x', {}^t d\varkappa(x')\eta') D_{z'}^{\alpha}\, e^{ih(z',x')\eta'}\Big|_{z'=x'}$$

$x' = \varkappa^{-1}(y'), h(z',x') = \varkappa(z') - \varkappa(x') - d\varkappa(x')(z'-x')$.

Let X be a C^∞ manifold with boundary ∂X. Then we define the classes

$\mathcal{W}_G^{\mu,(1)}(\,.\,)^{(-k)}$ on $(X,\partial X)$ by demanding that the operator belongs to the class in arbitrary local coordinates. In view of Theorem 14 an operator belongs to $\mathcal{W}_G^{\mu,(1)}(\,.\,)^{(-k)}$ on $(X,\partial X)$ if this is true in coordinates of an arbitrary atlas. We use the notations $\mathcal{W}_G^{\mu,(1)}(Q,P)^{(-k)}(X,X)$, $\mathcal{W}_G^{\mu,(1)}(Q)^{(-k)}(X,\partial X)$ etc. Then the assertions of Theorems 11,12 remain valid, where the symbols are to be understood in local coordinates.

In view of the isomorphisms

$$\mathcal{K}^{s,s}(\mathbb{R}^n_+) = \begin{cases} H^s(\mathbb{R}^n_+) & \text{if } s<\frac{1}{2} \\ H^s_0(\mathbb{R}^n_+) & \text{if } s>-\frac{1}{2} \end{cases}$$

the spaces $\mathcal{K}^{s,s}_{comp}(\mathbb{R}^n_+)$ and $\mathcal{K}^{s,s}_{loc}(\mathbb{R}^n_+)$ do not depend on the coordinates. The same is true for $\mathcal{K}^{s,s}(\mathbb{R}^n_+)$ provided the coordinate change is linear outside a compact set. For arbitrary $s,\gamma\in\mathbb{R}$ we have an isomorphism

$$Op_\partial([\xi']^{s-\gamma}) : \mathcal{K}^{s,\gamma}(\mathbb{R}^n_+) \longrightarrow \mathcal{K}^{\gamma,\gamma}(\mathbb{R}^n_+)$$

which is a ψDO with operator-valued symbol (values in scalar multiples of the identity). Here $[\xi']$ means a strictly positive C^∞ function of ξ' which equals $|\xi'|$ for $|\xi'|\geqq$ const. Under a change of coordinates in $\mathbb{R}^{n-1}=\partial\mathbb{R}^n_+$ which is linear outside a compact set we get another ψDO (with constant coefficients outside a compact set). Its inverse acts from $\mathcal{K}^{\gamma,\gamma}(\mathbb{R}^n_+)$ into $\mathcal{K}^{s,\gamma}(\mathbb{R}^n_+)$ continuously, which proves the invariance of $\mathcal{K}^{s,\gamma}_{comp}(\mathbb{R}^n_+)\,\mathcal{K}^{s,\gamma}_{loc}(\mathbb{R}^n_+)$, and $\mathcal{K}^{s,\gamma}(\mathbb{R}^n_+)$ provided the coordinate change is linear outside a compact set.

For each $P=(\gamma,\varphi)\in\underline{P}$, $s\in\mathbb{R}$ we have the exact sequence of Hilbert spaces

$$0 \longrightarrow \mathcal{K}^{s,\gamma}(\mathbb{R}^n_+) \hookrightarrow \mathcal{K}^s_P(\mathbb{R}^n_+) \xrightarrow{\bigoplus_j T_j} \bigoplus_j H^s(\mathbb{R}^{n-1}) \longrightarrow 0$$

where $1\leqq j\leqq \dim\varphi$ and T_j are the trace operators corresponding to φ from 3.2.1. A coordinate change which is linear near ∞ does not change neither $\mathcal{K}^{s,\gamma}(\mathbb{R}^n_+)$ nor $H^s(\mathbb{R}^{n-1})$. The orthogonal projection of $\mathcal{K}^s_P(\mathbb{R}^n_+)$ onto $\mathcal{K}^{s,\gamma}(\mathbb{R}^n_+)$ belongs to $1+\mathcal{W}^0_G(P,P)$, hence the complementary projection is in $\mathcal{W}^o_G(P,P)$ and has the orthocomplement of $\mathcal{K}^{s,\gamma}(\mathbb{R}^n_+)$ in $\mathcal{K}^s_P(\mathbb{R}^n_+)$ as its image. According to Theorem 14 we obtain from the invariance of $\mathcal{K}^{s,\gamma}(\mathbb{R}^n_+)$ that $\varkappa^*\mathcal{K}^s_P(\mathbb{R}^n_+)$, $\varkappa$ the change of coordinates, can be described as

$$\varkappa^*\mathcal{K}^s_P(\mathbb{R}^n_+) = \varkappa^*\mathcal{K}^{s,\gamma}(\mathbb{R}^n_+) + \varkappa^*\mathcal{G}(\bigoplus_j H^s(\mathbb{R}^{n-1}))$$
$$= \mathcal{K}^{s,\gamma}(\mathbb{R}^n_+) + \varkappa_*\mathcal{G}(\bigoplus_j H^s(\mathbb{R}^{n-1}))$$

where $\mathcal{G} \in \mathcal{W}^o_G(P)(\mathbb{R}^n_+,\mathbb{R}^{n-1})$ map $\bigoplus_j H^s(\mathbb{R}^{n-1})$ into the orthocomplement of $\mathcal{K}^{s,\gamma}(\mathbb{R}^n_+)$ in $\mathcal{K}^s_P(\mathbb{R}^n_+)$. Then, clearly $\varkappa^*\mathcal{K}^s_P(\mathbb{R}^n_+) = \mathcal{K}^s_P(\mathbb{R}^n_+)$. Now it is clear that the spaces $\mathcal{K}^{s,\gamma}, \mathcal{K}^s_P, \dots$ on manifolds can be characterized by demanding the properties in any fixed atlas (and not by arbitrary coordinates).

15. Remark. It is possible to show also that each change of coordinates preserving the boundary preserves $\mathcal{K}^{s,\gamma}_{comp}$ and $\mathcal{K}^{s,\gamma}_{loc}$, i.e. even if the coordinate transversal to the boundary is changed. For linear transformations it is simple to check this using the expression for the norm. The general case can be reduced to a superposition of linear changes and changes of coordinates on the boundary modulo an error term which behaves like $O(t)$. Provided that $P = (\gamma, \mathfrak{p})$ has the property that $\overline{\tau}(\gamma)(T^{-j}\mathfrak{p}) \subset \mathfrak{p}$ for $j = 0,1,\dots$ the same is true for $\mathcal{K}^s_P$.

Now we define the classes $\mathcal{MO}^{\mu,(1)}_G$ of Green operators with continuous asymptotics. The related classes of boundary symbols take values in the Green operators with continuous asymptotics on the half-line, which were introduced in 2.3.3.

16. Definition. Let $\mu \in \mathbb{R}, k \in \mathbb{Z}_+, P = (\gamma, \mathfrak{p}) \in \underline{P}$, $Q = (\lambda, \mathfrak{q}) \in \underline{P}$. A family of operators $g(y,\xi), y \in \Omega, \xi \in \mathbb{R}^d \setminus 0$, is said to belong to $\mathcal{R}^{(\mu)}((1)-\mathcal{G}_G(Q,P)^{(-k)}(\mathbb{R}_+,\mathbb{R}_+))(\Omega \times (\mathbb{R}^d\setminus 0))$ if it is a C^∞ family on $\Omega \times (\mathbb{R}^d \setminus 0)$ with values in $(1)-\mathcal{G}_G(Q,P)^{(-k)}(\mathbb{R}_+,\mathbb{R}_+)$ and the homogeneity condition (1) is satisfied. In the same way, by replacing $\mathcal{C}$ by $\mathcal{G}$ we define

$$\mathcal{R}^{(\mu)}(\mathcal{G}_G(Q,P)^{(-k)}(\mathbb{R}_+,\mathbb{R}_+)),\ \mathcal{R}^{(\mu)}(\mathcal{G}_G(Q)^{(-k)}(\mathbb{R}_+,\partial\mathbb{R}_+)),$$

$$\mathcal{R}^{(\mu)}(\mathcal{G}_G(P)^{(-k)}(\partial\mathbb{R}_+,\mathbb{R}_+)).$$

Then Proposition 4 and (5) remain valid if $\mathcal{L}$ is replaced everywhere by $\mathcal{G}$. We associate operators with boundary symbols in $\mathcal{R}^{(\mu)}(\mathcal{G}_G(\,.\,)^{(-k)})(V\times V\times(\mathbb{R}^{n-1}\setminus 0))$ according to (3). Then the Propositions 2,5,6,7,8 remain valid if $\mathcal{L}$ is replaced by $\mathcal{G}$ and $\mathcal{W}$ by $\mathcal{MW}$. The operator classes $\mathcal{MW}^{\mu,(1)}_G$ are defined by replacing in Definition 10 $\mathcal{L}$ by $\mathcal{G}$ and $\mathcal{W}$ by $\mathcal{MW}$, where the replacements of $\mathcal{L}$ by $\mathcal{G}$ in Definition 9 yield $\mathcal{R}^{\mu,(1)}(\mathcal{G}_G^{(-k)})$. Then the results of the calculus (Theorems 11,12,14) remain valid if the mentioned modifications are carried out.

Let X be a C^∞ manifold with boundary $Y = \partial X$. Then we introduce the corresponding classes of Green operators $\mathcal{W}^{\mu,(1)}_G(Q,P)^{(-k)}(X,X)$ and so on by demanding that in each coordinates the induced mapping (obtained by the diagram

$$\begin{array}{ccc}
\mathcal{K}^s_P(X) & \xrightarrow{\;\mathcal{G}\;} & \mathcal{K}^{s-\mu}_Q(X) \\
\uparrow & & \downarrow r_U \\
C^\infty_0(U') & \xrightarrow{\;\mathcal{G}_{U'}\;} & C^\infty(U') \\
\downarrow\cong & & \downarrow\cong \\
C^\infty_0(U) & \xrightarrow{\;\varkappa_*(\mathcal{G}_{U'})\;} & C^\infty(U)
\end{array}$$

$U'\subset X$ a coordinate patch, $U' \xrightarrow{\varkappa} U\subset\mathbb{R}^n$ the charte homeomorphism) belongs to $\mathcal{W}^{\mu,(1)}_G(Q,P)^{(-k)}(U,U)$. We shall use the above notations together with the appropriate simplifications because in all concrete cases it is clear where the given operator acts.

3.2.3. Mellin operators

Admitting boundary symbols with values in $\mathcal{L}_{M+G}$ or $\mathcal{G}_{M+G}$ we obtain a larger class of operators which are called Mellin operators in $U\subset\mathbb{R}^n_+$. We continue to use the notations from 3.2.2.

1. Definition. Let $\mu \in \mathbb{R}, k \in \mathbb{Z}_+, P = (\gamma, \mathcal{P}),\ Q = (\lambda, q) \in \underline{P},\ \mu - \gamma + \lambda \in \mathbb{Z}_+$.
A family of operators $m(y,\xi),\ y \in \Omega,\ \xi \in \mathbb{R}^d \setminus 0$, is said to belong to $\mathcal{R}^{(\mu)}(\mathcal{L}_{M+G}(Q,P)^{(-k)})(\Omega \times (\mathbb{R}^d \setminus 0))$ if it is a C^∞ family on $\Omega \times (\mathbb{R}^d \setminus 0)$ with values in $\mathcal{L}_{M+G}(Q,P)^{(-k)}$, homogeneous in ξ of order μ

$$m(y, \delta\xi) = \delta^\mu \varkappa_\delta m(y,\xi) \varkappa_\delta^{-1} \quad \text{for each } \delta > 0$$

and all $y \in \Omega$, $\xi \in \mathbb{R}^d \setminus 0$, and such that the Mellin symbols $\sigma_M^{-\lambda+\gamma-l}(m(y,\xi))$, $l = 0,1,\ldots,k-1$, (which are defined by 2.3.4. for arbitrarily fixed y,ξ), are homogeneous polynomials in ξ of order $\mu-\gamma+\lambda+l$.

In the same way $\mathcal{R}^{(\mu)}(\mathcal{L}_{M+G}(Q,P))$ is defined.

If we set

$$h_l(y,\xi,z) = \sigma_M^{-\lambda+\gamma-l}(m(y,\xi))(z), \quad l=0,1,\ldots,k-1,$$

we get from the Definition 1

$$h_l \in C^\infty(\Omega, \mathcal{M}_{\mathcal{J}_l})[\xi]^{(\mu-\gamma+\lambda+l)} \tag{1}$$

for certain $\mathcal{J}_l \in \underline{\mathcal{J}}$, where $A[\xi]^{(k)}$ denotes the space of all polynomials in ξ, homogeneous of order k with coefficients in A. Choose $\gamma_l : \gamma - l \leq \gamma_l \leq \gamma$ such that $\mathcal{J}_l \in \underline{\mathcal{J}}[\gamma_l]$, i.e. $\pi_{\mathbb{C}}\mathcal{J}_l \cap \Gamma_{\frac{1}{2}-\gamma_l} = \emptyset$. Then

$$m(y,\xi) - t^\lambda \sum_{l=0}^{k-1} t^l \omega(t|\xi|) op_M^{\gamma_l}(h_l(y,\xi,z))\, \omega(t|\xi|) t^{-\gamma} \in \mathcal{R}^{(\mu)}(\mathcal{L}_G(Q,P)^{(-k)}). \tag{2}$$

We consider $\mathcal{R}^{(\mu)}(\mathcal{L}_{M+G}{}^{(-k)})$ in the topology induced from the Fréchet space topology on $C^\infty(\Omega, \mathcal{M}_{\mathcal{J}_l})[\xi]^{(\mu-\gamma+\lambda+l)}$ and the topology on $\mathcal{R}^{(\mu)}(\mathcal{L}_G(Q,P)^{(-k)})$. Then $\mathcal{R}^{(\mu)}(\mathcal{L}_{M+G}{}^{(-k)})$ becomes an (LF)-space.

Let $V \subset \mathbb{R}^{n-1}$ be open, $U := V \times \mathbb{R}_+ \subset \mathbb{R}^n_+$. We associate with $m(x',y',\xi') \in \mathcal{R}^{(\mu)}(\mathcal{L}_{M+G}(Q,P)^{(-k)})(V \times V \times (\mathbb{R}^{n-1} \setminus 0))$ an operator U in the following way: First we treat the Mellin terms and set for

$$h_l(x',y',\xi',z) = \sigma_M^{-\lambda+\gamma-l}(m(x',y',\xi'))(z), \ l=0,1,\ldots,k-1,$$

$$\mathrm{Op}'(\{h_l\})u(x',t) = \sum_{l=0}^{k-1} t^{l-\gamma+\lambda} \iint e^{i(x'-y')\xi'} \omega(t[\xi'])\, \times$$

$$\times\, \mathrm{op}_M^{\gamma_l}(h_l(x',y',\xi'))(\omega(s[\xi'])u(y',s))(t)dy'\,đ\xi' \quad , \tag{3}$$

where γ_l and ω are chosen as in (1) and the action of op_M refers to the variables s. If g denotes the Green boundary symbol defined by (2), we set

$$\mathrm{Op}(m) = \mathrm{Op}'(\{\sigma_M^{\gamma-\lambda-l}(m)\}_{l=0}^{k-1}) + \mathrm{Op}_\partial(\chi g) \tag{4}$$

where Op_∂ is defined as in 3.2.2., χ an excision function.

2. Proposition. Let $m \in \mathcal{R}^{(\mu)}(\mathcal{L}_{M+G}(Q,P)^{(-k)})(V\times V\times(\mathbb{R}^{n-1}\setminus 0))$ and $\mathrm{Op}(m)$ defined as above. Then $\mathrm{Op}(m)$ defines a continuous operator

$$\mathrm{Op}(m) : \mathcal{K}^s_{P,\mathrm{comp}}(U) \longrightarrow \mathcal{K}^{s-\mu}_{Q,\mathrm{loc}}(U)$$

for each $s \in \mathbb{R}$ and for each $P' = (\gamma+k,\varphi') \in \underline{P}$, $\Gamma(\gamma)\varphi' \leq \varphi$ there exists $R = (\lambda+k,\mathcal{N}) \in \underline{P}$, $R = R(m,\varphi')$ such that

$$\mathrm{Op}(m) : \mathcal{K}^s_{P',\mathrm{comp}}(U) \longrightarrow \mathcal{K}^{s-\mu}_{R^+,\mathrm{loc}}(U).$$

The choices made in the definition of Op(m) do not affect the resulting operator modulo $\mathcal{D}_G^{\mu,(l)}(Q,P)^{(-k)}$ for each $l \in \mathbb{Z}_+$.

Proof. We conclude exactly as in the proof of 3.2.2. Proposition 2. First we use the decomposition

$$m(x',y',\xi') = \sum_{l=0}^{\infty} a_i(x')m_i(\xi')b_i(y'),$$

$$a_i, b_i \in C^\infty(V), m_i \in \mathcal{R}^{(\mu)}(\mathcal{L}_{M+G}(Q,P)^{(-k)}(\mathbb{R}^{n-1}\setminus 0),$$

which reduces the problem to the constant coefficient case. Then the homogeneity and the mapping properties of $m_i(\xi')$ on $\mathbb{R}_+$ for fixed ξ' give the estimates

$$\|\kappa^{-1}(\xi')(t^{-\gamma+\lambda+l}\omega(t[\xi'])\mathrm{op}_M^{\gamma_l}(h_l(\xi'))(\omega(s[\xi']))\,\kappa(\xi')\|_{\mathcal{L}(\mathcal{K}^s_P(\mathbb{R}_+),\mathcal{K}^{s-\mu}_{R^+}(\mathbb{R}_+))}$$

$$\leq c\langle\xi'\rangle^\mu .$$

This gives the asserted continuity of $Op'(\{\sigma^{\gamma-\lambda-1}(m)\}_{l=0}^{k-1})$. For the second term we use 3.2.2. Proposition 2. Finally, in 2.3.4. it was proved that different choices of γ_1 and ω give for fixed ξ' only a difference of Green type on $\mathbb{R}_+$. In view of the second part of 3.2.2. Proposition 2 the last assertion follows. □

It is easy to check that Mellin operators defined by (4) possess a certain pseudolocality property. Let $\varphi, \psi \in C_0^\infty(V)$, $\varphi\psi = 0$. Then $\varphi Op(m)\psi \in \mathcal{D}_G^{\mu-k,(o)}(Q,P)^{(-k)}$. In fact, in $\varphi Op'(\{\sigma_M^{\gamma-\lambda-1}\}_{l=0}^{k-1})\psi$ we can introduce

$$e^{i(x'-y')\xi'} = \frac{1}{|x'-y'|^{2N}} (-\Delta_{\xi'})^N e^{i(x'-y')\xi'}.$$

For $N \geq \mu-\gamma+\lambda+k$ we get by partial integration a sum of integrals containing terms of the form

$$c(\xi')\,\omega^{(k')}(t[\xi'])\,op_M^\gamma(h(x',y',\xi'))\,\omega^{(k'')}(t[\xi']),$$

where $c \in S^{-l'}(\mathbb{R}^{n-1})$ and h is a polynomial in ξ' of order $\leq \mu-\gamma+\lambda+k$, $k'+k''+l' \leq N$, $k'+k''>0$. All these terms yield Green operators since $\omega^{(k')}(t) \in C_0^\infty(\mathbb{R}_+)$ for $k'>0$. Then the result follows from 3.2.2.

Obviously, $m(y,\xi) \in \mathcal{R}^{(\mu)}(\mathcal{L}_{M+G}{}^{(-k)})$ implies

$$D_y^\alpha D_\xi^\beta\, m(y,\xi) \in \mathcal{R}^{(\mu-|\beta|)}(\mathcal{L}_{M+G}{}^{(-k)}).$$

3. <u>Proposition.</u> (i) Let $m_j(y,\xi) \in \mathcal{R}^{(\mu_j)}(\mathcal{L}_{M+G}{}^{(-k)})$, $j=1,2$, be such that for arbitrary y,ξ the composition $m(y,\xi) := m_1(y,\xi)m_2(y,\xi)$ is defined. Then $m(y,\xi) \in \mathcal{R}^{(\mu_1+\mu_2)}(\mathcal{L}_{M+G}{}^{(-k)})$.
(ii) Let $P = (\gamma,0)$, $Q = (\lambda,0)$ and $m(y,\xi) \in \mathcal{R}^{(\mu)}(\mathcal{L}_{M+G}(Q,P)^{(-k)})$. Then the formal adjoint, taken pointwise, belongs to $\mathcal{R}^{(\mu)}(\mathcal{L}_{M+G}(P^*,Q^*)^{(-k)})$.

<u>Proof.</u> The smoothness and homogeneity are obvious. Moreover, $\mathcal{L}_{M+G}{}^{(-k)}$ is closed under compositions and taking adjoints provided they are defined. The symbolic calculus on $\mathbb{R}_+$ shows that also the property of

the Mellin symbols to be a polynomial in ξ is preserved. □

Besides Mellin boundary symbols which are homogeneous we need sums with different homogeneity orders.

4. <u>Définition.</u> A family of operators $m(y,\xi)$, $y\in\Omega$, $\xi\in\mathbb{R}^d\setminus 0$, is said to belong to $\mathcal{R}^{\mu,(1)}(\mathcal{L}_{M+G}(Q,P)^{(-k)})$ if it is a C^∞ family with values in $\mathcal{L}_{M+G}(Q,P)^{(-k)}$ and there exist homogeneous families $m_{\mu-j}(y,\xi)\in \mathcal{R}^{(\mu-j)}(\mathcal{L}_{M+G}(Q,P)^{(-k)})$, $j=0,1,\dots,l-1$, such that

$$m(y,\xi)-\sum_{j=0}^{l-1} m_{\mu-j}(y,\xi)\in\mathcal{R}^{\mu,(1)}(\mathcal{L}_G(Q,P)^{(-k)}).$$

Each $m(y,\xi)\in\mathcal{R}^{\mu,(1)}(\mathcal{L}_{M+G}(Q,P)^{(-k)})$ can be represented modulo $\mathcal{R}^{\mu,(1)}(\mathcal{L}_G(Q,P)^{(-k)})$ as

$$t^\lambda \sum_{\substack{0\leq j\leq 1\\ j\leq i\leq k}} t^i\,\omega(t[\xi])\operatorname{op}_M^{\gamma_{ij}}(h_{ij}(y,\xi,z))\omega(t[\xi])t^{-\gamma}, \tag{5}$$

where $0\geq\gamma_{ij}\geq -i$, $h_{ij}\in C^\infty(\Omega,\mathcal{M}_{\mathfrak{Z}_{ij}})[\xi]^{(1-j)}$, $\mathfrak{Z}_{ij}\in\underset{\sim}{\mathfrak{Z}}[\gamma_{ij}]$. Then

$$\sigma_M^{\gamma-\lambda}(m(y,\xi))(z)=h_{00}(y,\xi,z)$$

is the leading Mellin symbol. It is a homogeneous polynomial in ξ of order $\mu-\gamma+\lambda$. In the case $\mu=\gamma-\lambda$ it is independent of ξ.

The lower order Mellin symbols

$$\sigma_M^{\gamma-\lambda-i}(m(y,\xi))(z)=\sum_{0\leq j\leq i} h_{ij}(y,\xi,z),\quad 0<i<k,$$

are polynomials in ξ of degree $\leq\mu-\gamma+\lambda+i$, in general non-homogeneous.

We associate with

$$m(x',y',\xi')=\sum_{j=0}^{l-1} m_{\mu-j}(x',y',\xi')$$

$m\in\mathcal{R}^{\mu,(1)}(\mathcal{L}_{M+G}{}^{(-k)})(V\times V\times(\mathbb{R}^{n-1}\setminus 0))$ an operator on U by

$$\operatorname{Op}(m)=\sum_{j=0}^{l=1}\left[\operatorname{Op}'(\{\sigma_M(m_{\mu-j})\})+\operatorname{Op}_\gamma(\chi g_j)\right] \tag{6}$$

according to (3). From Proposition 2 we see immediately that the arbitrary choices in the definition of $\operatorname{Op}(m)$ do not affect the result modulo $\mathcal{W}_G^{\mu,(l)}$.

5. Definition. An operator $\mathcal{A}$ is said to belong to $\mathcal{W}_{M+G}^{\mu,(l)}(Q,P)^{(-k)}$ if there exists a boundary symbol $m \in \mathcal{R}^{\mu,(l)}(\mathcal{L}_{M+G}(Q,P)^{(-k)})$ such that

$$\mathcal{A} - \mathrm{Op}(m) \in \mathcal{W}_G^{\mu,(l)}(Q,P)^{(-k)}.$$

Considering $\mathcal{A} \in \mathcal{W}_{M+G}^{\mu,(l)}(Q,P)^{(-k)}$ as ψDO with operator-valued symbol we get well-defined mappings

$$\sigma_\partial^{(\mu-j)} : \mathcal{W}_{M+G}^{\mu,(l)}(Q,P)^{(-k)} \longrightarrow \mathcal{R}^{(\mu-j)}(\mathcal{L}_{M+G}(Q,P)^{(-k)}),\ j=0,1,\dots,l-1,$$

which evaluate the homogeneous components of the boundary symbol. Then the sequence

$$0 \longrightarrow \mathcal{W}_{M+G}^{\mu-1,(l-1)}(Q,P)^{(-k)} \longrightarrow \mathcal{W}_{M+G}^{\mu,(l)}(Q,P)^{(-k)} \longrightarrow$$

$$\xrightarrow{\sigma_\partial^{(\mu)}} \mathcal{R}^{(\mu)}(\mathcal{L}_{M+G}(Q,P)^{(-k)}) \longrightarrow 0 \qquad (7)$$

is exact. Restricting ourselves to the leading Mellin symbol we get the exact sequence

$$0 \longrightarrow \mathcal{W}_{M+G}^{\mu-1,(l-1)}(Q,P)^{(-k)} + \mathcal{W}_G^{\mu,(l)}(Q,P)^{(-k)} \hookrightarrow \mathcal{W}_{M+G}^{\mu,(l)}(Q,P)^{(-k)}$$

$$\xrightarrow{\sigma_M^{-\mu}} \mathfrak{M}_{\underline{z}[0]} \longrightarrow 0. \qquad (8)$$

This follows from the exactness of the sequence

$$0 \longrightarrow \mathcal{L}_G(Q,P)^{(-k)} \hookrightarrow \mathcal{L}_{M+G}(Q,P)^{(-k)} \xrightarrow{\sigma_M^{-\mu}} \mathfrak{M}_{\underline{z}[0]} \longrightarrow 0.$$

By induction we get also the exactness of the sequence

$$0 \longrightarrow \mathcal{W}_G^{\mu-1,(o)}(Q,P)^{(-k)} \hookrightarrow \mathcal{W}_{M+G}^{\mu,(l)}(Q,P)^{(-k)}$$

$$\xrightarrow{\overset{l-1}{\underset{j=0}{\times}} \sigma_\partial^{(\mu-j)}} \overset{l-1}{\underset{j=0}{\times}} \mathcal{R}^{(\mu-j)}(\mathcal{L}_{M+G}(Q,P)^{(-k)}) \longrightarrow 0 \quad .$$

Repeating the arguments of the proof of 3.2.2. Proposition 3 we get the following mapping properties.

6. Proposition. Each $\mathcal{A} \in \mathcal{W}^{\mu,(1)}_{M+G}(Q,P)^{(-k)}$ defines a continuous linear mapping

$$\mathcal{A} : \mathcal{K}^{s}_{P,\mathrm{comp}}(U) \longrightarrow \mathcal{K}^{s-\mu}_{Q,\mathrm{loc}}(U)$$

for each $s \in \mathbb{R}$. Moreover, the restriction to $\mathcal{K}^{s}_{P',\mathrm{comp}}(U) \subset \mathcal{K}^{s}_{P,\mathrm{comp}}(U)$ gives a continuous operator

$$\mathcal{A} : \mathcal{K}^{s}_{P',\mathrm{comp}}(U) \longrightarrow \mathcal{K}^{s-\mu}_{R^{+},\mathrm{loc}}(U) ,$$

where $P' = (\gamma', \mathcal{P}')$, $R = (r, \mathcal{R})$, $\gamma' - \gamma = r - \lambda \geq -k$.

From Proposition 3 we get that the $\#$ product defined by 3.2.2. (8) defines a mapping

$$\# : \mathcal{R}^{\mu_1,(1)}(\mathcal{L}_{M+G}{}^{(-k)}) \times \mathcal{R}^{\mu_2,(1)}(\mathcal{L}_{M+G}{}^{(-k)}) \longrightarrow \mathcal{R}^{\mu_1+\mu_2,(1)}(\mathcal{L}_{M+G}{}^{(-k)}).$$

Moreover, the mapping $*$ defined by 3.2.2. (9) yields

$$* : \mathcal{R}^{\mu,(1)}(\mathcal{L}_{M+G}(Q,P)^{(-k)}) \longrightarrow \mathcal{R}^{\mu,(1)}(\mathcal{L}_{M+G}(P^{*},Q^{*})^{(-k)}).$$

This follows immediately from 3.2.2. and the fact that derivatives adjoints and compositions of homogeneous boundary symbols of Mellin type are again of this type (provided that they are defined).

Generalizing 3.2.2. Theorem 12 we get

7. Theorem. (i) Let $\mathcal{A}_j \in \mathcal{W}^{\mu_j,(1)}_{M+G}(\,.\,)^{(-k)}$ with boundary symbols $a_j \in \mathcal{R}^{\mu_j,(1)}(\mathcal{L}_{M+G}{}^{(-k)})$, $j=1,2$, such that $a_1 \# a_2$ is defined. Then the operator $\mathcal{A} := \mathcal{A}_1 \varphi \mathcal{A}_2 \in \mathcal{W}^{\mu_1+\mu_2,(1)}_{M+G}(\,.\,)^{(-k)}$ for each $\varphi \in C^{\infty}_{o}(V)$ and $\mathcal{A}$ has the boundary symbol $a_1 \# (\varphi a_2)$.

(ii) Let $P = (\gamma, 0)$, $Q = (\lambda, 0)$ and $\mathcal{A} \in \mathcal{W}^{\mu,(1)}_{M+G}(Q,P)^{(-k)}$ with boundary symbol $a \in \mathcal{R}^{\mu,(1)}(Q,P)^{(-k)}$. Then $\mathcal{A}^{*} \in \mathcal{W}^{\mu,(1)}_{M+G}(P^{*},Q^{*})^{(-k)}$ and $a^{*} \in \mathcal{R}^{\mu,(1)}(P^{*},Q^{*})^{(-k)}$ is its boundary symbol.

Proof. (i) As noted above $a := a_1 \# (\varphi a_2) \in \mathcal{R}^{\mu_1+\mu_2,(1)}(\mathcal{L}_{M+G}{}^{(-k)})$ and we set $\mathcal{A}_1 := \mathrm{Op}(a)$. By the general theory of ψDO's with operator-valued symbols we have $\mathcal{A} - \mathcal{A}_1 \in \mathcal{W}^{\mu_1+\mu_2-1,(o)}_{G}(\,.\,)^{(-k)}$, which proves (i). (ii) is treated in the same way. $\square$

In the same way the invariance of $\mathcal{M}^{\mu,(1)}_{M+G}(\,.\,)^{(-k)}$ under tangential changes of coordinates is proved. Therefore the class on a manifold with boundary can by defined by means of an arbitrary atlas.

8. Remark. By inserting $\mathcal{G}_{M+G}(Q,P)^{(-k)}$ (cf. Section 2.3.4.) in Definition 5, we also can define the space $\mathcal{M}^{\mu,(1)}_{M+G}(Q,P)^{(-k)}$ of Mellin and Green operators with continuous asymptotics.

The details are obvious and left to the reader.

3.2.4. The class $\mathcal{M}^{\mu}$

Now we extend the class $\mathcal{M}^{\mu}_{M+G}$ by ψDO's of order μ and define in this way $\mathcal{M}^{\mu}$.

First we specify the admitted ψDO symbols.

1. Definition. Denote by $S^{\mu}_{cl}(\overline{\mathbb{R}^n_+}\times\mathbb{R}^n)$ the subspace of $S^{\mu}_{cl}(\mathbb{R}^n_+\times\mathbb{R}^n)$ of all symbols $a(x',t,\xi',\tau)$ such that the homogeneous components $a_{\mu-j}(x',t,\xi',\tau)$ belong to $C^{\infty}(\overline{\mathbb{R}^n_+}\times(\mathbb{R}^n\setminus 0))$ and for all $K\subset\subset\overline{\mathbb{R}^n_+}$, $\alpha,\beta\in\mathbb{Z}^n_+$, $N\in\mathbb{Z}_+$ there are constants $c=c(K,\alpha,\beta,N)$ such that

$$|D^{\alpha}_{x',t}D^{\beta}_{\xi',\tau}(a(x',t,\xi',\tau)-\sum_{j=0}^{N-1}a_{\mu-j}(x',t,\xi',\tau)\chi(\xi',\tau))|\leq c\langle\xi',\tau\rangle^{\mu-N-|\beta|}$$

for all $(x',t)\in K$, $(\xi',\tau)\in(\mathbb{R}^n\setminus 0)$.

Note that we do not assume the transmission property.

Each symbol in $S^{\mu}_{cl}(\mathbb{R}^n_+\times\mathbb{R}^n)$ gives rise to a continuous operator in $\mathcal{L}(C^{\infty}_0(\mathbb{R}^n_+),C^{\infty}(\mathbb{R}^n_+))$ defined by the usual formula composed with extension by zero of the argument function and restriction of the result to the open half space. Based on the discussion in 2.3.5. of ψDO's on the half line we indicate now how to define an operator in $\mathcal{L}(\mathcal{K}^s_P(\mathbb{R}^n_+),$ $\mathcal{K}^s_Q(\mathbb{R}^n_+))$ having the given interior symbol in $S^{\mu}_{cl}(\overline{\mathbb{R}^n_+}\times\mathbb{R}^n)$.

First we consider symbols independent of t. The general case will be treated later. Consider $a(x',\xi',\tau)\in S^{\mu}_{cl}(\overline{\mathbb{R}^n_+}\times\mathbb{R}^n)$ and let $P=(\gamma,\varphi)\in\underline{P}$,

$Q = (\lambda, \eta) \in \underline{P}$ be given. Assume that $\lambda = \gamma - \mu$ and γ is admissible, i.e. $\gamma - \mu = \frac{1}{2} \bmod \mathbb{Z}_+, -\gamma = \frac{1}{2} \bmod \mathbb{Z}_+$. The smooth mapping

$$\mathbb{R}^{n-1} \times \mathbb{R}^{n-1} \ni (x', \xi') \longrightarrow a(x', \xi', .) \in S^{\mu}_{cl}(\mathbb{R})$$

composed with a (non-canonical) operator convention

$$op_{\psi} : S^{\mu}_{cl}(\mathbb{R}) \longrightarrow \mathcal{L}(\mathcal{K}^s_P(\mathbb{R}_+), \mathcal{K}^{s-\mu}_Q(\mathbb{R}_+))$$

defines a smooth family $op_{\psi}(a)(x', \xi')$ of operators in $\mathcal{L}(\mathcal{K}^s_P(\mathbb{R}_+), \mathcal{K}^{s-\mu}_Q(\mathbb{R}_+))$ which plays the role of a boundary symbol. Note that the choice of op_{ψ} is unique modulo Green operators on $\mathbb{R}_+$.

Assuming in addition that $a(x', \xi', \tau)$ vanishes for x' outside a compact set we define

$$Op(a)u(x',t) = \int e^{ix'\xi'} op_{\psi}(a)(x', \xi')(F'u)(\xi', .)(t)\, đ\xi'. \qquad (1)$$

It is easily seen that $Op(a) \in \mathcal{L}(\mathcal{K}^s_P(\mathbb{R}^n_+), \mathcal{K}^{s-\mu}_Q(\mathbb{R}^n_+))$. In fact, (1) means that $Op(a)$ is a ψDO on $\mathbb{R}^{n-1}$ with operator-valued symbol $op_{\psi}(a)(x', \xi')$. Then the estimates

$$\| D^{\alpha}_{x'} D^{\beta}_{\xi'} op_{\psi}(a)(x', \xi') \|_{\mathcal{L}(\mathcal{K}^s_P(\mathbb{R}_+)(\xi'), \mathcal{K}^{s-\mu}_Q(\mathbb{R}_+)(\xi'))} \le c \langle \xi' \rangle^{\mu - |\beta|} \qquad (2)$$

(here $\mathcal{K}^s_P(\mathbb{R}_+)(\xi')$ denotes the space $\mathcal{K}^s_P(\mathbb{R}_+)$ with the ξ'-depending norm introduced in 3.1.1.) and the compact support in x' imply the asserted continuity (cf. the proof of 3.2.2. Proposition 2). Clearly, different choices of op_{ψ} change $Op(a)$ by an operator in $\mathcal{D}^{\mu}_G$.

The estimates (2) follow from the homogeneity of $op_{\psi}(a_{\mu-j})$ and the estimates of the remainder in Definition 1. In view of 2.3.5. we observe that the action of $Op(a)$ as operator in $\mathcal{L}(\mathcal{K}^s_P(\mathbb{R}^n_+), \mathcal{K}^{s-\mu}_Q(\mathbb{R}^n_+))$ can be considered as the extension from C^{∞}_0 of the canonical ψDO modified by a Green operator in $\mathcal{D}^{\mu}_G$.

Let $a \in S^{\mu}_{cl}(\overline{\mathbb{R}^n_+} \times \mathbb{R}^n)$, now possibly depending on t. Then we have $op_{\psi}(a)(t, x', \xi')$, which denotes the smooth family of operators

$$\mathbb{R}^{n-1}\times\mathbb{R}^{n-1} \ni (x', \xi') \longrightarrow op_\psi a(t,x',\xi') \in \mathcal{L}(\mathcal{K}^s_P(\mathbb{R}_+), \mathcal{K}^{s-\mu}_Q(\mathbb{R}_+))$$

by means of $a(t,x',\xi',\tau)$ (cf. 2.3.5.). Again we get that if a vanishes for x' outside a compact set,

$$Op(a)u(x',t) = \int e^{ix'\xi'} op_\psi(a)(t,x',\xi')(F'u)(\xi',.)(t)\, đ\xi' \qquad (1')$$

yields an operator in $\mathcal{L}(K^s_P(\mathbb{R}^n_+), \mathcal{J}^{s-\mu}_Q(\mathbb{R}^n_+))$ which is a classical ψDO on $\mathbb{R}^n_+$ with a as its symbol. In contrast to the t-independent case it is not reasonable to look at the family $op_\psi(a)(t,x',\xi')$ as a boundary symbol. The latter should not depend on the values of a for $t \geq c > 0$, $\forall c > 0$. Moreover, boundary symbols split into homogeneous components which will be influenced only by the homogeneous components of the interior symbol.

2. <u>Definition.</u> Denote by $\mathcal{O}^{(\mu)}(\mathcal{L}^\mu(Q,P)^{(-k)})(\Omega\times(\mathbb{R}^{n-1}\setminus 0))$ the space of all families $\alpha(y,\xi')$, $y\in\Omega$, $\xi'\in\mathbb{R}^{n-1}\setminus 0$, that are C^∞ with values $\mathcal{L}^\mu(Q,P)^{(-k)}(\Omega\times(\mathbb{R}^{n-1}\setminus 0))$, homogeneous with respect to ξ' of degree μ and such that there exists an $a\in S^{(\mu)}(\Omega\times(\mathbb{R}^n\setminus 0))$ with the property

$$op_\psi(a)(y,\xi') - \alpha(y,\xi') \in \mathcal{O}^{(\mu)}(\mathcal{L}_{M+G}(Q,P)^{(-k)})(\Omega\times(\mathbb{R}^{n-1}\setminus 0)),$$

where op_ψ denotes any operator convention.

It is clear that the choice of op_ψ does not influence whether or not α belongs to $\mathcal{O}^{(\mu)}(\mathcal{L}^\mu(Q,P)^{(-k)})$. Remark that each smooth family $\alpha(y,\xi')$ with values in $\mathcal{L}^\mu(Q,P)^{(-k)}$ gives rise to a ψDO symbol $a(y,\xi',t,\tau)$. Definition 2 demands in addition t-independence. Then the homogeneity of the operator family implies that $a(y,\xi',\tau)$ is homogeneous of degree μ in (ξ',τ). Another assumption is that this function is smooth including $\xi'=0$, $\tau=\pm 1$.

For $a\in S^\mu_{cl}(\overline{\mathbb{R}^n_+}\times\mathbb{R}^n)$ we obtain $op_\psi(a_\mu)(x',0,\xi') \in \mathcal{O}^{(\mu)}(\mathcal{L}^\mu(Q,P)(\mathbb{R}^{n-1}\times(\mathbb{R}^{n-1}\setminus 0))$. Denote the mapping $a \longrightarrow op_\psi(a_\mu)(x',0,\xi')$ by

$$\sigma^{(\mu)}_\partial : S^\mu_{cl}(\overline{\mathbb{R}^n_+}\times\mathbb{R}^n) \longrightarrow \mathcal{O}^{(\mu)}(\mathcal{L}^\mu(Q,P))(\mathbb{R}^{n-1}\times(\mathbb{R}^{n-1}\setminus 0))$$

which depends, of course, on the choice of op_ψ.
Conversely, given $\sigma \in \mathcal{R}^{(\mu)}(\mathcal{L}^\mu(Q,P)^{(-k)})(\Omega \times (\mathbb{R}^{n-1} \setminus 0))$ implies that $a_\mu \in S^{(\mu)}(\Omega \times (\mathbb{R}^n \setminus 0))$ is uniquely determined by the condition of Definition 2. This follows since the interior symbol of an operator in $\mathcal{L}^\mu$ is uniquely determined provided it is t-independent.

Denote by $\mathcal{V}^{(\mu)}(Q,P)^{(-k)}(\Omega \times (\mathbb{R}^{n-1} \setminus 0))$ the space of all pairs $(a_\mu, \sigma_\mu) \in S^{(\mu)}(\Omega \times (\mathbb{R}^n \setminus 0)) \times \mathcal{R}^{(\mu)}(\mathcal{L}^\mu(Q,P)^{(-k)})(\Omega \times (\mathbb{R}^{n-1} \setminus 0))$ being compatible in the sense that they appear together in Definition 2.

According to the above remarks both projections to the first and to the second factor

$$pr_1 : \mathcal{V}^{(\mu)} \longrightarrow S^{(\mu)}, \quad pr_2 : \mathcal{V}^{(\mu)} \longrightarrow \mathcal{R}^{(\mu)}$$

are surjective.

With any given $\sigma \in \mathcal{V}^{(\mu)}(Q,P)^{(-k)}(\mathbb{R}^{n-1} \times (\mathbb{R}^{n-1} \setminus 0))$ we associate an operator $Op(\sigma)$ with σ in the following way: Above it was explained how to get an extension of the ψDO with given interior symbol (take $a_\mu(x',t,\xi',\tau)\chi(\xi',\tau)$, χ an excision function and choose op_ψ). Then it remains to work with $\sigma - op_\psi(a_\mu \chi) \in \mathcal{R}^{(\mu)}(\mathcal{L}_{M+G}(Q,P)^{(-k)})$. The operator convention for such boundary symbols was introduced in 3.2.3. Provided that the interior and the boundary symbol stabilize for large (x',t) and x', respectively, we get

$$Op(\sigma) : \mathcal{K}_P^s(\mathbb{R}_+^n) \longrightarrow \mathcal{K}_Q^{s-\mu}(\mathbb{R}_+^n).$$

3. <u>Definition.</u> An operator $\mathcal{A}$ is said to belong to $\mathcal{W}^{\mu,(1)}(Q,P)^{(-k)}$ if there exists $\sigma \in \mathcal{V}^{(\mu)}(Q,P)^{(-k)}$ such that $\mathcal{A} - Op(\sigma) \in \mathcal{W}_G^{\mu-1,(0)}(Q,P)^{(-k)}$ for some choice of the Op-convention. An operator $\mathcal{A}$ is said to belong to $\mathcal{W}^{\mu,(l)}(Q,P)^{(-k)}$ if there exists $\sigma \in \mathcal{V}^{(\mu)}(Q,P)^{(-k)}$ such that $\mathcal{A} - Op(\sigma) \in \mathcal{W}^{\mu-1,(l-1)}(Q,P)^{(-k)}$.

It is clear that the choice of the Op-convention has no influence on the question whether σ is a symbol of $\mathcal{A}$ because both the principal interior and the principal boundary symbol are uniquely determined by $\mathcal{A}$. In fact, for the interior symbol we can apply the usual procedure, since in the interior Mellin and Green operators are smoothing. For the principal boundary symbol we apply the modification for operator-valued symbols.

Operators from the class $\mathcal{M}^{\mu,(l)}(Q,P)^{(-k)}$ possess l homogeneous boundary symbols which can be recovered successively. The compatibility conditions between interior and boundary symbol components are more complicated for the lower order terms. The boundary symbol of homogeneity order $\mu-l$ is affected by all $\partial_t^j a_{\mu-i}|_{t=0}$, $i+j \leq l$. By induction over l we can avoid to work explicitly with the mentioned compatibility conditions.

We have the exact sequence

$$0 \longrightarrow \mathcal{M}^{\mu-1,(l-1)}(Q,P)^{(-k)} \hookrightarrow \mathcal{M}^{\mu,(l)}(Q,P)^{(-k)} \xrightarrow{\sigma^{(\mu)}} \mathcal{S}^{(\mu)}(Q,P)^{(-k)} \to 0.$$

If $\mathcal{A} \in \mathcal{M}^{\mu,(l)}$ and $a_{\mu-j} \in S^{(\mu-j)}$, $\alpha_{\mu-j} \in \mathcal{R}^{(\mu-j)}$, $j=0,1,\ldots,l-1$, are the homogeneous components of the interior and boundary symbol, then we denote the mapping $\mathcal{A} \longrightarrow \{(a_{\mu-j}, \alpha_{\mu-j})\}_{j=0}^{l-1}$ by $\sigma^{\mu,(l)}$. Let $\mathcal{S}^{\mu,(l)}$ be the image of $\sigma^{\mu,(l)}$

$$\sigma^{\mu,(l)} : \mathcal{M}^{\mu,(l)}(Q,P)^{(-k)} \longrightarrow \mathcal{S}^{\mu,(l)}(Q,P)^{(-k)}.$$

Then the sequence

$$0 \longrightarrow \mathcal{M}_G^{\mu-1,(0)}(Q,P)^{(-k)} \hookrightarrow \mathcal{M}^{\mu,(l)}(Q,P)^{(-k)} \xrightarrow{\sigma^{\mu,(l)}} \mathcal{S}^{\mu,(l)}(Q,P)^{(-k)} \longrightarrow 0$$

is exact, too.

The mapping properties of operators in $\mathcal{M}^{\mu,(l)}(Q,P)^{(-k)}$ are similar to those of Mellin operators.

4. Proposition. Each $\mathcal{A} \in \mathcal{M}^{\mu,(l)}(Q,P)^{(-k)}$ defines a continuous linear mapping

$$\mathcal{A} : \mathcal{K}^{s}_{P,\mathrm{comp}}(U) \longrightarrow \mathcal{K}^{s-\mu}_{Q,\mathrm{loc}}(U)$$

for each $s \in \mathbb{R}$. Moreover, the restriction to $\mathcal{K}^s_{P',comp}(U) \subset \mathcal{K}^s_{P,comp}(U)$ gives a continuous operator

$$\mathcal{A} : \mathcal{K}^s_{P',comp}(U) \longrightarrow \mathcal{K}^{s-\mu}_{R^+,loc}(U)$$

where $P' = (\gamma', \varphi'), R = (r, \mathfrak{r}), \gamma' - \gamma = r - \lambda \geq -k$.

Proof. In view of 3.2.3. Proposition 6 it is sufficient to show the stated properties for $Op(\sigma)$, $\sigma \in \mathcal{S}^{(\mu)}(Q,P)^{(-k)}$. The first statement is fulfilled by construction. The second one follows in view of the Mellin expansion of op_ψ. $\square$

5. Theorem. (i) Let $\mathcal{A} \in \mathcal{D}^{\mu,(1)}(Q,P)^{(-k)}$, $\mathcal{B} \in \mathcal{D}^{\nu,(1)}(Q',Q)^{(-k)}$, one of them property supported. Then $\mathcal{B}\mathcal{A} \in \mathcal{D}^{\mu+\nu,(1)}(Q',P)^{(-k)}$ and for the principal symbols we have

$$\sigma_\psi^{(\mu+\nu)}(\mathcal{B}\mathcal{A}) = \sigma_\psi^{(\nu)}(\mathcal{B}) \cdot \sigma_\psi^{(\mu)}(\mathcal{A}),$$

$$\sigma_\partial^{(\mu+\nu)}(\mathcal{B}\mathcal{A}) = \sigma_\partial^{(\nu)}(\mathcal{B}) \circ \sigma_\partial^{(\mu)}(\mathcal{A}),$$

($\circ$ denotes the composition of operators on $(\mathbb{R}_+, \partial\mathbb{R}_+)$).

(ii) Let $\mathcal{A} \in \mathcal{D}^{\mu,(1)}(Q,P)^{(-k)}$, $P = (\gamma, 0)$, $Q = (\gamma - \mu, 0)$. Then $\mathcal{A}^* \in \mathcal{D}^{\mu,(1)}(P^*,Q^*)^{(-k)}$ and

$$\sigma_\psi^{(\mu)}(\mathcal{A}^*) = \overline{\sigma_\psi^{(\mu)}(\mathcal{A})},$$

$$\sigma_\partial^{(\mu)}(\mathcal{A}^*) = \sigma_\partial^{(\mu)}(\mathcal{A})^*.$$

Proof. (ii) In view of 3.2.3. Theorem 7 (ii) it remains to consider a ψDO. Since $S^\mu_{cl}(\overline{\mathbb{R}^n_+} \times \mathbb{R}^n)$ is closed under the $*$ operation, the formal adjoint is a ψDO in the considered class. It possesses the formal adjoint principal boundary symbol, which proves that $\mathcal{A}^* - Op(\overline{\sigma_\psi^{(\mu)}(\mathcal{A})}, \sigma_\partial^{(\mu)}(\mathcal{A})^*) \in \mathcal{D}^{\mu-1,(1-1)}(P^*,Q^*)^{(-k)}$. Iterating this conclusion we get the assertion.

(i) First note that the composition rule of principal symbols preserves the compatibility between interior and boundary principal symbols.

In fact, the leading interior symbol at the boundary can be obtained as the (constant coefficient) complete symbol of the leading boundary symbol. Then $\mathcal{BA} - \mathrm{Op}(\sigma_\psi^{(\nu)}(\mathcal{B})\sigma_\psi^{(\mu)}(\mathcal{A}), \sigma_\partial^{(\nu)}(\mathcal{B})\sigma_\partial^{(\mu)}(\mathcal{A})) \in$ $\in \mathcal{D}^{\mu+\nu-1,(l-1)}(Q',Q)^{(-k)}$. Repeating the conclusion $l-1$ times we get the statement. □

Note that the composition rule for the Mellin symbols is implicitly contained in the composition of boundary symbols.

Let X be a C^∞ manifold with boundary $Y = \partial X$. Then the class $\mathcal{D}^{\mu,(l)}(Q,P)^{(-k)}(X,X)$ is defined in the usual way by means of a partition of unity and local coordinates. We continue to use again the above notations from the local situation indicating, if necessary, the manifold X. Remark that the corresponding classes $\mathcal{D}^{\mu,(l)}(.)(X,\partial X)$, $\mathcal{D}^{\mu,(l)}(.)(\partial X,X)$ and $\mathcal{D}^{\mu,(l)}(.)(\partial X,\partial X)$ coincide with the corresponding Green classes.

6. Remark. The definition of our operator classes on a manifold with boundary, the boundary symbols and so on depend on a choice of a normal to the boundary. It can be checked that the classes do not depend on this choice. Of course, the homogeneous components of the boundary symbol depend on this choice. However, the transformation of the leading boundary symbol is the same as in Boutet de Monvel's class.

7. Remark. A simple modification of Definition 3 admits also to define the version of the class with continuous asymptotics $\mathcal{MD}^{\mu,(l)}(Q,P)^{(-k)}$, and we then get an analogous calculus as for $\mathcal{D}^{\mu,(l)}(Q,P)^{(-k)}$ (cf. also Remark 8).

3.2.5. Ellipticity and Fredholm property

After the long preparations in the previous sections we come now to the main results. It turns out that the theory of the $\mathcal{D}$-classes (with arbitrary orders) is as comfortable as the Boutet de Monvel class of boundary value problems with the transmission property.

We consider (systems of) operators acting between sections of vector bundles over X and $Y = \partial X$. For the sake of simplicity we assume that the matrix of ψDO's in the interior consists of elements of the same orders. The orders of the boundary conditions can easily be transformed also to this order by a composition by an invertible ψDO on Y.

Let $E, F \in \mathrm{Vect}(X), J, G \in \mathrm{Vect}(Y)$, $\mu \in \mathbb{R}, k, l \in \mathbb{N}, P, Q \in \underline{P}$ and consider $\mathcal{D}^{\mu,(1)}(Q,P)^{(-k)}(X, Y; E, F, J, G)$ (below we shall indicate the bundles only if a confusion is to be avoided).

1. <u>Definition.</u> A symbol $\sigma = (a, \alpha) \in \mathcal{S}^{(\mu)}(Q,P)^{(-k)}$ is called elliptic iff a is bijective outside the zero section and in addition

$$\alpha(x', \xi') : \begin{matrix} E_{x'} \otimes \mathcal{K}_P^s(\mathbb{R}_+) \\ \oplus \\ J_{x'} \end{matrix} \longrightarrow \begin{matrix} F_{x'} \otimes \mathcal{K}_Q^{s-\mu}(\mathbb{R}_+) \\ \oplus \\ G_{x'} \end{matrix} \qquad (1)$$

is bijective for all $(x', \xi') \in T^*Y \setminus 0$.

Note that the assumption about the boundary symbol α implies that the left upper corner is a family of Fredholm operators. Then it follows that the leading Mellin symbol $\sigma_M^{\mu}(\alpha)(x', z) : E_{x'} \to F_{x'}$ is bijective for each $z \in \Gamma_{\frac{1}{2}-\gamma}$, $P = (\gamma, \mathcal{P})$.

An operator $\mathcal{A} \in \mathcal{D}^{\mu,(1)}(Q,P)^{(-k)}$ is called elliptic if its principal symbol $\sigma^{(\mu)}(\mathcal{A}) \in \mathcal{S}^{(\mu)}(Q,P)^{(-k)}$ is elliptic.

2. <u>Proposition.</u> Let $\sigma = (a, \alpha) \in \mathcal{S}^{(\mu)}(Q,P)^{(-k)}$ be elliptic and set $\sigma^{-1} := (a^{-1}, \alpha^{-1})$ (i.e. the inverse mappings for fixed parameter values). Then $\sigma^{-1} \in \mathcal{S}^{(-\mu)}(P,Q)^{(-k)}$.

<u>Proof.</u> Clearly, $a^{-1} \in S^{(-\mu)}$ is smooth up to the boundary. For fixed x', ξ' we have that $\alpha(x', \xi')$ belongs to $\mathcal{L}^{\mu}(Q,P)^{(-k)}$ and is bijective as mapping (1) for each $s \in \mathbb{R}$. It was proved in 2.3.6. that $\alpha^{-1}(x', \xi') \in \mathcal{L}^{-\mu}(P,Q)^{(-k)}$. Moreover, $\sigma_M^{-\mu}(\alpha^{-1}(x', \xi'))(z)$

$= [\sigma_M^{\mu}(\mathfrak{a}(x', \xi'))]^{-1}(z)$ is independent of ξ', smoothly depending on x' with values in $\sigma_M^{-\mu}(op_\psi(a^{-1}(x',0, \xi',.))) + \mathfrak{M}_{\bar{\gamma}[\gamma]}$, which shows at the same time the compatibility condition. In the same way we get for the lower Mellin symbols also a smooth dependence on x' and a polynomial dependence on ξ'. Then $\mathfrak{g}(x', \xi') := \mathfrak{a}^{-1}(x', \xi')$ $- op_\psi(a^{-1}(x',0, \xi',.)) - m_1(x', \xi')$ is a smooth family with values in $\bigcap_s \mathcal{L}(F_{x'}\otimes\mathcal{K}_Q^{s-\mu}(\mathbb{R}_+)\oplus G_{x'}, E_{x'}\otimes\mathcal{K}_P^{s}(\mathbb{R}_+)\oplus J_{x'})$ and it belongs to $\mathcal{L}_G^{(-k)}$ for all x', ξ'. Here $m_1(x', \xi')$ denotes a family of Mellin operators obtained from the inverse sequence of Mellin symbols. Then

$$\mathfrak{g}(x', \xi')\mathfrak{a}(x', \xi') = 1-(op_\psi(a^{-1}(x',0, \xi',.))+m_1(x', \xi'))$$

is a smooth family with values in $\mathcal{L}_G^{(-k)}$. Together with the smoothness of $\mathfrak{a}^{-1}(x', \xi')$ as a family of operators in $\bigcap_s \mathcal{L}(F_{x'}\otimes\mathcal{K}_Q^{s-\mu}(\mathbb{R}_+)\oplus G_{x'}, E_{x'}\otimes\mathcal{K}_P^{s}(\mathbb{R}_+)\oplus J_{x'})$ it follows that $\mathfrak{g}$ is a smooth family with values in $(1)-\mathcal{L}_G^{(-k)}$. Taking into account the adjoint (of the regular part) we get that $\mathfrak{g}$ is smooth with values in $\mathcal{L}_G^{(-k)}$. Then $\mathfrak{a}^{-1}$ is smooth with values in $\mathcal{L}^{-\mu}(Q,P)^{(-k)}$. $\square$

Remark that another variant of proving Proposition 2 can be given in the spirit of the proof of [R1] for Boutet de Monvel's class.

In fact, it is not hard to show that there is a smooth family $\mathfrak{g}_1(x', \xi')$ with values in $\mathcal{L}_G^{(-k)}$ such that $\mathfrak{a}_1(x', \xi') := op_\psi(a^{-1}(x',0, \xi')) + m_1(x', \xi') + \mathfrak{g}_1(x', \xi')$ is bijective for all x', ξ'. Then it is sufficient to show that the inverse of $\mathfrak{a}(x', \xi')\mathfrak{a}_1(x', \xi') = 1+\mathfrak{g}_2(x', \xi')$ has the form $1 + \mathfrak{g}_3(x', \xi')$ where $\mathfrak{g}_3(x', \xi')$ is a smooth family with values in $\mathcal{L}_G^{(-k)}$, too, because then $\mathfrak{a}^{-1}(x', \xi') = \mathfrak{a}_1(x', \xi')(1 +\mathfrak{g}_2(x', \xi'))^{-1}$ $= \mathfrak{a}_1(x', \xi')(1 + \mathfrak{g}_3(x', \xi'))$, (cf. Section 2.3.2).

3. <u>Theorem.</u> Let $\mathcal{A}\in\mathcal{W}^{\mu,(1)}(Q,P)^{(-k)}$ be elliptic. Then, for each $s\in\mathbb{R}$ the mapping

$$\mathcal{A} : \mathcal{K}^s_P(X,E) \oplus H^s(Y,J) \longrightarrow \mathcal{K}^{s-\mu}_Q(X,F) \oplus H^{s-\mu}(Y,G)$$

is Fredholm and $\mathcal{A}^{(-1)} = \mathrm{Op}(\sigma^{(\mu)}(\mathcal{A})^{-1}) \in \mathcal{D}^{-\mu,(1)}(P,Q)^{(-k)}$ is a parametrix.

Proof. By construction we have $1-\mathcal{A}\mathcal{A}^{(-1)}, 1-\mathcal{A}^{(-1)}\mathcal{A} \in \mathcal{D}^{-1,(1-1)}$, which gives a compact operator in view of the compact embeddings

$$\mathcal{K}^{s+1}_{R^+}(X) \hookrightarrow \mathcal{K}^s_Q(X), R = (\lambda,\mathcal{r}),\ Q = (\lambda',q),\ \lambda > \lambda', H^{s+1}(Y) \hookrightarrow H^s(Y).\ \square$$

In the same way as in the case of the class with transmission property one can prove that the ellipticity is also necessary for the Fredholm property in the spaces under consideration.

As usually the Fredholm property depends only on the principal interior and boundary symbol. Inverting the principal symbols only we get a parametrix $\mathcal{P}$ in the sense that $\mathcal{K}_1 := 1-\mathcal{A}\mathcal{P}$ and $\mathcal{K}_2 := 1-\mathcal{P}\mathcal{A}$ has order -1. We can easily find a finer parametrix $\mathcal{R} \in \mathcal{D}^{-\mu,(1)}(P,Q)^{(-k)}$ such that $1-\mathcal{A}\mathcal{R}$ and $1-\mathcal{R}\mathcal{A}$ belong to $\mathcal{D}_G^{-1,(o)}$.

Set $\mathcal{R} = \mathcal{P} \sum_{j=0}^{l-1} \mathcal{K}_1^j$. Clearly $\mathcal{R} \in \mathcal{D}^{-\mu,(1)}(P,Q)^{(-k)}$ and the components of the interior and the boundary symbol of homogeneity degree $\mu, \mu-1, \dots, \mu-l+1$ of $1-\mathcal{R}\mathcal{A}$ and $1-\mathcal{A}\mathcal{R}$ vanish. We call $\mathcal{R}$ a parametrix modulo Green operators without boundary symbols.

Clearly, the Fredholm property does not depend on the smoothness order s. This is in sharp contrast to the case when usual Sobolev spaces are considered (cf. [R1]).

On the other hand, a realization of the given boundary symbol in other spaces (e.g. by shrinking the domain of definition or enlarging the target space e.g., by modifying the weights) will affect, in general, the Fredholm property, since even the action on the half-line may change the index.

Now we can formulate the regularity theorem for solutions of elliptic boundary value problems.

4. Theorem. Let $\mathcal{A} \in \mathcal{D}^{\mu,(1)}(Q,P)^{(-k)}$, $P = (\gamma,p)$, $Q = (\gamma-\mu,q)$, be elliptic and $u \in \mathcal{K}^s_P(X,E) \oplus H^s(Y,J)$ be a solution of $\mathcal{A}u = f$, where

$f \in \mathcal{K}_{Q_1}^{t-\mu}(X,F) \oplus H^{t-\mu}(Y,G)$, $s < t < s+1$, $Q_1 = (\lambda, q_1)$, $\gamma-\mu < \lambda < \gamma-\mu+k$, $r(\gamma-\mu)q_1 \leq q$, then there exists a $P_1 = (\lambda+\mu, p_1)$, $r(\gamma)p_1 \leq p$ such that $u \in \mathcal{K}_{P_1^+}^{t}(X,E) \oplus H^t(Y,J)$.

Note that the formulation explains the different roles of the parameters s,k and l in the solvability theory.

Proof. Let $\mathcal{R}$ be a parametrix of $\mathcal{A}$ modulo Green operators without boundary symbols. Then $\mathcal{G} := 1-\mathcal{R}\mathcal{A} \in \mathcal{D}_G^{-1,(o)}$ and

$$u = \mathcal{R}f + \mathcal{G}u.$$

Thus the assertion follows from the mapping properties of $\mathcal{R}$ and $\mathcal{G}$. $\square$

The parametrix construction by means of a symbolic calculus implies, in particular, an information to which extent the coefficients of the asymptotics of a solution are determined locally or globally.

Of course, in the proof it was used only that $\mathcal{R}$ is a left parametrix. Moreover, we get in this way an a-priori estimate

$$\|u\|_{\mathcal{K}_{P_1^+}^{t}(X,E)\oplus H^t(Y,J)} \leq c\left(\|f\|_{\mathcal{K}_Q^{t-\mu}(X,F)\oplus H^t(Y,G)} + \|u\|_{\mathcal{K}_P^{s}(X,E)\oplus H^s(Y,J)}\right)$$

where $\|.\|_{\mathcal{K}_{P_1^+}^{t}(X,E)\oplus H^t(Y,J)}$ denotes any norm on $\mathcal{K}_{P_1^+}^{t}(X,E)\oplus H^t(Y,J)$.

As a special case we get that $u \in \mathcal{K}_P^s(X,E)\oplus H^s(Y,J)$ satisfying

$$\mathcal{A}u = 0$$

belongs to $\bigcap_t \mathcal{K}_{P_1^+}^{t}(X,E)\oplus H^t(Y,J)$, where now P_1 corresponds to $Q_1 = (\lambda, q)$.

In other words, $\ker_s \mathcal{A}$ does not depend on s. Similarly, one can find a finite-dimensional subspace of $\bigcap_t \mathcal{K}_Q^t(X,F)\oplus H^t(Y,G)$ complementing the image of $\mathcal{A}_s$ to the whole image space.

For the class $\mathcal{MO}^{\mu,(l)}(Q,P)^{(-k)}$ of operators with continuous asymptotics we adopt the notion of ellipticity from Definition 1 (note that there is no reference to the specific nature of the boundary symbols). Proposition 2 is also valid for elliptic elements of $\mathcal{G}^{(\mu)}(\mathcal{MO})(Q,P)^{(-k)}$. As corollary we get that an elliptic operator in

$\mathcal{MO}^{\mu,(1)}(Q,P)^{(-k)}$ admits a parametrix in $\mathcal{MO}^{-\mu,(1)}(P,Q)^{(-k)}$. The regularity result reads as

4'. Theorem. Let $\mathcal{A} \in \mathcal{MO}^{\mu,(1)}(Q,P)^{(-k)}$, $P = (\gamma, \varphi)$, $Q = (\gamma-\mu, \eta)$, be elliptic and $u \in \mathcal{K}^s_P(X,E) \oplus H^s(Y,J)$ be a solution of $\mathcal{A}u = f$, where

$$f \in \mathcal{K}^{t-\mu}_D(X,F) \oplus H^{t-\mu}(Y,G),\ s < t < s+1,\ D = (\lambda, \Delta), \gamma-\mu < \lambda < \gamma-\mu+k,$$

$r(\gamma-\mu)\Delta \leqq \eta$, then there exists $D_1 = (\lambda+\mu, \Delta')$, $r(\gamma)\Delta' \leqq \varphi$ such that $u \in \mathcal{K}^t_{D_1^+}(X,E) \oplus H^t(Y,J)$.

The proof follows from the mapping properties of the parametrix $\mathcal{MO}^{-\mu,(1)}(P,Q)^{(-k)}$, which gives in addition the corresponding a-priori estimates (now in the spaces with continuous asymptotics). In particular we obtain the regularity of solutions of the equation $\mathcal{A}u = 0$.

Having in mind mainly the question of the Fredholm property in given spaces it is natural to pass to the norm closure of the operator classes. Then, for the interior symbols only continuity and homogeneity remains. For the Mellin symbols all analytic extensions from the line $\mathrm{Re}\, z = \frac{1}{2} - \gamma$ disappear (cf. [R2]).

Note that the continuity of

$$\mathrm{Op} : \sigma^{(\mu)}(Q,P)^{(-k)} \longrightarrow \bigcap_s \mathcal{L}(\mathcal{K}^s_P, \mathcal{K}^{s-\mu}_Q)$$

implies that a homotopy through elliptic principal symbols gives rise to a continuous deformation through Fredholm operators. In particular, the index being $\mathbb{Z}$-valued remains constant under such deformations. Naturally the problem arises of describing the Fredholm index in topological terms of the manifold and the principal symbol. It can be treated along the lines of [R2], [R1]. The main missing tool is the existence of order reducing operators in the class. Provided they are found, for instance as powers of simple prototype operators, a homotopy classification as in [R 2] can be expected.

3.3. Notes

The theory of elliptic boundary value problems (bvp's) is one of the essential parts of the classical analysis with a long tradition and a deep influence to other parts of mathematics. Basic problems are the Fredholm property and the elliptic regularity in suitable distribution spaces, the construction of parametrices and the calculation of the index.

The pseudo-differential approach has emphasized the symbolic structure, and in K-theoretic terms it was a starting point in the global analysis and the topological index theory.

All this is well-understood today for those problems that are close enough to differential operators. In the case of bvp's these are just the ψDO's with the transmission property with respect to the boundary. They are obtained by constructing parametrices of elliptic differential boundary value problems.

Remember that the transmission property in our notation [R1] (and that of Boutet de Monvel [B1]) is a condition in local coordinates to a complete symbol on the conormal bundle of the boundary. It ensures the natural mapping properties of the associated ψDO's that are obvious for differential operators, namely to preserve C^{∞} up to the boundary and the Sobolev space smoothness s modulo the order shift (under the condition $s > -1/2$ in case of proper ψDO's).
One may ask quite formally whether it is necessary to restrict the consideration to operators with the transmission property. Very simple operators do not belong to this class, for instance those with $|\xi|^{\mu}$ as the homogeneous principal symbol, for $\mu \notin 2\mathbb{Z}$, ξ being the covector in local coordinates.

Moreover, this type of ψDO's appears in the parametrix constructions of mixed elliptic bvp's. A classical and simple example is the Zaremba problem

$$\Delta u = f \quad \text{in } \Omega,$$
$$T_{\pm}\, u = g_{\pm} \quad \text{on } X_{\pm}\,,$$

where Δ is the Laplacian in a bounded domain Ω, $\partial\Omega$ being C^∞, $\partial\Omega = \overline{X}_+ \cup \overline{X}_-$, $\overline{X}_\pm$ submanifolds of $\partial\Omega$ with C^∞ boundary $Y = \overline{X}_+ \cap \overline{X}_-$ of codimension 1 in $\partial\Omega$, $T_\pm$ Dirichlet and Neumann conditions, respectively. Mixed elliptic problems (for scalar operators as well as for systems) are not less classical than the "usual" ones, also if $\partial\Omega$ has an edge along Y. They arise in many models of the classical mathematical physics, e.g. in elasticity theory.

Simple manipulations (reduction to the boundary, local reflection along Y) lead to bvp's for ψDO's on X_+ (or likewise on X_-) without the transmission property with respect to Y. For the Zaremba problem we get an operator with homogeneous principal symbol $|\xi|$. All interesting mixed problems (in the case of smooth $\partial\Omega$) give rise to this general class of ψDO's.

In order to perform a parametrix construction we have to add extra boundary (and potential) conditions along Y. Their number depends on the choice of the weights in the distribution spaces (it may happen that no conditions are necessary). The extra data have to satisfy the Shapiro-Lopatinskij condition. Then we can establish the elliptic regularity.

By the mentioned manipulations, in the reverse direction we obtain the same information for the original problems under corresponding additional conditions of boundary and potential type along Y.

In other words, the essential question for mixed bvp's is to analyze bvp's for ψDO's without the transmission property. This was just the program of the preceding sections.

The necessity of studying bvp's in the case of violated transmission property has been recognized very early in the literature. Above all there is a series of papers of Višik, Eskin [V1], [V2], [V3], [V4]. They are based on a higher dimensional variant of the Wiener-Hopf technique, which works well for scalar operators with constant coefficients. It yields explicit solution formulas in the half space. By a localization argument ("freezing of coefficients") then also follows a parametrix construction for variable coefficients and on manifolds with boundary. Of course, the special case of operators with the transmission property is also included. The technical tools of the theory have been used later on by many authors for solving several types of singular problems. The approach of Višik, Eskin belongs to the essential achievements of the theory of boundary value problems. But in spite of its culture and beauty it has some disadvantages that become evident if we pose the same questions as for ψDO's on closed compact C^{∞} manifolds. In a well-organized calculus of ψDO's we have to expect the following elements

(i) an algebra of operators with symbolic structures (principal, complete), if possible with $*$ operation,

(ii) an ideal of smoothing operators such that the complete (principal) symbols determine the operator modulo a smoothing one (someone of lower order)

(iii) scales of adequate Sobolev spaces and C^{∞} subspaces with asymptotics

(iv) the concept of ellipticity, which is equivalent to the Fredholm property, and a parametrix construction within the algebra.

Clearly the concept of ellipticity should not make an essential difference between scalar and matrix-valued operators.

For pseudo-differential bvp's it needs certain new ideas to carry out a theory with all these features.

It is remarkable that even the smoothing operators (which are the trivial ones in a calculus on closed compact manifolds) are not at all obvious. Their structure may be conjectured by knowing the elliptic regularity. Solutions of elliptic bvp's with vanishing data are C^{∞} in the interior and have conormal asymptotics close to the boundary. Such functions are the substitute of usual C^{∞} functions. Then the smoothing operators should be characterized by the corresponding smoothing property: they map distributions to C^{∞} functions with asymptotics, and the adjoints do the same.

In case of the transmission property the asymptotics is rather special, it is the smoothness up to the boundary, the "Taylor asymptotics".

It is known that Boutet de Monvel's algebra [B1] represents a ψDO calculus with the above elements, and they are new compared with Višik, Eskin's work. In particular the Green operators are of interest in this class. These are ψDO's along the boundary with operator-valued symbols that act normal to the boundary as Green operators on $\mathbb{R}_+$ (with the transmission property). The Green operators in higher dimensions are not the smoothing ones in the algebra. They are of some order and not compact in the Sobolev spaces, though the associated boundary symbols are compact. Green operators appear as extra elements in compositions of ψDO's on manifolds with boundary, i.e. the resulting operators are ψDO + Green. Such sums then constitute an algebra and the Green operators form a two-sided ideal.

From the boundary symbolic calculus we already know that Boutet de Monvel's class has some properties which are not typical for the general case. The $\mathbb{R}_+$ calculus shows that also Mellin operators are generated in compositions. But then we get an algebra of sums of the form ψDO + Mellin + Green with Mellin + Green and Green operators as two-sided ideals. This is true indeed also in higher dimensions, as we have seen in the preceding sections. Moreover, the Sobolev spaces in higher dimensions have been replaced by $\mathcal{K}^{s,\gamma}$ and $\mathcal{K}^{s,\gamma}$ + {singular terms} -spaces. Thus the symbolic structure with respect to the Mellin contribution and also the specific sort of spaces could not be derived from the particular class with the transmission property. On the other hand, we do not only recover Boutet de Monvel's class as a special case but also see further natural scales of spaces where the elliptic regularity holds for this class. We can move the weights γ as we want and may drop in particular the "mysterious" condition $s > -1/2$. The weights are now arbitrary up to negative half integers if we only modify the operator by a smoothing one of the Green class, whereas the interior smoothness $s \in \mathbb{R}$ remains arbitrary and is shifted by the order of the operators.

Analogous results hold in general, including the parametrix construction or inverses within the class. From this point of view the higher dimensional theory of bvp's of Chapter 3 is an extension of Višik, Eskin's work in the sense of the mentioned "standard program" of a ψDO calculus.

The general algebra of pseudo-differential bvp's (with constant discrete as well as continuous asymptotics) has further interesting subalgebras, for instance, those which are generated by a single elliptic bvp. In such algebras we have a smaller variety of asymptotic types in the Green and Mellin operators. They are worth being studied in the future in more detail. Another subalgebra consists of the zero order operators realized in L^2 spaces. The principal symbolic calculus of this class has been investigated by Rempel, Schulze in [R2] under the point of view of stable homotopy classes of elliptic principal symbols and the analogue of the Atiyah-Singer index theorem. This investigation also has explained the topological role of the principal Mellin symbol as a family of morphisms parametrized by the conormal vectors to the boundary, of length ≤ 1. For the subclass with the transmission property this part disappears. Geometrically it means equality of the values of the homogeneous principal symbol on the north and south pole of the unit cospheres, restricted to the boundary.

The principal symbolic calculus allows to pass to continuous symbols which correspond to a closure of the algebra. The corresponding higher order analogue has been used by Rempel, Schulze [R5] for studying general mixed elliptic boundary value problems, in particular the parametrix construction on symbolic level.

There are also other motivations for considering ψDO's without the transmission property. For instance, a necessary condition for the transmission property is that the order is an integer. Thus the symbols with non-integer orders, in particular the fractional or complex powers, do not belong to this class, up to some exceptions. In case of ψDO's on closed compact manifolds it is very natural to study the

complex powers of ψDO's which are ψDO's again, cf. Seeley [S7]. For bvp's things are more complicated. Nevertheless, on the principal symbolic level one can obtain a number of analogous results, cf. Rempel, Schulze [R3]. It would be interesting to return to these question once again under the point of view of the complete symbolic structure of this chapter.
A related problem is that of a parameter-dependent calculus of ψD bvp's as an analogue of that occurring in the theory of complex powers on closed compact manifolds, cf. Šubin [S3]. Also this should be elaborated in future, in particular for obtaining convenient reductions of orders within the algebra. Note that a parameter-dependent calculus for operators with the transmission property has been established by Grubb [G6], cf. also the papers quoted there. Reductions of orders in terms of parameter-dependent families were constructed in [R3], and may also be found in the Russian translation of [R1].
It is not possible here to mention all merits and steps in the developement of the idea to use the ψDO calculus in bvp's. Further references may be found in the monographs of Rempel, Schulze [R1] and Grubb [G6]. As mentioned, the ψDO ideas are motivated by those on closed compact manifolds and they are linked to them by the principle of reduction to the boundary. In this direction fundamental achievements go back to Hörmander [H1], [H4], [H5], Kohn, Nirenberg [K2] and other authors on the techniques of the standard calculus. Clearly the Fredholm property and the elliptic regularity for differential problems in smooth domains were known much earlier, cf. for instance Agmon, Douglis, Nirenberg [A1], Lions, Magenes [L4].

Let us return once again to some technical aspects of our theory. The idea of establishing the higher-dimensional theory of bvp's is to perform a "lifting" of a corresponding operator calculus on $\mathbb{R}_+$ with parameters x', ξ', by applying the ψDO convention in terms of the Fourier transform along the boundary. A non-straightforward point is to keep in mind a ξ' depending action $\varkappa(\xi')$ along the fibre spaces (which are the spaces on $\mathbb{R}_+$, i.e. normal to the boundary). This is involved both in the definition of the distribution spaces on $\mathbb{R}^n_+$, in particular of the notion of conormal asymptotics, and the operator-valued symbol classes. We have an adequate notion of homogeneity of operator-valued symbols. This leads to a rather individual interaction in the Mellin symbols between conormal orders and degree in ξ'. It is remarkable that the Mellin symbols are polynomials in ξ'. In virtue of the mentioned homogeneity their degree increases with decreasing conormal orders, and in a complete symbolic calculus it would tend to infinity. For limiting the size of the exposition we control an arbitrary finite number N of symbol components. This requires a calculus of remainders, characterized by those mapping properties that would be satisfied by asymptotic sums of operators of orders $> -N$, tending to minus infinity. Clearly the calculus also admits a variant with infinitely many symbol components. Another typical structure in the higher dimensional theory is that the operators can be characterized by their symbolic hierarchies. The symbols (even the leading ones) are pairs with a compatibility condition, and the components which are operator-valued have again a symbolic structure. The leading symbol consists of the leading interior ψDO symbol and the boundary symbol. From that we can derive the leading Mellin symbol which reflects a specific information about the asymptotics. In case of the ellipticity the inversion of the leading components preserves the compatibility condition. The various components of elliptic symbols are responsible for different kinds of properties of the parametrix B of an operator A, which altogether imply $AB-I$, $BA-I$ compact. The leading Mellin symbol gives us the main contribution to the conormal asymptotics of solutions.

The methods of Chapter 3 can also be considered under a more general point of view, which will be discussed in Chapter 4, where we realize the idea of "edgefication" of a calculus with symbolic structures (ψDO, Mellin, boundary,...). The edge in Chapter 3 was just the boundary (locally $\mathbb{R}^{n-1}$) and the mentioned lifting was the edgefication of the calculus over $\mathbb{R}_+$. It is an exciting program to do the same for a cone instead, represented by $\mathbb{R}_+ \times \{\text{manifold}\}$, which is the subject of the theory below.

4. Mixed boundary value problems on manifolds with edges

In applications boundary value problems with discontinuous (mixed) boundary conditions and (or) geometric singularities of the boundary occur. The simplest geometric singularity of isolated conical points is now well-understood, also the corresponding symbol structure. The next more difficult case is that of an edge. Since mixed boundary value problems on smoothly bounded manifolds behave quite similar like mixed problems on manifolds with edges (where the boundary conditions change when crossing the edge) we shall consider immediately the edge case.

Let $\tilde{X}$ be a C^∞ manifold without boundary, $\mathring{X} \subset \tilde{X}$ an open submanifold. Denote by X the completion of $\mathring{X}$ in $\tilde{X}$, $\partial X = X \setminus \mathring{X}$. We call $(X, \partial X)$ a <u>manifold with edges</u> if for each point $x \in \partial X$ there exists a coordinate chart (U, φ) on $\tilde{X}, x \in U$, $\varphi : U \longrightarrow \mathbb{R}^n$, $n = \dim X$, with one of the two following properties:

(i) $\varphi|U \cap X : U \cap X \longrightarrow V \subset \mathbb{R}^n_+$,

$\quad\;\;\varphi|U \cap \partial X : U \cap \partial X \longrightarrow V' \subset \partial\mathbb{R}^n_+$,

(ii) $\varphi|U \cap X : U \cap X \longrightarrow \mathbb{R}^k \times C$, C a closed cone in $\mathbb{R}^{n-k}$,

$\quad\;\;\varphi|U \cap \partial X : U \cap \partial X \longrightarrow \mathbb{R}^k \times \partial(C \setminus 0)$.

We denote by $\partial_1 X$ the subset of ∂X of all points admitting coordinates with the property (i) and we set $\partial_2 X = \partial X \setminus \partial_1 X$. Clearly, $\partial_1 X$ is open in ∂X and dense or empty. Since $\mathbb{R}^k$ is also a cone, it is reasonable to choose the maximal possible K for a given $x \in \partial_2 X$. In general, this depends on x and is upper semi-continuous. For simplicity, we assume k to be (locally) constant.

It is natural to consider a manifold with edges as a stratified object. The zeroth stratum $\partial_0 X$ is the interior $\mathring{X}$, the 1-stratum the "smooth part" of the boundary $\partial_1 X$ and the 2-stratum the "edge" $\partial_2 X$.

The C^∞ structure on X is specified by a choice of an atlas of coordinate charts such that, for two coordinate homeomorphisms φ, ψ restricted to their common domain, $\varphi \circ \psi^{-1}$ is a diffeomorphism in the interior, extends as a diffeomorphism across $\partial\mathbb{R}^n_+$ and has the form $\mathbb{R}^k \times C \ni (y, \tilde{x}) \longrightarrow (\varkappa(y), \tilde{x}) \in \mathbb{R}^k \times C$, $\varkappa$ a diffeomorphism $\varkappa : \mathbb{R}^k \longrightarrow \mathbb{R}^k$, and the same on $\mathbb{R}^k \times \partial(C \setminus 0)$.

We shall consider vector bundles over X as vector bundles over $\partial_0 X$, the smooth structure of which is described by a cocycle $g_{UV}: U \cap V \longrightarrow GL(l)$, l the dimension of the fibre which is smooth in $\partial_0 X$ up to $\partial_1 X$ and is independent of $\tilde{x}$ near $\partial_2 X$.

Let us mention two particular cases of special interest:

Example 1. Let f,g be real valued smooth functions on X with the properties:

(i) $df = 0$ if $f = 0$, $dg = 0$ if $g = 0$

(ii) df and dg are linearly independent when $f = g = 0$.

In view of (i) both $\{x \in \tilde{X} : f(x) \leqq 0\}$ and $\{x \in \tilde{X} : g(x) \leqq 0\}$ are submanifolds with smooth boundary. According to (ii) their intersection $X := \{x \in \tilde{X} : f(x) \leqq 0, g(x) \leqq 0\}$ is a manifold with edges. In fact, near each point where f and g vanish the local model is $\mathbb{R}^{n-2} \times \{(x_1, x_2) \in \mathbb{R}^2 : x_1 \geqq 0, x_2 \geqq 0\}$. Then $\partial X = \{x \in \tilde{X} : f(x)g(x) = 0\}$, $\partial_2 X = \{x \in \tilde{X} : f(x) = g(x) = 0\}$.

Example 2. Let $\partial_2 X$ be an arbitrary closed submanifold of $\tilde{X}$ of any codimension, $\mathring{X} = \tilde{X} \setminus \partial_2 X$, $\partial_1 X = \emptyset$. Here $\partial_2 X$ is an "interior boundary".

It should be stressed that the geometric assumptions on the manifold must be completed by assumptions on the behaviour of the operator (or coefficients) in order to be precise which classes of problems are covered by a certain calculus. They will be given in terms of the local model. Before going into details let us sketch some of the main features of the theory. Our principal goal is the description of operator classes which are closed under composition (provided it is defined) and parametrix construction of elliptic problems. On $\partial_0 X$ and $\partial_1 X$ the structure is the same as for pseudo-differential boundary problems. The additional difficulty due to the edge gives rise to new operator classes which are smoothing outside the edge. They are controlled by the edge symbol $\partial_2 \sigma$, which takes values in an algebra of operators on a cone - the transversal cross section of the edge. The bijectivity of $\partial_2 \sigma$ is the additional ellipticity condition besides the usual conditions of bijectivity of the interior symbol $\partial_0 \sigma$ and of the boundary symbol $\partial_1 \sigma$. The elements of the triple $(\partial_0 \sigma, \partial_1 \sigma, \partial_2 \sigma)$ arising from operators are not independent but subject to natural compatibility conditions, which express the influence of $\partial_0 \sigma$ on $\partial_1 \sigma$ and $\partial_2 \sigma$ and of $\partial_1 \sigma$ on $\partial_2 \sigma$.

4.1. Function spaces

4.1.1. Function spaces on a cone

A subset C of $\mathbb{R}^n$ is called a <u>cone</u> if it is invariant under homoteties: $\lambda C = C$ for all $\lambda \geq C$. The intersection with the unit sphere $N = C \cap S^{n-1}$ is called <u>base</u> of C. Conversely, C is the flow out of N under homoteties and we write $C = C(N)$. It is convenient to use polar coordinates $\pi(x) = (r,\varphi) \in \mathbb{R}_+ \times S^{n-1}$, $r = |x|$, $\varphi = \frac{x}{x}$, besides the Cartesian coordinates. Then

$$\pi : \mathbb{R}^n \setminus 0 \longrightarrow \mathbb{R}_+ \times S^{n-1}$$

is a diffeomorphism. If N is a smooth manifold of dimension $n-1$ without boundary, then $\overline{\mathbb{R}}_+ \times N$ is a C^∞ manifold with boundary $\partial(\overline{\mathbb{R}}_+ \times N) = \partial\overline{\mathbb{R}}_+ \times N$. If N itself has a smooth boundary, then $\overline{\mathbb{R}}_+ \times N$ has edges. We resolve this singularity by "moving $\partial\overline{\mathbb{R}}_+$ to $-\infty$". Consider a diffeomorphism $s : \mathbb{R}_+ \longrightarrow \mathbb{R}$, given by a smooth function $r \longrightarrow s(r)$,

$$s(r) = \begin{cases} \log r & \text{for } 0 < r < 1, \\ r & \text{for } 2 < r, \end{cases}$$

where $s'(r)$ is nowhere vanishing. Note, that $s(r) \longrightarrow -\infty$ as $r \longrightarrow +0$. Denote by

$$\varepsilon : \mathbb{R}_+ \times N \longrightarrow \mathbb{R} \times N$$

the diffeomorphism given by $\varepsilon(r,\varphi) = (s(r),\varphi)$.

Let N be a smooth manifold. Consider the half-cylinder $\overline{\mathbb{R}}_+ \times N$. We obtain a topological cone $C(N)$ by contracting $N \times \partial\overline{\mathbb{R}}_+$ to one point, $C(N) = \overline{\mathbb{R}}_+ \times N/\sim$, where two points $x, x' \in \overline{\mathbb{R}}_+ \times N$ are identified by $\sim$ iff $x, x' \in \partial\overline{\mathbb{R}}_+ \times N$ or $x = x'$. We denote by $\partial_2 C(N)$ the vertex of the cone. If $\partial N \neq \emptyset$, then $C(N)$ has as boundary $\partial C(N) = C(\partial N)$. The smooth part $\partial C(N) \setminus \partial_2 C(N)$ of the boundary is denoted by $\partial_1 C(N)$. The interior $\partial_0 C(N)$ is an open C^∞ manifold and $\partial_0 C(N) \cup \partial_1 C(N)$ is a (non-closed) C^∞ manifold with boundary (i.e. the local model is an open subset of $\mathbb{R}^n$ or $\overline{\mathbb{R}^n_+}$). Near the vertex $\partial_2 C(N)$ the admissible coordinates arise in the following way. Consider local coordinates on N

$$\psi : U \longrightarrow S^{n-1}$$

$n-1 = \dim N$, $U \subset N$ open. Consider (r,ψ), $r \in \mathbb{R}_+$, as polar coordinates of $C(\psi(U)) \subset \mathbb{R}^n$. For technical reasons below we assume that N admits a finite covering $\{U_j\}$ by coordinate neighbourhoods such that the transition diffeomorphisms $C(\psi_j(U_j \cap U_k)) \longrightarrow C(\psi_k(U_j \cap U_k))$ are restrictions of affine transformations of $\mathbb{R}^n$. This is a topological condition and satisfied, for instance, for spheres.

We recall the definitions of some function spaces from [R6]. Let N be a smooth manifold without boundary. Introducing local coordinates on N we may regard $\mathbb{R}_+ \times \Omega$, $\Omega \subset \mathbb{R}^{n-1}$ open, as local model of $\mathbb{R}_+ \times N$. Consider the completion of $C_0^\infty(\mathbb{R}_+ \times \Omega)$ with respect to the norm

$$\| u \|_{\mathcal{H}^s} = \left\{ \frac{1}{2\pi i} \int_{\Gamma_{\frac{1}{2}}} \int_{\mathbb{R}^{n-1}} (1+|z|^2+|\xi|^2)^s \, |MFu(z,\xi)|^2 dz d\xi \right\}^{1/2}, \qquad (1)$$

where

$$MFu(z,\xi) = \int_0^\infty t^{z-1} \int_\Omega e^{-ix\xi} u(t,x) dx dt. \qquad (2)$$

If N is compact, we define $\mathcal{H}^s(\mathbb{R}_+ \times N)$ by a finite partition of unity $\{\varphi_j\}$ and a finite sum of norms (1) of $\varphi_j(x)u(t,x)$. For non-compact N we get $\mathcal{H}^s_{loc}(\mathbb{R}_+ \times N)$ by demanding finiteness of the norms (1) for $\varphi(x)u(t,x)$ when φ runs over an (infinite) partition of unity. $\mathcal{H}^s_{comp}(\mathbb{R}_+ \times N)$ is the subspace of $\mathcal{H}^s_{loc}(\mathbb{R}_+ \times N)$ with support, having a compact projection into N. One checks easily that for $s \in \mathbb{Z}_+$ (1) is equivalent to

$$\sum_{j+|\alpha| \leq s} \| \vartheta_t^j D_x^\alpha u(t,x) \|_{L^2(\mathbb{R}_+ \times \Omega)} .$$

For each $\gamma \in \mathbb{R}$ we set for compact N

$$\mathcal{H}^{s,\gamma}(\mathbb{R}_+ \times N) = t^\gamma \mathcal{H}^s(\mathbb{R}_+ \times N) . \qquad (3)$$

We have the obvious embeddings

$$\mathcal{H}^{s,\gamma}(\mathbb{R}_+ \times N) \hookrightarrow \mathcal{H}^{t,\gamma}(\mathbb{R}_+ \times N) \text{ for } s \geq t.$$

The same is true for non-compact N with the subscripts comp or loc.

Denote by I_o the interval $[0,2] \subset \overline{\mathbb{R}}_+$. Let $\mathcal{H}_i^{s,\gamma}(I_o \times N)$ be the subspace of $\mathcal{H}_i^{s,\gamma}(\mathbb{R}_+ \times N)$, $i \in \{\text{comp}, \text{loc}\}$, of all elements supported by I_o with respect to t. Set

$$\mathcal{K}_i^{s,\gamma}(\mathbb{R}_+ \times N) = \mathcal{H}_i^{s,\gamma}(I_o \times N) + (\pi^{-1})^* H_i^s(\pi^{-1}(I_\infty \times N)), \qquad (4)$$

where $I_\infty = [1,\infty)$, $(\pi^{-1})^* H_i^s(\pi^{-1}(I_\infty \times N))$ is defined by localization on $U \subset N$, $\pi^{-1}(I_\infty \times U) = C(U) \cap \{|\tilde{x}| \geq 1\} \subset \mathbb{R}^n$, the H^s norm refers to the standard coordinates of $\mathbb{R}^n$, $(\pi^{-1})^*$ is the pull back $(\pi^{-1})^* u(\tilde{x}) = u(|\tilde{x}|, \frac{\tilde{x}}{|\tilde{x}|})$ which makes sense, since supp $u \cap \{0\} = \emptyset$. The subspace of $(\pi^{-1})^* H_i^s(\pi^{-1}(I_\infty \times N))$ of distributions supported in $I_{c_1,c_2} = \{t \in \mathbb{R}_+ : 0 < c_1 \leq t \leq c_2 < \infty\}$ coincides with the subspace of $H_i^s(\mathbb{R} \times N)$ with t-support in I_{c_1,c_2}. The same is true for the corresponding subspace of $\mathcal{H}_i^{s,\gamma}(I_o \times N)$. In fact, we have

<u>1. Lemma.</u> Let $\varepsilon : \mathbb{R}_+ \longrightarrow \mathbb{R}$ be the diffeomorphism given by $\varepsilon(t) = \log t$ for $t \in \mathbb{R}_+$. Then

$$\varepsilon^*[\mathcal{H}^{s,\gamma}(\mathbb{R}_+ \times N)] = g^{\gamma - 1/2} H^s(\mathbb{R} \times N)$$

as topological vector spaces, where the weight g is given by e^s, $s \in \mathbb{R}$.

<u>Proof.</u> Substituting $t = e^s$ in

$$MFu(\tfrac{1}{2} - i\tau, \xi) = \int_0^\infty t^{-i\tau - 1/2} Fu(t,\xi)\, dt$$

we get

$$\int_{-\infty}^{\infty} e^{-is\tau} e^{\frac{s}{2}} Fu(e^s, \xi) ds,$$

i.e. the Fourier transform $\tilde{F}$ with respect to x and s of $g^{1/2} \varepsilon^* u$. Then

$$\| u\|^2_{\mathcal{H}^s} = \int_{-\infty}^{\infty}\int_{\mathbb{R}^{n-1}} (1+|\tau|^2+|\xi|^2)^s \left|MFu(\tfrac{1}{2}-i\tau,\xi)\right|^2 d\tau d\xi$$

$$= \int_{\mathbb{R}^n} (1+|\tau|^2+|\xi|^2)^s \left|\tilde{F}(g^{\frac{1}{2}}\varepsilon^* u)(\tau,\xi)\right|^2 d\tau d\xi,$$

which proves the assertion for $\gamma = 0$. For arbitrary γ we use the definition of $\mathcal{H}^{s,\gamma}$ and $\varepsilon^*(t^\gamma) = e^{s\gamma}$, $0<t<c$. $\square$

As a consequence the definition (4) of the $\mathcal{K}_i^{s,\gamma}(\mathbb{R}_+ \times N)$ spaces does not change if we replace I_0 by I_{0,c_2} and I_∞ by $I_{c_1,\infty}$, $0<c_1<c_2<\infty$ arbitrary. The definition (4) is related to the intuitive picture that the polar coordinates are useful and necessary only in an arbitrary small neighbourhood of the vertex.

The spaces $\mathcal{K}^{s,\gamma}$ (pulled back to the cone) play a role similar to $\mathcal{K}^{s,\gamma}(\mathbb{R}_+)$ introduced in 2.3.1. They must be distinguished from the function spaces in the half-space introduced in 3.1.1. because here the vertex of a cone is distinguished and there the boundary $\partial\mathbb{R}^n_+$.

Note the continuous embeddings

$$\mathcal{K}_i^{s,\gamma}(\mathbb{R}_+ \times N) \hookrightarrow \mathcal{K}_i^{s',\gamma'}(\mathbb{R}_+ \times N)$$

for $s \geq s'$, $\gamma \geq \gamma'$.

Assume from now on that N is compact. We choose a Riemannian metric g on N and consider $\mathbb{R}\times N$ with the natural product metric $ds^2 + g$. This yields the corresponding L^2 space $L^2(\mathbb{R}\times N)$.

It is well-known that the dual of $g^{\gamma-1/2}H^s(\mathbb{R}\times N)$ with respect to the L^2 scalar product is $g^{-\gamma+1/2}H^{-s}(\mathbb{R}\times N)$. In view of

$$\int_{-\infty}^{\infty} f(s)\overline{g(s)}ds = \int_0^{\infty} f(\log t)\overline{g(\log t)}\,\frac{dt}{t}\ , \quad f,g \in L^2(\mathbb{R}),$$

we get from Lemma 1 that the dual of $\mathcal{H}^{s,\gamma}(\mathbb{R}_+ \times N)$ with respect to $L^2(\mathbb{R}_+ \times N, dtdx)$, dx denotes the density on N induced by the metric g, can be identified with $\mathcal{H}^{-s,-\gamma}(\mathbb{R}_+ \times N)$.

The metric on $\pi^{-1}(\mathbb{R}_+ \times N)$ is given by $dt^2 + t^2 g$. It is equivalent to the metric induced by the euclidean metric in $\mathbb{R}^n$ on the conical subsets $\pi^{-1}(\mathbb{R}_+ \times U)$. The corresponding L^2-scalar product on $\mathbb{R}_+ \times N$, which gives $L^2(\mathbb{R}_+ \times N,\ t^{n-1}dtdx)$, generates the duality between $(\pi^{-1})^* H^s(\pi^{-1}(\mathbb{R}_+ \times N))$ and $(\pi^{-1})^* H^{-s}(\pi^{-1}(\mathbb{R}_+ \times N))$. Note that on compact parts of $\mathbb{R}_+$ all mentioned metrics are equivalent and define the same space L^2. The difference is precisely in the behaviour near $t = 0$ and $t = \infty$.

As a consequence we get that $\mathcal{K}^{s,\gamma}(\mathbb{R}_+ \times N)$ and $\mathcal{K}^{-s,-\gamma}(\mathbb{R}_+ \times N)$ are dual with respect to the scalar product of the Hilbert space $L^2(\mathbb{R}_+ \times N, \rho(t)dtdx)$, $\rho(t) = t^{n-1}$ near ∞, $\rho(t) = 1$ near 0, $n = \dim N+1$.

Besides the weighted Sobolev spaces $\mathcal{K}^{s,\gamma}$ we shall also need spaces with asymptotics. They arise as extensions of $\mathcal{K}^{s,\gamma}$ by singular functions of the form

$$\omega(t)t^{-p_j} \log t^k \varphi_{jk}(x) \tag{5}$$

where $p_j \in \mathbb{C}, k \in \mathbb{Z}_+, \varphi_{jk} \in C^\infty(N)$, $0 \leq k \leq m_j$. In addition to the data p_j and m_j we have to fix a finite-dimensional space $L_j \subset C^\infty(N)$ where φ_{jk} belongs to L_j for all k.

__2. Definition.__ Denote by $\underline{\mathcal{P}}(N)$ the set of all countable sets $\mathcal{P} = \{(p_j, m_j, L_j)\}$ where $p_j \in \mathbb{C}$, Re $p_j \to -\infty$, $m_j \in \mathbb{Z}_+$, $L_j \subset C^\infty(N)$ a finite-dimensional subspace. Again we have the operations $r(\gamma), \bar{r}(\gamma), l(\gamma), \bar{l}(\gamma)$, $\gamma \in \mathbb{R}$, on $\underline{\mathcal{P}}(N)$. We denote by $\underline{P}(N)$ the set of all $P = (\gamma, \mathcal{P})$, $\gamma \in \mathbb{R}$, $\mathcal{P} \in \bar{r}(\gamma)\underline{\mathcal{P}}(N)$.

Given $P = (\gamma, \mathcal{P}) \in \underline{P}(N)$ we define

$$\mathcal{K}^s_P(\mathbb{R}_+ \times N) = \mathcal{K}^{s,\gamma}(\mathbb{R}_+ \times N) + \mathcal{E}(\mathcal{P})(\mathbb{R}_+ \times N),$$

where $\mathcal{E}(\mathcal{P})(\mathbb{R}_+ \times N)$ denotes the linear span of all functions of the form (5), $(p_j, m_j, L_j) \in \mathcal{P}$, $\varphi_{jk} \in L_j$. $\mathcal{E}(\mathcal{P})(\mathbb{R}_+ \times N)$ depends, of course, on the choice of the cut-off function ω, but $\mathcal{K}^s_P$ is independent of this choice. In fact, another cut-off function in (5) yields a difference which is smooth and has a compact support in $\mathbb{R}_+$ with respect to t. Clearly $\mathcal{K}^s_P$ is a "finite-dimensional perturbation" of the Hilbert space $\mathcal{K}^{s,\gamma}$. Each element u of $\mathcal{K}_P^s$ determines a linear combination of the singular terms (5) u_s such that $u - u_s \in \mathcal{K}^{s,\gamma}$. Here u_s is uniquely determined.

Let $\varphi \in \underline{\varphi}(N)$ be given. We define a space $\mathcal{K}_\varphi^s$ describing an asymptotics modulo an arbitrary good weight γ in the following way. Consider a sequence $P_j \in \underline{P}(N)$, $j = 1,2,\ldots P_j = (\gamma_j, \varphi_j), \gamma_j \longrightarrow \infty$ as $j \longrightarrow \infty$, with the property $\bar{r}(\gamma_j)\varphi = \varphi_j$. Then

$$K_{P_1}^s \hookleftarrow K_{P_2}^s \hookleftarrow \ldots \hookleftarrow K_{P_j}^s \hookleftarrow \ldots$$

and we set

$$\mathcal{K}_\varphi^s(\mathbb{R}_+ \times N) = \varprojlim \mathcal{K}_{P_j}^s(\mathbb{R}_+ \times N).$$

It is clear that $\mathcal{K}_\varphi^s$ is independent of the choice of the sequence P_j.

Consider now the case when the base N of the cone (or half-cylinder) is a compact C^∞-manifold with boundary. Then the function spaces include a specification of the behaviour near $\mathbb{R}_+ \times \partial N$. Among the different possibilities the most simple choices arise from $\dot{H}_o^s(N)$ and $H^s(N)$, which are special cases of the $\mathcal{K}_P^s(N)$ introduced in 3.1.2. Fix $P \in \underline{P}$, which controls the behaviour near ∂N (cf.3.1.2.), and $\gamma \in \mathbb{R}$ as the weight for $t \longrightarrow 0$. Since $\mathbb{R}_+ \times N$ has corners, we define first spaces on $\mathbb{R} \times N$ to be pulled back by a diffeomorphism $\varepsilon: \mathbb{R}_+ \times N \longrightarrow \mathbb{R} \times N$ to $\mathbb{R}_+ \times N$. Let $\{U_j\}$ be a covering of N by coordinate neighbourhoods, $\{\varphi_j\}$ a partition of unity, supp $\varphi_j \subset U_j$. Define the norm of $\mathcal{K}_P^s(\mathbb{R} \times N)$ by

$$\|u\|_{\mathcal{K}_P^s(\mathbb{R} \times N)} = \sum_j \|\varphi_j(x) u(s,x)\|_s ,$$

where x represents local coordinates of U_j and $\|\cdot\|_s$ is the $H^s(\mathbb{R}^n)$ norm if $U_j \cap \partial N = \emptyset$ and the $\mathcal{K}_P^s(\mathbb{R}_+^n)$ norm otherwise. Then $\varepsilon^*(\mathcal{K}_P^s(\mathbb{R} \times N))$ is a space of distributions on $\mathbb{R}_+ \times N$ and $\varepsilon^*(\mathcal{K}_P^s(\mathbb{R} \times N))(I_o \times N)$ denotes the subspace with t-support in I_o.

Define the norm of $\mathcal{K}_P^s(\pi^{-1}(I_\infty \times N))$ by

$$\| u \|_{\mathcal{K}_P^s(\pi^{-1}(I_\infty \times N))} = \sum_j \| \varphi_j(\tfrac{\tilde{x}}{|\tilde{x}|}) u(|\tilde{x}|, \tfrac{\tilde{x}}{|\tilde{x}|}) \|_s ,$$

where $\tilde{x} \in \mathbb{R}^n, \frac{\tilde{x}}{|\tilde{x}|}$ is a coordinate $U_j \longrightarrow S^{n-1}$ and $\| \cdot \|_s$ is the $H^s(\mathbb{R}^n)$ norm if $U_j \cap \partial N = \emptyset$ and the $\mathcal{K}_P^s(\mathbb{R}^n_+)$ norm otherwise.

Note that if u has a t-support in a fixed compact in $\mathbb{R}_+$, the above norms are equivalent to the $\mathcal{K}_P^s(\mathbb{R} \times N)$ norm.

Now we set

$$\mathcal{K}_P^{s,\gamma}(\mathbb{R}_+ \times N) = t^\gamma \, \varepsilon^*(\mathcal{K}_P^s(\mathbb{R} \times N))(I_0 \times N) + (\pi^{-1})^* \mathcal{K}_P^s(\pi^{-1}(I_\infty \times N)).$$

It is clear that I_0 and I_∞ can be replaced by I_{0,c_2} and $I_{c_1,\infty}$.

<u>3. Definition.</u> Denote by $\underline{P}(N, \partial N)$ the set of all $P = (P_1, P_2), P_1 \in \underline{P}$ and $P_2 = (\gamma_2, \mathcal{P}_2), \mathcal{P}_2 = \{(p_j, m_j, L_j)\}, (p_j, m_j)$ as in Definition 2 and L_j are finite-dimensional subspaces of $\mathcal{K}_{P_1}(N)$.

If $P \in \underline{P}(N, \partial N)$, we introduce now $\mathcal{K}_P^s(\mathbb{R}_+ \times N)$ by

$$\mathcal{K}_P^s(\mathbb{R}_+ \times N) = \mathcal{K}_{P_1}^{s,\gamma_2}(\mathbb{R}_+ \times N) + \mathcal{E}(\mathcal{P}_2)$$

if $P_2 = (\gamma_2, \mathcal{P}_2)$, $\mathcal{K}_{P_1}^{s,\gamma_2}$ was introduced above and $\mathcal{E}(\mathcal{P}_2)$ is the linear span of singular functions (5) with $\varphi_{jk} \in L_j \subset \mathcal{K}_{P_1}^\infty(N)$. Again we have a "finite-dimensional perturbation" of a Hilbert space possessing a Hilbert space structure itself. Clearly

$$\mathcal{K}_P^s(\mathbb{R}_+ \times N) \longrightarrow \mathcal{K}_{P'}^{s'}(\mathbb{R}_+ \times N)$$

if $s \geq s'$, $P \leq P'$ which means

$$\gamma_1 \geq \gamma_1', \mathcal{P}_1 \leq \mathcal{P}_1', \gamma_2 \geq \gamma_2', \mathcal{P}_2 \leq \mathcal{P}_2' ,$$

if $P = ((\gamma_1, \mathcal{P}_1), (\gamma_2, \mathcal{P}_2))$, $P' = ((\gamma_1', \mathcal{P}_1'), (\gamma_2', \mathcal{P}_2'))$.

Several connections between the spaces $\mathcal{K}_P^s(\mathbb{R}_+ \times N)$ and $\mathcal{K}_{P_2}^s(\mathbb{R}_+ \times \partial N)$ are given by trace and potential operators to be studied in 4.2.1.

<u>4.1.2. Function spaces on a wedge</u>

In 3.1.1. we generalized the $\mathcal{K}^{s,\gamma}$ spaces from the half-line $\mathbb{R}_+$ to

the half-space $\mathbb{R}^{n-1}\times\mathbb{R}_+$. In an analogous manner we introduce here spaces on $\mathbb{R}^d\times C$, C a cone. Often we shall work in polar coordinates on conical subsets of C as in 4.1.1.

Denote by y the standard coordinate in $\mathbb{R}^d$ and by η the dual variable. Let F' be the (partial) Fourier transform with respect to y. For a conical open subset C(U) of $\mathbb{R}^n$ we have a parametrization by standard euclidean coordinates $z \in \mathbb{R}^n$. We set $t = |z|$ and $x = \frac{z}{|z|}$ for the corresponding polar coordinates.

The $\mathbb{R}_+$ action by homoteties on $\mathbb{R}^n$ gives rise to the following definition

$$ {}^n\varkappa_\lambda v(z) = \lambda^{n/2} v(\lambda z), \ \lambda > 0, v \in C_0^\infty(\mathbb{R}^n \setminus 0), $$

being a unitary representation in $L^2(\mathbb{R}^n, dz)$.

In polar coordinates we get

$$ {}^n\varkappa_\lambda u(t,x) = \lambda^{n/2} u(\lambda t, x), u \in C_0^\infty(\mathbb{R}_+ \times S^{n-1}), $$

which is unitary in $L^2(\mathbb{R}_+\times S^{n-1}, t^{n-1}dtdx) = \mathcal{K}^{0,(1-n)/2}(\mathbb{R}_+\times S^{n-1})$.
A characterization of $H^s(\mathbb{R}^d\times\mathbb{R}^n)$ is given by the following generalization of 3.1.1. Lemma 1.

<u>1. Lemma.</u> The norm

$$ \| u\|_{H^s(\mathbb{R}^d\times\mathbb{R}^n)} = \left\{ \int_{\mathbb{R}^d\times\mathbb{R}^n} (1+|\eta|^2+|\zeta|^2)^s |Fu(\eta,\zeta)|^2 d\eta d\zeta \right\}^{1/2} $$

(F denotes the Fourier transform in $\mathbb{R}^{d+n}$) is equal to

$$ \left\{ \int_{\mathbb{R}^d} \langle\eta\rangle^{2s} \| {}^n\varkappa^{-1}_{\langle\eta\rangle} F'u(\eta,\cdot)\|^2_{H^s(\mathbb{R}^n)} d\eta \right\}^{1/2} . $$

This can be proved exactly as 3.1.1. Lemma 1. From the definition of the spaces $\mathcal{K}^s_p$ on cones (or cylinders) in 4.1.1. it is clear that ${}^n\varkappa_\lambda$ acts continuously in these spaces, of course, in general not norm-preserving. We introduce η -depending norms

$$ \| u\|_{\mathcal{K}(\eta)} = \| {}^n\varkappa^{-1}_{\langle\eta\rangle} u\|_{\mathcal{K}} $$

which are equivalent for different η but not uniformly for arbitrary large γ. In particular, this makes sense for spaces on $\mathbb{R}_+ \times N$ defined by means of a partition of unity on N.

2. Definition. Let N be a compact C^∞-manifold without boundary, $n - 1 = \dim N$, $s \in \mathbb{R}$, $P \in \underline{P}(N)$. Denote by $\mathcal{K}_P^s(\mathbb{R}^d \times \mathbb{R}_+ \times N)$ the space defined by the existence of a finite norm

$$\| u \|_{\mathcal{K}_P^s} = \left\{ \int_{\mathbb{R}^d} \langle \eta \rangle^{2s} \| F' u(\eta, \cdot\,\cdot) \|^2_{\mathcal{K}_P^s(\mathbb{R}_+ \times N)(\eta)} d\eta \right\}^{1/2} .$$

Clearly $\mathcal{K}_P^s(\mathbb{R}^d \times \mathbb{R}_+ \times N)$ is a Hilbert space with scalar product induced by a scalar product on $\mathcal{K}_P^s(\mathbb{R}_+ \times N)$. If K is a compact subset of $\mathbb{R}_+ \times N$, then $\mathcal{K}_P^s(\mathbb{R}_+ \times N) \cap \mathcal{E}'(K) = H^s(\mathbb{R}_+ \times N) \cap \mathcal{E}'(K)$ implies in view of Lemma 1 that

$$\mathcal{K}_P^s(\mathbb{R}^d \times \mathbb{R}_+ \times N) \cap \mathcal{E}'(\mathbb{R}^d \times K) = H^s(\mathbb{R}^d \times \mathbb{R}_+ \times N) \cap \mathcal{E}'(\mathbb{R}^d \times K).$$

The natural half-order in $\underline{P}(N)$ yields a continuous embedding

$$\mathcal{K}_P^s(\mathbb{R}^d \times \mathbb{R}_+ \times N) \hookrightarrow \mathcal{K}_{P'}^{s'}(\mathbb{R}^d \times \mathbb{R}_+ \times N)$$

for $s \geq s'$ and $P \leq P'$. It can easily be proved (cf.3.1.2. Lemma 5) that the multiplication by smooth functions in $C_0^\infty(\mathbb{R}^d)$ preserves the space $\mathcal{K}_P^s(\mathbb{R}^d \times \mathbb{R}_+ \times N)$. Moreover, $\mathcal{K}_P^s(\mathbb{R}_+ \times N)$ and $\mathcal{K}_P^s(\mathbb{R}^d \times \mathbb{R}_+ \times N)$ are stable under multiplication by functions $\varphi(t,x) \in C^\infty(\mathbb{R}_+ \times N)$ which are constant for $0 < t < \varepsilon$, $\varepsilon > 0$.
In fact, outside a neighbourhood of $t = 0$ the $\mathcal{K}_P^s$ spaces coincide locally with H^s. In particular, multiplication by functions from (special) partitions of unity are possible.

The decomposition

$$\mathcal{K}_P^s(\mathbb{R}_+ \times N) = \mathcal{K}^{s,\gamma}(\mathbb{R}_+ \times N) + \mathcal{E}(\mathcal{P}),$$

$P = (\gamma, \mathcal{P})$ gives rise to a splitting of an arbitrary $u \in \mathcal{K}_P^s(\mathbb{R}^d \times \mathbb{R}_+ \times N)$ into a sum of singular functions of the form

$$u_{jk}(y,t,x) = \int_{\mathbb{R}^d} e^{iy\eta} \langle \eta \rangle^{n/2} \omega(\langle \eta \rangle t)(\langle \eta \rangle t)^{-p_j} \log^k(\langle \eta \rangle t) \varphi_{jk}(x) \hat{v}(\eta) d\eta , \quad (1)$$

where $0 \leq k \leq m_j$, $(p_j, m_j) \in \mathcal{P}$, $\varphi_{jk} \in L_j$, $v \in H^s(\mathbb{R}^d)$ and ω is a cut-off function, and a remainder

$$u - \sum_j \sum_{k=0}^{m_j} u_{jk} \in \mathcal{K}^{s,\gamma}(\mathbb{R}^d \times \mathbb{R}_+ \times N).$$

In fact, according to Definition 2 we have

$${}^n\varkappa_{\langle\eta\rangle}^{-1} F'u(\eta,.) \in \mathcal{K}_P^s(\mathbb{R}_+ \times N).$$

Choosing ω we get an isomorphism

$$\mathcal{K}_P^s(\mathbb{R}_+ \times N) = \mathcal{K}^{s,\gamma}(\mathbb{R}_+ \times N) \oplus \bigoplus_j \bigoplus_k \mathbb{C}.$$

Applying ${}^n\varkappa_{\langle\eta\rangle}$ and the inverse Fourier transform we get $\mathcal{K}^{s,\gamma}(\mathbb{R}^d \times \mathbb{R}_+ \times N)$ for the remainder and (1) for the singular functions (cf. 3.1.2.).

Note that our singular functions are not simply a product of functions depending on y and on (t,x). They are to be understood as potentials of densities v on $\mathbb{R}^d$. The operation $v \longrightarrow u_{jk}$ given by (1) will turn out to be a special Green operator acting like a potential from the edge of the wedge into its interior. This is in complete analogy to the constructions in Section 3.1.2..

Let now N be a compact C^∞-manifold with boundary ∂N. Then the above notions make sense for $\mathbb{R}^d \times \mathbb{R}_+ \times \partial N$. Function spaces on the interior incorporate a control for $t \longrightarrow 0$ and near ∂N. Let $P \in \underline{P}(N, \partial N)$. Then we can repeat Definition 2 and get a space again denoted by $\mathcal{K}_P^s(\mathbb{R}^d \times \mathbb{R}_+ \times N)$. Outside a neighbourhood of $t = 0$ it reduces simply to $\mathcal{K}_{P_1}^s(\mathbb{R}^d \times \mathbb{R} \times N)$, $P = (P_1, P_2)$. Now the notion of singular expansions must be refined into a two-step procedure. First the expansion with respect to $\partial_1(\mathbb{R}^d \times \mathbb{R}_+ \times N) = \mathbb{R}^d \times \mathbb{R}_+ \times \partial N$, which is determined by P_1, yields densities on $\mathbb{R}^d \times \mathbb{R}_+ \times \partial N$ and a term with good weight, both with P_2-asymptotics for $t \longrightarrow 0$. The t-asymptotics is the second step. We postpone the discussion in detail until the corresponding trace and potential operators will be prepared (cf. 4.3.2.).

4.2. Operators on a cone

Algebras of operators on a cone play a twofold role for problems on singular manifolds: they appear as the local models near conical points and constitute the operator classes where the edge symbols take values. The first case was treated in [R 6] and [S 5]. Here we are interested in the second interpretation.

At the same time we will formulate a more precise variant of the calculus of [R 6], [S 5], i.e. we study proper subclasses of those in the earlier papers. We shall see that the present cone theory is an immediate analogue of that on $\mathbb{R}_+$ in the Chapters 1,2.

4.2.1. Green operators

Green operators on a cone are characterized by their mapping properties. Like in the case of $\mathbb{R}_+$ (cf.1.2.2., 2.3.3.) they do not possess any symbols. Away from the vertex $\partial_2 C$ of the cone C they are smoothing.

First we assume that the base N of the cone $C = \mathbb{R}_+ \times N/\sim$ has no boundary, i.e. $\partial_1 C = \emptyset$. Let $P = (\gamma, \varphi) \in \underline{P}(N)$ be given, $\varphi = \{(p_j, m_j, L_j)\}$. Set $P_o = (\gamma, 0)$, then $\mathcal{K}^s_{P_o}(\mathbb{R}_+ \times N) \subset \mathcal{K}^s_P(\mathbb{R}_+ \times N)$ has finite codimension. Choosing a cut-off function ω we obtain an embedding

$$\bigoplus_{k=o}^{m_j} L_j \ni (v_o, \ldots, v_{m_j}) \longrightarrow \omega(t) \sum_{k=o}^{m_j} t^{-p_j} \log^k t \, v_k \in \mathcal{K}^s_P(\mathbb{R}_+ \times N),$$

the image of which is a complement of $\mathcal{K}^s_{P_o}(\mathbb{R}_+ \times N)$, i.e. we have an isomorphism

$$\mathcal{K}^s_P(\mathbb{R}_+ \times N) \cong \mathcal{K}^s_{P_o}(\mathbb{R}_+ \times N) \oplus_j \bigoplus_{k=o}^{m_j} L_j. \tag{1}$$

We shall use the representation of an arbitrary operator $\mathcal{G} \in \mathcal{L}(\mathcal{K}^s_P(\mathbb{R}_+ \times N), \mathcal{K}^t_Q(\mathbb{R}_+ \times N))$ in the form

$$\mathcal{G} = \begin{pmatrix} \mathcal{G}_{11} & \mathcal{G}_{12} \\ \mathcal{G}_{21} & \mathcal{G}_{22} \end{pmatrix} : \begin{matrix} \mathcal{K}^s_{P_o}(\mathbb{R}_+ \times N) \\ \oplus \\ \bigoplus_j \bigoplus_k L_j \end{matrix} \longrightarrow \begin{matrix} \mathcal{K}^t_{Q_o}(\mathbb{R}_+ \times N) \\ \oplus \\ \bigoplus_j \bigoplus_k L'_j \end{matrix}, \tag{2}$$

where $Q = (\lambda, q), Q_o = (\lambda, 0)$, L'_j are the finite-dimensional spaces occurring in q. Introducing finite bases in L_j and L'_j we may replace $\bigoplus_j \bigoplus_k L_j$ $(\bigoplus_j \bigoplus_k L'_j)$ by $\mathbb{C}^{\dim P}$ $(\mathbb{C}^{\dim Q})$. (2) is used for describing formal adjoints. The scalar product of $L^2(\mathbb{R}_+ \times N, t^{n-1} dt dx)$ induces the duality of $\mathcal{K}^{s,\gamma}(\mathbb{R}_+ \times N)$ and $\mathcal{K}^{-s,\gamma'}(\mathbb{R}_+ \times N)$, where $\gamma' = -\gamma + 1 - n$. Then the formal adjoint of $\mathcal{G}$ takes the form

$$g^* = \begin{pmatrix} g_{11}^* & g_{21}^* \\ g_{12}^* & g_{22}^* \end{pmatrix} : \begin{matrix} \mathcal{K}^{-s}_{P^*_o}(\mathbb{R}_+ \times N) \\ \oplus \\ \mathbb{C}^{\dim P} \end{matrix} \longrightarrow \begin{matrix} \mathcal{K}^{-t}_{Q^*_o}(\mathbb{R}_+ \times N) \\ \oplus \\ \mathbb{C}^{\dim Q} \end{matrix} , \quad (3)$$

where g_{ij}^* is the formal adjoint of g_{ij}.

<u>1. Definition.</u> Let $k \in \mathbb{Z}_+$, $P = (\gamma, \wp)$, $Q = (\lambda, \sigma) \in \underline{P}(N)$. Denote by $\mathcal{L}_G(Q,P)^{(-k)}(C, \partial_2 C)$, $C = \mathbb{R}_+ \times N/\sim$ the subspace of $\bigcap_s \mathcal{L}(\mathcal{K}^s_P \oplus \mathbb{C}, \mathcal{K}^{-\infty}_Q \oplus \mathbb{C})$ of all operators g such that:

(i) there exists an $R \in \underline{P}(N)$, $R = (\lambda + k, \mathcal{N})$, $\bar{F}(\lambda)\mathcal{N} = 0$ such that for all $j \in \mathbb{Z}_+$, $\varepsilon > 0$ the mapping

$$(1+t)^j g : \begin{matrix} \mathcal{K}^s_{P_o}(\mathbb{R}_+ \times N) \\ \oplus \\ \mathbb{C}^{\dim P+1} \end{matrix} \longrightarrow \begin{matrix} \mathcal{K}_{R_\varepsilon}(\mathbb{R}_+ \times N) \\ \oplus \\ \mathbb{C}^{\dim Q+1} \end{matrix} ,$$

$R_\varepsilon = (\lambda + k - \varepsilon, r(\lambda + k - \varepsilon)\mathcal{N})$, is continuous

(ii) the same for g^* (possibly with another R' instead of R).

The extra dimension has an analogous meaning as the boundary symbols of the matrix elements in 3.1.2. formula (5) related to ∂X; cf. also the remarks given there.

Outside a neighbourhood of the vertex $\partial_2 C$ each operator in $\mathcal{L}_G(Q,P)^{(-k)}(C, \partial_2 C)$ is smoothing. For $t \longrightarrow \infty$ the kernel is rapidly decreasing. The same is true for the formal adjoint. The index $-k$ stands for the conormal order.

From the definition it is obvious that $\mathcal{L}_G(Q,P)^{(-k)}(C, \partial_2 C)$ is a two-sided ideal in the sense that the composition with any operator in $\mathcal{L}(\mathcal{K}^s_P, \mathcal{K}^t_P)$ from the right and any operator in $\mathcal{L}(\mathcal{K}^s_Q, \mathcal{K}^t_Q)$ from the left is again in the class of Green operators $\mathcal{L}_G(Q,P)^{(-k)}(C, \partial_2 C)$.

If R and R' in the definition are fixed, we have a natural Fréchet topology induced by the norms of operators in $\mathcal{L}(\mathcal{K}^s_{P_o}(\mathbb{R}_+ \times N) \oplus \mathbb{C}^{\dim P+1}, \mathcal{K}^t_{R_\varepsilon}(\mathbb{R}_+ \times N) \oplus \mathbb{C}^{\dim Q+1})$, where s, t, ε run over a countable set. Therefore $\mathcal{L}_G(Q,P)^{(-k)}(C, \partial_2 C)$ is an (LF) space.

Sometimes it is not convenient to keep the conormal order $-k$ fixed. We can pass to the projective limit for $k \longrightarrow \infty$ and denote the resulting class by $\mathcal{L}_G(Q,P)(C,\partial_2 C)$.

Consider now the case of a cone C with boundary $\partial_1 C$, i.e. the base N is a compact smooth manifold with boundary ∂N. Let $P = (P_1, P_2) \in \underline{P}(N, \partial N)$. Consider the half-space $\mathbb{R}^n_+ = \{(x', x_n) \in \mathbb{R}^n : x_n > 0\}$ as the local model of a boundary neighbourhood on N. By definition

$$\mathcal{K}^s_{(P_1,P_2)}(\mathbb{R}_+ \times \mathbb{R}^n_+) = \mathcal{K}^s_{(P_1,P_{2,0})}(\mathbb{R}_+ \times \mathbb{R}^n_+) + \mathcal{E}(\varphi_2)(\mathbb{R}_+ \times \mathbb{R}^n_+),$$

where $P_2 = (\gamma_2, \varphi_2)$, $P_{2,0} = (\gamma_2, 0)$ and $\mathcal{E}(\varphi_2)$ is a finite-dimensional space spanned by

$$\omega(t) t^{-p_j} \log^k t \; \varphi_{jk}(x', x_n),$$

with $\varphi_{jk} \in L_j \subset \mathcal{K}^\infty_{P_1}(\mathbb{R}^n_+)$. The space $\mathcal{K}^s_{(P_1,P_{2,0})}$ was introduced in 4.1.1. by means of the pull back from the infinite cylinder $\mathbb{R} \times \mathbb{R}^n_+$ in the following way

$$\mathcal{K}^s_{(P_1,P_{2,0})}(\mathbb{R}_+ \times \mathbb{R}^n_+) = \mathfrak{z}(t)^{\gamma_2 - \frac{n}{2}} \varepsilon^* \mathcal{K}^s_{P_1}(\mathbb{R} \times \mathbb{R}^n_+).$$

Now we insert the decomposition of $\mathcal{K}^s_{P_1}(\mathbb{R}_+ \times \mathbb{R}^n_+)$ into $\mathcal{K}^s_{P_{1,0}}(\mathbb{R} \times \mathbb{R}^n_+)$ and a sum of potentials of densities at $x_n = 0$.

Denote by $K(P_1)$ the vector of $\dim P_1$ potential operators of the form

$$v(s, x') \longrightarrow \int_{\mathbb{R}} \int_{\mathbb{R}^{n-1}} e^{i(x'\xi' + s\sigma)} \omega(x_n \langle \xi', \sigma \rangle)(x_n \langle \xi', \sigma \rangle)^{-p_j}$$

$$\log^k (x_n \langle \xi', \sigma \rangle) v(\xi', \sigma) \, d\xi' d\sigma ,$$

where $0 \le k \le m_j, (p_j, m_j) \in \varphi_1, v \in H^s(\mathbb{R} \times \mathbb{R}^{n-1})$.

Then

$$\mathcal{K}^s_{P_1}(\mathbb{R} \times \mathbb{R}^n_+) = \mathcal{K}^s_{P_{1,0}}(\mathbb{R} \times \mathbb{R}^n_+) + K(P_1)(H^s(\mathbb{R} \times \mathbb{R}^{n-1})^{\dim P_1})$$

(cf. 3.2.2.) and we get

$$\mathcal{K}^s_{(P_1,P_{2,0})}(\mathbb{R}_+ \times \mathbb{R}^n_+) = \mathcal{K}^s_{(P_{1,0},P_{2,0})}(\mathbb{R}_+ \times \mathbb{R}^n_+) + \varepsilon_* K(P_1)(\mathcal{K}^s_{P_{2,0}}(\mathbb{R}_+ \times \mathbb{R}^{n-1})^{\dim P_1}),$$

where

$$\varepsilon_* K(P_1) = \varepsilon^* K(P_1)(\varepsilon^{-1})^* .$$

The decomposition

$$\mathcal{K}^s_{(P_1,P_2)}(\mathbb{R}_+\times\mathbb{R}^n_+) = \mathcal{K}^s_{(P_{1,0},P_{2,0})}(\mathbb{R}_+\times\mathbb{R}^n_+) + \varepsilon_* K(P_1)(\mathcal{K}^s_{P_{2,0}}(\mathbb{R}_+\times\mathbb{R}^{n-1})^{\dim P_1}) + \qquad (4)$$
$$+ \mathcal{E}(\varphi_2)(\mathbb{R}_+\times\mathbb{R}^n_+)$$

implies an isomorphism

$$\mathcal{K}^s_{(P_1,P_2)}(\mathbb{R}_+\times\mathbb{R}^n_+) = \mathcal{K}^s_{(P_{1,0},P_{2,0})}(\mathbb{R}_+\times\mathbb{R}^n_+)\oplus\mathcal{K}^s_{P_{2,0}}(\mathbb{R}_+\times\mathbb{R}^{n-1})^{\dim P_1}$$
$$\oplus\ \mathbb{C}^{\dim P_2} \qquad (5)$$

Using this isomorphism we can replace an operator

$$\mathcal{G}: \begin{matrix}\mathcal{K}^s_P(\mathbb{R}_+\times N)\\ \oplus \\ \mathcal{K}^s_{P'}(\mathbb{R}_+\times\partial N)\end{matrix} \longrightarrow \begin{matrix}\mathcal{K}^t_Q(\mathbb{R}_+\times N)\\ \oplus \\ \mathcal{K}^t_{Q'}(\mathbb{R}_+\times\partial N)\end{matrix},$$

$P,Q\in\underline{P}(N,\partial N)$, $P',Q'\in\underline{P}(\partial N)$ in local coordinates of $U\subset N, U\cap\partial N = \emptyset$ by a mapping

$$\mathcal{G}_U: \begin{matrix}\mathcal{K}^s_{P_o}(\mathbb{R}_+\times\mathbb{R}^n_+)\\ \oplus \\ \mathcal{K}^s_{P_{2,0}}(\mathbb{R}_+\times\mathbb{R}^{n-1})^{\dim P_1}\\ \oplus \\ \mathcal{K}^s_{P'_o}(\mathbb{R}_+\times\mathbb{R}^{n-1})\\ \oplus \\ \mathbb{C}^{\dim P_2+\dim P'}\end{matrix} \longrightarrow \begin{matrix}\mathcal{K}^t_{Q_o}(\mathbb{R}_+\times\mathbb{R}^n_+)\\ \oplus \\ \mathcal{K}^t_{Q_{2,0}}(\mathbb{R}_+\times\mathbb{R}^{n-1})^{\dim Q_2}\\ \oplus \\ \mathcal{K}^t_{Q'_o}(\mathbb{R}_+\times\mathbb{R}^{n-1})\\ \oplus \\ \mathbb{C}^{\dim Q_2+\dim Q'}\end{matrix} \qquad (6)$$

where $P_o = (P_{1,0},P_{2,0})$.

If U does not intersect ∂N, the terms containing P_1,Q_1,P',Q' do not appear and we get

$$\mathcal{G}_U: \begin{matrix}\mathcal{K}^s_{P_{2,0}}(\mathbb{R}_+\times\mathbb{R}^n)\\ \oplus\ \mathbb{C}^{\dim P_2}\end{matrix} \longrightarrow \begin{matrix}\mathcal{K}^t_{Q_{2,0}}(\mathbb{R}_+\times\mathbb{R}^n)\\ \oplus\ C^{\dim Q_2}\end{matrix}$$

as before. Similarly as (2) the representation (6) has the advantage that the formal adjoints act in spaces of the same kind. Note that starting from

$$\mathcal{G} \in \mathcal{L}(\mathcal{K}^s_P(\mathbb{R}_+ \times N) \oplus \mathcal{K}^s_{P'}(\mathbb{R}_+ \times \partial N) \oplus \mathbb{C},\ \mathcal{K}^t_Q(\mathbb{R}_+ \times N) \oplus \mathcal{K}^t_{Q'}(\mathbb{R}_+ \times \partial N) \oplus \mathbb{C})$$

we get again a representation (6) with $\dim P_2 + \dim P' + 1$ ($\dim Q_2 + \dim Q' + 1$) instead of $\dim P_2 + \dim P$ ($\dim Q_2 + \dim Q'$). The class $\mathcal{L}_G(\ldots)^{(-k)}(C, \partial_1 C, \partial_2 C)$ is defined in terms of mapping properties of the local representations.

2. Definition. Let $k \in \mathbb{Z}_+$, $P = (P_1, P_2)$, $Q = (Q_1, Q_2) \in \underline{P}(N, \partial N)$, $P', Q' \in \underline{P}(\partial N)$, $C = \mathbb{R}_+ \times N/\sim$, $\partial_1 C = \mathbb{R}_+ \times \partial N/\sim$. Denote by $\mathcal{L}_G(Q, Q', P, P')^{(-k)}(C, \partial_1 C, \partial_2 C)$ the subspace of all $\mathcal{G}$ in $\mathcal{L}(\mathcal{K}^s_P \oplus \mathcal{K}^s_{P'} \oplus \mathbb{C}, \mathcal{K}^{-\infty}_Q \oplus \mathcal{K}^{-\infty}_{Q'} \oplus \mathbb{C})$ such that:

(i) there exist $R = ((\lambda_1 + k, \mathfrak{r}_1), (\lambda_2 + k, \mathfrak{r}_2)) \leq Q_o = ((\lambda_1, 0), (\lambda_2, 0))$ and $R' = (\lambda' + k, \mathfrak{r}') \leq Q'_o = (\lambda', 0)$ such that for all $\varepsilon > 0$, $j \in \mathbb{Z}_+$ the local representation of $(1+t)^j \mathcal{G}$ is in fact a continuous operator in

$$\mathcal{L}(\mathcal{K}^s_{P_o} \oplus (\mathcal{K}^s_{P_{2,o}})^{\dim P_1} \oplus \mathcal{K}^s_{P'_o} \oplus \mathbb{C}^{\dim P_2 + \dim P' + 1},\ \mathcal{K}_{R_\varepsilon} \oplus (\mathcal{K}^\infty_{R_{2,\varepsilon}})^{\dim Q_2} \oplus \mathcal{K}^\infty_{R'_\varepsilon} \oplus \mathbb{C}^{\dim Q_2 + \dim Q'}),$$

where $R_\varepsilon = ((\lambda_1 + k - \varepsilon, r(\lambda_1 + k - \varepsilon)\mathfrak{r}_1), (\lambda_2 + k - \varepsilon, r(\lambda_2 + k - \varepsilon)\mathfrak{r}_2)$,
$R'_\varepsilon = (\lambda' + k - \varepsilon, r(\lambda' + k - \varepsilon)\mathfrak{r}')$,

(ii) the same for the formal adjoint of the local representation, possibly with other R and R'.

In the interior these operators are smoothing. Near $\partial_1 C$ outside a neighbourhood of $\partial_2 C$ they reduce to Green operators considered in 3.2.1. Their kernels are rapidly decreasing near infinity.

The class of Green operatorsis atwo-sided ideal in the sense that the composition with any operator in $\mathcal{L}(\mathcal{K}^s_P \oplus \mathcal{K}^s_{P'} \oplus \mathbb{C}, \mathcal{K}^t_P \oplus \mathcal{K}^t_{P'} + \mathbb{C})$ from the right and any operator in $\mathcal{L}(\mathcal{K}^s_Q \oplus \mathcal{K}^s_{Q'} \oplus \mathbb{C}, \mathcal{K}^t_Q \oplus \mathcal{K}^t_{Q'} \oplus \mathbb{C})$ from the left is again in this class.

The natural topology arising from the definition as in the case $\partial N = \emptyset$ turns $\mathcal{L}_G(\ldots)^{(-k)}(C, \partial_1 C, \partial_2 C)$ into an (LF) space.

4.2.2. Mellin operators

Mellin operators on a cone are defined by operator-valued symbols on the base. Compared with [R6] we mainly restrict ourselves here to the case of rapidly decreasing Mellin symbols with values in the smoothing operators. This will be sufficient for describing the Mellin part of the edge symbols which is not due to the ψDO.

We consider first the case $\partial_1 C = \emptyset$, i.e. the base N of the cone has no boundary.

1. Definition. Let N be a compact C^∞ manifold without boundary. Denote by $\underline{\mathfrak{z}}(N)$ the set of all countable sets $\mathfrak{z} = \{(p_j, m_j, L_j)\}$ such that $\{(p_j, m_j)\} \in \underline{\mathfrak{z}}$ (cf. 1.3.1.) and L_j are finite-dimensional subspaces of $L^{-\infty}(N)$, the space of smoothing operators on N.

Again the operations $r(\gamma)$, $\bar{r}(\gamma)$ etc. and the shifts T^γ make sense.

2. Definition. Denote by $\mathfrak{M}_\mathfrak{z}(N)$, $\mathfrak{z} \in \underline{\mathfrak{z}}(N)$, the space of all meromorphic functions $z \longrightarrow m(z)$ with values in $L^{-\infty}(N)$ having poles at most in p_j, $\mathfrak{z} = \{(p_j, m_j, L_j)\}$ and admitting near each p_j a representation

$$m(z) = \sum_{k=0}^{m_j - 1} \frac{m_{jk}}{(z-p_j)^{k+1}} + m_j(z), \tag{1}$$

where m_j is holomorphic near p_j with values in $L^{-\infty}(N)$ and $m_{jk} \in L_j$. Moreover, for the kernel $m(x,y,z)$ of $m(z)$ all semi-norms

$$\sup_{\substack{x,y \in N \\ z \in S}} \left| D_x D_y D_z^l (\chi_\mathfrak{z}(z) m(x,y,z) \right| (1+|z|)^M \tag{2}$$

are finite, where D_x and D_y are differential operators on N of arbitrary order (applied with respect to the first and the second argument of $m(x,y,z)$), $l, M \in \mathbb{Z}_+$, S is a strip given by

$$S = \{ z \in \mathbb{C} : c_1 \le \operatorname{Re} z \le c_2 \}$$

and $\chi_\mathfrak{z}$ a $\mathfrak{z}$-excision function.

It is not hard to endow $\mathfrak{M}_\mathfrak{z}(N)$ with a topology turning it into a Fréchet space. For a fixed strip we may restrict ourselves to a sequence of $\mathfrak{z}$-excision functions (with decreasing complement of the support) and to powers of the Laplace-Beltrami operator in the semi-norms (2). Consider, moreover, the standard norm of L_j applied to m_{jk} as semi-norm on $\mathfrak{M}_{\mathfrak{z}}(N)$ and the countable number of semi-norms on $C(U_j, L^{-\infty}(N))$, U_j a closed neighbourhood of p_j in $\mathbb{C}$ applied to $m_j(z)$. Taking a sequence of increasing strips, $c_1 \rightarrow -\infty$, $c_2 \rightarrow +\infty$, we obtain a countable system of semi-norms on $\mathfrak{M}_{\mathfrak{z}}(N)$ giving a Fréchet space.

<u>3. Proposition.</u> $\mathfrak{M}_{\mathfrak{z}}(N)$ is nuclear.

<u>Proof.</u> We have

$$\mathfrak{M}_0(N) = \mathfrak{M}_0 \otimes_\pi L^{-\infty}(N) \cong \mathfrak{M}_0 \otimes_\pi C^\infty(N \times N)$$

where $\mathfrak{M}_0$ are the (scalar) holomorphic functions in the whole $\mathbb{C}$ which are in all strips S rapidly decreasing (cf. 1.3.1.). Both factors of the tensor product are nuclear and so is $\mathfrak{M}_0(N)$. Then $\mathfrak{M}_0(N)$ is the projective limit of Hilbert spaces H_j of operator-valued functions holomorphic in a strip S_j with nuclear embeddings $H_{j+1} \hookrightarrow H_j$. Then

$$\mathfrak{M}_{\mathfrak{z}}(N) = \varprojlim \, (H_j + M_j)$$

with finite-dimensional spaces M_j (build from the L_j) and nuclear embeddings $H_{j+1} + M_{j+1} \hookrightarrow H_j + M_j$. $\square$

For abbreviation we simply write $\mathfrak{M}_{\mathfrak{z}}$ instead of $\mathfrak{M}_{\mathfrak{z}}(N)$.

We define compositions and adjoints of Mellin symbols in $\mathfrak{M}_{\underline{\mathfrak{z}}} = \varinjlim_{\mathfrak{z} \in \underline{\mathfrak{z}}} \mathfrak{M}_{\mathfrak{z}}$ pointwise for each fixed z (where the given functions are defined). Then we have the obvious

<u>4. Lemma.</u> (i) Let $m_j \in \mathfrak{M}_{\mathfrak{z}_j}$, j=1,2. Then there exists $\mathfrak{z} \in \underline{\mathfrak{z}}$ depending on $\mathfrak{z}_1, \mathfrak{z}_2$ and the Taylor expansion of m_1 at $\pi_{\mathbb{C}}(\mathfrak{z}_2)$ and of m_2 at $\pi_{\mathbb{C}}(\mathfrak{z}_1)$ such that $m_1(z) m_2(z) \in \mathfrak{M}_{\mathfrak{z}}$.
(ii) If $m \in \mathfrak{M}_{\mathfrak{z}}$, then $m^*(z) = (m(z))^* \in \mathfrak{M}_{\mathfrak{z}^*}$ where $\mathfrak{z}^* = \{(p_j, m_j, L_j^*)\}$ if $\mathfrak{z} = \{(p_j, m_j, L_j)\}$.

Now we associate operators on $\mathbb{R}_+ \times N$ with Mellin symbols. We shall use the notation $\underline{\mathfrak{Z}}(N)[\gamma]$ for the subset of $\underline{\mathfrak{Z}}(N)$ of all $\mathfrak{z}$ satisfying $\pi_{\mathbb{C}}\mathfrak{z} \cap \Gamma_{\frac{1}{2}-\gamma} = \emptyset$. Given $m \in \mathfrak{M}_{\mathfrak{z}}$, $\mathfrak{z} \in \underline{\mathfrak{Z}}(N)[\gamma]$ we define $op_M^{\gamma}(m)$ by

$$op_M^{\gamma}(m)u(t,x) = \frac{1}{2\pi i} \int_{\Gamma_{\frac{1}{2}-\gamma}} t^{-z} m(z) Mu(z,\cdot)dz. \qquad (3)$$

The integral in (3) certainly makes sense if $u \in C_0^{\infty}(\mathbb{R}_+ \times N)$. The estimates (2) in Definition 2 yield immediately the continuity of $op_M^{\gamma}(m)$ as a mapping

$$op_M^{\gamma}(m) : \mathcal{H}^{s,\gamma}(\mathbb{R}_+ \times N) \longrightarrow \mathcal{H}^{\infty,\gamma}(\mathbb{R}_+ \times N)$$

for every $s \in \mathbb{R}$, and also

$$\omega op_M^{\gamma}(m)\omega : \mathcal{K}^{s,\gamma}(\mathbb{R}_+ \times N) \longrightarrow \mathcal{K}^{\infty,\gamma}(\mathbb{R}_+ \times N),$$

where ω is an arbitrary cut-off function. Consider now spaces with asymptotics

$$\mathcal{K}_P^s(\mathbb{R}_+ \times N) \cong \mathcal{K}^{s,\gamma}(\mathbb{R}_+ \times N) \oplus \mathcal{E}(\mathfrak{p})$$

where $P = (\gamma, \mathfrak{p})$. For all $\delta < \delta_0(P)$ we have $\mathcal{K}_P^s(\mathbb{R}_+ \times N) \subset \mathcal{K}^{s,\delta}(\mathbb{R}_+ \times N)$. Assume that δ is chosen in such a way that $\mathfrak{z} \in \underline{\mathfrak{Z}}(N)[\delta]$. Consider the restriction of $\omega op_M^{\delta}(m)\omega$ to $\mathcal{K}_P^s(\mathbb{R}_+ \times N)$.

<u>5. Lemma.</u> (i) $\omega op_M^{\delta}(m)\omega$ induces a continuous operator

$$\omega op_M^{\delta}(m)\omega \quad : \mathcal{K}_P^s(\mathbb{R}_+ \times N) \longrightarrow \mathcal{K}_Q^{\infty}(\mathbb{R}_+ \times N)$$

where $Q = (\gamma, \mathfrak{q})$, $\mathfrak{q} = \mathfrak{q}(\mathfrak{p}, m)$.

(ii) The difference $\omega[op_M^{\delta}(m) - op_M^{\gamma}(m)]\omega$ on $\mathcal{K}^{s,\gamma}(\mathbb{R}_+ \times N)$, $\gamma < \delta$, is a finite range Green operator $\mathcal{G} \in \mathcal{L}_G$.

<u>Proof.</u> (ii) Using $(p_j, m_j, L_j) \in \mathfrak{z}$ with $\frac{1}{2} - \gamma < \operatorname{Re} p_j < \frac{1}{2} - \delta$ we find a finite range Green operator $\mathcal{G} \in \mathcal{L}_G$ such that $P(\ker L_j)\frac{d^k}{dz^k} M(u - \mathcal{G}u)(z_j, \cdot) = 0$, $0 \leq k \leq m_j$, where $P(\ker L_j)$ denotes a projector onto the

null-space of all operators in L_j. Then

$$\omega(\mathrm{op}_M^{\delta}(m) - \mathrm{op}_M^{\gamma}(m))\omega - \mathcal{G} = 0$$

because all poles of $m(z)M(u-\mathcal{G}u)(z,\cdot)$ in $\frac{1}{2}-\gamma < \mathrm{Re}\, z < \frac{1}{2}-\delta$ disappear.

(i) The projection $\mathcal{K}_P^s(\mathbb{R}_+ \times N) \longrightarrow \mathcal{K}^{s,\gamma}(\mathbb{R}_+ \times N)$ has the form $1-\mathcal{G}$, $\mathcal{G} \in \mathcal{L}_G$. Up to a finite range operator with values in $\mathcal{K}_Q^{\infty}(\mathbb{R}_+ \times N)$ we may consider $\omega\, \mathrm{op}_M^{\delta} \omega(1-\mathcal{G})$, which is modulo $\mathcal{L}_G$ equal to $\omega\, \mathrm{op}_M^{\gamma}(m)\omega$. □

Let $k \in \mathbb{Z}_+$, $\gamma \in \mathbb{R}$, be given and $m = \{m_0, m_1, \dots, m_{k-1}\} \in \mathop{\times}\limits_{j=0}^{k-1} \mathfrak{M}_{\mathfrak{z}^j}$, where $\mathfrak{z}^0 \in \underline{\mathfrak{z}}(N)[\gamma]$, $\mathfrak{z}^j \in \underline{\mathfrak{z}}(N)$, $j>0$. We associate with $\{m_j\}$ an operator by

$$\omega\mathrm{op}_M^{\gamma}(m)\omega = \omega \sum_{j=0}^{k-1} t^j \mathrm{op}_M^{\gamma_j}(m_j)\omega , \qquad (4)$$

where $\gamma \le \gamma_j \le \gamma + j$ are chosen in such a way that $\mathfrak{z}^j \in \underline{\mathfrak{z}}(N)[\gamma_j]$. Of course, this definition depends on the choice of γ_j. It is independent of this choice modulo Green operators. Then

$$\omega\, \mathrm{op}_M^{\gamma}(m)\omega \;: \mathcal{K}^{s,\gamma}(\mathbb{R}_+ \times N) \longrightarrow \mathcal{K}^{\infty,\gamma}(\mathbb{R}_+ \times N)$$

is continuous.

<u>6. Definition.</u> An operator $\mathfrak{A}$ is said to belong to $\mathcal{L}_{M+G}(Q,P)^{(-k)}(C)$ if there exists $m \in \mathop{\times}\limits_{j=0}^{k-1} \mathfrak{M}_{\mathfrak{z}^j}$, $\mathfrak{z}^0 \in \underline{\mathfrak{z}}(N)[\gamma]$, $P = (\gamma, \mathfrak{p})$, $Q = (\lambda, \mathfrak{q})$, $\mathfrak{z}^j \in \underline{\mathfrak{z}}(N)$ such that with appropriate γ_j we have $\mathfrak{A} - \omega t^{\gamma-\lambda}\mathrm{op}_M(m)\omega \in \mathcal{L}_G(Q,P)^{(-k)}(C)$.

From Lemma 5 it is clear that the choice of γ_j has no influence whether or not a given $\mathfrak{A}$ belongs to our class.

As in 1.3.1. and 2.3.5. it can be proved that the operator $\mathfrak{A}$ possesses uniquely determined Mellin symbols $m_j \in \mathfrak{M}_{\mathfrak{z}^j}$. If we denote the mapping $\mathfrak{A} \longrightarrow m = \{m_j\}_{j=1}^{k}$ by $\sigma_M^{(k)}$ we get the exact sequence

$$0 \longrightarrow \mathcal{L}_G(Q,P)^{(-k)}(C) \hookrightarrow \mathcal{L}_{M+G}(Q,P)^{(-k)}(C) \xrightarrow{\sigma_M^{(k)}} \mathfrak{M}_{\underline{\mathfrak{z}}(N)[\gamma]} \mathop{\times}\limits_{j=1}^{k-1} \mathfrak{M}_{\underline{\mathfrak{z}}(N)} \longrightarrow 0 . \qquad (5)$$

<u>7. Proposition.</u> $\mathfrak{A} \in \mathcal{L}_{M+G}(Q,P)^{(-k)}(C)$ defines a continuous mapping

$$\mathfrak{A} : \mathcal{K}_P^s(\mathbb{R}_+ \times N) \longrightarrow \mathcal{K}_Q^{\infty}(\mathbb{R}_+ \times N) .$$

The restriction to $\mathcal{K}_{P'}^s(\mathbb{R}_+ \times N) \subset \mathcal{K}_P^s(\mathbb{R}_+ \times N)$, $P' = (\gamma', \mathfrak{p}')$, $\gamma < \gamma' < \gamma + k$, gives a continuous operator

$$\mathcal{O}\!\iota : \mathcal{K}^{s}_{P'}(\mathbb{R}_+ \times N) \longrightarrow \mathcal{K}^{s}_{R}(\mathbb{R}_+ \times N),$$

where $R = (r, \varkappa)$, $r - \lambda < \gamma' - \gamma$, $Q = (\lambda, q)$.

This follows from the mapping properties of Green operators and the above remarks about elementary Mellin operators.

As for scalar Mellin symbols we have the same rules of compositions and adjoints which preserve the class of Mellin symbols. In particular, the analogue of 1.3.2. (5) now reads as follows.

Let $m = (m_o, m_1, \ldots, m_{k-1}) \in \mathop{\times}\limits_{j=0}^{k-1} \mathfrak{M}_{\underline{\gamma}}$ and $m' = (m'_o, m'_1, \ldots, m'_{k-1}) \in \mathop{\times}\limits_{j=0}^{k-1} \mathfrak{M}_{\underline{\gamma}}$.

Then we set

$$(m \# m')_j(z) = \sum_{k+l=j} m_k(z - l) m_l(z), \qquad (6)$$

which belongs to our class of Mellin symbols again.

8. Theorem. (i) Let $\mathcal{O}\!\iota_j \in \mathcal{L}_{M+G}^{(-k)}$, $j = 1,2$, be such that $\mathcal{O}\!\iota_1 \mathcal{O}\!\iota_2$ is defined. Then $\mathcal{O}\!\iota_1 \mathcal{O}\!\iota_2 \in \mathcal{L}_{M+G}^{(-k)}$ and the Mellin symbols are evaluated by

$$\sigma^{(k)}_M(\mathcal{O}\!\iota_1 \mathcal{O}\!\iota_2) = \sigma^{(k)}_M(\mathcal{O}\!\iota_1) \# \sigma^{(k)}_M(\mathcal{O}\!\iota_2).$$

(ii) Let $\mathcal{O}\!\iota \in \mathcal{L}_{M+G}(Q,P)^{(-k)}$, $P = (\gamma, 0)$, $Q = (\lambda, 0)$. Then $\mathcal{O}\!\iota^* \in \mathcal{L}_{M+G}(P^*, Q^*)^{(-k)}$, $P^* = (-\gamma+n-1, 0)$, $Q^* = (-\lambda+n-1, 0)$ and for the leading Mellin symbol we have

$$\sigma^{o}_M(\mathcal{O}\!\iota^*)(z) = \sigma^{o}_M(\mathcal{O}\!\iota)^*(n-z).$$

Proof. Since Green operators are a two-sided $*$-ideal (cf. 4.2.1.), we may restrict ourselves to elementary Mellin operators (modulo Green operators). Using Lemma 5 we may commute t^l and Mellin operators causing an argument shift by $-l$ in the Mellin symbol. This implies (i). (ii) is proved as 1.3.1. Lemma 6, noting that $t^{n-1}\,dtdx$ is the measure involved in the reference L^2 space. □

Without difficulties one may pass to the case when $k = \infty$ and an infinite number of Mellin symbols is present. The only non-trivial point is the construction of an operator for a given sequence of Mellin symbols. It is done as usually by a suitable choice of the cut-off functions ω_j.

Let us now treat the case $\partial_1 C \neq 0$, i.e. the base N of the cone is a compact manifold with non-empty boundary ∂N. Fix $P_1, Q_1 \in \underline{P}$. Mellin symbols will be defined as meromorphic families with values in $\mathcal{W}^{-\infty}(Q_1,P_1)(N,\partial N)$ (cf. 3.2.1.). First we introduce the corresponding class of singularity types.

1'. Definition. Denote by $\mathfrak{z}(N,\partial N)$ the set of all countable sets $\mathfrak{z} = \{(p_j, m_j, L_j)\}$ such that $\{(p_j, m_j)\} \in \underline{\mathfrak{z}}$ and L_j are finite-dimensional subspaces of $\mathcal{W}^{-\infty}(Q_1,P_1)(N,\partial N)$.

Clearly the specialization to smoothing operators on ∂N gives Definition 1.

The space $\mathcal{W}_G^{-\infty}(Q_1,P_1)(N,\partial N)$ is endowed with an (LF) topology in the same way as done in 2.3.4. for $\mathcal{L}_G(Q,P)$. In fact, keeping a singularity type $v \in \underline{\mathcal{P}}^{\gamma_1}$ fixed, the subspace of $\mathcal{W}_G^{-\infty}$ acting continuously as

$$\begin{matrix} \mathcal{K}^s_{P_1}(N) \\ \oplus^1 \\ H^s(\partial N) \end{matrix} \longrightarrow \begin{matrix} \mathcal{K}^\infty_{v}(N) \\ \oplus \\ H^\infty(\partial N) \end{matrix}$$

is Fréchet. Restricting the adjoint by choosing v' we get a Fréchet space $\mathcal{W}_G^{-\infty}(Q_1,P_1)(v,v')$ and

$$\mathcal{W}_G^{-\infty}(Q_1,P_1) = \varinjlim_{v,v'} \mathcal{W}_G^{-\infty}(Q_1,P_1)(v,v').$$

2'. Definition. Denote by $\mathfrak{M}_{\mathfrak{z}}(N,\partial N)$, $\mathfrak{z} \in \mathfrak{z}(N,\partial N)$, the space of all meromorphic functions $z \longrightarrow m(z)$ with values in $\mathcal{W}_G^{-\infty}(Q_1,P_1)$ having poles at most at p_j, $\mathfrak{z} = \{(p_j, m_j, L_j)\}$, admitting a Laurent expansion near each p_j

$$m(z) = \sum_{k=0}^{m_j - 1} \frac{m_{jk}}{(z-p_j)^{k+1}} + m_j(z), \qquad (1')$$

where m_j is holomorphic near p_j with values in $\mathcal{W}_G^{-\infty}(Q_1,P_1)$ and $m_{jk} \in L_j$. Moreover, for each semi-norm q on

$\mathcal{D}_G^{-\infty}(Q_1,P_1)(\mathfrak{N},\mathfrak{N}')$, $l,M\in\mathbb{Z}_+$, $\chi_{\mathfrak{Z}}$ a $-\mathfrak{Z}$ excision function

$$\sup_{z\in S} q\left[D_z^l(\chi_{\mathfrak{Z}}(z)m(z))\right](1+|z|)^M \qquad (2')$$

is finite, $S=\{z\in\mathbb{C} : c_1\le \operatorname{Re} z\le c_2\}$.

Its complete analogy to the case without boundary $\mathcal{M}_{\mathfrak{Z}}(N,\partial N)$ is endowed with an (LF) space topology.

<u>3: Proposition.</u> $\mathcal{M}_{\mathfrak{Z}}(N,\partial N)$, $\mathfrak{Z}\in\underline{\mathfrak{Z}}(N,\partial N)$ is nuclear.

<u>Proof.</u> Use that

$$\mathcal{M}_O(N,\partial N)=\mathcal{M}_O\hat{\otimes}_\pi\mathcal{D}_G^{-\infty}(Q_1,P_1)$$

and the arguments of the proof of Proposition 3. $\square$

As above, for abbreviation we also write $\mathcal{M}_{\mathfrak{Z}}$ instead of $\mathcal{M}_{\underline{\mathfrak{Z}}}(N,\partial N)$ and similarly for

$$\mathcal{M}_{\underline{\mathfrak{Z}}(N,\partial N)}(N,\partial N):=\mathcal{M}_{\underline{\mathfrak{Z}}(N,\partial N)}=\varinjlim_{\mathfrak{Z}\in\underline{\mathfrak{Z}}(N,\partial N)}\mathcal{M}_{\mathfrak{Z}}$$

<u>4: Lemma.</u> (i) Let $m_j\in\mathcal{M}_{\underline{\mathfrak{Z}}(N,\partial N)}$, $j=1,2$, and assume that $m(z):=m_1(z)m_2(z)$ is defined for almost all z. Then $m(z)\in\mathcal{M}_{\underline{\mathfrak{Z}}(N,\partial N)}$.

(ii) Let $m\in\mathcal{M}_{\underline{\mathfrak{Z}}(N,\partial N)}$ and assume that a formal adjoint makes sense for almost all z, then $m^*(z)=m(z)^*\in\mathcal{M}_{\underline{\mathfrak{Z}}(N,\partial N)}$.

Consider now $P=(P_1,P_2)$, $Q=(Q_1,Q_2)\in\underline{P}(N,\partial N)$. Assume first $P_2=(\gamma,0)$, $Q_2=(\gamma,0)$. For $m\in\mathcal{M}_{\mathfrak{Z}}$, $\mathfrak{Z}\in\underline{\mathfrak{Z}}(N,\partial N)[\gamma]$ we define $op_M^\gamma(m)$ by (3) where $u\in\mathcal{K}_P^s(\mathbb{R}_+\times N)\oplus\mathcal{K}_{P_2}^s(\mathbb{R}_+\times\partial N)$ and get by (2') a continuous mapping

$$\omega\, op_M^\gamma(m)\,\omega : \begin{array}{c}\mathcal{K}_P^s(\mathbb{R}_+\times N)\\ \oplus\\ \mathcal{K}_{P_2}^s(\mathbb{R}_+\times\partial N)\end{array}\longrightarrow\begin{array}{c}\mathcal{K}_Q^\infty(\mathbb{R}_+\times N)\\ \oplus\\ \mathcal{K}_{Q_2}^\infty(\mathbb{R}_+\times\partial N)\end{array}$$

The case of general $P_2'=(\gamma,\mathfrak{P}_2)$ is included by embedding into a space with weight $\mathcal{K}_P^s$, $P=(P_1,(\delta,0))$. The choice of δ is not essential modulo Green operators.

<u>5: Lemma.</u> (i) The restriction of $\omega\, op_M^\delta(m)\,\omega$ to $\mathcal{K}_P^s$ gives a continuous operator

$$\omega \operatorname{op}_M^{\delta}(m)\,\omega : \begin{matrix} \mathcal{K}^{s}_{P'}(\mathbb{R}_+ \times N) \\ \oplus \\ \mathcal{K}^{s}_{P_2'}(\mathbb{R}_+ \times \partial N) \end{matrix} \longrightarrow \begin{matrix} \mathcal{K}^{\infty}_{Q'}(\mathbb{R}_+ \times N) \\ \\ \mathcal{K}^{\infty}_{Q_2'}(\mathbb{R}_+ \times \partial N) \end{matrix}$$

where $Q_2' = (\gamma, \mathcal{Q}_2), Q' = (Q_1, Q_2')$.

(ii) The difference $\omega[\operatorname{op}_M^{\delta}(m) - \operatorname{op}_M^{\gamma}(m)]\omega$, $\mathfrak{Z} \in \underline{\mathfrak{Z}}(N,\partial N)[\gamma] \cap \underline{\mathfrak{Z}}(N,\partial N)[\delta]$ on $\mathcal{K}^s_P(\mathbb{R}_+ \times N) \oplus \mathcal{K}^s_{P_2}(\mathbb{R}_+ \times \partial N)$, $P = (P_1, P_2)$, $P_2 = (\gamma, O)$, $\gamma < \delta$, is a finite-range Green operator.

The proof is completely analogous to the proof of Lemma 5.

With given $m = \{m_o, m_1, \dots, m_{k-1}\} \in \mathop{X}\limits_{j=0}^{k-1} \mathfrak{M}_{\mathfrak{Z}^j}, \mathfrak{Z}^j \in \underline{\mathfrak{Z}}(N,\partial N)$ we associate an operator $\omega \operatorname{op}_M^{\gamma}(m)\omega, \mathfrak{Z}^o \in \underline{\mathfrak{Z}}(N,\partial N)[\gamma]$, by (4). The choice of the γ_j has no influence modulo Green operators and

$$\omega \operatorname{op}_M^{\gamma}(m)\omega : \begin{matrix} \mathcal{K}^{s}_{P}(\mathbb{R}_+ \times N) \\ \oplus \\ \mathcal{K}^{s}_{P_2}(\mathbb{R}_+ \times \partial N) \end{matrix} \longrightarrow \begin{matrix} \mathcal{K}^{\infty}_{Q}(\mathbb{R}_+ \times N) \\ \oplus \\ \mathcal{K}_{Q_2}(\mathbb{R}_+ \times \partial N) \end{matrix}$$

$P = (P_1, P_2), P_2 = (\gamma, O), Q = (Q_1, Q_2), Q_2 = (\gamma, O)$, is continuous.

6: <u>Definition</u>. An operator $\mathfrak{a}$ is said to belong to $\mathcal{L}_{M+G}(Q', P')^{(-k)}$ $(C, \partial_1 C)$, $P' = (P_1, P_2')$, $P_2' = (\gamma, \mathcal{P}_2)$, $Q' = (Q_1, Q_2')$, $Q_2' = (\lambda, \mathcal{Q}_2)$, if there exists $m \in \mathop{X}\limits_{j=0}^{k-1} \mathfrak{M}_{\mathfrak{Z}^j}$, $\mathfrak{Z}^j \in \underline{\mathfrak{Z}}(N,\partial N)$, $\mathfrak{Z}^o \in \underline{\mathfrak{Z}}(N,\partial N)[\gamma]$ such that

$$\mathfrak{a} - \omega t^{\gamma-\lambda} \operatorname{op}_M^{\gamma}(m)\,\omega \in \mathcal{L}_G(Q', P')^{(-k)}(C, \partial_1 C).$$

Since Mellin operators of the form (4) belong to $\mathcal{L}_G^{(-k)}$ only if all m_j vanish, we have k uniquely determined Mellin symbols of $\mathfrak{a} \in \mathcal{L}_{M+G}^{(-k)}$ and the sequence

$$0 \longrightarrow \mathcal{L}_G(Q',P')^{(-k)}(C,\partial_1 C) \hookrightarrow \mathcal{L}_{M+G}(Q',P')^{(-k)}(C,\partial_1 C) \xrightarrow{\sigma_M^{(k)}} \tag{5'}$$

$$\longrightarrow \mathfrak{M}_{\underline{\mathfrak{Z}}(N,\partial N)[\gamma]} \mathop{X}\limits_{j=1}^{k-1} \mathfrak{M}_{\underline{\mathfrak{Z}}(N,\partial N)} \longrightarrow 0$$

is exact, where $\sigma_M^{(k)}(\mathfrak{a}) = \{m_o, m_1, \dots, m_{k-1}\}$ is the Mellin symbol mapping.

7'. Proposition. Let $\sigma \in \mathcal{L}_{M+G}(Q',P')^{(-k)}(C,\partial_1 C)$. Then σ defines a continuous mapping

$$\sigma : \begin{array}{c} \mathcal{K}^s_{P'}(\mathbb{R}_+ \times N) \\ \oplus \\ \mathcal{K}^s_{P_2'}(\mathbb{R}_+ \times \partial N) \end{array} \longrightarrow \begin{array}{c} \mathcal{K}^\infty_{Q'}(\mathbb{R}_+ \times N) \\ \oplus \\ \mathcal{K}^\infty_{Q_2'}(\mathbb{R}_+ \times \partial N). \end{array}$$

The restriction to $\mathcal{K}^s_{P''} \oplus K^s_{P_2''}$, $P'' = (P_1,P_2'')$, $P_2'' = (\gamma',\mathfrak{p}_2')$, $\gamma < \gamma' < \gamma + k$, gives a continuous operator

$$\sigma : \begin{array}{c} \mathcal{K}^s_{P''}(\mathbb{R}_+ \times N) \\ \oplus \\ \mathcal{K}^s_{P_2''}(\mathbb{R}_+ \times \partial N) \end{array} \longrightarrow \begin{array}{c} \mathcal{K}^\infty_{Q''}(\mathbb{R}_+ \times N) \\ \oplus \\ \mathcal{K}^\infty_{Q_2''}(\mathbb{R}_+ \times \partial N) \end{array}$$

where $Q'' = (Q_1'',Q_2'')$, $Q_2'' = (\lambda',\mathfrak{q}_2')$, $\lambda' - \lambda < \gamma' - \gamma$, Q_1'' has an arbitrarily large weight.

The composition rule (6) makes sense also in the case with boundary (provided the domains and ranges fit together).

Then Theorem 8 remains valid where in (ii) the singularity types $P' = (P_1,P_2)$ and $Q' = (Q_1,Q_2)$ have to be of the form $P_1 = (\gamma_1,0)$, $P_2 = (\gamma_2,0)$, $Q_1 = (\lambda_1,0)$, $Q_2 = (\lambda_2,0)$.

4.2.3. The algebra $\mathcal{L}^\infty$ on a cone

The goal of this section is the extension of the class $\mathcal{L}_{M+G}$ by pseudo-differential operators. For the resulting class $\mathcal{L}^\infty$ we derive a precise symbolic calculus. The operator classes here are proper subclasses of the algebras of operators on a cone introduced in [R 6] and [R 8]. In this class our wedge symbols take values. The most important step here is the proof that arbitrary (classical) pseudo-differential symbols give rise to a Mellin expansion with meromorphic operator valued Mellin symbols. As in the one-dimensional case (cf. 2.3.5.) it follows that a ψDO (possibly after a modification by a Green operator) admits an extension to $\mathcal{K}^s_{\underline{P}}$ spaces and that the class is closed under composition and adjoints.

Let us recall briefly some well-known facts about homogeneous functions and distributions in $\mathbb{R}^n$. Let $u \in C^\infty(\mathbb{R}^n \setminus 0)$ be homogeneous of degree $z \in \mathbb{C}$, i.e. $u(\lambda\tilde{x}) = \lambda^z u(\tilde{x})$, $\lambda \in \mathbb{R}_+$, $\tilde{x} \in \mathbb{R}^n \setminus 0$. Then, for each $v \in C_0^\infty(\mathbb{R}^n \setminus 0)$

$$\int_{\mathbb{R}^n} u(\tilde{x})\, {}^n\varkappa_\lambda v(\tilde{x})d\tilde{x} = \int_{\mathbb{R}^n} u(\tilde{x})\lambda^{n/2} v(\lambda\tilde{x})d\tilde{x} =$$

$$= \lambda^{-n/2}\int_{\mathbb{R}^n} u(\lambda^{-1}\tilde{x})v(\tilde{x})d\tilde{x} = \lambda^{-z-n/2}\int_{\mathbb{R}^n} u(\tilde{x})v(\tilde{x})d\tilde{x}.$$

A distribution $u \in \mathcal{D}'(\mathbb{R}^n \setminus 0)(\mathcal{D}'(\mathbb{R}^n))$ is called homogeneous of degree $z \in \mathbb{C}$ if for each $\lambda \in \mathbb{R}_+, v \in C_0^\infty(\mathbb{R}^n \setminus 0)(C_0^\infty(\mathbb{R}^n))$

$$\langle u, {}^n\varkappa_\lambda v\rangle = \lambda^{-z-n/2}\langle u, v\rangle.$$

Denote by $\mathcal{D}'(\mathbb{R}^n \setminus 0)^{(z)}(\mathcal{D}'(\mathbb{R}^n)^{(z)})$ the space of all homogeneous distributions on $\mathbb{R}^n \setminus 0(\mathbb{R}^n)$ of degree z. Let $C^\infty(\mathbb{R}^n \setminus 0)^{(z)}$ be the subspace of smooth functions in $\mathcal{D}'(\mathbb{R}^n \setminus 0)^{(z)}$. Moreover, we set

$$\mathcal{D}'(\mathbb{R}^n)^{(z)}_\infty = \left\{u \in \mathcal{D}'(\mathbb{R}^n)^{(z)} : u\big|_{\mathbb{R}^n \setminus 0} \in C^\infty(\mathbb{R}^n \setminus 0)\right\}$$

with the topology induced by the embedding $\mathcal{D}'(\mathbb{R}^n)^{(z)}_\infty \subset \mathcal{D}'(\mathbb{R}^n)^{(z)}$ and the restriction $\mathcal{D}'(\mathbb{R}^n)^{(z)} \longrightarrow C^\infty(\mathbb{R}^n \setminus 0)$. For given z each smooth function $v \in C^\infty(S^{n-1})$ gives rise to a function $u^{(z)}(v) \in C^\infty(\mathbb{R}^n \setminus 0)^{(z)}$ given by

$$u^{(z)}(v)(\tilde{x}) = |\tilde{x}|^z v\left(\frac{\tilde{x}}{|\tilde{x}|}\right).$$

If $\pi : \mathbb{R}^n \setminus 0 \longrightarrow \mathbb{R}_+ \times S^{n-1}$, $\pi(\tilde{x}) = (t, x)$, is the diffeomorphism corresponding to polar coordinates, then

$$(\pi^{-1})^* u^{(z)}(v)(t, x) = u^{(z)}(v)(tx) = t^z v(x).$$

Interpreting $u^{(z)}(v)$ as a density of a distribution

$$C_0^\infty(\mathbb{R}^n \setminus 0) \ni \varphi \longrightarrow \langle u^{(z)}(v), \varphi\rangle = \int_{\mathbb{R}^n} u^{(z)}(v)(\tilde{x})\varphi(\tilde{x})d\tilde{x}$$

the push forward defined by

$$\langle \pi_* u, \psi\rangle = \langle u, \pi^*\psi|d\pi|\rangle, \psi \in C_0^\infty(\mathbb{R}_+ \times S^{n-1})$$

gives

$$\langle \pi_* u^{(z)}(v),\psi\rangle = \int_{\mathbb{R}^n} |\tilde{x}|^z v\left(\frac{\tilde{x}}{|\tilde{x}|}\right) \psi\left(|\tilde{x}|,\frac{\tilde{x}}{|\tilde{x}|}\right) d\tilde{x}$$

$$= \int_0^\infty dt \int_{S^{n-1}} t^{z+n-1} v(x)\,\psi(t,x)\,dx$$

where dx is the density on S^{n-1} induced by the standard metric.

By regularization at $\tilde{x}=0$ one gets distributions on $\mathbb{R}^n$. The canonical regularization $\dot{u}^{(z)}(v)$ by analytic extension with respect to z from $\operatorname{Re} z > -n$ gives homogeneous distributions in $\mathbb{R}^n$ for all $z\in\mathbb{C}, z\neq -n,-n-1,\dots,-n-j,\dots$ It is expressed in terms of the distributions $t_+^z \in \mathcal{D}'(\mathbb{R})$

$$\langle \dot{u}^{(z)}(v),\varphi\rangle = \int_{S^{n-1}} v(x)\,\langle t_+^{z+n-1},\varphi(tx)\rangle\,dx,$$

$\varphi\in C_0^\infty(\mathbb{R}^n)$, which can be written for $\varphi\in C_0^\infty(\mathbb{R}^n\setminus 0)$ as

$$\int_0^\infty dt \int_{S^{n-1}} v(x)\,t^{z+n-1}\,\varphi(tx)\,dx.$$

The family $z \longrightarrow t_+^{z+n-1} \in \mathcal{D}'(\mathbb{R})$ of distributions depends meromorphically on z with simple poles at $z=-n,-n-1,\dots,-n-j,\dots$ and residues at $-n-j$

$$\operatorname*{Res}_{z=-n-j} t_+^{z+n-1} = \frac{(-1)^j}{j!}\,\delta^{(j)}.$$

Normalizing by the Γ-function we get that

$$\frac{1}{\Gamma(z+n)}\,t_+^{z+n-1}$$

is an entire function with values in $\mathcal{D}'(\mathbb{R})$ and the value $\delta^{(j)}$ at $z=-n-j$. It follows that

$$\frac{1}{\Gamma(z+n)}\,\dot{u}^{(z)} : C^\infty(S^{n-1}) \longrightarrow \mathcal{D}'(\mathbb{R}^n)^{(z)}_\infty$$

is a holomorphic family of continuous operators. At $z=-n-j$ the value is the finite-range operator

$$v \longrightarrow \sum_{|\alpha|=j} \frac{(-1)^j}{\alpha!} \int_{S^{n-1}} v(x)x^\alpha\,dx\,\partial_x^\alpha\delta \quad .$$

Recall the following

1. <u>Proposition</u>. Each homogeneous distribution in $\mathbb{R}^n$ belongs to $\mathcal{S}'$. The Fourier transform yields topological isomorphisms

$$\mathcal{D}'(\mathbb{R}^n)^{(z)} \xrightarrow{F} \mathcal{D}'(\mathbb{R}^n)^{(-n-z)} \quad \text{and} \quad \mathcal{D}'(\mathbb{R}^n)^{(z)}_\infty \xrightarrow{F} \mathcal{D}'(\mathbb{R}^n)^{(-n-z)}_\infty .$$

A proof is given in [H 3], vol. I.

Now we turn to ψDO's on a conical manifold $C = \mathbb{R}_+ \times N/\sim$ where $\partial N = \emptyset$. Let $U \subset N$ be open and $U \longrightarrow \Omega \subset S^{n-1}$ be a coordinate homeomorphism. Besides the coordinates $(t,x) \in \mathbb{R}_+ \times S^{n-1}$ we also use euclidean coordinates $\tilde{x} = tx \in \mathbb{R}^n$. Among all classical ψDO's on $C \setminus \partial_2 C$ we are going to distinguish a subclass with a special behaviour with respect to t. We start from x-independent homogeneous symbols. Then we consider the more general case of classical symbols, all homogeneous components of which have an asymptotic expansion when $\tilde{x} \longrightarrow 0$.

Let $u \in C_0^\infty(\mathbb{R}_+ \times \Omega)$, $\Omega \subset S^{n-1}$. Then the inversion formula for the Mellin transform

$$u(t,x) = \frac{1}{2\pi i} \int_{\Gamma_{1/2-\gamma}} t^{-z} Mu(z,x)dz, \ \gamma \in \mathbb{R},$$

gives a decomposition of u into functions $t^{-z}Mu(z,x)$ homogeneous of degree $-z$. Assume now that a given operator A obeys the property that $t^\mu A$ maps $\mathcal{D}'(\mathbb{R}^n)^{(-z)}_\infty$ into itself. Then we get formally

$$t^\mu Au(t,x) = \frac{1}{2\pi i} \int_{\Gamma_{1/2-\gamma}} t^{-z}(t^{z+\mu}A(s^{-z}Mu(z,.))(t,x))dz$$

and $v(x) \longrightarrow t^{z+\mu}A(s^{-z}v(.))(t,x)$ can be viewed as operator-valued Mellin symbol of conormal order 0 of $t^\mu A$.

Let $a_\mu(\tilde{\xi}) \in C^\infty(\mathbb{R}^n \setminus 0)^{(\mu)}$ and $\chi(\tilde{\xi})$ an excision function. Denote by A_μ^χ the ψDO on $\mathbb{R}^n F^{-1}\chi a_\mu F$. The homogeneity of a_μ implies that $|\tilde{x}|^\mu A_\mu^\chi$ commutes with ${}^n\varkappa_\lambda$ modulo smoothing operators. In fact,

$$(F\, {}^n\varkappa_\lambda u)(\tilde{\xi}) = ({}^n\varkappa_\lambda^{-1} Fu)(\tilde{\xi}), \ \lambda > 0$$

and

$${}^n\varkappa_\lambda^{-1} |\tilde{x}|^\mu A_\mu^\chi \, {}^n\varkappa_\lambda = |\tilde{x}|^\mu A_\mu^{\chi_\lambda} ,$$

where $\chi_\lambda(\tilde{\xi}) = \chi(\lambda\tilde{\xi})$ is another excision function. Obviously, $A_\mu^\chi - A_\mu^{\chi_1}$

is smoothing.

Assume $\mu > -n$ and consider $A_\mu = F^{-1} a_\mu F$. Since $a_\mu \in L^1_{loc}(\mathbb{R}^n)$, we get a continuous mapping

$$A_\mu \;:\; C_0^\infty(\mathbb{R}^n \setminus 0) \longrightarrow C^\infty(\mathbb{R}^n \setminus 0)$$

and in view of the homogeneity also

$$A_\mu \;: \mathcal{D}'(\mathbb{R}^n)_\infty^{(-z)} \longrightarrow \mathcal{D}'(\mathbb{R}^n)_\infty^{(-z-\mu)},$$

where Proposition 1 is used.

Provided that $\operatorname{Re} z < n$, the extension with homogeneity $-z$ of $v \in C^\infty(S^{n-1})$ belongs to $\mathcal{D}'(\mathbb{R}^n)_\infty^{(-z)}$. Define the operator-valued Mellin symbol $\sigma_M^\mu(a_\mu)$ of a_μ of conormal order μ by

$$\sigma_M^\mu(a_\mu)(z)(v)(x) = |\tilde{x}|^{z+\mu} A_\mu(|\tilde{y}|^{-z} v(\frac{\tilde{y}}{|\tilde{y}|}))(\tilde{x}), \tag{1}$$

where $x = \frac{\tilde{x}}{|\tilde{x}|}$ and the right-hand side is homogeneous of degree 0 in $\tilde{x}$. Then we have for each γ, $-\mu < \frac{1}{2} - \gamma < n$, and $u(t,x) \in C_0^\infty(\mathbb{R}_+ \times S^{n-1})$

$$\pi_* A_\mu u(t,x) = op_M^\gamma(\sigma_M^\mu(a_\mu))u(t,x) =$$

$$= \frac{1}{2\pi i} \int_{\Gamma_{1/2-\gamma}} t^{-z} \sigma_M^\mu(a_\mu)(z)(Mu(z,.))(x)dz.$$

2. Theorem. (i) Let ω be a cut-off function, $-\mu < \frac{1}{2} - \gamma < n$. Then

$$G_\omega = \omega(\pi_* A_\mu^\chi - t^{-\mu} op_M^\gamma(\sigma_M^\mu(a_\mu)))\omega \tag{2}$$

is a Green operator with kernel in $\mathcal{K}^\infty_{\mathfrak{P}_0}(\mathbb{R}_+ \times S^{n-1}) \hat{\otimes}_\pi \mathcal{K}^\infty_{-n-1_{\mathfrak{P}_0}}(\mathbb{R}_+ \times S^{n-1})$, where $\mathfrak{P}_0 = \{(j,0,L_j)\}_{j=0}^\infty \in \underline{P}(S^{n-1})$ and $L_j \subset C^\infty(S^{n-1})$ is the finite-dimensional space of traces on S^{n-1} of homogeneous polynomials of degree j in $\mathbb{R}^n$.

(ii) The Mellin symbol $\sigma_M^\mu(a_\mu)(z)$ is a meromorphic family on the whole complex plane with values in $L^\mu_{cl}(S^{n-1})$. At most simple poles can appear in $\{(n+j),(-\mu-j)\}_{j=0}^\infty \subset \mathbb{C}$ and the residues are operators of finite range. For each fixed $\gamma \in \mathbb{R}$ the family of operators $\rho \longrightarrow \sigma_M^\mu(a_\mu)(\gamma + i\rho)$, $\rho \in \mathbb{R} \setminus 0$, is a classical ψDO with parameter ρ.

Proof. (i) The kernel of $G := A_\mu^\chi - A_\mu$ is a smooth function

$g\in C^\infty(\mathbb{R}^n\times\mathbb{R}^n) = C^\infty(\mathbb{R}^n)\otimes_\pi C^\infty(\mathbb{R}^n)$. Then $\omega\pi_* G\omega$ has the kernel

$$\omega(t)g(tx,t'x')t'^{n-1}\omega(t')\in \mathcal{K}^\infty_{p_o}(\mathbb{R}_+\times S^{n-1})\otimes_\pi \mathcal{K}^\infty_{T^{n-1}p_o}(\mathbb{R}_+\times S^{n-1})$$

with respect to the density $dt'dx$ on $\mathbb{R}_+\times S^{n-1}$, dx the density on S^{n-1} coming from the metric in $\mathbb{R}^n$.

(ii) Let $\varepsilon : \mathbb{R}_+\times S^{n-1} \longrightarrow \mathbb{R}\times S^{n-1}$ be the diffeomorphism given by $\varepsilon(t,x) = (\log t, x)$. The invariance of L^μ_{cl} under coordinate changes implies that $\varepsilon_*\pi_* A^\gamma_\mu \in L^\mu_{cl}(\mathbb{R}\times S^{n-1})$. The same is true for $\varepsilon_*\pi_* A_\mu$ because the difference is smoothing. Since $|\tilde{x}|^\mu A$ commutes with homotetics, the operator

$$L_\gamma(r,D_r,x,D_x) = e^{r(\frac{1}{2}-\gamma+\mu)}\varepsilon_*\pi_* A\, e^{-r(\frac{1}{2}-\gamma)}$$

on $\mathbb{R}\times S^{n-1}$ commutes with r-translations, i.e. $L_\gamma(r,D_r,x,D_x) = L_\gamma(D_r,x,D_x)$. It admits a representation

$$L_\gamma(D_r,x,D_x)u(r,x) = \int e^{ir\varrho}\sigma(L_\gamma)(\varrho,x,D_x)F_1u(\varrho,x)\, d\varrho \qquad (3)$$

where

$$F_1u(\varrho,x) = \int e^{-ir\varrho}u(r,x)dr$$

and

$$\sigma(L_\gamma)(\varrho,x,D_x)(v)(x) = e^{-ir\varrho}L_\gamma(D_r,x,D_x)(e^{ir\varrho}v(.))(x)$$

is a classical ψDO with parameter ϱ.

Putting $z = \frac{1}{2}-\gamma+i\varrho$ we have

$$\sigma^\mu_M(a_\mu)(\tfrac{1}{2}-\gamma+i\varrho)(v)(x) = t^{\frac{1}{2}-\gamma+i\varrho+\mu}\pi_* A_\mu(t^{-\frac{1}{2}+\gamma-i\varrho}v(\cdot))(x)$$

$$= e^{ir\varrho}L_\gamma(D_r,x,D_x)(e^{-ir\varrho}v)(x)$$

$$= \sigma(L_\gamma)(\varrho,x,D_x)v(x).$$

In particular, $\sigma^\mu_M(a_\mu)$ belongs to $L^\mu_{cl}(S^{n-1})$ for each fixed z. Obviously, all homogeneous components of the complete ψDO symbol of $\sigma^\mu_M(a_\mu)$ (in local coordinates on S^{n-1}) depend continuously on z, therefore $z \longrightarrow \sigma^\mu_M(a_\mu)(z)$ is a continuous $L^\mu_{cl}(S^{n-1})$-valued function.

$$\frac{1}{\Gamma(n-z)}\,\dot{u}^{(-z)} : C^\infty(S^{n-1}) \longrightarrow \mathcal{D}'(\mathbb{R}^n)_\infty^{(-z)}$$

is a holomorphic family of continuous operators and $\Gamma(n-z)^{-1}\Gamma(z+\mu)^{-1}\sigma_M^\mu(a_\mu)(z)$ is the composition

$$\begin{array}{ccccc} C^\infty(S^{n-1}) & \xrightarrow{\Gamma(n-z)^{-1}\dot{u}^{(-z)}} & \mathcal{D}'(\mathbb{R}^n)_\infty^{(-z)} & \xrightarrow{F} & \mathcal{D}'(\mathbb{R}^n)_\infty^{(z-n)} \\ & & & & \downarrow \Gamma(z+\mu)^{-1}a_\mu \\ C^\infty(S^{n-1}) & \xleftarrow{r'_s} & \mathcal{D}'(\mathbb{R}^n)_\infty^{(-z-\mu)} & \xleftarrow{F^{-1}} & \mathcal{D}'(\mathbb{R}^n)_\infty^{(z+\mu-n)}, \end{array} \quad (4)$$

where r'_s denotes the restriction to S^{n-1} and $\Gamma(z+\mu)^{-1}a_\mu$ is the operator of multiplication by $\Gamma(z+\mu)^{-1}a_\mu(\tilde{\xi})$. This yields the meromorphy of $\sigma_M^\mu(a_\mu)(z)$ with values in $\mathcal{L}(C^\infty(S^{n-1}),C^\infty(S^{n-1}))$, poles at the points where $\Gamma(n-z)$ and $\Gamma(z+\mu)$ have their poles and finite-range residues. Since $\sigma_M^\mu(a_\mu)(z)$ is outside the poles continuous with values in $L_{cl}^\mu(S^{n-1})$ we conclude by Cauchy's formula that it is also meromorphic with values in $L_{cl}^\mu(S^{n-1})$. □

Note some essential consequences of (ii).

3. <u>Corollary.</u> Let $\varphi,\psi\in C^\infty(S^{n-1}),\varphi\psi=0$. Then $\varphi\sigma_M^\mu(a_\mu)(z)\psi$ is smoothing for each z (outside the poles). Moreover, its kernel and all derivatives are rapidly decreasing in each strip $S=\{z\in\mathbb{C} : c_1<\mathrm{Re}\, z<c_2, |\mathrm{Im}\, z|>c_3>0\}$, c_j arbitrary constants.

<u>Proof.</u> The first assertion is the pseudo-locality of operators in $L_{cl}^\mu(S^{n-1})$. The second one is a consequence of the structure of ψDO's with parameter $\mathrm{Im}\, z$. □

Corollary 3 can be reformulated as

$$\varphi\sigma_M^\mu(a_\mu)\psi\in \mathfrak{M}_{\mathfrak{z}}(S^{n-1}),\ \pi_{\mathbb{C}}\mathfrak{z}=\{(n+j),(-\mu-j)\}_{j=0}^\infty$$

(cf. 4.2.2.).

Another consequence is the invariance of the class of Mellin symbols with respect to diffeomorphisms of domains on S^{n-1}.

Let $\Omega,\Omega'\subsetneq S^{n-1}$ be open and $\varkappa:\Omega\longrightarrow\Omega'$ a diffeomorphism. Regard the realization of $\sigma_M^\mu(a_\mu)$ as operator

$$\sigma_M^\mu(a_\mu)_\Omega : C_0^\infty(\Omega)\longrightarrow C^\infty(\Omega)$$

defined as composition of $\sigma_M^\mu(a_\mu)$ with the natural embedding $C_0^\infty(\Omega) \hookrightarrow C^\infty(S^{n-1})$ and the restriction $C^\infty(S^{n-1}) \longrightarrow C^\infty(\Omega)$. Then $\varkappa_*(\sigma_M^\mu(a_\mu)_\Omega)$ is defined by the commutativity of the diagram

$$\begin{array}{ccc} C_0^\infty(\Omega) & \xrightarrow{\ \sigma_M^\mu(a_\mu)_\Omega\ } & C^\infty(\Omega) \\ {\scriptstyle \varkappa^*}\big\downarrow & & \big\downarrow{\scriptstyle \varkappa^*} \\ C_0^\infty(\Omega') & \xrightarrow{\ \varkappa_*(\sigma_M^\mu(a_\mu)_\Omega)\ } & C^\infty(\Omega') \end{array}$$

The diffeomorphism $\varkappa$ induces a diffeomorphism $\tilde{\varkappa} : C(\Omega) \longrightarrow C(\Omega')$ of the conical subsets of $\mathbb{R}^n$ given by

$$\begin{array}{ccc} C(\Omega) & \xrightarrow{\ \tilde{\varkappa}\ } & C(\Omega') \\ \big\downarrow{\scriptstyle \pi} & & \big\downarrow{\scriptstyle \pi} \\ \mathbb{R}_+ \times \Omega & \xrightarrow{\ \mathrm{id}\times\varkappa\ } & \mathbb{R}_+ \times \Omega' \end{array} .$$

It is clear that in general $\tilde{\varkappa}$ is not the restriction of a diffeomorphism of domains containing the origin.

Then $\tilde{\varkappa}_*(A_\mu^\chi) = \tilde{\varkappa}^* A_\mu^\chi (\tilde{\varkappa}^{-1})^*$ is a classical ψDO, again. Since $\pi_* \tilde{\varkappa}_* = (\mathrm{id}\times\varkappa_*)\pi_*$ we get that

$$\omega(\pi_* \tilde{\varkappa}_*(A_\mu^\chi) - t^{-\mu}\mathrm{op}_M^\gamma(\varkappa_* \sigma_M^\mu(a_\mu)))\omega$$

is a Green operator with kernel in

$$(\mathrm{id}\times\varkappa^*)\left[\mathcal{K}_{P_0}(\mathbb{R}_+\times\Omega)\right] \hat{\otimes}_\pi (\mathrm{id}\times\varkappa^*)\left[\mathcal{K}^\infty_{T^{n-1}P_0}(\mathbb{R}_+\times\Omega)\right] .$$

Let b_μ be the principal symbol of $\tilde{\varkappa}_*(A_\mu^\chi)$. The corresponding Mellin symbol $\sigma_M^\mu(b_\mu)$ is equal to $\varkappa_* \sigma_M^\mu(a_\mu)$ because the Mellin symbol is uniquely determined by the operator. Moreover, in view of Theorem 2 (ii) it is again a meromorphic family with values in $L_{cl}^\mu(\Omega')$ with at most simple poles at $\{(n+j),(-\mu-j)\}_{j=0}^\infty$ and (transformed) finite-range operators as residues and a family of classical ψDO's with parameter. Therefore it is proved

4. <u>Proposition.</u> The transformed ψDO $\tilde{\varkappa}_*(A_\mu^\chi)$ has $\varkappa_* \sigma_M^\mu(a_\mu)$ as its Mellin symbol, which possesses the properties encountered in Theorem 2 (ii).

Note that the position and the multiplicity of the poles of the Mellin symbols of homogeneous ψDO symbols do not change under diffeomorphisms. The finite-range operator residues change, however. By this reason one can not expect to get a simple explicit expression like in the case of $\tilde{x}$-independent symbols.

The reformulation of a ψDO as Mellin operator modulo Green operators in Theorem 2 is the crucial point in order to get extensions in the $\mathcal{K}^{s,\gamma}$ $\mathcal{K}_P^s$ etc. spaces.

Let $m(z) = \sigma_M^\mu(a_\mu)(z)$ be given and $\gamma \in \mathbb{R}$ such that $-\mu < 1/2-\gamma < n$. Define the operator-valued symbol $\sigma_\gamma(\rho)$ by

$$\sigma_\gamma(\rho) = m(1/2-\gamma-i\rho)$$

and set

$$L_\gamma = Op(\sigma_\gamma) = F_1^{-1}\sigma_\gamma F_1$$

(cf.(3)), which is a ψDO of order μ on $\mathbb{R}xS^{n-1}$. It extends from $C_0^\infty(\mathbb{R}xS^{n-1})$ to a continuous mapping

$$L_\gamma : H^s(\mathbb{R}xS^{n-1}) \longrightarrow H^{s-\mu}(\mathbb{R}xS^{n-1})$$

for each $s \in \mathbb{R}$. The diagram

$$\begin{array}{ccc}
H^s(\mathbb{R}xS^{n-1}) & \xrightarrow{L_\gamma} & H^{s-\mu}(\mathbb{R}xS^{n-1}) \\
\downarrow \varepsilon^* & & \downarrow \varepsilon^* \\
\mathcal{H}^{s,1/2}(\mathbb{R}_+xS^{n-1}) & & \mathcal{H}^{s-\mu,1/2}(\mathbb{R}_+xS^{n-1}) \\
\downarrow t^{\gamma-1/2} & & \downarrow t^{\gamma-1/2} \\
\mathcal{H}^{s,\gamma}(\mathbb{R}_+xS^{n-1}) & \xrightarrow{op_M^\gamma(m)} & \mathcal{H}^{s-\mu,\gamma}(\mathbb{R}_+xS^{n-1})
\end{array}$$

commutes, hence

$$\omega op_M^\gamma(m)\omega : \mathcal{K}^{s,\gamma}(\mathbb{R}_+xS^{n-1}) \longrightarrow \mathcal{K}^{s-\mu,\gamma}(\mathbb{R}_+xS^{n-1}) \qquad (5)$$

is continuous.

It is easy to describe the difference

$$\omega(op_M^\gamma(m) - op_M^\delta(m))\omega \text{ on } \mathcal{K}^{s,\gamma}(\mathbb{R}_+xS^{n-1}) \cap \mathcal{K}^{s,\delta}(\mathbb{R}_+xS^{n-1}).$$

If there are no poles of m in the strip $S = \{z\in\mathbb{C} : 1/2-\max(\gamma,\delta) < \operatorname{Re} z < 1/2-\min(\gamma,\delta)\}$, the difference is zero, since the integration contour can be moved from $\Gamma_{1/2-\gamma}$ to $\Gamma_{1/2-\delta}$. If there are poles in the strip S we can find a Green operator $G = G(m,\gamma,\delta)\in\mathcal{L}_G(\min(\gamma,\delta), \max(\gamma,\delta))$ with finite range such that

$$m(z)M((1-G)u)(z)$$

has no poles in the strip S. Therefore

$$\omega(\mathrm{op}_M^{\gamma}(m) - \mathrm{op}_M^{\delta}(m))\omega = \omega(\mathrm{op}_M^{\gamma}(m) - \mathrm{op}_M^{\delta}(m))G\,\omega$$

is a Green operator with finite range in $\mathcal{L}_G(\min(\gamma,\delta), \max(\gamma,\delta))$. So we have proved the following

5. <u>Lemma.</u> Let $m(z)$ be a Mellin symbol of a homogeneous ψDO symbol $a_\mu(\tilde{\xi})$ and $\gamma,\delta\in\mathbb{R}$ such that $1/2-\gamma, 1/2-\delta\notin\{-\mu-j, n+j\}_{j=0}^{\infty}$. Then

$$\omega(\mathrm{op}_M^{\gamma}(m) - \mathrm{op}_M^{\delta}(m))\omega\in\mathcal{L}_G(\min(\gamma,\delta), \max(\gamma,\delta))$$

and it has finite-dimensional range.

Moreover, G_ω acts continuously

$$G_\omega : \mathcal{K}^{s,\gamma}(\mathbb{R}_+ \times S^{n-1}) \longrightarrow \mathcal{K}^{\infty}_{P_0}(\mathbb{R}_+\times S^{n-1}) \subset \mathcal{K}^{\infty,\gamma-\mu}(\mathbb{R}_+\times S^{n-1}),$$

where $1/2-\gamma< n$ guarantees an action on $\mathcal{K}^{s,\gamma}$ and $-\mu< 1/2-\gamma$ the inclusion.

Since $C_0^\infty(\mathbb{R}_+\times S^{n-1})$ is dense in the spaces with weights $\mathcal{K}^{s,\gamma}(\mathbb{R}_+\times S^{n-1})$, the realization

$$\omega\pi_* A_\mu^{\gamma}\omega : \mathcal{K}^{s,\gamma}(\mathbb{R}_+\times S^{n-1}) \longrightarrow \mathcal{K}^{s-\mu,\gamma-\mu}(\mathbb{R}_+\times S^{n-1})$$

is the extension from C_0^∞ by continuity. It is proved

6. <u>Proposition.</u> For each $s\in\mathbb{R}, -\mu< 1/2-\gamma< n$ the operator $\omega\pi_* A_\mu^{\gamma}\omega$ extends from C_0^∞ to a continuous operator

$$\omega\pi_* A_\mu^{\gamma}\omega : \mathcal{K}^{s,\gamma}(\mathbb{R}_+\times S^{n-1}) \longrightarrow \mathcal{K}^{s-\mu,\gamma-\mu}(\mathbb{R}_+\times S^{n-1}) \quad . \tag{6}$$

The limitations on μ and γ which follow from $-\mu < 1/2-\gamma < n$ can be overcome as follows. Given $\delta \in \mathbb{R}$ such that $1/2-\delta \notin \{-\mu-j, n+j\}_{j=0}^{\infty}$. Then

$$\omega\, op_M^{\delta}(m)\omega : \mathcal{K}^{s,\delta}(\mathbb{R}_+ \times S^{n-1}) \longrightarrow \mathcal{K}^{s-\mu,\delta}(\mathbb{R}_+ \times S^{n-1})$$

is continuous because no pole of $m(z)$ is situated on the integration line $\Gamma_{1/2-\delta}$ in the inverse Mellin transform. This realization of a Mellin operator with Mellin symbol m is in general different from $\omega op_M^{\gamma}(m)\omega|_{\mathcal{K}^{s,\gamma} \cap \mathcal{K}^{s,\delta}}$.

Consider now a weight $\delta > \gamma$, γ as in Proposition 6 and $P = (\delta, \wp)$ such that

$$\mathcal{K}_P^{s}(\mathbb{R}_+ \times S^{n-1}) \subset \mathcal{K}^{s,\gamma}(\mathbb{R}_+ \times S^{n-1}).$$

Assume that $1/2-\delta+\mu \notin -\mathbb{Z}_+$. Then there exists a Green operator $G \in \mathcal{L}_G((\delta-\mu, \wp_0), \delta)$

$$G : \mathcal{K}^{s,\delta}(\mathbb{R}_+ \times S^{n-1}) \longrightarrow \mathcal{K}^{\infty}_{(\delta-\mu, \wp_0)}(\mathbb{R}_+ \times S^{n-1})$$

such that

$$\omega \pi_* A_{\mu}^{\chi} \omega = \omega op_M^{\delta}(\sigma_M^{\mu}(a_{\mu}))\omega + G.$$

The restriction of $\omega \pi_* A_{\mu}^{\chi}\omega$ (6) to $\mathcal{K}^{s,\delta}(\mathbb{R}_+ \times S^{n-1})$ acts continuously

$$\omega \pi_* A_{\mu}^{\chi}\omega : \mathcal{K}^{s,\delta}(\mathbb{R}_+ \times S^{n-1}) \longrightarrow \mathcal{K}^{s-\mu}_{(\delta-\mu, \wp_0)}(\mathbb{R}_+ \times S^{n-1}). \quad (7)$$

This is another extension of $\omega \pi_* A_{\mu}^{\chi}\omega$ from C_0^{∞}. In this way the limitation $1/2-\gamma > -\mu$ disappeared.

Consider now a weight $\delta < \gamma$, γ as in Proposition 6, $1/2-\delta \notin \{n+j\}_{j=0}^{\infty}$. The Green operator G_{ω} in Theorem 2 does not admit an extension to $\mathcal{K}^{s,\delta}$. Such an extension is possible after a modification by a finite-range Green operator.

In fact, the decomposition

$$\mathcal{K}_{\wp}^{\infty} = \mathcal{K}_{\wp_{\delta}}^{\infty} + \mathcal{E}(\delta)$$

where $P_\delta \in \underline{P}(N), \pi_{\mathbb{C}} P_\delta \subset \{z \in \mathbb{C} : \mathrm{Re}\, z < 1/2+\delta\}$ and $\mathcal{E}(\delta)$ is a finite-dimensional space, applied to the kernel of G in the second argument gives rise to a decomposition

$$G = G(\delta)' + G(\delta)''$$

where $G(\delta)' \in \mathcal{L}_G(\gamma-\mu,\delta)$ and $G(\delta)''$ is an operator of finite range. Then

$$\omega \pi_* A_\mu^\gamma \omega = \omega t^{-\mu} op_M^\gamma(m)\omega + G_\omega$$

$$= \omega t^{-\mu} op_M^\delta(m)\omega + G_1 + G_\omega$$

and

$$\omega \pi_* A_\mu^\gamma \omega - (G_1+G_\omega)(\delta)' : \mathcal{K}^{s,\delta}(\mathbb{R}_+ \times S^{n-1}) \longrightarrow \mathcal{K}^{s-\mu,\delta-\mu}(\mathbb{R}_+ \times S^{n-1})$$

is continuous.

In the case $-\mu < n$ we can admit therefore all weights $\gamma : \{1/2-\gamma \notin \{-\mu-j, n+j\}_{j=0}^{\infty}$. Consider finally the case $\mu \geq -n$. Note that the homogeneity μ can be assumed to be complex. For $-\mathrm{Re}\,\mu \geq n$ we obtain from $a_\mu \in C^\infty(\mathbb{R}^n \setminus 0)$ a homogeneous distribution in $\mathcal{D}'(\mathbb{R}^n \setminus 0)_\infty^{(\mu)}$ by analytic extension from $-\mathrm{Re}\,\mu < n$ and the diagram

$$\begin{array}{ccc} \mathcal{D}'(\mathbb{R}^n \setminus 0)_\infty^{(z)} & \xrightarrow{A_\mu} & \mathcal{D}'(\mathbb{R}^n \setminus 0)_\infty^{(z-\mu)} \\ \downarrow F & & \downarrow F \\ \mathcal{D}'(\mathbb{R}^n \setminus 0)_\infty^{(-n-z)} & \xrightarrow{a_\mu} & \mathcal{D}'(\mathbb{R}^n \setminus 0)_\infty^{(-n-z+\mu)} \end{array}$$

makes sense for $z \notin \{-\mu-j, n+j\}_{j=0}^{\infty}$ and gives the analytic extension of A_μ. By (1) we get the operator-valued Mellin symbol. It preserves the properties from Theorem 2 (ii) with the only exception that for $\mu \leq -n$ integer poles of order 2 appear. The representation (2) of $\omega \pi_* A_\mu^\gamma \omega$ remains valid. Then $\omega \pi_* A_\mu^\gamma \omega$ extends to a continuous operator

$$\omega \pi_* A_\mu^\gamma \omega : \mathcal{K}^{s,\gamma}(\mathbb{R}_+ \times S^{n-1}) \longrightarrow \mathcal{K}^{s-\mu}_{(\gamma-\mu, P_0)}(\mathbb{R}_+ \times S^{n-1})$$

if $1/2-\gamma < n, 1/2-\gamma \notin \{-\mu-j\}_{j=0}^{\infty}$.

For $1/2-\gamma > n, 1/2-\gamma \notin \{-\mu-j, n+j\}_{j=0}^{\infty}$ we find a Green operator G such that

$$\omega \pi_* A_\mu^\chi \omega + G : \mathcal{K}^{s,\gamma}(\mathbb{R}_+ \times S^{n-1}) \longrightarrow \mathcal{K}^{s-\mu}_{(\gamma-\mu,\wp_0)}(\mathbb{R}_+ \times S^{n-1})$$

is continuous.

Let us summarize the conclusions of the above discussion.

7. <u>Theorem.</u> Let $a_\mu(\tilde{\xi})$ be positively homogeneous of degree μ and $A_\mu^\chi = F^{-1}\chi a_\mu F$. Let $\gamma \in \mathbb{R}$ such that $1/2-\gamma \notin \{-\mu-j, n+j\}_{j=0}^\infty$. Then $\pi_* A_\mu^\chi$ extends to a continuous operator

$$\pi_* A_\mu^\chi : \mathcal{K}^{s,\gamma}(\mathbb{R}_+ \times S^{n-1}) \longrightarrow \mathcal{K}^{s-\mu}_{(\gamma-\mu,\wp_0)}(\mathbb{R}_+ \times S^{n-1}) \tag{8}$$

for $1/2-\gamma < n$, and for $\gamma : 1/2-\gamma > n$ there exists a finite-range Green operator $G_1 \in \mathcal{L}_G((\gamma-\mu,\wp_0),0)$ depending on γ and a_μ such that $\pi_* A_\mu^\chi + G_1$ yields a continuous mapping

$$\pi_* A_\mu^\chi + G_1 : \mathcal{K}^{s,\gamma}(\mathbb{R}_+ \times S^{n-1}) \longrightarrow \mathcal{K}^{s-\mu}_{(\gamma-\mu,\wp_0)}(\mathbb{R}_+ \times S^{n-1}). \tag{9}$$

For each γ there exists another Green operator G_2 such that

$$\pi_* A_\mu^\chi + G_2 : \mathcal{K}^{s,\gamma}(\mathbb{R}_+ \times S^{n-1}) \longrightarrow \mathcal{K}^{s-\mu,\gamma-\mu}(\mathbb{R}_+ \times S^{n-1}) \tag{10}$$

is continuous. If $\delta \in \mathbb{R}, 1/2-\delta \notin \{-\mu-j, n+j\}_{j=0}^\infty$, $\delta > \gamma$, and $P = (\delta, \wp)$ such that $\mathcal{K}_P^s(\mathbb{R}_+ \times S^{n-1}) \subset \mathcal{K}^{s,\gamma}(\mathbb{R}_+ \times S^{n-1})$ then for the restrictions of the operators (9) and (10) to $\mathcal{K}_P^s(\mathbb{R}_+ \times S^{n-1})$ there exist $Q_1 = (\delta-\mu, q_1)$, $Q_2 = (\delta-\mu, q_2)$ and we get continuous mappings

$$\pi_* A_\mu^\chi + G_1 : \mathcal{K}_P^s(\mathbb{R}_+ \times S^{n-1}) \longrightarrow \mathcal{K}_{Q_1}^{s-\mu}(\mathbb{R}_+ \times S^{n-1}) \subset \mathcal{K}^{s-\mu}_{(\gamma-\mu,\wp_0)}(\mathbb{R}_+ \times S^{n-1}),$$

$$\pi_* A_\mu^\chi + G_2 : \mathcal{K}_P^s(\mathbb{R}_+ \times S^{n-1}) \longrightarrow \mathcal{K}_{Q_2}^{s-\mu}(\mathbb{R}_+ \times S^{n-1}) \subset \mathcal{K}^{s-\mu,\gamma-\mu}(\mathbb{R}_+ \times S^{n-1}).$$

There is no problem to generalize Theorems 2 and 7 to symbols of the form

$$a_\mu(\tilde{x},\tilde{\xi}) = \sum b_j(\tilde{x}) a_{\mu j}(\tilde{\xi}), \tag{11}$$

where b_j are homogeneous of degree 0 and belong to a finite-dimensional subspace of $C^\infty(\mathbb{R}^n \setminus 0)$. Consider now symbols which are not necessarily homogeneous of degree zero with respect to $\tilde{x}$ but have asymptotics when $\tilde{x} \to 0$.

Let $\gamma_0 = \{(-j,0,L_j)\}_{j=0}^{\infty} \in \gamma(N)$ be fixed in the following.

Assume

$$a_\mu(\tilde{x},\tilde{\xi}) \in (\pi^{-1})^* \mathcal{K}_{\gamma_0}^{\infty}(\mathbb{R}_+ \times S^{n-1}) \otimes_\pi C^\infty(\mathbb{R}^n \setminus 0)^{(\mu)}. \tag{12}$$

Then we have an asymptotic expansion (for $\tilde{x} \to 0$)

$$a_\mu(\tilde{x},\tilde{\xi}) = a^0(x,\tilde{\xi}) + ta_\mu^1(x,\tilde{\xi}) + \dots + t^j a_\mu^j(x,\tilde{\xi}) + \dots$$

$(t = |\tilde{x}|, x = \tilde{x}/|\tilde{x}|)$, where a_μ^j have the form (11). Each term gives rise to a Mellin symbol. The conormal order of $t^j A_\mu^j$ is $\mu - j$. Therefore, only a_μ^0 contributes to σ_M^μ, and $\sigma_M^{\mu-j}(A_\mu) = \sigma^\mu(a_\mu^j)$.

The assumption (12) contains also a certain behaviour of the symbols for $\tilde{x} \to \infty$, namely like $H^\infty(\mathbb{R}^n) \otimes_\pi C^\infty(\mathbb{R}^n \setminus 0)^{(\mu)}$. It is not hard to take into account also an asymptotic expansion for $\tilde{x} \to \infty$. However, for the sake of simplicity we make the simpler, but more restrictive assumption that the symbols are $\tilde{x}$-independent outside a compact set.

The generalization of Theorem 7 to $\tilde{x}$-dependent symbols of the described class follows immediately from the expansion

$$a_\mu(\tilde{x},\tilde{\xi}) = \sum_{j=0}^{\infty} c_j(\tilde{x}) b_j(\tilde{\xi})$$

$c_j \in (\pi^{-1})^* \mathcal{K}_{\gamma_0}^{\infty}(\mathbb{R}_+ \times S^{n-1}), b_j \in C^\infty(\mathbb{R}^n \setminus 0)^{(\mu)}$, which converges absolutely in $(\pi^{-1})^* \mathcal{K}_{\gamma_0}^{\infty}(\mathbb{R}_+ \times S^{n-1}) \otimes_\pi C^\infty(\mathbb{R}^n \setminus 0)^{(\mu)}$. Then we apply Theorem 7 to each b_j. Note that the norms of the arising mappings (8),(9) and (10) depend continuously on the homogeneous symbols. Then the series of compositions of multiplication by c_j and the ψDO with symbol b_j converges in the corresponding operator norm. Of course, the multiplication by c_j gives rise to a new singularity type. The finite-dimensional space of singular functions can be changed in an arbitrary way by adding a suitable Green operator. In general, the resulting weight is $\gamma - \mu$. However, if we know in addition that

$$a^j_\mu = 0 \quad \text{for} \quad j=0,\dots,N-1$$

i.e. a_μ has a zero at $\tilde{x}=0$ of order N, then the weight of the image space can be chosen arbitrarily, $< \gamma - \mu + N$.

Finally, we take into consideration classical symbols, all homogeneous components of which have the above properties, i.e.

$$a(\tilde{x},\tilde{\xi}) \sim \sum_{j=0}^{\infty} a_{\mu-j}(\tilde{x},\tilde{\xi})$$

and $a_{\mu-j}$ are stabilizing for $\tilde{x} \longrightarrow \infty$ for all j and

$$a_{\mu-j}(\tilde{x},\tilde{\xi}) \in \pi^* \mathcal{K}^{\infty}_{\varphi_0}(\mathbb{R}_+ \times S^{n-1}) \otimes_\pi C^\infty(\mathbb{R}^n \setminus 0)^{(\mu-j)} \tag{13}$$

$\varphi \in \underline{\varphi}(N)$ as above. Denote this symbol space by $S^\mu_{cl,\varphi_0,st}(\mathbb{R}^n \times \mathbb{R}^n)$. Then, for arbitrarily given $P = (\gamma,\varphi) \in \underline{P}(N)$, $1/2-\gamma \notin \{-\mu-j, n+j\}^\infty_{j=0}$ and $Q = (\gamma-\mu, \varphi) \in \underline{P}(N)$ we get for each $a_{\mu-j}$ an operator in $\mathcal{L}(\mathcal{K}^s_P, \mathcal{K}^{s-\mu}_Q)$ which is canonically defined modulo Green operators.

It remains to treat

$$a(\tilde{x},\tilde{\xi}) - \sum_{j=0}^{N-1} \chi(\tilde{\xi}) a_{\mu-j}(\tilde{x},\tilde{\xi}) \in \pi^* \mathcal{K}^\infty_{\varphi_0}(\mathbb{R}_+ \times S^{n-1}) \otimes_\pi S^{\mu-N}(\mathbb{R}^n) \, . \tag{14}$$

Consider $b(\tilde{\xi}) \in S^{\mu-N}(\mathbb{R}^n)$. Then $F^{-1}b \in C^1(\mathbb{R}^n)$ for $1 < N-\mu-n$. The kernel of the ψDO with symbol b is given by $F^{-1}b(\tilde{x}-\tilde{y})$. Adding a suitable Green operator we obtain a kernel vanishing to the order 1 at $\tilde{y}=0$ and an operator in $\mathcal{L}(\pi^*\mathcal{K}^{s,\gamma}, C^1)$ for all $\gamma : 1/2-\gamma < n+1$. For a given γ we choose N and 1 so large that $1/2-\gamma < n+1$ and $C^1 \subset \pi^* \mathcal{K}^{s-}_{Q_1,loc}, Q_1 = (\gamma-\mu, \varphi_1)$ with suitable φ_1. Then we repeat the arguments of composition with multiplication and series expansion.

If we consider $S^\mu_{cl,\varphi_0,st}$ in the topology defined by the countable set of semi-norms induced by (13) and (14), we get that the operator in the norm topology depends continuously on the symbol.

8. <u>Theorem.</u> Let $\mu, \gamma \in \mathbb{R}$, $1/2-\gamma \notin \{-\mu-j, n+j\}^\infty_{j=0}$, $P = (\gamma, \varphi) \in \underline{P}(N)$, $Q = (\gamma-\mu, \varphi) \in \underline{P}(N)$. Then there exists a continuous mapping

$$G : S^{\mu}_{cl,\varphi_0,st} \longrightarrow \mathcal{L}_G$$

such that $F^{-1}aF + G(a), a \in S^{\mu}_{cl,\varphi_0}$, admits an extension from $C_0^{\infty}(C(N)\setminus\partial_2 C(N))$ to an operator in $\mathcal{L}(\pi^*\mathcal{K}^s_P, \pi^*\mathcal{K}^{s-\mu}_Q)$. If $a \in S^{\mu}_{cl,\varphi_0,st}$ has a zero of order 1 at $\partial_2 C(N)$, then the same is true for each $Q = (\lambda,\eta) \in \underline{P}(N), \lambda < \gamma - \mu + 1$.

Let us denote the mapping $a \longrightarrow \pi_*(F^{-1}aF+G(a))$ by

$$op_{\psi}^{Q,P} : S_{cl,\varphi_0,st} \longrightarrow \mathcal{L}(\pi^*\mathcal{K}^s_P, \pi^*\mathcal{K}^{s-\mu}_Q) \qquad (15)$$

and call it a (Q,P) operator convention.

It is clear that we can admit also symbols $a(\tilde{x},\tilde{\xi})$ which are not stabilizing for $\tilde{x} \to \infty$ but have a more general form.

Denote by $S^{\mu}_{cl,\varphi_0,(o)}$ the subspace of $S^{\mu}_{cl,\varphi_0,loc} = \pi^*\mathcal{K}^{\infty}_{\varphi_0,loc} \otimes_{\pi} S^{\mu}$ of all symbols $a(\tilde{x},\tilde{\xi})$ that are decreasing for $\tilde{x} \to \infty$, i.e. for all multi-indices α, β

$$\sup_{|\tilde{x}| \geq R} |D^{\alpha}_{\tilde{x}} D^{\beta}_{\tilde{\xi}} a(\tilde{x},\tilde{\xi})| \langle \tilde{\xi} \rangle^{-\mu+|\beta|} \longrightarrow 0 \quad \text{as} \quad R \to \infty.$$

Then we define S^{μ}_{cl,φ_0} as the subspace of all $a(\tilde{x},\tilde{\xi}) \in S^{\mu}_{cl,\varphi_0,loc}$ of the form

$$a(\tilde{x},\tilde{\xi}) = a_0(\tilde{x},\tilde{\xi}) + a_1(\tilde{x},\tilde{\xi}),$$

where $a_0(\tilde{x},\tilde{\xi}) \in S^{\mu}_{cl,\varphi_0,st}$ and $a_1(\tilde{x},\tilde{\xi}) \in S^{\mu}_{cl,\varphi_0,(o)}$. Each symbol $a \in S^{\mu}_{cl,\varphi_0}$ possesses a unique symbol $b(\tilde{\xi}) = a(\infty,\tilde{\xi})$ inducing a mapping

$$\sigma_{\psi,\infty} : S^{\mu}_{cl,\varphi_0} \longrightarrow S^{\mu}_{cl}(\mathbb{R}^n).$$

Then

$$0 \longrightarrow S^{\mu}_{cl,\varphi_0,(o)} \hookrightarrow S^{\mu}_{cl,\varphi_0} \xrightarrow{\sigma_{\psi,\infty}} S^{\mu}_{cl}(\mathbb{R}^n) \longrightarrow 0$$

is exact.

Theorem 8 remains valid for symbols $a \in S^{\mu}_{cl,\varphi_0}$.

Consider now the global situation of a cone $C(N)$ with base N being a compact C^∞ manifold without boundary. Let $\{U_j\}$ be a finite covering of N by coordinate neighbourhoods and $\{\varphi_j\}$ a partition of unity, supp $\varphi_j \subset U_j$. Consider homeomorphisms

$$\chi_j : U_j \longrightarrow \Omega_j \subset S^{n-1},$$

$n-1 = \dim N$ and

$$\tilde{\chi}_j : C(U_j) \longrightarrow C(\Omega_j) \subset \mathbb{R}^n,$$

homogeneous of degree 1, such that

$$\tilde{\chi}_j(\tilde{x}) = |\tilde{x}|\chi_j\left(\frac{\tilde{x}}{|\tilde{x}|}\right) \text{ for } 0<|\tilde{x}|\leq c \tag{16}$$

$$d\tilde{\chi}_j(\tilde{x}) = \text{const.} \quad \text{for } |\tilde{x}|\geq 2c \tag{17}$$

for a suitable constant $c>0$.

Consider a collection of symbols $a_j \in S^{\mu}_{cl,\gamma_0}(C(\Omega_j)\times\mathbb{R}^n)$. Assume that the following compatibility conditions are satisfied:

$$\sigma_{\psi,\infty}(a_j)(\tilde{\xi}) = \sigma_{\psi,\infty}(a_i)({}^t d\tilde{\varkappa}^{ij}\tilde{\xi})$$

$$a_j = \sigma_\psi(\tilde{\varkappa}^{ij}_* \mathrm{op}_\psi(a_i)) \bmod S^{-\infty}_{\gamma_0},$$

where σ_ψ denotes the complete ψDO symbol. We denote the space of all such symbols by $S^{\mu}_{cl,\gamma_0}(C(N))$.

Let $\psi_j \in C_0^\infty(\Omega_j)$ be such that $\psi_j\varphi_j = \varphi_j$. Denote by $\tilde{\psi}_j$ and $\tilde{\varphi}_j$ the extension to $C(\Omega_j)\subset\mathbb{R}^n$ with homogeneity zero. Define the Mellin symbol of conormal order μ by

$$\sigma^{\mu}_M(a)(z) = \sum_j \psi_j \sigma^{\mu}_M(a_j)\varphi_j$$

(where, as usual, the pull backs by the coordinate mappings χ_j are not indicated). It is a meromorphic family of operators in $L^{\mu}_{cl}(N)$ with poles at most at $\{(-\mu-j, n+j)\}_{j=0}^{\infty}$ and finite-dimensional operators as coefficients of the principal part of the Laurent expansion.

9. <u>Lemma.</u> The choice of coordinates, partition of unity φ_j and the

functions ψ_j do not change the resulting Mellin symbol modulo $\mathfrak{M}_{\mathfrak{Z}}(N)$.

Proof. The assertion follows immediately from Corollary 3 and Proposition 4. □

Note that the principal Mellin symbol depends only on the principal interior symbol at $t=0$. In the same way we define lower order Mellin symbols of conormal order $\mu-k$. They are influenced by the interior symbol components $\sigma_\psi^{(\mu-j)}$, $0 \leq j \leq k$.

For given $a \in S^\mu_{cl,\varphi_0}(C(N))$ we define a (Q,P) operator convention by patching together the local expressions

$$op_\psi^{Q,P}(a) = \sum \psi_j op_\psi^{Q,P}(a_j)\varphi_j .$$

which gives an operator in $\mathcal{L}(\sigma^*\mathcal{K}^s_P, \pi^*\mathcal{K}^{s-\mu}_Q)$. From Theorem 8 and Lemma 9 we get that the arbitrary choices made in the (Q,P) operator convention have no influence to the resulting operator modulo $\mathcal{L}_{M+G}$. From Theorem 7 we see immediately that the restriction to spaces with better weight amounts to the same improvement of the weight in the image space.

We define now the class $\mathcal{L}^\mu(Q,P)$ as extension of $\mathcal{L}_{M+G}(Q,P)$ by ψDO's with symbol in S^μ_{cl,φ_0}.

10. Definition. An operator $A \in \mathcal{L}(\sigma^*\mathcal{K}^s_P, \pi^*\mathcal{K}^{s-\mu}_Q)$ is said to belong to $\mathcal{L}^\mu(Q,P)$ if there exists $a \in S^\mu_{cl,\varphi_0}$ and an operator convention $op_\psi^{Q,P}$ such that $A - op_\psi^{Q,P}(a) \in \mathcal{L}_{M+G}(Q,P)$.

The definition does not depend on the choice of the operator convention $op_\psi^{Q,P}$. Moreover, it is easily checked that the sequence

$$0 \longrightarrow \mathcal{L}_{M+G}(Q,P) \hookrightarrow \mathcal{L}^\mu(Q,P) \xrightarrow{\sigma_\psi} S^\mu_{cl,\varphi_0}(C(N)) \longrightarrow 0 \qquad (18)$$

is exact. In fact, if $A \in \mathcal{L}^\mu(Q,P)$ yields a continuous mapping (modulo $\mathcal{L}_{M+G}(Q,P)$)

$$A \in \mathcal{L}(\sigma^*\mathcal{K}^s_P, \pi^*\mathcal{K}^{s-\mu+1}_Q),$$

then $\sigma_\psi^\mu(A) = 0$. By induction over the homogeneity order μ we get

$\sigma_\psi(A) = 0$, i.e. $A \in \mathcal{L}_{M+G}(Q,P)$.

The main point is, of course, that compositions, adjoints and parametrices, if defined, belong to our class. For $\mathcal{L}_{M+G}$ this was discussed already in 4.2.2.

Consider $A \in \mathcal{L}^\mu(\gamma-\mu,\gamma)$ and the duality between $\mathcal{K}^{s,\gamma}$ and $\mathcal{K}^{-s,-\gamma}$ described in 4.1.1. Then the adjoint to $\pi_* A$

$$(\pi_* A)^* : \mathcal{K}^{\mu-s,\mu-\gamma}(\mathbb{R}_+ \times N) \longrightarrow \mathcal{K}^{-s,-\gamma}(\mathbb{R}_+ \times N)$$

has $\overline{\sigma_\psi^\mu(\pi_* A)}$ as principal ψDO symbol, which is in our class $S^{(\mu)}_{\mathcal{P}_0}$. Proceeding by induction we get that $\sigma_\psi((\pi_* A)^*) \in S^\mu_{cl,\mathcal{P}_0}$ and $A^* \in \mathcal{L}^\mu(-\gamma,\mu-\gamma)$ in view of 4.2.2. Theorem 8 (ii). Thus it is proved

11. Proposition. If $A \in \mathcal{L}^\mu(\gamma-\mu,\gamma)$, then $A^* \in \mathcal{L}^\mu(-\gamma,\mu-\gamma)$ for all admissible $\gamma \in \mathbb{R}$.

In case $A \in \mathcal{L}^\mu(Q,P)$, we use the isomorphisms

$$\mathcal{K}^s_P = \mathcal{K}^{s,\gamma} \underset{\mathcal{E}(\mathcal{P})}{\oplus} \;,\; \mathcal{K}^{s-\mu}_Q = \mathcal{K}^{s-\mu,\gamma-\mu} \underset{\mathcal{E}(\mathcal{Q})}{\oplus}$$

and get a 2x2 matrix of operators possessing an adjoint in the class.

Now we treat compositions. First it is clear that $\mathcal{L}_G$ is a two-sided ideal in $\mathcal{L}^\mu$. Let $A \in \mathcal{L}^\mu(P',P), B \in \mathcal{L}^\nu(Q,P')$ and assume from now on that the finite-dimensional spaces involved in $\mathcal{P}_0$ which controls the asymptotics of the symbol for $t \longrightarrow 0$ may depend on the operator.

12. Theorem. If $A \in \mathcal{L}^\mu(P',P), B \in \mathcal{L}^\nu(Q,P')$, then $BA \in \mathcal{L}^{\mu+\nu}(Q,P)$, $\sigma_\psi^{\mu+\nu}(BA) = \sigma_\psi^\nu(B)\sigma_\psi^\mu(A)$, $\sigma_M^{\mu+\nu}(BA)(z) = \sigma_M^\nu(B)(z-\mu)\sigma_M^\mu(A)(z)$.

Proof. According to 4.2.2. Theorem 8 we may restrict ourselves to compositions $op_\psi^{Q,P'}(b)op_\psi^{P',P}(a)$ and compositions of Mellin operators with $op_\psi^{P',P}(a)$ and $op_\psi^{Q,P'}(b)$. In the second case we get Mellin operators by inserting the Mellin representation of Theorem 2. Note that the space of Mellin symbols $\mathfrak{M}_{\mathfrak{z}}(N)$ is a two-sided ideal in all spaces of Mellin symbols $\mathfrak{M}_{\mathfrak{z}}(N) + \sigma_M^{\mu-k}(S^\mu_{cl,\mathcal{P}_0})$. The composition of two ψDO's yields in the interior another ψDO with principal symbol, being the product of the principal symbols.

Denote by $b \#_\psi a$ the complete symbol of BA (mod $S^{-\infty}$) where a and b are the complete ψDO symbols of A and B. If a has constant coefficients, then

$$\mathrm{op}_\psi^{Q,P}(b \#_\psi a) = \sum \psi_j \mathrm{op}_\psi^{Q,P}(b_j a_j)\varphi_j$$
$$= \mathrm{op}_\psi^{Q,P'}(b)\mathrm{op}_\psi^{P',P}(a) \mod \mathcal{L}_{M+G}.$$

For general a we take the adjoint of the ψDO A^* which can be represented as an absolutely converging sequence of composition of multiplication by functions and constant coefficient ψDO's. $\square$

Our final goal is the treatment of the concept of ellipticity for operators in $\mathcal{L}^\mu(Q,P)(C(N),\partial_2 C(N))$. It is quite similar to that used in 1.5.3. and 2.3.6. with the main difference that here the Mellin symbols are operator-valued.

We shall exploit the interior principal symbol $\sigma_\psi^\mu(A)(\tilde{x},\tilde{\xi})$, the principal conormal symbol $\sigma_M^\mu(A)(z), z \in \Gamma_{1/2-\gamma}, P = (\gamma,\varphi)$, $A \in \mathcal{L}^\mu(Q,P)$. In addition we need the complete symbol near infinity $\sigma_{\psi,\infty}(A)(\tilde{\xi})$ which is well-defined in view of our assumption of stabilization of the symbols modulo decreasing ones.

13. <u>Definition.</u> An operator $A \in \mathcal{L}^\mu(Q,P), P = (\gamma,\varphi)$, $Q = (\gamma-\mu,\eta)$ is called <u>elliptic</u> if

(i) $\sigma_\psi^\mu(A)(\tilde{x},\tilde{\xi}) \neq 0$ for all $\tilde{\xi} \neq 0$, $\tilde{x} = tx, t \geq 0$,

(ii) $\sigma_M^\mu(A)(z) : H^s(N) \longrightarrow H^{s-\mu}(N)$ is bijective for some $s \in \mathbb{R}$ and all $z \in \Gamma_{1/2-\gamma}$.

(iii) $\sigma_{\psi,\infty}(A)(\tilde{\xi}) \neq 0$ for all $\tilde{\xi} \in \mathbb{R}^n$.

An operator $B \in \mathcal{L}^{-\mu}(P,Q)$ is called a parametrix of A if

$$1 - BA \in \mathcal{L}_G(P,P), 1 - AB \in \mathcal{L}_G(Q,Q).$$

14. <u>Theorem.</u> For $A \in \mathcal{L}^\mu(Q,P)$ the following conditions are equivalent:

(i) A is elliptic,

(ii) A possesses a parametrix B in $\mathcal{A}^{-\infty}(P,Q)$

(iii) $A \in \bigcap_s \mathcal{L}(\pi^* \mathcal{K}_P^s(C(N)), \pi^* \mathcal{K}_Q^{s-\mu}(C(N)))$ is Fredholm.

Proof. The main point to show is that the inverse principal interior symbol and the inverse principal conormal symbols fit together, i.e. $\sigma_M^{-\mu}(a_\mu^{-1}) = (\sigma_M^\mu(a_\mu))^{-1} \bmod \mathfrak{M}_\partial$. Clearly, $(\sigma_M^\mu(a_\mu))^{-1}$ is a meromorphic family with values in $L_{cl}^{-\mu}(N)$ and

$$\sigma_M^{-\mu}(a_\mu^{-1}) - (\sigma_M^\mu(a_\mu))^{-1} = (\sigma_M^{-\mu}(a_\mu^{-1})\sigma_M^\mu(a_\mu) - 1)(\sigma_M^\mu(a_\mu))^{-1}$$

belongs to $\mathfrak{M}_\partial$ since this is true for $1 - \sigma_M^{-\mu}(a_\mu^{-1})\sigma_M^\mu(a_\mu)$ (cf. Theorem 12). Repeating the arguments of the proof of 2.3.6. Theorem 8 we get the assertion. □

As standard consequences we get a-priori estimates (cf. 2.3.6. Proposition 9) and the ellipticity of bijective operators in $\mathcal{A}^\mu(Q,P)$. Moreover, the generalization to systems of operators acting as

$$\begin{array}{ccc} \pi^*\mathcal{K}_P^s(C(N)) \otimes \mathbb{C}^M & & \pi^*\mathcal{K}_Q^{s-\mu}(C(N)) \otimes \mathbb{C}^N \\ \oplus & \longrightarrow & \oplus \\ \mathbb{C}^{M'} & & \mathbb{C}^{N'} \end{array}$$

is straightforward.

4.2.4. The case of a cone with boundary

Consider now the case of a cone C with boundary $\partial_1 C = \mathbb{R}_+ \times \partial N/\sim$ if the base N has a non-empty boundary ∂N, which is assumed to be a smooth manifold without boundary. Besides the operators on $\partial_1 C$, which were discussed in 4.2.3., we get operators of trace type acting from C to $\partial_1 C$, of potential type acting from $\partial_1 C$ to C. The operators on C contain besides ψDO's also Mellin and Green parts near $\partial_1 C$. Away from the vertex of the cone it is the structure of operators of the class $\mathfrak{D}$ from 3.2. Our attention is concentrated especially on the behaviour near the vertex where new types of operators appear. The approach from 4.2.3. will be the guiding line of the exposition.

Let $P = (P_1, P_2) \in \underline{P}(N)$, $Q = (Q_1, Q_2) \in \underline{P}(N)$. Consider an operator

$$A = \begin{pmatrix} A_{11} & A_{12} \\ A_{21} & A_{22} \end{pmatrix} : \begin{array}{c} \mathcal{K}_P^\infty(C) \\ \oplus \\ \mathcal{K}_{P_2}^\infty(\partial_1 C) \end{array} \longrightarrow \begin{array}{c} \mathcal{K}_Q^\infty(C) \\ \oplus \\ \mathcal{K}_{Q_2}^\infty(\partial_1 C) \end{array} , \qquad (1)$$

which belongs to the class $\mathcal{W}^\mu(Q_1, P_1)$ outside a neighbourhood of the vertex $\partial_2 C$. The assumptions about the structure near $\partial_2 C$ will be described below.

The operator A_{22} acts on $\partial_1 C$ and is assumed to have the structure described in 4.2.3., i.e. it consists of a ψDO (with certain assumptions about the coefficients near the vertex) and Mellin and Green operators on $\partial_1 C$. In order to distinguish them from the Mellin and Green operators near $\partial_1 C$ involved in the class $\mathcal{W}^\mu$ (cf. 3.2.4.) we shall speak about ∂_2-Mellin and ∂_2-Green operators on $\partial_1 C$.

The operators A_{12} and A_{21} are outside $\partial_2 C$ ∂_1-Green operators of potential and trace type, respectively, of the class $\mathcal{W}^\mu$. Modulo $\mathcal{W}^{\mu-1}$ they are controlled by their principal boundary symbols: $\partial_1^{(\mu)}\sigma(A_{12})$ taking values in $\mathcal{L}_G(Q_1)(\mathbb{R}_+, \partial\mathbb{R}_+)$ and $\partial_1^{(\mu)}\sigma(A_{21})$ in $\mathcal{L}_G(P_1)(\partial\mathbb{R}_+, \mathbb{R}_+)$. We assume that they have a conormal asymptotics for the argument $x \in \partial_1 C$ approaching $\partial_2 C$. The value at the vertex contributes to the $\partial_2\sigma$-symbol which is the principal ∂_2-Mellin symbol with values in $\mathcal{W}_G^\mu(Q_1)(N, \partial N)$ and $\mathcal{W}_G^\mu(P_1)(\partial N, N)$, respectively.

As in 4.2.3. we start from the local situation of $\mathbb{R}_+^n = \{\tilde{x} \in \mathbb{R}^n : \tilde{x}_n > 0\}$, $\tilde{x} = (\tilde{x}', \tilde{x}_n)$, and boundary symbols

$$\partial_1^{(\mu)}\sigma(A_{12})(\tilde{\xi}') : \mathbb{C} \longrightarrow \mathcal{K}_{Q_1}^\infty(\mathbb{R}_+),$$

$$\partial_1^{(\mu)}\sigma(A_{21})(\tilde{\xi}') : \mathcal{K}_{P_1}^\infty(\mathbb{R}_+) \longrightarrow \mathbb{C},$$

(first $\tilde{x}'$-independent) being homogeneous in the sense that

$$\partial_1^{(\mu)}\sigma(A_{12})(\lambda\tilde{\xi}') = \lambda^\mu \varkappa_\lambda \partial_1^{(\mu)}\sigma(A_{12})(\tilde{\xi}'), \qquad (2)$$

$$\partial_1^{(\mu)}\sigma(A_{21})(\lambda\tilde{\xi}') = \lambda^\mu \partial_1^{(\mu)}\sigma(A_{21})(\tilde{\xi}')\varkappa_\lambda^{-1}. \qquad (3)$$

The Mellin symbols are evaluated by applying $(F')^{-1}\partial_1^{(\mu)}\sigma F'$, F' the partial Fourier transform with respect to $\tilde{x}'$, to homogeneous functions. Denote by $\mathcal{K}^{\infty,\gamma}(\mathbb{R}^n_+)^{(z)}$ the subspace of $\mathcal{K}^{\infty,\gamma}(\mathbb{R}^n_+)$ (cf. 3.1.1., here the weight γ corresponds to $\tilde{x}_n \to 0$) of all distributions possessing an extension to $\mathbb{R}^n$ belonging to $\mathcal{S}'(\mathbb{R}^n \setminus 0)^{(z)}$. The restriction to the unit half sphere $S^{n-1}_+ = \{\tilde{x} \in \mathbb{R}^n, |\tilde{x}| = 1, \tilde{x}_n > 0\}$ gives

$$r' : \mathcal{K}^{\infty,\gamma}(\mathbb{R}^n_+)^{(z)} \longrightarrow \mathcal{K}^{\infty,\gamma}(S^{n-1}_+),$$

which is an isomorphism for $z \in \mathbb{C} \setminus \{-n-j\}^{\infty}_{j=0}$. Its inverse is the extension by homogeneity z (and regularization at the origin).
Let us fix a neighbourhood of ∂S^{n-1}_+ on S^{n-1}_+ not containing the north pole $\{\tilde{x}' = 0, \tilde{x}_n = 1\}$ and take here, for convenience, $\tilde{x}_n$ as coordinate transversal to the boundary (instead of the geodesic distance, which is, in general, more natural). Then $\mathcal{K}^{\infty}_{P_1}(S^{n-1}_+)$ is well-defined and we obtain $\mathcal{K}^{\infty}_{P_1}(\mathbb{R}^n_+)^{(z)}$ as inverse of the restriction (applied to a suitable weight). Let us introduce the notation

$$Eu(\tilde{\xi}', \tilde{x}_n) = u(\tilde{\xi}', |\tilde{\xi}'|^{-1}\tilde{x}_n).$$

1. __Lemma.__ For all $z \in \mathbb{C} \setminus \{-n-j, j\}^{\infty}_{j=0}$

$$EF' : \mathcal{K}^{\infty}_{P_1}(\mathbb{R}^n_+)^{(z)} \longrightarrow \mathcal{S}'(\mathbb{R}^{n-1})^{(1-n-z)}_{\infty} \otimes_{\pi} \mathcal{K}^{\infty}_{P_1}(\mathbb{R}_+) \tag{4}$$

is a continuous isomorphism.

__Proof.__ By definition we have $F'u(\tilde{\xi}', |\tilde{\xi}'|^{-1} \tilde{x}_n) \in \mathcal{S}'(\mathbb{R}^{n-1}) \otimes_{\pi} \mathcal{K}^{\infty}_{P_1}(\mathbb{R}_+)$ iff $u \in \mathcal{K}^{\infty}_{P_1}(\mathbb{R}^n_+)$ and the homogeneities are simply checked. $\square$
Let $\varphi, \psi \in C^{\infty}(S^{n-1})$ be functions having their supports in $S^{n-1} \setminus \{\tilde{x}_n = \pm 1\}$ and $\varphi = 1, \psi = 1$ near ∂S^{n-1}_+.
We associate a Mellin symbol with $\partial^{(\mu)}_1 \sigma(A_{12})(\tilde{\xi}')$ by the diagram

$$\begin{array}{ccccc} C^{\infty}(S^{n-2}) & \xrightarrow{\dot{u}^{(-z)}} & \mathcal{S}'(\mathbb{R}^{n-1}\setminus 0)^{(-z)}_{\infty} & \xrightarrow{F'} & \mathcal{S}'(\mathbb{R}^{n-1}\setminus 0)^{(1-n+z)}_{\infty} \\ & & & & \big\downarrow E\partial^{(\mu)}_1 \sigma(A_{12}) \\ \mathcal{K}^{\infty}_{Q_1}(S^{n-1}_+) & \xleftarrow{r'} \mathcal{K}^{\infty}_{Q_1}(\mathbb{R}^n_+)^{(-z-\mu+1/2)} & \xleftarrow{\psi(EF')^{-1}} & & \mathcal{S}'(\mathbb{R}^{n-1}\setminus 0)^{(1-n+z+\mu+1/2)}_{\infty} \otimes_{\pi} \mathcal{K}^{\infty}_{Q_1}(\mathbb{R}_+). \end{array} \tag{5}$$

Since the conormal order is $\mu + 1/2$, we call it $m_{12}(z) = \sigma^{\mu+1/2}_M(\partial^{(\mu)}_1 \sigma(A_{12}))(z)$. The vertical mapping is to be interpreted in an analogous manner as in 4.2.3.(4). Obviously, it is a meromorphic family of operators with poles at $\{(-\mu-1/2-j),\ (n-1+j)\}_{j=0}^{\infty}$. Remark that the shift by 1/2 ($\mu+1/2$ instead of μ) is caused by our definition of homogeneity by means of $\varkappa_{\lambda}v(t) = \lambda^{1/2}v(\lambda t)$.

Let χ be an excision function and consider

$$\mathrm{Op}'_{\chi}(\partial_1^{(\mu)}\sigma(A_{12})) = (F')^{-1}\chi(|\tilde{\xi}'|)\partial_1^{(\mu)}\sigma(A_{12})(\tilde{\xi}')F' \in \mathcal{W}_G^{\mu}(Q_1)(\mathbb{R}_+^n,\partial\mathbb{R}_+^n).$$

In polar coordinates this operator is modulo a ∂_2-Green operator represented by the ∂_2-Mellin operator

$$\omega(t)t^{-\mu-1/2}\mathrm{op}_M^{\gamma_2}(m_{12}(z))\omega(t'),$$

$t = |\tilde{x}|$, $t' = |\tilde{x}'|$, i.e.

$$\omega(\pi_*\mathrm{Op}'_{\chi}(\partial_1^{(\mu)}\sigma(A_{12})) - t^{-\mu-1/2}\mathrm{op}_M^{\gamma_2}(m_{12}(z)))\omega$$

is an operator with kernel in $\mathcal{K}_{\varphi}^{\infty}(\mathbb{R}_+\times S_+^{n-1})\hat{\otimes}_{\pi}\mathcal{K}_{\varphi_0}^{\infty}(\mathbb{R}_+\times S^{n-2})$ (with respect to $t^{n-2}dtdx$), $\varphi\in\underline{\varphi}(N,\partial N)$, i.e. it is a ∂_2-Green operator of potential type. Introducing logarithmic coordinates on $\mathbb{R}_+$ (moving $t=0$ to $-\infty$) we stay in the class $\mathcal{W}^{\mu}$. Then $m_{12}(z)$ is a family of operators in $\mathcal{W}^{\mu}(Q_1)(S_+^{n-1},S^{n-2})$ obtained by a testing against $e^{-ir\rho}$ (cf. the proof of 4.2.3. Theorem 2). Also the influence of the choice of the function ψ is negligible modulo $\mathcal{W}_G^{-\infty}(Q_1)(S_+^{n-1},S^{n-2})$. Another consequence is that the choice of other parametrizations of S^{n-2} and S_+^{n-1} will not lead to operators outside our class.

Now we can repeat the arguments of 4.2.3., which cover also $\tilde{x}'$-dependent symbols and the global situation on $(N,\partial N)$.

Let $\varphi_0 = \{(-j,L_j)\}_{j\in\mathbb{Z}_+}\in\underline{\varphi}(\partial N)$ be fixed. Denote by $\mathcal{T}_1^{(\mu)}(Q_1)(C,\partial_1 C)(\varphi_0)$ the subspace of all μ-homogeneous boundary symbols of potential type on $\partial_1 C$ with φ_0-asymptotics for $\tilde{x}'\to\partial_2 C$, i.e. a collection of $\tilde{x}',\tilde{\xi}'$- dependent families of operators belonging to $\pi^*\mathcal{K}_{\varphi_0}^{\infty}(\mathbb{R}_+\times S^{n-2})_{x'}\hat{\otimes}_{\pi}C^{\infty}(\mathbb{R}^{n-1}\setminus 0)_{\xi'}\hat{\otimes}_{\pi}\mathcal{L}_G(Q_1)(\mathbb{R}_+,\partial\mathbb{R}_+)$ with the homogeneity (2) which are compatible over intersections of coordinate neighbourhoods. We regard also $\mathcal{T}_1^{(\mu)}(Q_1)(C,\partial_1 C)$ as the inductive limit over all choices of L_j. For the sake of simplicity, we assume moreover that the symbols are stabilizing for $|\tilde{x}'| \longrightarrow \infty$.

Let $z \in \mathbb{C}$, $z \notin \{(-\mu-1/2-j),(n-1+j)\}_{j=0}^{\infty}$ be fixed. By the above construction applied in local coordinates we get a mapping

$$\mathcal{B}_1^{(\mu)}(Q_1)(C,\partial_1 C) \longrightarrow \mathcal{D}_G^{\mu}(Q_1)(N,\partial N) \qquad (6)$$

which gives the value of the principal Mellin symbol at the point z. Note that only the leading term of the expansion of the boundary symbol when $|\tilde{x}'| \longrightarrow 0$ contributes to this Mellin symbol. The family of operators is meromorphic in z and has finite range operator residues for each given boundary symbol. Modulo $\mathcal{D}_G^{-\infty}(Q_1)(N,\partial N)$ there is no effect of the arbitrary choices (of coordinates, partition of unity and so on.)

Denote by $\mathfrak{M}_{\underline{\partial}}^{\mu}(N,\partial N)$ the space of sums of Mellin symbols in $\mathfrak{M}_{\underline{\partial}}(N,\partial N)$ (cf.4.2.2.) and Mellin symbols of order μ arising as above from $\mathcal{B}_1^{(\mu)}$ boundary symbols. Then the mapping

$$\mathcal{B}_1^{(\mu)}(C,\partial_1 C) \longrightarrow \mathfrak{M}_{\underline{\partial}}^{\mu}(N,\partial N) \qquad (7)$$

induced by (6) assigns the leading Mellin symbol to a given boundary symbol. Here $\mathcal{B}_1^{(\mu)}(C,\partial_1 C)$ is the inductive limit of $\mathcal{B}_1^{(\mu)}(Q_1)(C,\partial_1 C)$ over all singularity types Q_1.

In the same way we can treat (asymptotic) sums of homogeneous boundary symbols, i.e. classical boundary symbols. Then the above assumptions about the behaviour for $|\tilde{x}'| \longrightarrow 0$ are made with respect to each $\tilde{\xi}'$-homogeneous component. We denote this class of boundary symbols by $\mathcal{B}_1^{\mu}(Q_1)(C,\partial_1 C)$. Then the leading Mellin symbol depends only on (the limit for $|\tilde{x}'| \longrightarrow 0$ of) the $\tilde{\xi}'$-homogeneous part of order μ. The lower order Mellin symbols are affected by the (coefficients of the $|\tilde{x}'|$-asmyptotics of the) $\tilde{\xi}'$-homogeneous components of order not less than the conormal order. All Mellin symbols belong to $\mathfrak{M}_{\underline{\partial}}$ iff all $\tilde{\xi}'$-homogeneous components are rapidly decreasing for $|\tilde{x}'| \longrightarrow 0$.

In a straightforward manner we introduce a topology on $\mathcal{S}_1^\mu(Q_1)(C,\partial_1 C)$ turning it into an (LF) space.

Given $P_2 \in \underline{P}(\partial N)$, $Q = (Q_1,Q_2) \in \underline{P}(N,\partial N)$ and $a_{12}(\tilde{x}',\tilde{\xi}') \in \mathcal{S}_1^\mu(Q_1)(C,\partial_1 C)$. As in 4.2.3. we get that the corresponding Green operator of potential type A_{12}(belonging to $\mathcal{W}^\mu$) admits an extension from $C_0^\infty(\partial_1 C \setminus \partial_2 C)$ to a continuous operator

$$A_{12} : \pi^* \mathcal{K}^s_{P_2}(\mathbb{R}_+ \times \partial N) \longrightarrow \pi^* \mathcal{K}^{s-\mu}_Q(\mathbb{R}_+ \times N) \tag{8}$$

possibly after a modification by a ∂_2-Green operator of potential type, $s \in \mathbb{R}$ arbitrary, $P_2 = (\gamma_2,\varphi_2)$, $1/2 - \gamma_2 \notin \{(-\mu-1/2-j),(n-1+j)\}_{j=0}^\infty$. Since the norm of this operator can be estimated by a linear combination of semi-norms of the symbol a_{12}, the mapping

$$\mathcal{S}_1^\mu(Q_1)(C,\partial_1 C) \longrightarrow \mathcal{L}(\pi^* \mathcal{K}^s_{P_2}(\mathbb{R}_+ \times \partial N), \pi^* \mathcal{K}^{s-\mu}_Q(\mathbb{R}_+ \times N))$$

is continuous. Of course, A_{12} possesses a_{12} as its complete boundary symbol $\partial_1\sigma(A_{12})$.

Now we define the class $\mathcal{L}^\mu(Q,P_2)(C,\partial_1 C)$ of operators in $\mathcal{L}(\pi^*\mathcal{K}^s_{P_2}(\mathbb{R}_+ \times \partial N), \pi^*\mathcal{K}^{s-\mu}_Q(\mathbb{R}_+ \times N))$ as sums of operators in $\mathcal{L}_{M+G}(Q,P_2)(C,\partial_1 C)$ and those arising as above from $\mathcal{S}_1^\mu(Q_1)(C,\partial_1 C)$. Then the sequence

$$0 \longrightarrow \mathcal{L}_{M+G}(Q,P_2) \hookrightarrow \mathcal{L}^\mu(Q,P_2) \xrightarrow{\partial_1\sigma} \mathcal{S}_1^\mu(Q_1) \longrightarrow 0 \tag{9}$$

is exact ($(C,\partial_1 C)$ is omitted here in the notation).

Besides the complete ∂_1-boundary symbol $\partial_1\sigma$ a ∂_2-symbol is given by the sequence of Mellin symbols of orders $\mu + 1/2, \mu - 1/2, \ldots,$ $\mu + 1/2 - j, \ldots$ taking values in $\mathcal{W}^\mu(Q_1)(N,\partial N)$. Let us denote by $\mathcal{S}_2^\mu(Q_1)(N,\partial N)$ the space of all those sequences and by $\partial_2\sigma$ the corresponding symbol mapping. Knowing both the ∂_1- and the ∂_2-symbol the operator is controlled modulo ∂_2-Green operators. Of course, the image of

$$(\partial_1\sigma,\partial_2\sigma) : \mathcal{L}^\mu(Q,P_2) \longrightarrow \mathcal{S}_1^\mu(Q_1) \times \mathcal{S}_2^\mu(Q_1) \tag{10}$$

is a proper subspace. It is characterized by the compatibility condition that the Mellin symbol of any conormal order arises from the ∂_1-symbol modulo the proper Mellin symbols of 4.2.2.

In the same way we introduce the class $\mathcal{L}^{\mu}(Q_2,P)(\partial_1 C,C)$ of operators of trace type acting from $\partial_1 C$ to C. There are only a few minor differences. In view of the definition of homogeneity μ by (3) we get a Mellin symbol associated with $\partial_1^{(\mu)}\sigma(A_{21})(\tilde{\xi}')$ by the diagram

$$\begin{array}{ccccc} \mathcal{K}^{\infty}_{P_1}(S^{n-1}_+) & \xrightarrow{\dot{u}(-z)} \mathcal{K}^{\infty}_{P_1}(\mathbb{R}^n_+)^{(-z)} & \xrightarrow{EF'\varphi} & \mathcal{D}'(\mathbb{R}^{n-1}\setminus 0)^{(1-n+z)}_{\infty}\otimes_{\pi}\mathcal{K}^{\infty}_{P_1}(\mathbb{R}_+) & \\ & & & \Big\downarrow \partial_1^{(\mu)}\sigma(A_{21})E^{-1} & (11) \\ C^{\infty}(S^{n-2}) & \xleftarrow{r'} D'(\mathbb{R}^{n-1}\setminus 0)^{(-z-\mu+1/2)}_{\infty} & \xleftarrow{(F')^{-1}} & \mathcal{D}'(\mathbb{R}^{n-1}\setminus 0)^{(1-n+z+\mu-1/2)}_{\infty} & \end{array}$$

It is a meromorphic family of operators with poles in $\{(-\mu+1/2-j),(n+j)\}_{j=0}^{\infty}$. The conormal order is here $\mu-1/2$ (in contrast to $\mu+1/2$ for potential operators). $\mathcal{T}_1^{\mu}(P_1)(\partial_1 C,C)$ denotes the classical boundary symbols of trace type (leading order μ) stabilizing at infinity with φ_0-asymptotics near the vertex of the cone $\partial_1 C$. We associate with $a_{21}(\tilde{x}',\tilde{\xi}')\in\mathcal{T}_1^{\mu}(P_1)(\partial_1 C,C)$ a trace operator A_{21} which acts continuously

$$A_{21}\ :\ \pi^*\mathcal{K}^s_P(\mathbb{R}_+\times N)\longrightarrow \pi^*\mathcal{K}^{s-\mu}_{Q_1}(\mathbb{R}_+\times\partial N). \qquad (12)$$

All previous notations are used with obvious modifications.

It remains to treat the operators in the left upper corner of (1). Away from $\partial_2 C$ they consist of ∂_1-Green, Mellin and ψDO's. The ∂_1-Green part is treated exactly as the potential and trace Green operators, since it possesses only a (complete) ∂_1-boundary symbol $a_{11}(\tilde{x}',\tilde{\xi}')$. The diagram

$$\begin{array}{ccccc} \mathcal{K}^{\infty}_{P_1}(S^{n-1}_+) & \xrightarrow{\dot{u}(-z)} \mathcal{K}^{\infty}_{P_1}(\mathbb{R}^n_+)^{(-z)} & \xrightarrow{EF'\varphi} & \mathcal{D}'(\mathbb{R}^{n-1}\setminus 0)^{(1-n+z)}_{\infty}\otimes_{\pi}\mathcal{K}^{\infty}_{P_1}(\mathbb{R}_+) & \\ & & & \Big\downarrow Ea_{11}E^{-1} & (13) \\ \mathcal{K}^{\infty}_{Q_1}(S^{n-1}_+) & \xleftarrow{r'} \mathcal{K}^{\infty}_{Q_1}(\mathbb{R}^n_+)^{(-z-\mu)} & \xleftarrow{\psi(EF')^{-1}} & \mathcal{D}'(\mathbb{R}^{n-1}\setminus 0)^{(1-n+z+\mu)}_{\infty}\otimes_{\pi}\mathcal{K}^{\infty}_{Q_1}(\mathbb{R}_+) & \end{array}$$

defines a Mellin symbol associated with a_{11}.

It is a meromorphic family of operators with values in $\mathcal{W}^{\mu}_{G}(Q_1, P_1)(N,N)$ and poles in $\{(-\mu-j),\ (n+j)\}_{j=0}^{\infty}$. Let us denote the corresponding operator class by $\mathcal{L}^{\mu}_{G}(Q,P)(C,C)$. It belongs to $\mathcal{W}^{\mu}_{G}$ away from $\partial_2 C$, but it contains, of course, ∂_2-Mellin and Green operators, too. As before, we have the ∂_1- and the ∂_2-symbolic level.

Next we consider the ∂_1-Mellin operators. They are also controlled by their ∂_1-boundary symbol, taking now values in $\mathcal{L}_{M+G}$ on the half-line and possessing, therefore, a sequence of ∂_1-Mellin symbols. By the diagram (13) we get again a ∂_2-Mellin symbol $m_{11}(z)$, now taking values in $\mathcal{W}^{\mu}_{M+G}(Q_1, P_1)(N,N)$ with poles in $\{(-\mu-j),\ (n+j)\}_{j=0}^{\infty}$. The ∂_1-Mellin operator is obtained from the homogeneous boundary symbol by the tangential Op'-convention applied to the boundary symbol, where $\omega(t\,|\tilde{\xi}'|)$ is replaced by $\omega(t\,[\tilde{\xi}'])$ (cf. 3.2.3.). For each $\tilde{\xi}' \neq 0$ the difference is in $\mathcal{L}_G$ on the half-line. Then we can apply the arguments for ∂_1-Green operators proving that (for homogeneous symbols with constant coefficients) the ∂_1-Mellin operator can be replaced modulo a ∂_2-Green operator by the ∂_2-Mellin operator with the ∂_2-Mellin symbol given by (13). This result extends to asymptotic sums with decreasing homogeneities. Admitting, moreover, $\tilde{x}'$-depending ∂_1-Mellin symbols with $\mathcal{P}_0$-asymptotics for $|\tilde{x}'| \to 0$ and stabilizing at ∞ we get operators in $\mathcal{L}(\pi^{*}\mathcal{K}^{s}_{P}(\mathbb{R}_+ \times N),\ \pi^{*}\mathcal{K}^{s-\mu}_{Q}(\mathbb{R}_+ \times N))$ which are (possibly after modification by ∂_2-Green operators) an extension of the ∂_1-Mellin operator given outside $\partial_2 C$. Denote the class by $\mathcal{L}^{\mu}_{M+G}(Q,P)(C,C)$.

The last component of A_{11} is the ψDO. Its ∂_2-Mellin symbol is obtained by localization on N. In the interior the results of 4.2.3. work. When localizing near ∂N we have to take into account both the interior ∂_0- and the boundary ∂_1-symbol. First we take the leading interior symbol (at $|\tilde{x}| = 0$) and get a ψDO on N. This is extended in an arbitrary manner to an operator in $\mathcal{L}(\mathcal{K}^{s}_{P_1}(N), \mathcal{K}^{s-\mu}_{Q_1}(N))$. The corresponding boundary symbol differs from that obtained from (13) applied to the given boundary symbol at most by a ∂_1-Mellin and

Green operator. Taking now into account also the ∂_2-Mellin and Green part in the left upper corner we can apply the above constructions of ∂_2-Mellin symbols, extensions to the $\mathcal{K}_P^s$ spaces and so on. Of course, in case of variable coefficients, the interior symbols are assumed to have a φ_0-asymptotics (cf. 4.2.3.) and stabilize at infinity. Denote the resulting operator class by $\mathcal{L}(Q,P)(C,C)$.

Now we have three symbolic levels:

(i) the (complete) interior symbol of the ψDO $\partial_0\sigma$

(ii) the (complete) boundary symbol $\partial_1\sigma$

(iii) the (sequence of) Mellin symbol(s) $\partial_2\sigma$

$\partial_2\sigma$ is now a meromorphic family of operators in $\mathcal{D}^\mu(Q_1,P_1)(N,N)$. Modulo the class of proper Mellin symbols (cf. 4.2.2.) it is determined by $\partial_0\sigma$ and $\partial_1\sigma$(near $\partial_2 C$). Denote by $\mathcal{S}^\mu$ the space of all compatible triples $(\partial_0\sigma, \partial_1\sigma, \partial_2\sigma)$. Then we can define an operator in $\mathcal{L}(\pi^*\mathcal{K}_P^s(\mathbb{R}_+ \times N), \pi^*\mathcal{K}_Q^{s-\mu}(\mathbb{R}_+ \times N))$ with the given symbol triple. We call this a P,Q-operator convention. Then the sequence

$$0 \longrightarrow \mathcal{L}_G(C,C) \hookrightarrow \mathcal{L}^\mu(C,C) \xrightarrow{\partial_0\sigma,\, \partial_1\sigma,\, \partial_2\sigma} \mathcal{S}^\mu \longrightarrow 0$$

is exact (singularity types omitted).

Sometimes it is more convenient to look only at the principal symbols $\partial_0^{(\mu)}, \partial_1^{(\mu)}, \partial_2^{(\mu)}$. If $\mathcal{S}^{(\mu)}$ denotes the space of triples of principal symbols, we get an exact sequence

$$0 \longrightarrow \mathcal{L}^{\mu-1}(C,C) \hookrightarrow \mathcal{L}^\mu(C,C) \xrightarrow{\partial_0^{(\mu)}\sigma,\, \partial_1^{(\mu)}\sigma,\, \partial_2^{(\mu)}\sigma} \mathcal{S}^{(\mu)} \longrightarrow 0.$$

We shall use the notation $\mathcal{L}^\mu(Q,P)(C,\partial_1 C)$ also for matrices of operators of the form (1), the components of which were characterized above. Similarly we shall also use the notations for the symbol components.

The class $\mathcal{L}^\mu$ on a cone with boundary was constructed in order to be able to carry out compositions and adjoints withinthe class (whenever they are defined). Away from $\partial_2 C$ this is clear from the results about the class $\mathcal{D}^\mu$. In particular, the principal interior and boundary symbol of the composition is equal to the composition of the principal interior and boundary symbols. The principal ∂_2-symbol of an operator in $\mathcal{L}^\mu$ is determined modulo proper Mellin symbols (in $\mathfrak{M}_z$) by the ∂_0- and the ∂_1-symbol. Since the class $\mathfrak{M}_z$ is an ideal in the class $\mathfrak{M}^\mu_z$, the product (possibly with a shift in the argument z) of symbols in $\mathfrak{M}^\infty_z = \varinjlim_\mu \mathfrak{M}^\mu_z$ is again in this class and the compatibility conditions are satisfied. In the case of constant coefficients it is clear that the composition of operators has the composition of the leading Mellin symbols (with suitable argument shift) as leading Mellin symbol. For instance, with the above notations and

$A \in \mathcal{L}^\mu(Q,P_1)(C,\partial_1 C), B \in \mathcal{L}^\nu(P_1,Q')(\partial_1 C,C)$, then

$$\partial_2^{(\mu+\nu)}\sigma(BA)(z) = \partial_2^{(\mu)}\sigma(B)(z-\nu+1/2)\partial_2^{(\nu)}\sigma(A)(z)$$

Repeating the arguments of the proof of 4.2.3. it is proved that this is valid also in the case of variable coefficients.

Thus it is shown the first part of

2. Theorem. (i) Let $A,B \in \mathcal{L}^\infty$ and assume that BA is defined, i.e. the image of A fits to the domain of definition of B. Then $BA \in \mathcal{L}^\infty$ has the composition of the leading symbol components as leading symbol components, where in the ∂_2-symbol of B the usual shift by the conormal order of A occurs.

(ii) Let $A \in \mathcal{L}^\infty$ be an operator acting between weighted spaces such that A^* is well-defined. Then $A^* \in \mathcal{L}^\infty$.

The second part is proved as follows: As shown in 3.2.3. the class is closed under taking adjoints (if possible). Moreover, the symbols of the adjoint operators have the same asymptotic properties near $\partial_2 C$. An operator with this interior and boundary symbol is equal to the adjoint modulo $\mathcal{L}_{M+G}$.

In the concept of ellipticity of operators in $\mathcal{L}^\mu(Q,P)(C,\partial_1 C)$ we have all elements from the case when $\partial_1 C = \emptyset$ (corresponding to the part $C \setminus \partial_1 C$ of our cone). Moreover, the usual ellipticity of boundary value problems (in case of smooth boundary) is involved near $\partial_1 C \setminus \partial_2 C$.

In addition a condition on the leading Mellin symbol with values in the boundary value problems on $(N,\partial N)$ is necessary.

3. Definition. An operator $A \in \mathcal{L}^{\mu}(Q,P)(C,\partial_1 C)$ is called elliptic if

(i) the principal interior and boundary symbols are bijective up to the boundary for $\tilde{\xi} \neq 0$ and $\tilde{\xi}' \neq 0$, respectively,

(ii) the complete interior and boundary symbols at ∞ are bijective for all $\tilde{\xi}$ and $\tilde{\xi}'$, respectively,

(iii) the principal Mellin symbol takes values in the intertible operators of $\mathcal{L}^{\mu}(Q_1,P_1)(N,\partial N)$ for all $z \in \Gamma_{1/2-\gamma_2}$.

An operator $B \in \mathcal{L}^{-\mu}(P,Q)(C,\partial_1 C)$ is called a parametrix of A if

$$1-BA \in \mathcal{L}_G(P,P)(C,\partial_1 C),\ 1-AB \in \mathcal{L}_G(Q,Q)(C,\partial_1 C).$$

4. Theorem. For $A \in \mathcal{L}^{\mu}(Q,P)(C,\partial_1 C)$ the following conditions are equivalent:

(i) A is elliptic,

(ii) A possesses a parametrix in $\mathcal{L}^{-\mu}(P,Q)(C,\partial_1 C)$,

(iii) A is a Fredholm operator as a mapping

$$A: \begin{array}{c} \mathcal{K}^s_P(C) \\ \oplus \\ \mathcal{K}^s_{P_1}(\partial_1 C) \end{array} \longrightarrow \begin{array}{c} \mathcal{K}^{s-\mu}_Q(C) \\ \oplus \\ \mathcal{K}^{s-\mu}_{Q_1}(\partial_1 C) \end{array}$$

$s \in \mathbb{R}$, cf. (1).

The proof follows the lines of the proof of 4.2.3. Theorem 14. The main point is that the pointwise inverse of the three symbol components has the desired structure and satisfies the compatibility conditions. For the interior symbol it is clear (the invertibility up to the boundary and including the limit values in the vertex is assumed). The ∂_1-boundary symbol and its compatibility with the interior symbol was treated in 3.2.5. for the ∂_2-symbol it follows also algebraically that its inverse must have the structure of a sum of a ∂_2-symbol associated with the inverse interior and inverse boundary symbol and a

proper $\mathcal{O}_2$-Mellin symbol. This contains both the asserted structure and the compatibility. Then we can continue step by step and determine all lower order symbol components. Having obtained the complete symbol triple we get a parametrix as the associated operator.
The Fredholm property implies the ellipticity, because the symbol components can be recovered from the operator by testing it against suitable sequences of functions. Inserting this into the a-priori estimate we get injectivity (and similarly surjectivity) of the symbol components.

Then it follows that the inverse of a bijective operator belongs to our class again.

4.2.5. Notes

The theory of ψDO's on wedges is an analogue of the theory of boundary value problems. The wedges are local models of manifolds with edges and of the form $\mathbb{R}^d \times \{\text{cone}\}$, where in the case of boundary value problems the cone degenerates to $\mathbb{R}_+$.

The previous chapter has shown the role of the Mellin symbolic calculus on $\mathbb{R}_+$ for the ellipticity of boundary value problems and the asymptotics of solutions. A crucial point was the Mellin representation of ψDO's on $\mathbb{R}_+$ that are first based on the Fourier transform. That was just the program of Chapter 1.

Now it is obvious that analogous constructions are needed for the general cones $\mathbb{R}_+ \times N/\sim$ for being able to treat the edge problems. In the simplest form the question is to reformulate the ψDO action in $\mathbb{R}^n$, restricted to $\mathbb{R}^n \setminus 0$, by Mellin actions, after introducing polar coordinates $\pi : \mathbb{R}^n \setminus 0 \to \mathbb{R}_+ \times S^{n-1}$. Such a reformulation was obtained by Plamenevskij in [P 3]. More details may also be found in [P 4].
The calculations in [P 3] (as well as those in Chapter 1) give no direct idea how to generalize this to boundary value problems.

For this reason we have developed in 4.2.3. another approach to derive the Mellin representations. The starting point here is that the Mellin inversion formula gives a decomposition of a distribution into a family of homogeneous ones (with respect to the natural homoteties of complex orders running over the weight line. A formal calculation then shows that an operator which respects the homogeneity of distributions up to an order shift $-\mu$ has an operator valued Mellin symbol of conormal order μ. The ψDO's in question are asymptotic sums of such simple ones, modulo smoothing operators. In the concrete situation we also have seen that the Mellin symbols satisfy the structure conditions of [R 6], namely to be parameter-depending ψDO's on the base with parameter on the weight line and meromorphic extensions, where the Laurent coefficients for the poles are finite-dimensional smoothing operators on the base. In addition the picture is local in a natural way, i.e. we may talk about restrictions to coordinate neighbourhoods of S^{n-1} and preserve the essential properties under coordinate diffeomorphisms. This admits us to globalize the constructions, i.e. to pass from S^n to arbitrary bases N. The construction gives an answer to the question of extending ψDO's to the totally characteristic Sobolev spaces $\mathcal{H}^s$ with respect to a vertex. As we have seen these actions are induced indeed, after removing smoothing operators with a controlled asymptotic structure. The Mellin representation and the meromorphy of the Mellin symbols imply, in addition, the actions on spaces with asymptotics.

For boundary value problems, i.e. when the base N of the cone has a boundary, the situation is much more complex, especially since we admit operators of the type of Chapter 3 on $(N,\partial N)$. In that case we have the problem of reformulating into Mellin actions not only for ψDO's but also for the Mellin and Green parts transversal to ∂N as well as for the potential and trace operators. Nevertheless the picture is finally analogous to that without boundary. The Mellin representations complete the classes of 4.2.1., 4.2.2. to an algebra with ψDO's in the same sense as in Chapter 1. In particular, we see again the role of the strongly decreasing Mellin symbols of 4.2.2. and of the Green kernels of 4.2.1. as the extra elements that appear in compositions of ψDO's (or boundary value problems) without those parts.

They are generated also in parametrix constructions for elliptic operators. We are now in a good position to talk about the analogous parameter-dependent operators, where the parameters are just the edge variables and covariables (y, η). Then we shall pass again to ψDO's along the edge, based on the corresponding edge symbolic calculus. This is the program of the following section.

Note that the algebra of 4.2.1. is a proper subalgebra of that in [R 6]. Here we observe a more precise structure of the non-decreasing Mellin symbols, and the ψDO symbols in the interior close to the vertex are induced by ψDO symbols in a neighbouring space, e.g. $\mathbb{R}^n$ in the case of the cone $\cong \mathbb{R}_+ \times S^n = \mathbb{R}^n \setminus 0$. An analogous remark holds for the boundary value problems in connection with [R 8] if we talk about the subalgebra of $\mathcal{L}^\infty$ in 4.2.4. of operators which are of Boutet de Monvel's type transversal to ∂N.

In spite of the position of the cone operators in this book as preliminary material for manifolds with edges, the cone theory is of independent interest and has its own tradition. So we briefly want to discuss some basic achievements in the literature though it is not our aim here to give a complete list of papers. As already mentioned there is a classical paper of Kondrat'ev [K 5] which contains a general theory of differential operators in domains with conical points. Certain ideas on parameter-dependent elliptic operators go back to Agranovič, Višik [A 2]. Essential elements of the symbolic structure of parametrices on manifolds with conical singularities in terms of meromorphic families of ψDO's were achieved in the work of Plamenevskij, cf. [P 7] and the bibliography there. Also the aspect of algebras of ψDO's in the Mellin setting was studied by Plamenevskij. The picture was further completed by Rempel, Schulze [R 6], [R 8] in the sense of a Mellin ψDO algebra containing the Green operators as negligible parts and the Mellin translation product of complete Mellin symbols. [R 6] considers the case of operators on manifolds with conical points whereas [R 8] contains boundary value problems with the transmission

property if the boundary has conical points. The results of Section 4.2.4. explain the symbolic structure of the parametrices from another point of view than [R 8]. But in 4.2.4. are also admitted boundary value problems without the transmission property. The Mellin translation product of Mellin symbols shows beautiful internal relations of the sequences of Mellin symbols under compositions.
Again the concept of Green operators is important for the algebra properties. It may be a challenge for the reader to understand their complex structure, but they are formally associated with the spaces with asymptotics in the same way as in the simplest cases of Chapter 1.

The functional analytic background can be further developed in many directions. First it is clear that the concept of spaces with continuous asymptotics for the cone can be studied in more detail, cf. also Schulze [S 5]. This leads to more general Mellin symbolic structures. Moreover, it is interesting to study the cone operators in spaces based on L^p and Hölder norms, cf. Baranowski [B 6]. There also exists a higher dimensional analogue of the Mellin transform and a corresponding calculus of ψDO's which are of Fuchs type in several variables, cf. Schulze [S 4]. Also other variants of the calculus are useful, such as an analogue of the theory of Section 4.2.4. for mixed and transmission boundary problems if the submanifold with the jump of the boundary conditions touches the conical points. The latter problem is not yet elaborated but is should be possible to apply the structures obtained in this book.

4.3. Operators on a wedge

Now we investigate classes of operators on a wedge containing ψDO's and also Green and Mellin operators arising in compositions. The local model near a point at the edge has the form $X = \mathbb{R}^k \times C$, C a cone possibly with boundary $\partial_1 C$. The simpler boundaryless case ($\partial_1 C = 0$) is considered first. Based on the operator classes on the cone we get the operators on the wedge as in 3.2. We can use notations similar to those in 3.2. because in the special case when the edge is absent they coincide.

4.3.1. ∂_2-Green operators

Let X be a manifold with edges, i.e. the local model of a neighbourhood of a point is an open subset of $\mathbb{R}^n$ or $\mathbb{R}^n_+$ or it is of the form $\mathbb{R}^k \times C$, C a closed cone of dimension $n-k$ possibly with boundary. First we assume $\partial_1 C = \emptyset$, i.e. if X has a smooth boundary $\partial_1 X$, it is separated from $\partial_2 X$.

We define now negligible operators. Let $C = C(N)$, N a compact manifold without boundary.

1. Definition. Let $k \in \mathbb{Z}_+$, $P = (\gamma, \varphi) \in \underline{P}(N)$, $Q = (\lambda, q) \in \underline{P}(N)$. Denote by $(1)\text{-}\mathcal{W}_G^{-\infty}(Q,P)^{(-k)}(X,X)$ the subspace of $\bigcap_{s,s' \in \mathbb{R}} \mathcal{L}(\mathcal{K}^s_{P,\mathrm{comp}}(X), \mathcal{K}^{s'}_{Q,\mathrm{loc}}(X))$ of all operators $\mathcal{A}$ such that there exists an $R = R(\mathcal{A})(\lambda+k, r) \in \underline{P}(N)$, and $\mathcal{A}$ is continuous

$$\mathcal{A} : \mathcal{K}^s_{P,\mathrm{comp}}(X) \longrightarrow \mathcal{K}^{s'}_{R^+,\mathrm{loc}}(X) \quad \text{for all } s,s' \in \mathbb{R},$$

where $\mathcal{K}^{s'}_{R^+,\mathrm{loc}} = \varprojlim_{\varepsilon > 0} \mathcal{K}^{s'}_{R_\varepsilon,\mathrm{loc}}$, $R_\varepsilon = (\lambda + k - \varepsilon, r(\lambda + r - \varepsilon) r)$.

The elements of this space are called smoothing (1)-Green operators of conormal order $-k$. Set

$$(1)\text{-}\mathcal{W}_G^{-\infty}(Q,P)(X,X) = \bigcap_k (1)\text{-}\mathcal{W}_G^{-\infty}(Q,P)^{(-k)}(X,X).$$

We define the classes of smoothing Green operators (without (1)) by similar assumptions about the adjoints, understood in a proper sense. Fix a covering of X by coordinate neighbourhoods and a partition of unity $\{\varphi_j\}_{j \in J}$. Then the operators $\mathcal{A}_{ij} := \varphi_i \mathcal{A} \varphi_j$ can be considered in local coordinates. If both neighbourhoods belong to the edge, we have

$$\mathcal{A}_{ij} : \mathcal{K}^s_{P,\mathrm{loc}}(\mathbb{R}^k \times C) \longrightarrow \mathcal{K}^{s'}_{Q,\mathrm{comp}}(\mathbb{R}^k \times C).$$

Introducing the isomorphisms

$$\mathcal{K}^s_P(\mathbb{R}^k \times C) \cong \mathcal{K}^{s,\gamma}(\mathbb{R}^k \times C) \oplus (H^s(\mathbb{R}^k))^{\dim P},$$

$$\mathcal{K}^{s'}_Q(\mathbb{R}^k \times C) \cong \mathcal{K}^{s',\lambda}(\mathbb{R}^k \times C) \oplus (H^{s'}(\mathbb{R}^k))^{\dim Q},$$

arising from

$$\mathcal{K}^s_P(\mathbb{R}_+\times N) = \mathcal{K}^{s,\gamma}(\mathbb{R}_+\times N) + \mathcal{E}(p)(\mathbb{R}_+\times N),$$

$$\mathcal{K}^{s'}_Q(\mathbb{R}_+\times N) = \mathcal{K}^{s',\lambda}(\mathbb{R}_+\times N) + \mathcal{E}(q)(\mathbb{R}_+\times N),$$

we may represent $\mathcal{A}_{ij}$ as a matrix of operators

$$\begin{pmatrix} \mathcal{A}^{11}_{ij} & \mathcal{A}^{12}_{ij} \\ \mathcal{A}^{21}_{ij} & \mathcal{A}^{22}_{ij} \end{pmatrix} : \begin{matrix} \mathcal{K}^{s,\gamma}(\mathbb{R}^k\times C) \\ \oplus \\ (H^s(\mathbb{R}^k))^{\dim P} \end{matrix} \longrightarrow \begin{matrix} \mathcal{K}^{s',\lambda}(\mathbb{R}^k\times C) \\ \oplus \\ (H^{s'}(\mathbb{R}^k))^{\dim Q} \end{matrix}$$

For the adjoints acting in the dual spaces we obtain in view of

$$(\mathcal{K}^{s,\gamma}(\mathbb{R}^k\times C))' = \mathcal{K}^{-s,-\gamma}(\mathbb{R}^k\times C),$$

$$(\mathcal{A}^{11}_{ij})^* : \mathcal{K}^{-s',-\lambda}(\mathbb{R}^k\times C) \longrightarrow \mathcal{K}^{-s,-\gamma}(\mathbb{R}^k\times C),$$

$$(\mathcal{A}^{21}_{ij})^* : (H^{-s'}(\mathbb{R}^k))^{\dim Q} \longrightarrow \mathcal{K}^{-s,-\gamma}(\mathbb{R}^k\times C).$$

Similarly we proceed in case only one of the coordinate neighbourhoods belongs to the edge.

2. <u>Definition.</u> Denote by $\mathcal{W}^{-\infty}_G(Q,P)(X,X)$ the subspace of $(1)\text{-}\mathcal{W}^{-\infty}_G(Q,P)(X,X)$ of all $\mathcal{A}$ such that the above adjoints are continuous operators:

$$(\mathcal{A}^{11}_{ij})^* : \mathcal{K}^{s,-\lambda}(\mathbb{R}^k\times C) \longrightarrow \mathcal{K}^{s'}_{R_1^+}(\mathbb{R}^k\times C),$$

$$(\mathcal{A}^{21}_{ij})^* : (H^s(\mathbb{R}^k))^{\dim Q} \longrightarrow \mathcal{K}^{s'}_{R_1^+}(\mathbb{R}^k\times C),$$

for arbitrary s, s', l and some $R_1 = R_1(\mathcal{A}) = (-\gamma+1, \mathcal{N}_1^{\cdot}) \in \underline{P}(N)$, and similarly in case only one of the coordinate neighbourhoods belongs to the edge.

From the definitions it is clear that the composition of smoothing Green operators is again in this class provided it is defined.

The same is true for the adjoint.

Since the spaces $\mathcal{K}^\infty_{P,\mathrm{comp}}(X)$ and $\mathcal{K}^\infty_{P,\mathrm{loc}}(X)$ are nuclear, smoothing Green operators may be characterized by their kernels.

In the same way we define smoothing Green operators acting from

X to $\partial_2 X$ (trace type), from $\partial_2 X$ to X (potential type). Since $\partial_2 X$ is a smooth manifold there, we have the usual smoothing operators. Classes of Green operators of order $\mu > -\infty$ without edge symbols are defined by their mapping properties.

3. Definition. Let $\mu \in \mathbb{R}, P, Q \in \underline{P}(N)$ as above. Denote by $(1)\text{-}\mathcal{W}_G^{\mu,(0)}(Q,P)^{(-k)}(X,X)$ the subspace of $\bigcap_{s \in \mathbb{R}} \mathcal{L}(\mathcal{K}^s_{P,\mathrm{comp}}(X), \mathcal{K}^{s-\mu}_{Q,\mathrm{loc}}(X))$ of all operators $\mathcal{A}$ such that there exists $R = R(\mathcal{A}) = (\lambda + k, \mathcal{N}) \in \underline{P}(N)$ and $\mathcal{A}$ is continuous

$$\mathcal{A} : \mathcal{K}^s_{P,\mathrm{comp}}(X) \longrightarrow \mathcal{K}^{s-\mu}_{R^+,\mathrm{loc}}(X) \quad \text{for all } s$$

and for arbitrary $\varphi, \psi \in C_0^\infty(\partial_2 X), \varphi\psi = 0$ we have $\psi\mathcal{A}\varphi \in (1)\text{-}\mathcal{W}_G^{-\infty}(Q,P)^{(-k)}(X,X)$.

It is clear that the classes are closed under compositions and adjoints provided they are defined.

The edge symbols constitute the new element which is necessary for an optimal control of operators on $(X, \partial_2 X)$ (cf. the analogy to the boundary symbols in 3.2.2., 3.2.3. and 3.2.4.). They are families of operators on the cone C transversal to the edge $\partial_2 X$ parametrized by points of $T^*(\partial_2 X)\setminus 0$ and take values in $\mathcal{L}^{(\mu)}$ introduced in 4.2.3. If they take values in $\mathcal{L}_{M+G}$ ($\mathcal{L}_G$), we speak of edge symbols of Mellin (Green) type. Similar as for boundary symbols we have a notion of homogeneity coming from the ${}^n\varkappa_\lambda$ -action on functions and distributions on C. We consider edge symbols which admit an asymptotic expansion into homogeneous ones.

In this section we only deal with Green edge symbols. Let $\Omega \subseteq \mathbb{R}^d$ be open, C a cone with base N, $\partial N = \emptyset$, n = dim C.

4. Definition. Let $\mu \in \mathbb{R}, P, Q \in \underline{P}(N)$. A family of operators $g(y,\eta), y \in \Omega, \eta \in \mathbb{R}^d \setminus 0$, is said to belong to $\mathcal{R}^{(\mu)}(\mathcal{L}_G(Q,P)(C,\partial_2 C))(\Omega \times (\mathbb{R}^d \setminus 0))$ if it is a C^∞ family on $\Omega \times (\mathbb{R}^d \setminus 0)$ with values in $\mathcal{L}_G(Q,P)(C,\partial_2 C)$ satisfying the homogeneity condition

$$g(y,\lambda\eta) = \lambda^{\mu}\, {}^n\varkappa_\lambda\, g(y,\eta)\, {}^n\varkappa_\lambda^{-1} \quad \text{for all } \lambda > 0, \qquad (1)$$

where

$$ {}^{n}\varkappa_{\lambda} u(\tilde{x}) = \lambda^{n/2} u(\lambda\tilde{x}),\ \tilde{x}\in C. $$

For fixed μ an edge symbol $g\in \mathcal{R}^{(\mu)}(\mathcal{L}_G(Q,P)(C,\partial_2 C))(\Omega\times\mathbb{R}^d\setminus 0)$ is uniquely determined by its values at $|\eta|=1$ and each smooth family on $|\eta|=1$ gives an edge symbol by extension by homogeneity. We consider the space of Green edge symbols in the topology given by

$$ C^{\infty}(\Omega\times S^{d-1}, \mathcal{L}_G(Q,P)(C,\partial_2 C)) = C^{\infty}(\Omega\times S^{d-1})\otimes_{\pi}\mathcal{L}_G(Q,P)(C,\partial_2 C). $$

If $g\in \mathcal{R}^{(\mu)}(\mathcal{L}_G(Q,P)(C,\partial_2 C))(\Omega\times(\mathbb{R}^d\setminus 0))$, then

$$ g_0(y,\eta) := {}^{n}\varkappa^{-1}_{|\eta|}\, g(y,\eta)\, {}^{n}\varkappa_{|\eta|} \tag{2} $$

is positively homogeneous with respect to η of degree μ in the usual sense, i.e.

$$ g_0 \in C^{\infty}(\Omega)\otimes_{\pi} C^{\infty}(\mathbb{R}^d\setminus 0)^{(\mu)}\otimes_{\pi}\mathcal{L}_G(Q,P)(C,\partial_2 C). \tag{3} $$

This motivates the definition of the space of non-homogeneous Green edge symbols

$$ \mathcal{R}^{\mu}(\mathcal{L}_G(Q,P)(C,\partial_2 C))(\Omega\times\mathbb{R}^d) := $$

$$:= ({}^{n}\varkappa)_{*}\,(S_{cl}(\Omega\times\mathbb{R}^d)\otimes_{\pi}\mathcal{L}(Q,P)(C,\partial_2 C)) $$

where $({}^{n}\varkappa)_{*}$ is given by

$$ (({}^{n}\varkappa)_{*}\, g_0)(y,\eta) = {}^{n}\varkappa_{[\eta]}\, g_0(y,\eta)\, {}^{n}\varkappa^{-1}_{[\eta]}\ , \tag{4} $$

and the topology is induced by $S^{\mu}_{cl}(\Omega\times\mathbb{R}^d)\otimes_{\pi}\mathcal{L}(Q,P)(C,\partial_2 C)$. Here $[\eta]$ is a strictly positive smooth function $\mathbb{R}^k\longrightarrow\mathbb{R}_+$ which is equal to $|\eta|$ outside a compact set.

In view of the stability of $\mathcal{L}_G$ under compositions and adjoints we get immediately that these operations (provided they are defined) give again Green edge symbols. Moreover, series of symbols of orders tending to $-\infty$ can be summed up asymptotically and the result is unique modulo $\mathcal{R}^{-\infty}(\mathcal{L}_G(Q,P)(C,\partial_2 C))(\Omega\times\mathbb{R}^d)$.

Let $g(y,y',\eta)\in\mathcal{R}^{\mu}(\mathcal{L}_G(Q,P)(C,\partial_2 C))(V\times V\times\mathbb{R}^d)$, $V\subseteq\mathbb{R}^d$ open.

g plays the role of an amplitude function of a ψDO with operator-valued symbol on $(U, \partial_2 U)$, $U := V \times C, \partial_2 U := V \times \partial_2 C$, i.e. it acts from U to U or U to $\partial_2 U$ or $\partial_2 U$ to U or $\partial_2 U$ to $\partial_2 U$. In the last case we have, of course, an ordinary ψDO on $V \cong \partial_2 U$ ($\partial_2 C$ is a point). The two preceding cases yield operators of trace and potential type. Consider, for example, the first case, i.e. g takes values in operators acting from C to C. Then we set

$$\mathrm{Op}(g)u(y) = \int_{\mathbb{R}^d} \int_V e^{i(y-y')\eta} g(y,y',\eta)u(y')dy'\,đ\eta, \tag{5}$$

$u \in C_0^\infty(V, \mathcal{K}_P(C))$. The result is a C^∞ function on V with values in $\mathcal{K}_Q^\infty(C)$. The decomposition of g

$$g(y,y',\eta) = \sum_{j=1}^{\infty} a_j(y) g_j(\eta) b_j(y')$$

following from the definition of $\mathcal{R}^\mu(\mathcal{L}_G)$ allows the reduction of continuity questions to the constant coefficient case. Exactly as in the proof of 3.2.2. Proposition 2 we obtain

5. <u>Proposition.</u> For $\mu \geq -\infty$ and $g \in \mathcal{R}^\mu(\mathcal{L}_G(Q,P)(C,\partial_2 C))(V \times V \times \mathbb{R}^d)$ we have $\mathrm{Op}(g) \in \mathcal{W}_G^{\mu,(0)}(Q,P)(V \times C, V \times C)$.

Note that in the same way the class of Green edge symbols $\mathcal{R}^\mu(\mathcal{L}_G(Q,P)^{(-k)})$ can be defined. The corresponding operators belong to $\mathcal{W}_G^{\mu,(0)}(Q,P)^{(-k)}$.

As in 3.2.2. we define the class $\mathcal{W}_G^{\mu,(l)}$ of Green operators of order μ possessing l homogeneous edge symbols of homogeneity $\mu-j$, $j=0, 1, \ldots, l-1$, by extending $\mathcal{W}_G^{\mu-l,(0)}$ by $\mathrm{Op}(\mathcal{R}^\mu(\mathcal{L}_G))$. Putting

$$\mathcal{R}^{\mu,(l)}(\mathcal{L}_G) = \bigoplus_{j=0}^{l} \mathcal{R}^{(\mu-j)}(\mathcal{L}_G)$$

we get the exact sequence

$$0 \longrightarrow \mathcal{W}_G^{\mu-l,(0)} \hookrightarrow \mathcal{W}_G^{\mu,(l)} \longrightarrow \mathcal{R}^{\mu,(l)}(\mathcal{L}_G) \longrightarrow 0. \tag{6}$$

The class $\mathcal{W}_G^\mu$ with infinite sequence of homogeneous edge symbols is obtained as projective limit

$$\mathcal{W}_G^{\mu} = \varinjlim_{l} \mathcal{W}_G^{\mu,(l)} .$$

Then the sequence

$$0 \longrightarrow \mathcal{W}_G^{-\infty} \longrightarrow \mathcal{W}_G^{\mu} \longrightarrow \mathcal{R}^{\mu}(\mathcal{L}_G) \longrightarrow 0 \quad . \qquad (7)$$

is exact.

Note that operators in $\mathcal{W}_G^{\mu,(1)}$ are concentrated near $\partial_2 X$ modulo negligible operators (in $\mathcal{W}_G^{\mu-1,(0)}$). In fact, if $\varphi(\tilde{x})$ is a smooth function on C with $\operatorname{supp}\varphi \cap \partial_2 C = \emptyset$ and $g \in \mathcal{R}^{\mu}(\mathcal{L}_G)$, then φg and $g\varphi$ belong to $\mathcal{R}^{-\infty}(\mathcal{L}_G)$.

Operators in $\mathcal{W}_G^{\mu,(1)}$ are ψDO's on $V \subseteq \mathbb{R}^k$ with symbols taking values in the class $\mathcal{L}_G$ of Green operators on $(C, \partial_2 C)$. Since this class is closed under compositions and taking adjoints (whenever they are defined), the same follows about the class $\mathcal{W}_G^{\infty,(1)}$. For the symbols the usual formulas are valid.

6. Theorem. (i) Let $\mathcal{G}_j \in \mathcal{W}_G^{\mu_j,(1)}$ such that $\mathcal{G}_1 \mathcal{G}_2$ is defined. Then $\mathcal{G}_1 \mathcal{G}_2 \in \mathcal{W}_G^{\mu_1+\mu_2,(1)}$.

(ii) Let $\mathcal{G} \in \mathcal{W}_G^{\mu,(1)}$ and $\mathcal{G}^*$ be well-defined. Then $\mathcal{G}^* \in \mathcal{W}_G^{\mu,(1)}$.

(iii) Let $\varkappa : V \longrightarrow V_1$ be a diffeomorphism and $\mathcal{G} \in \mathcal{W}_G^{\mu,(1)}$ defined on $V \times C$ (or $V \times \partial_2 C$). Then $\varkappa_* \mathcal{G} \in \mathcal{W}_G^{\mu,(1)}$ is an operator on $V_1 \times C$ (or $V_1 \times \partial_2 C$).

The invariance (iii) of the class $\mathcal{W}_G^{\mu,(1)}$ under coordinate changes implies that the class is well-defined on a manifold with edges referring to an arbitrary atlas.

7. Example. Let X be now a compact manifold with wedge $\partial_2 X$, $P \in \underline{P}(N)$, N the base of the cone transversal to $\partial_2 X$. Consider

$$1 + \mathcal{A} : \begin{array}{c} \mathcal{K}_P^s(X) \\ \oplus \\ H^t(\partial_2 X) \end{array} \longrightarrow \begin{array}{c} \mathcal{K}_P^s(X) \\ \oplus \\ H^t(\partial_2 X), \end{array}$$

$s, t \in \mathbb{R}$, $\mathcal{A} \in \mathcal{W}_G^{0,(1)}(P,P)$. Provided that the principal edge symbol $\partial_2\sigma^{(0)}(\mathcal{A})$ vanishes, i.e. $\mathcal{A} \in \mathcal{W}_G^{-1,(0)}(P,P)$, then $\mathcal{A}$ is compact and $1+\mathcal{A}$ is Fredholm with index zero.

Consider now the case $\partial_1 X \neq 0$, i.e. for the cone $C = C(N)$ transversal to the boundary we have $\partial_1 C = C(\partial N) \neq 0$. For the triple $(X, \partial_1 X, \partial_2 X)$ we have to consider 9 types of operators, all of them occurring in 3x3 matrices of the form

$$\mathcal{A} = (\mathcal{A}_{ij}) : \begin{matrix} \mathcal{K}_P^{\infty}(X) \\ \oplus \\ \mathcal{K}^{\infty}_{\partial_1 P_2}(\partial_1 X) \\ \oplus \\ C^{\infty}(\partial_2 X) \end{matrix} \longrightarrow \begin{matrix} \mathcal{K}_Q^{\infty}(X) \\ \oplus \\ \mathcal{K}^{\infty}_{\partial_1 Q_2}(\partial_1 X) \\ \oplus \\ C^{\infty}(\partial_2 X) \end{matrix} , \tag{8}$$

$P = (P_1, P_2)$, $Q = (Q_1, Q_2) \in \underline{P}(N, \partial N)$, $i,j = 1,2,3$. $\partial_1 P_2$ and $\partial_1 Q_2$ are obtained from P_2, Q_2 by taking the trace on ∂N. Here $\mathcal{A}_{22}, \mathcal{A}_{23}, \mathcal{A}_{32}, \mathcal{A}_{33}$ belong to the class $\mathcal{W}_G^{\infty,(1)}$ described above. The remaining five types will be described now.

In the first step we describe smoothing ∂_2-Green operators. We consider, for example, $\mathcal{A}_{13}$.

8. Definition. Denote by $\mathcal{W}_G^{-\infty}(P)(X, \partial_2 X)$ the subspace of $\bigcap_{s,s'} \mathcal{L}(H^s_{comp}(\partial_2 X), K^{s'}_{Q,loc}(X))$ of all operators $\mathcal{A}$ acting continuously

$$\mathcal{A} : H^s_{comp}(\partial_2 X) \longrightarrow \mathcal{K}^{s'}_{R^+, loc}(X)$$

for arbitrary $s, s', l, R = R(\mathcal{A}) = ((\lambda_1 + l, \mathfrak{r}_1), (\lambda_2 + l, \mathfrak{r}_2)) \in \underline{P}(N, \partial N)$
$Q = ((\lambda_1, q_1), (\lambda_2, q_2))$.

Similarly, $\mathcal{A}_{11}, \mathcal{A}_{12}, \mathcal{A}_{31}, \mathcal{A}_{21}$ are defined by their mapping properties and by the same properties of the adjoints (in the proper sense).

In the second step we define Green operators of a finite order μ without edge symbols demanding the same mapping properties with $s-\mu$ instead of s' and pseudolocality. Denote the corresponding classes again by $\mathcal{W}_G^{\mu,(0)}$.

In the third step we include also edge symbols. Given a smooth family $g(y, \eta)$ with values in $\mathcal{L}_G(C, \partial_1 C, \partial_2 C)$. The ${}^n\varkappa_\lambda$-action on C and the ${}^{n-1}\varkappa_\lambda$-action on $\partial_1 C$ yield the notion of homogeneity of order μ

$$g(y,\lambda\eta) = \lambda^{\mu} ({}^{n}\varkappa_{\lambda} \oplus {}^{n-1}\varkappa_{\lambda} \oplus 1) g(y,\eta) ({}^{n}\varkappa_{\lambda} \oplus {}^{n-1}\varkappa_{\lambda} \oplus 1)^{-1} .$$

Then

$$g_0(y,\eta) := ({}^{n}\varkappa_{[\eta]} \oplus {}^{n-1}\varkappa_{[\eta]} \oplus 1)^{-1} g(y,\eta) ({}^{n}\varkappa_{[\eta]} \oplus {}^{n-1}\varkappa_{[\eta]} \oplus 1)$$

is homogeneous for large $|\eta|$ in the usual sense, i.e. $\chi(\eta) g_0 \in C^{\infty}(\Omega)_y \otimes_{\pi} C^{\infty}(\mathbb{R}^d \setminus 0)^{(\mu)} \otimes_{\pi} \mathcal{L}_G(C,\partial_1 C,\partial_2 C)$, χ an excision function. Classical edge symbols arise in this way from $S^{\mu}_{cl}(\Omega \times \mathbb{R}^d) \otimes_{\pi} \mathcal{L}_G (C,\partial_1 C,\partial_2 C)$. Since $\mathcal{L}_G(C,\partial_1 C,\partial_2 C)$ is stable under compositions and taking adjoints in the class of classical edge symbols $\mathfrak{R}^{\infty}(\mathcal{L}_G(C,\partial_1 C,\partial_2 C)) = \bigcup_{\mu} \mathfrak{R}^{\mu}(\mathcal{L}_G(C,\partial_1 C,\partial_2 C))$, all operations of the calculus can be carried out.

We associate with edge symbols in $\mathfrak{R}^{\mu}(\mathcal{L}_G(C,\partial_1 C,\partial_2 C))$ ∂_2-Green operators by (5), i.e. again we get ψDO's with operator-valued symbols. They belong to $\mathfrak{W}_G^{\mu,(0)}$ on $(V\times C, V\times\partial_1 C, V\times\partial_2 C)$.
Then we define the class $\mathfrak{W}_G^{\mu,(1)}$ on $(V\times C, V\times\partial_1 C, V\times\partial_2 C)$ as an extension of $\mathfrak{W}_G^{\mu-1,(0)}$ by operators with edge symbols in $\mathfrak{R}^{\mu}(\mathcal{L}_G(C,\partial_1 C,\partial_2 C))$. We get again the exact sequences (6) and (7) (now with a more general meaning). Finally, the operator class remains closed under compositions, adjoints and coordinate changes, i.e. Theorem 6 is valid also in the more general situation.

In view of the extension of our operator class below (cf.4.3.3.) we shall introduce other notations. Up to now we were guided by the class on a manifold with smooth boundary. Our classes of ∂_2-Green operators are concentrated near $\partial_2 X$. Later on we take into account also ∂_1-Green and Mellin operators and ψDO's, both on $\partial_1 X$ and on $\partial_0 X$. We shall write $\mathcal{T}^{(\mu)}_{2,G}$ instead of $\mathfrak{R}^{(\mu)}(\mathcal{L}_G)$ for the μ-homogeneous ∂_2-symbols of Green type and $\mathcal{T}^{\mu}_{2,G}$ instead of $\mathfrak{R}^{\mu}(\mathcal{L}_G)$. Denote by $\mathfrak{W}^{\mu,(1)}_{2,G}$ the above defined classes of ∂_2-Green operators of order $\leq \mu$ with $1 \leq \infty$ homogeneous symbol components. Then the sequence

$$0 \longrightarrow \mathfrak{W}^{\mu-1,(1-1)}_{2,G} \hookrightarrow \mathfrak{W}^{\mu,(1)}_{2,G} \xrightarrow{\partial_2^{(\mu)}\sigma} \mathcal{T}^{(\mu)}_{2,G} \longrightarrow 0$$

is exact. Moreover, there is a right inverse Op_2 of $\partial_2^{(\mu)}\sigma$.

4.3.2. ∂_2-Mellin operators

Now we are going to define the class $\mathcal{W}^{\mu,(1)}_{2,M+G}$ obtained by extending $\mathcal{W}^{\mu,(1)}_{2,G}$ by ∂_2-Mellin operators. It is obtained by admitting the edge symbols to take values in $\mathcal{L}_{M+G}$ on the cone instead of $\mathcal{L}_G$. As in 3.2.3. we assume the conormal symbols to be polynomials in η, the covariable in $T^*(\partial_2 X)$.

First let us consider the simpler case if the cone has a base N without boundary. Let $P = (\gamma, \varphi) \in \underline{P}(N)$, $Q = (\lambda, q) \in \underline{P}(N)$ be given. At first we specify $\sigma_M \mathcal{T}^{(\mu)}_{2,M+G}$ - the space of ∂_2 Mellin symbols of μ-homogeneous edge symbols with their dependence on $y \in V \subset \mathbb{R}^k$ and $\eta \in \mathbb{R}^k$. We set

$$\sigma_M \mathcal{T}^{(\mu)}_{2,M+G}(Q,P)(C,\partial_2 C)(V \times \mathbb{R}^k) = C^\infty(V \times \mathcal{M}_{\underline{\delta}(N)}[\gamma])[\eta]^{(\mu-\delta)} \bigoplus_{1-1}^{\infty} C^\infty(V, \mathcal{M}_{\underline{\delta}(N)})[\eta]^{(\mu-\delta+1)} \tag{1}$$

$\delta = \gamma - \lambda$, and

$$\sigma_M \mathcal{T}^{\mu}_{2,M+G}(Q,P)(C,\partial_2 C)(V \times \mathbb{R}^k) = C^\infty(V, \mathcal{M}_{\delta(N)}[\gamma])[\eta]^{\mu-\delta} \bigoplus_{l=1}^{\infty} C^\infty(V, \mathcal{M}_{\delta(N)})[\eta]^{\mu-\delta+1}, \tag{2}$$

where $A[\eta]^{(\nu)}$ denotes the polynomials in η with coefficients in A and homogeneity ν and $A[\eta]^{\nu} = \dot{+}_{j=0}^{\nu} A[\eta]^{(j)}$. Note that not the singularity types P,Q, but only their weights enter the definition. We consider the elements of $\sigma_M \mathcal{T}^{(\mu)}_{2,M+G}$ and $\sigma_M \mathcal{T}^{\mu}_{2,M+G}$ (which are in fact sequences of functions depending on $y \in V$, $z \in \mathbb{C}$ and $\eta \in \mathbb{R}^k$) in the topology of componentwise convergence. As usually we shall also use abbreviated notations where the specification of unessential data is suppressed.

1. Definition. A family $m(y,\eta)$, $y \in V$, $\eta \in \mathbb{R}^k \setminus 0$, is said to belong to $\mathcal{T}^{(\mu)}_{2,M+G}(Q,P)(V \times (\mathbb{R}^k \setminus 0))$ if it is a C^∞ family with values in $\mathcal{L}_{M+G}(Q,P)(C,\partial_2 C)$ and its sequence of Mellin symbols

$\sigma_M^{\delta-1}(m(y,\eta))$, $1=0,1,2,\ldots$ forms an element

of $\sigma_M \mathcal{T}^{(\mu)}_{2,M+G}(Q,P)(V \times \mathbb{R}^k)$.

From 4.2.2. we have a (non-canonical) operator-convention op_M

assigning an operator in $\mathcal{A}_{M+G}$ to any given sequence of Mellin symbols. Then

$$\mathcal{Y}^{(\mu)}_{2,M+G} = op_M(\sigma_M \mathcal{Y}^{(\mu)}_{2,M+G}) + \mathcal{Y}^{(\mu)}_{2,G} \qquad (3)$$

and the sequence

$$0 \longrightarrow \mathcal{Y}^{(\mu)}_{2,G} \hookrightarrow \mathcal{Y}^{(\mu)}_{2,M+G} \xrightarrow{\sigma_M} \sigma_M \mathcal{Y}^{(\mu)}_{2,M+G} \longrightarrow 0 \qquad (4)$$

is exact (σ_M denotes the evaluation of the sequence of Mellin symbols for fixed y and η). In fact, the Mellin operator convention is a left inverse to σ_M.

Next we associate with a given Mellin symbol in $\sigma_M \mathcal{Y}^{(\mu)}_{2,M+G}$ (or $\sigma_M \mathcal{Y}^{\mu}_{2,M+G}$) a Mellin operator on $V \times C$. For homogeneous symbols (in η) the neighbourhood of $\eta = 0$ is not simply treated by η-excision (as for Green symbols). We proceed in the spirit of 3.2.3.

Let $\{h_l(y,y',\eta)\}_{l=0}^{\infty} \in \sigma_M \mathcal{Y}^{(\mu)}_{2,M+G}(Q,P)$ and let $\gamma_l : \gamma - 1 - 1/2 \leq \gamma_l \leq \gamma - 1/2$ be admissible weights for h_l. Consider the parametrization of C by $(t,x) \in \mathbb{R}_+ \times N$. We define Mellin operators on $V \times C$ by

$$\mathrm{Op}(h_l)u(y,t,x) = \qquad (5)$$

$$= t^{l-\delta} \int e^{i(y-y')\eta} \omega_l(t[\eta]) op_M^{\gamma_l}(h_l(y,y',\eta))(\omega_l(\cdot[\eta])u(y',\cdot,\cdot))(t,x)dy' đ\eta$$

where the cut-off functions ω_l are to be chosen below and $[\eta]$ denotes a strictly positive smooth function equal to $|\eta|$ outside a neighbourhood of the origin. The arbitrary choices of ω_l, γ_l and the function $[\cdot]$ in this definition do not affect the operator modulo $\mathcal{W}^{\mu}_{2,G}$.

In the usual way we get that the operator $\mathrm{Op}(h_l)$ defines a continuous mapping

$$\mathrm{Op}(h_l) : \mathcal{K}^{s}_{P,comp}(V \times C) \longrightarrow \mathcal{K}^{s-\mu}_{Q,loc}(V \times C), \text{ for each } s \in \mathbb{R}, \qquad (6)$$

(comp and loc with respect to V) and for each $P' = (\gamma + k, \varphi') \in \underline{P}(N)$ $\overline{r}(\gamma)\varphi' \leq \varphi$ there exists $R = (\lambda + k + l, \sigma) \in \underline{P}(N)$, $\overline{r}(\lambda)\sigma \leq q$ such that

$$\mathrm{Op}(h_l) : \mathcal{K}^{s}_{P',comp}(V \times C) \longrightarrow \mathcal{K}^{s-\mu}_{R^+,loc}(V \times C) \qquad (7)$$

is continuous. It follows from similar mapping properties of the adjoint that $Op(h_l)$ is a ∂_2-Green operator in $\mathfrak{W}_G^{\mu}(Q,P)^{(-1)}$. For fixed $k \in \mathbb{N}$ we associate with $m(y,y',\eta) \in \mathcal{S}_{2,M+G}^{(\mu)}(Q,P)$ the operator

$$Op_k(m) = \sum_{l=0}^{k-1} Op(h_l) + Op(g_k) \tag{8}$$

where

$$g_k = m - \sum_{l=0}^{k-1} t^{l-\delta} \omega op_M^{\gamma_l}(h_l)\omega \in \mathcal{R}^{(\mu)}(\mathcal{L}_G(Q,P)^{(-k)})$$

and $Op(g_k)$ is defined as in 4.3.1. by means of an η-excision function. The result is independent of the choices mod $\mathfrak{W}_G^{-\infty}(Q,P)$.
Passing to the limit $k \longrightarrow \infty$ we put

$$Op(m) = \sum_{l=0}^{\infty} Op(h_l) + Op(g), \tag{9}$$

where

$$g = m - \sum_{l=0}^{\infty} t^{l-\delta} \omega_l op_M^{\gamma_l}(h_l)\omega_l \in \mathcal{R}^{(\mu)}(\mathcal{L}_G(Q,P))$$

and ω_l are chosen in a suitable way, which gives a unique result mod $\mathfrak{W}_G^{-\infty}(Q,P)$.
Now we treat the case of non-homogeneous ∂_2-symbols. Given a sequence $m = \{m_j\}_{j=0}^{\infty}$, $m_j \in \mathcal{S}_{2,M+G}^{(\mu-j)}(Q,P)$. Then we get $Op(m_j)$ by (9). Note that $Op(m_j) \in \mathfrak{W}_G^{\mu-j,(0)}(Q,P)^{(-j)}$. For a calculus modulo $\mathfrak{W}_G^{\mu',(0)}(Q,P)$ with given μ' we take

$$\sum_{\mu-j>\mu'} Op(m_j)$$

as operator convention. Then a certain finite number of homogeneous edge symbols is controlled (cf. the calculus in 3.2.3).For the finer calculus with an infinite number of homogeneous edge symbols we introduce a device of asymptotic summation. Returning to the definition of $Op(m_j)$ by (9) we introduce other cut-off functions ω_j

$$Op(m_j) = \omega_j \sum_{l=0}^{\infty} Op(h_l(m_j))\,\omega_j + Op(g_j) \tag{10}$$

where $g_j \in \mathcal{R}^{(\mu-j)}(\mathcal{L}_G(Q,P))$. We can choose ω_j such that for each $k \in \mathbb{Z}_+$ there is a $k' \in \mathbb{Z}_+$ such that

$$\sum_{j>k'} \omega_j \sum_{l=0}^{\infty} \mathrm{op}(h_l(m_j))\omega_j$$

converges in $\mathcal{W}_G^{\mu-k,(0)}(Q,P)^{(-k)}$. Then we choose in addition η-excision functions χ_j such that

$$\sum_j \mathrm{op}_G(\chi_j g_j)$$

converges in the sense of Green operators.

Put

$$\mathrm{Op}(m) = \sum_{j=0}^{\infty} \omega_j \sum_{l=0}^{\infty} \mathrm{op}(h_l(m_j))\omega_j + \sum_{j=0}^{\infty} \mathrm{op}_G(\chi_j g_j). \tag{11}$$

Clearly the result is independent of our choices modulo $\mathcal{W}_G^{-\infty}$. Moreover, the constructed operator obeys the given sequence of homogeneous edge symbols.

Denote the operator convention assigning a ∂_2-Mellin operator on $V \times C$ to a sequence of Mellin symbols by

$$\mathrm{op}_M : \sigma_M \mathcal{T}^{\mu}_{2,M+G} \longrightarrow \mathcal{L}(\mathcal{K}_P^s, \mathcal{K}_Q^{s-\mu}). \tag{12}$$

We define the class $\mathcal{W}^{\mu}_{2,M+G}$ by

$$\mathcal{W}^{\mu}_{2,M+G} = \mathrm{op}_M(\sigma_M \mathcal{T}^{\mu}_{2,M+G}) + \mathcal{W}^{\mu}_{2,G}. \tag{13}$$

Hence, each operator in $\mathcal{W}^{\mu}_{2,M+G}$ possesses a series of homogeneous edge symbols with values in $\mathcal{A}_{M+G}$ on the cone C.

The sequences

$$0 \longrightarrow \mathcal{W}^{\mu}_{2,G} \hookrightarrow \mathcal{W}^{\mu}_{2,M+G} \xrightarrow{\sigma_M} \sigma_M \mathcal{T}^{\mu}_{2,M+G} \longrightarrow 0 \tag{14}$$

and

$$0 \longrightarrow \mathcal{W}^{\mu-1}_{2,M+G} \hookrightarrow \mathcal{W}^{\mu}_{2,M+G} \xrightarrow{\partial_2^{(\mu)}\sigma} \mathcal{T}^{(\mu)}_{2,M+G} \longrightarrow 0 \tag{15}$$

are exact, where $\partial_2^{(\mu)}\sigma$ denotes the mapping evaluating the μ-homogeneous (principal) edge symbol.

Let us summarize the most important facts about the mapping properties of operators in the class $\mathcal{W}^{\mu}_{2,M+G}$.

2. Proposition. Let $\mathcal{A} \in \mathcal{L}(\mathcal{K}^s_{P,comp}(V\times C), \mathcal{K}^{s-\mu}_{Q,loc}(V\times C))$ belong to $\mathcal{W}^\mu_{2,M+G}(Q,P)$. For each $P' = (\gamma', \mathcal{P}') \in \underline{P}(N)$, $\gamma' \geq \gamma$, $\bar{r}(\gamma)\mathcal{P}' \leq \mathcal{P}$ there exists $R = (\lambda + \gamma' - \gamma, \mathcal{R}) \in \underline{P}(N)$, $\bar{r}(\lambda)\mathcal{R} \leq \mathcal{Q}$ such that the restriction of $\mathcal{A}$ to $\mathcal{K}^s_{P';comp}(V\times C)$ is continuous as operator

$$\mathcal{A} : \mathcal{K}^s_{P';comp}(V\times C) \longrightarrow \mathcal{K}^{s-\mu}_{R^+,loc}(V\times C). \tag{16}$$

Proof. The statement follows in the usual way from the corresponding mapping property of Mellin operators on the cone C. □

By construction our class of ∂_2-Mellin operators is stable under compositions, adjoints and changes in the y-coordinate.

3. Theorem. (i) Let $\mathcal{A} \in \mathcal{W}^\mu_{2,M+G}(P',P)$, $\mathcal{B} \in \mathcal{W}^{\mu'}_{2,M+G}(Q,P')$ and $\varphi \in C^\infty_0(V)$. Then $\mathcal{B}\varphi\mathcal{A} \in \mathcal{W}^{\mu+\mu'}_{2,M+G}(Q,P)$ and

$$\partial_2^{(\mu+\mu')}\sigma(\mathcal{B}\varphi\mathcal{A}) = \varphi\partial_2^{(\mu')}\sigma(\mathcal{B}) \cdot \partial_2^{(\mu)}\sigma(\mathcal{A}).$$

(ii) Let $\mathcal{A} \in \mathcal{W}^\mu_{2,M+G}(Q,P)$, $P = (\gamma, 0)$, $Q = (\lambda, 0)$.
Then $\mathcal{A}^* \in \mathcal{W}^\mu_{2,M+G}(Q^*,P^*)$ and

$$\partial_2^{(\mu)}\sigma(\mathcal{A}^*) = (\partial_2^{(\mu)}\sigma(\mathcal{A}))^*.$$

(iii) Let $\varkappa : V \longrightarrow V'$, $V, V' \subset R^k$ open, be a diffeomorphism. Then $\mathcal{A} \in \mathcal{W}^\mu_{2,M+G}(V\times C)$ implies $\varkappa_*\mathcal{A} \in \mathcal{W}^\mu_{2,M+G}(V'\times C)$.

Proof. Considering ∂_2-Mellin operators as ψDO's with operator-valued symbol (acting on C) we have to show only that the operations on symbolic level are carried out in our symbol class $\mathcal{T}^\infty_{2,M+G} = \bigcup_\mu \mathcal{T}^\mu_{2,M+G}$. This is simply checked on the level of $\sigma_M\mathcal{T}^\infty_{2,M+G}$, since only differentiations with respect to y and η, compositions and adjoints are involved. Taking into account (15) we get that the usual rules of composition and so on of symbols do not lead beyond the symbol class $\mathcal{T}^\infty_{2,M+G}$. □

According to (iii) the class $\mathcal{W}^\mu_{2,M+G}$ is well-defined on a manifold X with edge $\partial_2 X$ referring to an arbitrarily chosen atlas. Clearly, the class $\mathcal{W}^\mu_{2,M+G}$ gives smoothing operators in $\mathcal{W}^{-\infty}_G$ when localized outside a neighbourhood of $\partial_2 X$.

In case $\partial_1 C \neq \emptyset$, we define ∂_2-Mellin operators similar to

∂_2-Green operators in 4.3.1. by admitting the edge symbols to take values in $\mathcal{L}_{M+G}$ on the cone C. Then, in the 3x3 matrix of operators 4.3.1. (7) we get a more general form of $\mathcal{A}_{11}, \mathcal{A}_{22}$, whereas all remaining elements are not changed. Here $\mathcal{A}_{22}$ is enlarged by ∂_2-Mellin operators described above. For $\mathcal{A}_{11}$ the operator class is enriched by Mellin operators defined by Mellin symbols in $\bigoplus_{l=0}^{\infty} C^{\infty}(V, \mathfrak{M}_{\mathfrak{Z}(N,\partial N)})[\eta]^{\mu-\delta+1}$. Since no essential new points arise, let us note only that the classes $\mathcal{W}^{\mu,(1)}_{2,M+G}$ obey the usual properties of being closed under compositions, adjoints and changes of coordinates in the wedge.

4.3.3. The class $\mathcal{W}^{\mu}: \partial_1 C = \emptyset$

The class $\mathcal{W}^{\mu}$ is obtained by extending $\mathcal{W}^{\mu}_{M+G}$ by ψDO's. In the interior we have polyhomogeneous ψDO symbols. The specific new feature is the behaviour near $\partial_2 X$, which can be described in terms of the local model $\mathbb{R}^k \times C$ (or $V \times C, V \subset \mathbb{R}^k$ open).

We shall treat symbols $a(y,\tilde{x},\eta,\tilde{\xi}) \in \mathcal{K}^{\infty}_{\underline{P}(N)}(\mathbb{R}^k \times C) \otimes_{\pi} S^{\mu}_{cl}(\mathbb{R}^n)$, $y, \eta \in \mathbb{R}^k$, $\tilde{x} \in C$, $\tilde{\xi} \in \mathbb{R}^{n-k}$. They admit an expansion

$$a(y,\tilde{x},\eta,\tilde{\xi}) = \sum_j b_j(y,\tilde{x}) c_j(\eta,\tilde{\xi}), b_j \in \mathcal{K}^{\infty}_{\underline{P}(N)}(\mathbb{R}^k \times C), c_j \in S^{\mu}_{cl}(\mathbb{R}^k) \qquad (1)$$

converging in the tensor product. By (1) we are reduced to multiplications by b_j and constant coefficient ψDO's. Let $c(\eta,\tilde{\xi}) \in S^{\mu}_{cl}(\mathbb{R}^n)$ be given. It induces a continuous mapping

$$\mathbb{R}^k \ni \eta \longrightarrow c(\eta,\cdot) \in S^{\mu}_{cl}(\mathbb{R}^{n-k}).$$

For given $P_2 = (\gamma_2, \mathcal{P}_2) \in \underline{P}(N)$, $Q_2 = (\gamma_2 - \mu, q_2) \in \underline{P}(N)$, $1/2 - \gamma_2 \notin \{-\mu - j, n-k+j\}_{j=0}^{\infty}$, we described in 4.2.3. the construction of a (non-canonical) operator convention

$$op_{\psi} : S^{\mu}_{cl}(\mathbb{R}^{n-k}) \longrightarrow \mathcal{L}(\mathcal{K}^{s}_{P_2}(C), \mathcal{K}^{s-\mu}_{Q_2}(C)), \qquad (2)$$

where the auxilary choices are immaterial modulo $\mathcal{L}_{M+G}$ on C. The η dependence may be organized in such a way that the resulting operator families behave as classical operator-valued symbols. In particular the η-dependent cut-offs are now $\omega(t[\eta])$.

For the sake of simplicity we assume that the asymptotics of b_j for $|\tilde{x}| \longrightarrow 0$ has the form

$$b_j(y,\tilde{x}) \sim \sum_{l=0}^{\infty} |\tilde{x}|^l b_{jl}(y, \frac{\tilde{x}}{|\tilde{x}|}).$$

Denote the corresponding class of interior symbols by $\mathcal{S}_0^\mu$.

Now we associate with interior symbols in $\mathcal{S}_0^\mu$ operators in $\mathcal{L}(\mathcal{K}^s_{P_2}(\mathbb{R}^k \times C), \mathcal{K}^{s-\mu}_{Q_2}(\mathbb{R}^k \times C))$. We assume the symbols to stabilize for y outside a compact set in $\mathbb{R}^k$ (otherwise we have to deal with $\mathcal{L}(\mathcal{K}^s_{P_2,comp}(\mathbb{R}^k \times C), \mathcal{K}^{s-\mu}_{Q_2,loc}(\mathbb{R}^k \times C))$, where comp and loc are understood with respect to $\mathbb{R}^k$). For a constant coefficient symbol $c(\eta,\tilde{\xi})$ we take $op_\psi(c(\eta,\cdot))$ as operator-valued symbol and define

$$Op(c)u(y,\tilde{x}) = \int e^{iy\eta} op_\psi(c(\eta,\cdot)) Fu(\eta,\cdot)(\tilde{x})\, d\eta, \tag{3}$$

F denotes here the partial Fourier transform along $\mathbb{R}^k$. In the usual way it is checked that $Op(c) \in \mathcal{L}(\mathcal{K}^s_{P_2}(\mathbb{R}^k \times C), \mathcal{K}^{s-\mu}_{Q_2}(\mathbb{R}^k \times C))$ and its norm depends continuously on c. Given $a(y,\tilde{x},\eta,\tilde{\xi}) \in \mathcal{S}_0^\mu$ with decomposition (1). Then we set

$$Op_\psi(a) = \sum_j Op_\psi(b_j) Op_\psi(c_j), \tag{4}$$

where $Op_\psi(b_j)$ is understood as multiplication operator acting in $\mathcal{K}^{s-\mu}_{Q_2}(\mathbb{R}^k \times C)$. This operator convention is independent of the arbitrary choices modulo $\mathcal{W}^\mu_{M+G}$.

1. <u>Definition.</u> An operator $\mathcal{A}$ is said to belong to $\mathcal{W}^\mu(Q_2,P_2)$ on $\mathbb{R}^k \times C$ if there exists a (complete) interior symbol $\sigma \in \mathcal{S}_0^\mu$ such that $\mathcal{A} - Op(\sigma) \in \mathcal{W}^\mu_{M+G}(Q_2,P_2)$ for some Op-convention.

Since away from $\partial_2 X$ the Mellin and Green operators are smoothing, we can recover from $\mathcal{A}$ the interior symbol in a unique way obtaining a mapping

$$\sigma_0\sigma : \mathcal{W}^\mu \longrightarrow \mathcal{S}_0^\mu \tag{5}$$

which is surjective. Moreover, the sequence

$$0 \longrightarrow \mathcal{W}^\mu_{M+G} \hookrightarrow \mathcal{W}^\mu \xrightarrow{\sigma_0\sigma} \mathcal{S}_0^\mu \longrightarrow 0 \tag{6}$$

is exact.

Besides the interior symbol we have an edge symbol being a smooth family parametrized by (y,η) with values in $\mathcal{L}^{\mu}$ on the cone C.

Let us focus the attention on the leading homogeneity μ. For $a \in \mathcal{S}_O^{\mu}$ we take the part $a_{\mu}(y,\tilde{x},\eta,\tilde{\xi})$ homogeneous of degree μ in $\eta,\tilde{\xi}$ and associate with

$$a_{\mu}^{0}(y,\frac{\tilde{x}}{|\tilde{x}|},\eta,\tilde{\xi}) = \lim_{|\tilde{x}|\to 0} a_{\mu}(y,\tilde{x},\eta,\tilde{\xi}) \tag{7}$$

a family of operators

$$op_{\psi}(a_{\mu}^{0})(y,\eta) \in \mathcal{L}(\mathcal{K}_{P_2}^{s}(C),\mathcal{K}_{Q_2}^{s-\mu}(C)).$$

It is homogeneous of degree μ in the sense of 4.3.1.(1). Note that lower order parts $a_{\mu-j}, j=1,2,\ldots$ and also terms with $O(|\tilde{x}|^{1}), 1 \geq 1$, of the expansion of a_{μ} do not contribute to this leading homogeneity. Then $op_{\psi}(a_{\mu})$ represents the edge symbol modulo μ-homogeneous edge symbols of Mellin and Green type, i.e. modulo $\mathcal{S}_{2,M+G}^{(\mu)}$. Introducing the notation

$$\mathcal{S}_2^{(\mu)} = op_{\psi}(\mathcal{S}_O^{(\mu)}) + \mathcal{S}_{2,M+G}^{(\mu)} \tag{8}$$

we shall speak of $\mathcal{S}^{(\mu)}$ as the space of couples (a,α) in $\mathcal{S}_O^{(\mu)} \times \mathcal{S}_2^{(\mu)}$ being compatible in the sense that $\alpha - op_{\psi}(a) \in \mathcal{S}_{2,M+G}^{(\mu)}$.

Each $\mathcal{L}^{\mu}$ operator on the cone C has a sequence of operator-valued Mellin symbols (with respect to the vertex). It is important to describe the influence of the interior symbol. We have the expansion

$$a_{\mu}(y,\frac{\tilde{x}}{|\tilde{x}|},\eta,\tilde{\xi}) \sim \sum_{\alpha} a_{\mu\alpha}(y,\frac{\tilde{x}}{|\tilde{x}|},\tilde{\xi})\, q_{\alpha}(\eta) \quad \text{for} \quad |\tilde{\xi}| \to \infty ,$$

where $a_{\mu\alpha}$ is homogeneous in $\tilde{\xi}$ of degree $\mu - |\alpha|$ and q_{α} is a homogeneous polynomial of degree α. In particular, $a_{\mu}(y,\frac{\tilde{x}}{|\tilde{x}|},0,\tilde{\xi})$ is the leading part which is the only contribution to the Mellin symbol of conormal order μ. The Mellin symbols of conormal order $\mu-j$ are homogeneous polynomials in η of degree j. This means that $\sigma_M \mathcal{S}_2^{(\mu)}$ – the space of sequences of Mellin symbols of edge symbols in $\mathcal{S}_2^{(\mu)}$ – are also j-homogeneous η-polynomials with coefficients in $C_y^{\infty} \otimes \mathfrak{M}_{\underline{\mathfrak{z}}(N)}$.

It follows that

$$\mathcal{Y}_{2,M+G}^{(\mu)} \cdot \mathcal{Y}_{2}^{(\nu)} \subset \mathcal{Y}_{2,M+G}^{(\mu+\nu)}$$

$$\mathcal{Y}_{2}^{(\mu)} \cdot \mathcal{Y}_{2,M+G}^{(\nu)} \subset \mathcal{Y}_{2,M+G}^{(\mu+\nu)}$$

and

$$\mathcal{Y}_{2}^{(\mu)} \cdot \mathcal{Y}_{2}^{(\nu)} \subset \mathcal{Y}_{2}^{(\mu+\nu)},$$

where $\cdot$ means composition of operators on the cone (provided it is defined) for fixed y,η. The last assertion is based on the fact that $\mathcal{Y}_{0}^{(\mu)} \cdot \mathcal{Y}_{0}^{(\nu)} \subset \mathcal{Y}_{0}^{(\mu+\nu)}$.

2. Proposition. Each $\mathcal{A} \in \mathcal{W}^{\mu}(Q,P)(V\times C)$ defines a continuous linear operator

$$\mathcal{A} : \mathcal{K}_{P,\mathrm{comp}}^{s}(V\times C) \longrightarrow \mathcal{K}_{Q,\mathrm{loc}}^{s-\mu}(V\times C) \tag{9}$$

for each $s \in \mathbb{R}$. Its restriction to $\mathcal{K}_{P',\mathrm{comp}}^{s}(V\times C) \subset \mathcal{K}_{P,\mathrm{comp}}^{s}(V\times C)$ is a continuous operator

$$\mathcal{A} : \mathcal{K}_{P',\mathrm{comp}}^{s}(V\times C) \longrightarrow \mathcal{K}_{R^{+},\mathrm{loc}}^{s-\mu}(V\times C), \tag{10}$$

where $R = (r,\sigma), r = \gamma'-\gamma+\lambda$, $P' = (\gamma',\varphi')$. $P = (\gamma,\varphi)$. $Q,(\lambda,q)$.

Proof. The first statement is fulfilled by construction. The second one is a consequence of the Mellin expansion for ψDO's on the cone C. $\square$

The most important facts about the class $\mathcal{W}^{\mu}$ on $V\times C$ are collected in the following

3. Theorem. (i) Let $\mathcal{A} \in \mathcal{W}^{\mu}(Q,P)$, $\mathcal{B} \in \mathcal{W}^{\nu}(Q',Q)$, $\varphi \in C_0^{\infty}(V)$. Then $\mathcal{B}\varphi\mathcal{A} \in \mathcal{W}^{\mu+\nu}(Q',P)$ and for the principal interior and edge symbol we have

$$\partial_0^{(\mu+\nu)}\sigma(\mathcal{B}\varphi\mathcal{A}) = \varphi\partial_0^{(\nu)}\sigma(\mathcal{B})\partial_0^{(\mu)}\sigma(\mathcal{A}),$$

$$\partial_2^{(\mu+\nu)}\sigma(\mathcal{B}\varphi\mathcal{A}) = \varphi\partial_2^{(\nu)}\sigma(\mathcal{B})\circ\partial_2^{(\mu)}\sigma(\mathcal{A}),$$

where $\circ$ denotes the composition of operators on C (for fixed y,η).

(ii) Let $\mathcal{A} \in \mathcal{W}^{\mu}(Q,P)$, $P = (\gamma,0)$, $Q = (\gamma-\mu,0)$. Then $\mathcal{A}^{*} \in \mathcal{W}^{\mu}(P^{*},Q^{*})$ and

$$\partial_0^{(\mu)}\sigma(\mathcal{A}^*) = \overline{\partial_0^{(\mu)}\ \sigma(\mathcal{A})},$$

$$\partial_2^{(\mu)}\sigma(\mathcal{A}^*) = \partial_2^{(\mu)}\ \sigma(\mathcal{A})^*.$$

Proof. First note that the operations on symbolic level remain in the class, since $\mathcal{L}^\infty$ on the cone is stable under compositions and taking adjoints and differentiation with respect to η and y is possible. Moreover, $\mathcal{B}\varphi\mathcal{A} - \mathrm{Op}(\partial_0\sigma(\mathcal{B}\varphi\mathcal{A})) \in \mathcal{W}^{\mu+\nu}_{M+G}(Q',P)$, where $\partial_0\sigma$ denotes the complete ∂_0-symbol. The composition law for the edge symbols follows from the general ψDO calculus with operator-valued symbol. In the same way we conclude for the adjoint. □

The invariance of $\mathcal{W}^\mu$ in $V\times C$ under diffeomorphisms $V \to V', V' \subset \mathbb{R}^k$, is proved by similar arguments. It is clear for the ψDO part. For the ∂_2-Mellin and Green operators we refer to the general calculus of ψDO's with operator-valued symbols, noting that the class of ∂_2-symbols of Mellin and Green type is stable.

Let X be a manifold with edge $\partial_2 X$ (but up to now without boundary $\partial_1 X$). Then we define the class $\mathcal{W}^\mu(Q,P)$ on $(X,\partial_2 X)$ as the space of mappings

$$\mathcal{A} = \begin{pmatrix} \mathcal{A}_{11} & \mathcal{A}_{12} \\ \mathcal{A}_{21} & \mathcal{A}_{22} \end{pmatrix} : \begin{matrix} \mathcal{K}^s_P(X) \\ \oplus \\ H^s(\partial_2 X) \end{matrix} \longrightarrow \begin{matrix} \mathcal{K}^{s-\mu}_Q(X) \\ \oplus \\ H^{s-\mu}(\partial_2 X) \end{matrix}, s \in \mathbb{R},$$

which belong in local coordinates to the class $\mathcal{W}^\mu$, i.e. away from $\partial_2 X$ we have simply a classical ψDO with symbol in Γ^μ_0 and near $\partial_2 X$ in the 2x2 matrix of operators also ∂_2-Mellin and Green operators are involved.

4.3.4. The class $\mathcal{W}^\mu$: $\partial_1 C \neq \emptyset$

Now we are going to describe the class $\mathcal{W}^\mu$ in full generality, i.e. we include the case that $\partial_1 X$ is not separated from $\partial_2 X$. In 4.3.1. and 4.3.2. we discussed already the ∂_2-Mellin and Green operators in this case. The new feature is the effect produced by the ∂_1-symbol on the ∂_2-symbol (and ∂_2-operators). Our approach is the same as in 4.3.3. First we specify our assumptions on the asymptotic behaviour of the ∂_0- and ∂_1-symbols near $\partial_2 X$. Then we associate with the couple $(\partial_0\sigma, \partial_1\sigma)$ an operation on $(V\times C, V\times\partial_1 C)$, $V\subset\mathbb{R}^k$, which is canonically defined modulo ∂_2-Mellin and Green operators considered in 4.3.1. and 4.3.2. As a result we get the class $\mathcal{S}_2$ of edge symbols $\partial_2\sigma$ taking values in $\mathcal{L}^\mu(C,\partial_1 C,\partial_2 C)$ and the compatibility condition between $(\partial_0\sigma, \partial_1\sigma)$.

Let $P = (P_1, P_2)\in \underline{P}(N,\partial N)$ and $Q = (Q_1, Q_2)\in\underline{P}(N,\partial N)$ be given and let α be a μ-homogeneous ∂_1-symbol on $V\times\partial_1 C$, i.e.

$$\alpha(y,\tilde{x}',\lambda\eta,\lambda\tilde{\xi}') = \lambda^\mu \varkappa_\lambda \alpha(y,\tilde{x}',\eta,\tilde{\xi}')\varkappa_\lambda^{-1}, \quad \lambda>0$$

with values in $\mathcal{L}_{M+G}(Q_1,P_1)(\mathbb{R}_+,\partial\mathbb{R}_+)$. Set

$$\alpha_0(y,\tilde{x}',\eta,\tilde{\xi}') = \varkappa^{-1}_{|\eta,\tilde{\xi}'|}\alpha(y,\tilde{x}',\eta,\tilde{\xi}')\varkappa_{|\eta,\tilde{\xi}'|}, \tag{1}$$

which is μ-homogeneous in the ordinary sense. α_0 belongs to $C^\infty(V\times\partial_1 C)\otimes_\pi \mathcal{S}'(\mathbb{R}^n\setminus 0)^{(\mu)}_\infty \otimes_\pi \mathcal{L}_{M+G}(Q_1,P_1)(\mathbb{R}_+,\partial\mathbb{R}_+)$. We assume that near $\partial_2(V\times\partial_1 C) = V\times\partial_2 C$ we have a $\mathcal{P}_0$-asymptotics, i.e.

$\alpha_0\in\mathcal{K}_{\mathcal{P}_0}(V\times\partial_1 C)\otimes_\pi\mathcal{S}'(\mathbb{R}^n\setminus 0)^{(\mu)}_\infty\otimes_\pi\mathcal{L}_{M+G}(Q_1,P_1)(\mathbb{R}_+,\partial\mathbb{R}_+)$, where $\mathcal{P}_0\in\mathcal{P}(N,\partial N)$ involves only non-negative integers and no logarithmic terms. Denote this class by $\mathcal{S}^{(\mu)}_{1,M+G}(V\times C, V\times\partial_1 C)$. Then we have an expansion

$$\alpha(y,\tilde{x}',\eta,\tilde{\xi}') = \sum_j b_j(y,\tilde{x}')c_j(\eta,\tilde{\xi}'), \tag{2}$$

where c_j are again μ-homogeneous but with constant coefficients and $b_j\in\mathcal{K}^\infty_{\mathcal{P}_0}(V\times\partial_1 C)$. Then we note that $\eta\to c_j(\eta,\cdot)$ is a smooth mapping

$$\mathbb{R}^k\setminus 0 \longrightarrow \mathcal{S}^\mu_{1,M+G}(C,\partial_1 C).$$

We described in 4.2.4. a mapping

$$\mathrm{op}: \mathcal{S}^\mu_{1,M+G}(C,\partial_1 C) \longrightarrow \mathcal{L}_{M+G}(C,\partial_1 C)$$

depending on P,Q and auxiliary choices which have no effect modulo ∂_2-Mellin and Green operators.

Put

$$\mathcal{S}^{(\mu)}_{2,M+G} = \mathcal{S}^{(\mu)}_{2,M+G} + \mathrm{op}(\mathcal{S}^{(\mu)}_{1,M+G}). \tag{3}$$

Note that each ∂_2-symbol in $\mathcal{S}^{(\mu)}_{2,M+G}$ as before gives rise to a sequence of Mellin symbols with values in $\mathcal{M}^{\mu}_{M+G}(N,\partial N)$ which is a polynomial in η. With $\mathfrak{a}\in\mathcal{S}^{(\mu)}_{1,M+G}$ we associate an operator in $\mathcal{L}(\mathcal{K}^{s}_{P,comp}(V\times C)\oplus\mathcal{K}^{s}_{\partial_1 P_2,comp}(V\times\partial_1 C),\ \mathcal{K}^{s-\mu}_{Q,loc}(V\times C)\oplus\mathcal{K}^{s-\mu}_{\partial_1 Q_2,loc}(V\times C))$ extending the action from the interior (possibly after a modification by a ∂_2-Mellin and Green operator).

In the same way we proceed with non-homogeneous ∂_1-symbols. If Op denotes the operator convention, we put

$$\mathcal{W}^{\mu}_{M+G} = \mathrm{Op}(\mathcal{S}^{\mu}_{1,M+G}) + \mathcal{W}^{\mu}_{2,M+G}. \tag{4}$$

Then the sequence

$$0 \longrightarrow \mathcal{W}^{\mu-1}_{M+G} + \mathcal{W}^{\mu}_{2,M+G} \hookrightarrow \mathcal{W}^{\mu}_{M+G} \xrightarrow{\partial_1^{(\mu)}\sigma} \mathcal{S}^{(\mu)}_{1,M+G} \longrightarrow 0 \tag{5}$$

is exact.

Let us take now into account also a ψDO in the interior. Let $a(y,\tilde{x},\eta,\tilde{\xi})$ be a smooth function in the interior, μ-homogeneous in $\eta,\tilde{\xi}$. We assume a $\mathcal{P}_0$-asymptotics near $V\times\partial_1 C$ and, in addition, a $\mathcal{P}_0$-asymptotics near $V\times\partial_2 C$. Denote by $\mathcal{S}^{(\mu)}_0$ this class of interior (principal) symbols. Let $\mathfrak{a}$ be a ∂_1-symbol, μ-homogeneous and satisfying the compatibility condition with a

$$\mathrm{op}_1(a) - \mathfrak{a}\in\mathcal{S}^{(\mu)}_{1,M+G}\ , \tag{6}$$

where $\mathrm{op}_1(a)$ denotes the μ-homogeneous ∂_1-symbol associated with a (cf. 3.2.4.). Denote this space of μ-homogeneous ∂_1-symbols by $\mathcal{S}^{(\mu)}_1$. Clearly, it is characterized by a $\mathcal{P}_0$-asymptotics near $V\times\partial_2 C$.

Next we associate with the ∂_0-symbol $a\in\mathcal{S}^{(\mu)}_0$ an edge symbol $\mathrm{op}_2(a)$ being a family of operators in $\mathcal{L}^{\mu}(C,C)$ parametrized by $(y,\eta)\in V\times(\mathbb{R}^k\setminus 0)$, homogeneous in η of degree μ. For fixed y,η the operator is patched together from conical subsets of C.

We apply the construction from 4.2. to $a_0(y,\frac{\tilde{x}}{|\tilde{x}|},\eta,\cdot) = \lim_{|\tilde{x}|\to 0} a_0(y,\tilde{x},\eta,\cdot)$. Denote this class of edge symbols by $\mathrm{op}_2(\mathcal{S}^{(\mu)}_0)$ and set

$$\mathcal{S}^{(\mu)}_2 = \mathrm{op}_2(\mathcal{S}^{(\mu)}_0) + \mathcal{S}^{(\mu)}_{2,M+G}\ . \tag{7}$$

As in 4.3.3. we obtain a sequence of ∂_2-Mellin symbols which are polynomials in η (η-independent for the leading Mellin symbol of conormal order μ).

In the same way we proceed also for non-homogeneous ∂_0-symbols which admit a $\mathcal{P}_0$-asymptotics for each homogeneous component. It follows as in 4.3.3. that (possibly after a suitable modification by a ∂_2-Mellin and Green operator) the ψDO with given symbol in $\mathcal{T}_0^{\mu}$ admits an extension as continuous operator in $\mathcal{L}(\mathcal{K}_P^s(V\times C), \mathcal{K}_Q^{s-\mu}(V\times C))$. Denote the resulting operator class by $\mathrm{Op}(\mathcal{T}_0^{\mu})$ and set

$$\mathfrak{D}^{\mu} = \mathrm{Op}(\mathcal{T}_0^{\mu}) + \mathfrak{D}^{\mu}_{M+G}. \tag{8}$$

Then the sequences are exact:

$$0 \longrightarrow \mathfrak{D}^{\mu}_{M+G} \hookrightarrow \mathfrak{D}^{\mu} \xrightarrow{\partial_0\sigma} \mathcal{T}_0^{\mu} \longrightarrow 0 \tag{9}$$

and

$$0 \longrightarrow \mathfrak{D}^{\mu-1} \hookrightarrow \mathfrak{D}^{\mu} \xrightarrow{\partial_0^{(\mu)}\sigma,\, \partial_1^{(\mu)}\sigma,\, \partial_2^{(\mu)}\sigma} \mathcal{T}^{(\mu)} \longrightarrow 0 \tag{10}$$

where $\partial_0\sigma$ denotes the complete ∂_0-symbol mapping, $\partial_j^{(\mu)}\sigma$ evaluates the μ-homogeneous ∂_j-symbol and $\mathcal{T}^{(\mu)}$ consists of triples $(a_0, \alpha_1, \alpha_2) \in (\mathcal{T}_0^{(\mu)}, \mathcal{T}_1^{(\mu)}, \mathcal{T}_2^{(\mu)})$ satisfying the compatibility conditions:

$$\alpha_1 - op_1(a_0) \in \mathcal{T}^{(\mu)}_{1,M+G},\quad \alpha_2 - op_2(a_0) - op_2(\alpha_1 - op_1(a_0)) \in \mathcal{T}^{(\mu)}_{2,M+G}. \tag{11}$$

Besides the principal homogeneous part all three symbolic levels possess a sequence of symbols with lower homogeneity. For them the compatibility conditions become more and more involved. It must be noted that compositions and other operations of the symbolic calculus preserve the classes $\mathcal{T}_j^{(\infty)}$, j=0,1,2, and also the compatibility conditions. This is clear for $\mathcal{T}_0^{(\infty)}$. The compatibility condition between $\mathcal{T}_0^{(\infty)}$ and $\mathcal{T}_1^{(\infty)}$ was proved in 3.2. Since $\mathcal{T}_{2,M+G}$ is a two-sided ideal in $\mathcal{T}_2^{(\infty)}$ it follows from (7) that also the compatibility between $\mathcal{T}_0^{(\infty)}, \mathcal{T}_1^{(\infty)}$ and $\mathcal{T}_2^{(\infty)}$ remains true after compositions.

By construction an operator $\mathcal{A} \in \mathfrak{D}^{\mu}(Q,P)$ defines a continuous mapping

$$\mathcal{A}:\begin{array}{c}\mathcal{K}^{s}_{P,comp}(V\times C)\\ \oplus\\ \mathcal{K}^{s}_{\partial_1 P_2,comp}(V\times\partial_1 C)\\ \oplus\\ H^{s}(V)\end{array}\longrightarrow\begin{array}{c}\mathcal{K}^{s-\mu}_{Q,loc}(V\times C)\\ \oplus\\ \mathcal{K}^{s-\mu}_{\partial_1 Q_2,loc}(V\times\partial_1 C)\\ \oplus\\ H^{s-\mu}(V)\end{array}\qquad(12)$$

Its restriction to a subspace with better weights acts continuously into a subspace with correspondingly improved weight, possibly with a loss of an arbitrarily small $\varepsilon>0$. This is a consequence of the Mellin expansion of the edge symbols near $\partial_2 C$.

We prove now that the class $\mathfrak{W}^{\infty}=\bigcup_{\mu}\mathfrak{W}^{\mu}$ on $(V\times C,V\times\partial_1 C,V\times\partial_2 C)$ is stable under compositions, adjoints and diffeomorphism $V\longrightarrow V'$.

1. Theorem. (i) Let $\mathcal{A}\in\mathfrak{W}^{\mu}(P',P)$, $\mathcal{B}\in\mathfrak{W}^{\nu}(Q,P')$, $\varphi\in C_0^{\infty}(V)$. Then $\mathcal{B}\varphi\mathcal{A}\in\mathfrak{W}^{\mu+\nu}(Q,P)$ and $\partial_j^{(\mu+\nu)}\sigma(\mathcal{B}\varphi\mathcal{A})=\partial_j^{(\nu)}\sigma(\mathcal{B})\varphi\partial_j^{(\mu)}\sigma(\mathcal{A})$, $j=0,1,2$.

(ii) Let $\mathcal{A}\in\mathfrak{W}^{\mu}(Q,P)$ and $P=((\gamma_1,0),(\gamma_2,0))$, $Q=((\gamma_1-\mu,0),(\gamma_2-\mu,0))$. Then $\mathcal{A}^*\in\mathfrak{W}^{\mu}(P^*,Q^*)$ and $\partial_j^{(\mu)}\sigma(\mathcal{A}^*)=\partial_j^{(\mu)}\sigma(\mathcal{A})^*$.

(iii) Let $\varkappa: V\longrightarrow V'$ be a diffeomorphism and $\mathcal{A}\in\mathfrak{W}^{\mu}$ defined on $(V\times C,V\times\partial_1 C,V\times\partial_2 C)$. Then $\varkappa_*\mathcal{A}\in\mathfrak{W}^{\mu}$ on $(V'\times C,V'\times\partial_1 C,V'\times\partial_2 C)$.

Proof. (i) The case if $\partial_0\sigma(\mathcal{A})=0,\partial_1\sigma(\mathcal{A})=0$ and $\partial_0\sigma(\mathcal{B})=0,\partial_1\sigma(\mathcal{B})=0$, i.e. $\mathcal{A},\mathcal{B}\in\mathfrak{W}^{\infty}_{2,M+G}$, was treated in 4.3.2. In case $\partial_0\sigma(\mathcal{A})=0$, $\partial_0\sigma(\mathcal{B})=0$, i.e. $\mathcal{A},\mathcal{B}\in\mathfrak{W}^{\infty}_{M+G}$, we have $\mathcal{A}=\mathrm{Op}(\partial_1\sigma(\mathcal{A}))$ mod $\mathfrak{W}^{\mu}_{2,M+G}$, $\mathcal{B}=\mathrm{Op}(\partial_1\sigma(\mathcal{B}))$ mod $\mathfrak{W}^{\nu}_{2,M+G}$. From $\mathcal{Y}_2^{\mu}\cdot\mathcal{Y}^{\nu}_{2,M+G}\subset\mathcal{Y}^{\mu+\nu}_{2,M+G}$, $\mathcal{Y}^{\mu}_{2,M+G}\cdot\mathcal{Y}_2^{\nu}\subset\mathcal{Y}^{\mu+\nu}_{2,M+G}$ we get that $\mathcal{B}\varphi\mathcal{A}-\mathrm{Op}(\partial_1\sigma(\mathcal{B})\circ\varphi\partial_1\sigma(\mathcal{A}))\in\mathfrak{W}^{\mu+\nu}_{2,M+G}$.

where $\circ$ denotes the composition of complete ∂_1-symbols by the Leibniz rule. In fact, we have $\mathrm{Op}(\partial_1^{(\nu)}\sigma(\mathcal{B})\varphi\,\partial_1^{(\mu)}\sigma(\mathcal{A}))=$ $\mathrm{Op}(\partial_1^{(\nu)}\sigma(\mathcal{B}))\varphi\,\mathrm{Op}(\partial_1^{(\mu)}\sigma(\mathcal{A}))$ mod $(\mathfrak{W}^{\mu+\nu-1}+\mathfrak{W}^{\mu+\nu}_{2,M+G})$ and we conclude by induction.

Finally, in the general case we include also $\mathrm{Op}(\partial_0\sigma(\mathcal{B}))$ and $\mathrm{Op}(\partial_0\sigma(\mathcal{A}))$ and use that $\mathrm{Op}(\partial_0^{(\nu)}\sigma(\mathcal{B}))\varphi\partial_0^{(\mu)}\sigma(\mathcal{A}))=\mathrm{Op}(\partial_0^{(\nu)}\sigma(\mathcal{B}))\varphi\,\mathrm{Op}(\partial_0^{(\mu)}\sigma(\mathcal{A}))$ mod $(\mathfrak{W}^{\mu+\nu-1}+\mathfrak{W}^{\mu+\nu}_{M+G})$ and $\mathrm{Op}(\partial_0^{(\nu)}\sigma(\mathcal{B}))\mathfrak{W}^{\mu}_{M+G}\subset\mathfrak{W}^{\mu+\nu}_{M+G}+\mathfrak{W}^{\mu+\nu-1}$ $\mathfrak{W}^{\nu}_{M+G}\mathrm{Op}(\partial_0^{(\mu)}\sigma(\mathcal{A}))\subset\mathfrak{W}^{\mu+\nu}_{M+G}+\mathfrak{W}^{\mu+\nu-1}$.

(ii) First note that the adjoint of a symbol in $\mathcal{S}^{(\mu)}$ is again in this class since this is clear for the components, and the compatibility condition is preserved, too. Then $\mathcal{A}^* - Op(\partial_0^{(\mu)}\sigma(\mathcal{A})^*, \partial_1^{(\mu)}\sigma(\mathcal{A})^*, \partial_2^{(\mu)}\sigma(\mathcal{A})^*) \in \mathcal{W}^{\mu-1}$.

(iii) Changes of coordinates of the considered form preserve the symbol class $\mathcal{S}^{(\mu)}$ modulo lower order symbols. Therefore, also the operator class is preserved. □

The definition of the class $\mathcal{W}^{\mu}$ guarantees that for $\varphi, \psi \in \mathcal{K}^{\infty}_{\varphi_0}$, $\varphi\psi = 0$, the operator $\varphi\mathcal{A}\psi$ is a smoothing Green operator. Moreover, the extension of ψDO's in the interior by certain ∂_1-boundary operators (of Mellin and Green type) does not affect the structure outside a neighbourhood of $\partial_1 X$ since there the new operators act as smoothing Green operators. In the same way the ∂_2-Mellin and Green operators are negligible outside an arbitrary neighbourhood of $\partial_2 X$.

In view of Theorem 1(iii) we can define the class $\mathcal{W}^{\mu}$ in the global situation by referring to any atlas. We shall continue to use the previous notations. It is clear that Theorem 1(i) and (ii) remain valid then. In the standard way we get also operators acting between sections of vector bundles.

4.3.5. Ellipticity and Fredholm property

Let X be a compact manifold with smooth boundary $\partial_1 X$ and edge $\partial_2 X$. First we interpret the principal symbols of operators in $\mathcal{W}^{\mu}$ as global bundle morphisms.

Let $E_j, E_j \in \mathrm{Vect}(\partial_j X)$, $j = 0,1,2$, and

$$\mathcal{A} : \begin{matrix} \mathcal{K}^{\infty}_{P}(\partial_0 X, E_0) \\ \oplus \\ \mathcal{K}^{\infty}(\partial_1 X, E_1) \\ \oplus \\ C^{\infty}(\partial_2 X, E_2) \end{matrix} \longrightarrow \begin{matrix} \mathcal{K}^{\infty}_{Q}(\partial_0 X, E_0) \\ \oplus \\ \mathcal{K}^{\infty}(\partial_1 X, E_1) \\ \oplus \\ C^{\infty}(\partial_2 X, E_2) \end{matrix} \qquad (1)$$

$\mathcal{A} \in \mathcal{W}^{\mu}$. Denote by $\sigma = (\partial_0\sigma, \partial_1\sigma, \partial_2\sigma)$ its principal symbol, which is evaluated in local coordinates and with respect to local frames of the bundles. The local representations give rise to the following bundle morphisms: The interior principal symbol $\partial_0\sigma$ yields a morphism

$$\partial_0\sigma : \pi_0^* E_0 \longrightarrow \pi_0^* E_0' \qquad (2)$$

where $\pi_j : S^*(\partial_j X) \longrightarrow \partial_j X$ and we have a specific asymptotic behaviour of $\partial_0\sigma$ near $\partial_1(S(\partial_0 X))$ and $\partial_2(S^*(\partial_0 X))$. Here we restrict ourselves to the sphere bundle (with respect to a given metric) in $T^*(\partial_0 X)$ since the homogeneity order μ is kept fixed.

The boundary symbol $\partial_1\sigma$ yields a bundle morphism

$$\partial_1\sigma : \begin{array}{c} \pi_1^* E_0 \otimes \mathcal{K}^\infty_{P_1}(\mathbb{R}_+) \\ \oplus \\ \pi_1^* E_1 \end{array} \longrightarrow \begin{array}{c} \pi_1^* E_0' \otimes \mathcal{K}^\infty_{Q_1}(\mathbb{R}_+) \\ \oplus \\ \pi_1^* E_1' \end{array} \qquad (3)$$

with specific asymptotic behaviour near $\partial_2(S^*(\partial_1 X))$. For fixed $(x',\xi') \in S^*(\partial_1 X)$ the operator $\partial_1\sigma(x',\xi')$ belongs to $\mathcal{L}^\mu(Q_1,P_1)$ and it acts as

$$\partial_1\sigma(x',\xi') : \begin{array}{c} E_{0,x'} \otimes \mathcal{K}^\infty_{P_1}(\mathbb{R}_+) \\ \oplus \\ E_{1,x'} \end{array} \longrightarrow \begin{array}{c} E'_{0,x'} \otimes \mathcal{K}_{Q_1}(\mathbb{R}_+) \\ \oplus \\ E'_{1,x'} \end{array} \qquad (4)$$

Finally, the edge symbol $\partial_2\sigma$ yields a bundle morphism

$$\partial_2\sigma : \begin{array}{c} \pi_2^* E_0 \otimes \mathcal{K}^\infty_P(C) \\ \oplus \\ \pi_2^* E_1 \otimes \mathcal{K}^\infty_{\partial_1 P_2}(\partial_1 C) \\ \oplus \\ \pi_2^* E_2 \end{array} \longrightarrow \begin{array}{c} \pi_2^* E_0' \otimes \mathcal{K}^\infty_Q(C) \\ \oplus \\ \pi_2^* E_1' \otimes \mathcal{K}^\infty_{\partial_1 Q_2}(\partial_1 C) \\ \oplus \\ \pi_2^* E_2' \end{array} \qquad (5)$$

and takes values in $\mathcal{L}^\mu(Q,P)$ on $(C,\partial_1 C)$ for each fixed $(y,\eta) \in S^*(\partial_2 X)$. The compatibility conditions between $\partial_0\sigma, \partial_1\sigma$ and $\partial_2\sigma$ make sense for the bundle morphisms since they are independent of the local representation.

Denote by $\mathcal{G}^{(\mu)}$ the space of compatible triples of $(\partial_0\sigma, \partial_1\sigma, \partial_2\sigma)$ as above, considered as bundle morphisms.

1. <u>Definition.</u> A symbol $\sigma \in \mathcal{G}^{(\mu)}$ is called <u>elliptic</u> if all its components are bijective linear mappings for arbitrary fixed arguments

including the limit values of $\partial_0\sigma$ at $\partial_1(S^*(\partial_0 X))$ and $\partial_2(S^*(\partial_0 X))$ and of $\partial_1\sigma$ at $\partial_2(S^*(\partial_1 X))$. An operator is called elliptic if its symbol is elliptic.

2. Proposition. Let $\sigma = (\partial_0\sigma, \partial_1\sigma, \partial_2\sigma) \in \mathcal{S}^{(\mu)}$ be elliptic and define σ^{-1} as the triple $\sigma^{-1} := ((\partial_0\sigma)^{-1}, (\partial_1\sigma)^{-1}, (\partial_2\sigma)^{-1})$ (under the assumption of y-independent zeros of the principal Mellin symbols). Then $\sigma^{-1} \in \mathcal{S}^{(-\mu)}$.

Proof. Concerning $\partial_0\sigma$ and $\partial_1\sigma$ and its compatibility we may refer to 3.2.5. Proposition 2. Moreover, we proved in 4.2.4. that the inverse of an operator in $\mathcal{L}^{\mu}(C, \partial_1 C, \partial_2 C)$ belongs to $\mathcal{L}^{-\mu}(C, \partial_1 C, \partial_2 C)$. The compatibility with $(\partial_0\sigma)^{-1}$ and $(\partial_1\sigma)^{-1}$ and the smooth dependence on $(y, \eta) \in S^*(\partial_2 X)$ is proved as in 3.2.5. Proposition 5. $\square$

3. Theorem. Let $\mathcal{A} \in \mathcal{D}^{\mu}$ be elliptic. Then, for each $s \in \mathbb{R}$ the mapping

$$\mathcal{A} : \begin{array}{c} \mathcal{K}^{s}_{P}(\partial_0 X, E_0) \\ \oplus \\ \mathcal{K}^{s}_{\partial_1 P_2}(\partial_1 X, E_1) \\ \oplus \\ H^{s}(\partial_2 X, E_2) \end{array} \longrightarrow \begin{array}{c} \mathcal{K}^{s-\mu}_{Q}(\partial_0 X, E_0') \\ \oplus \\ \mathcal{K}^{s-\mu}_{\partial_1 Q_2}(\partial_1 X, E_1') \\ \oplus \\ H^{s-\mu}(\partial_2 X, E_2') \end{array} \qquad (6)$$

is Fredholm and $\mathcal{A}^{(-1)} = \mathrm{Op}(\sigma(\mathcal{A})^{-1})$ is a parametrix. It belongs to $\mathcal{D}^{-\mu}$ if the zeros of the principal Mellin symbols are y-independent.

Proof. By construction we have $1-\mathcal{A}\mathcal{A}^{(-1)}$, $1-\mathcal{A}^{(-1)}\mathcal{A} \in \mathcal{D}^{-1}$, which gives a compact operator in view of the compact embeddings of spaces with strictly larger smoothness index and better weight, since the manifold X is assumed to be compact. $\square$

The usual procedure of reconstructing the symbol components from the given operator yields immediately the proof of the converse assertion that the Fredholm property of the realization (6) can hold only if the given operator is elliptic. Note that this is not the case for the realization (1).

Since in our class a complete symbolic calculus is available, we can find for an elliptic operator also an improved parametrix $\mathcal{A}'$ such that $1-\mathcal{A}\mathcal{A}'$, $1-\mathcal{A}'\mathcal{A}$ are smoothing Green operators. Put $\mathcal{A}' \sim \mathcal{A}^{(-1)} \sum_{j=0}^{\infty} (1-\mathcal{A}\mathcal{A}^{(-1)})^j$ in the sense that we evaluate the complete symbol of the right-hand side and associate the operator $\mathcal{A}'$.

As a consequence we get a regularity result about solutions to elliptic boundary value problems on manifolds with edges.

4. Corollary. Consider an elliptic operator $\mathcal{A} \in \mathcal{W}^{\mu}$ in the realization (6). Let u be a solution to $\mathcal{A}u = f$, where $f \in \mathcal{K}^{t-\mu}_{Q'}(\partial_0 X, E_0') \oplus \mathcal{K}^{t-\mu}_{\partial_1 Q_2'}(\partial_1 X, E_1') \oplus H^{t-\mu}(\partial_2 X, E_2')$, $Q' = ((\gamma_1-\mu+k, q_1'), (\gamma_2-\mu+k, q_2)), Q' \leq Q$. Then there exists $P' = ((\gamma_1+k, p_1'), (\gamma_2+k, p_2')) \leq P$ such that $u \in \mathcal{K}^{t}_{P',+}(\partial_0 X, E_0) \oplus \mathcal{K}^{t}_{\partial_1 P_2',+}(\partial_1 X, E_1) \oplus H^t(\partial_2 X, E_2)$.

Proof. Let $\mathcal{A}'$ be a parametrix of $\mathcal{A}$ modulo smoothing Green operators, i.e. $\mathcal{G} = 1 - \mathcal{A}'\mathcal{A} \in \mathcal{W}_G^{-\infty}$. Then

$$u = \mathcal{A}'f + \mathcal{G}u$$

and the assertion follows from the mapping properties of $\mathcal{A}'$ and $\mathcal{G}$. $\square$

The continuity of $\mathcal{A}'$ and $\mathcal{G}$ yields the a-priori estimate

$$\|u\|_{t,P'} \leq c(\|f\|_{t,Q'} + \|u\|_{s,P}),$$

where we indicated in the norms the only relevant information about the smoothness order and the singularity type.

In particular, the null space $\ker_s \mathcal{A}$ of the realization (6) does not depend on s and belongs to $\mathcal{K}^{\infty}_{\underline{P}}(\partial_0 X, E_0) \oplus \mathcal{K}^{\infty}_{\underline{P}}(\partial_1 X, E_1) \oplus C^{\infty}(\partial_2 X, E_2)$, i.e. we have asymptotic expansions up to an arbitrary good weight. In this space we can find also a finite-dimensional space which is a complement of the image of (6) for each $s \in \mathbb{R}$.

Naturally the question arises about operators $\mathcal{A} \in \mathcal{W}^{\mu}$ satisfying not all of the ellipticity conditions. If $\partial_2 X = \emptyset$, it is well-known that the ellipticity of $\partial_0 \sigma$ implies in view of the compatibility conditions that $\partial_1 \sigma$ is a Fredholm operator at each point. Provided that a certain topological condition on the difference element in $K(S^*(\partial_1 X))$ is satisfied, one can enlarge the given ψDO by ∂_1-Mellin and Green operators (including trace and potential operators) so that the result is an elliptic operator in the class. Here ($\partial_2 X \neq \emptyset$) the situation is similar: The ellipticity of $\partial_0\sigma$ and $\partial_1\sigma$ (including the limits at $\partial_1(S^*(\partial_0 X))$ and $\partial_2(S^*(\partial_1 X)))$ implies that $\partial_2\sigma(y,\eta)$ is a Fredholm operator in $\mathcal{L}(E_{0,y}\otimes\mathcal{K}_P^s(C)\oplus E_{1,y}\otimes\mathcal{K}^s_{\partial_1 P_2}(C_1)\oplus E_{2,y}, E_{0,y}\otimes\mathcal{K}_Q^{s-\mu}(C) \oplus E'_{1,y}\otimes\mathcal{K}^{s-\mu}_{\partial_1 Q_2}(\partial_1 C)\oplus E'_{2,y})$. In fact, in view of the compatibility condition between $\partial_0\sigma, \partial_1\sigma$ and $\partial_2\sigma$ we get that $\partial_2\sigma$ is determined by the limits of $\partial_0\sigma$ and $\partial_1\sigma$ at $\partial_2(S^*(\partial_0 X))$ and $\partial_2(S^*(\partial_1 X))$ modulo compact operators in the above space of operators on the cone $(C, \partial_1 C, \partial_2 C)$ and $(\partial_0\sigma)^{-1}$, $(\partial_1\sigma)^{-1}$ induce a parametrix of $\partial_2\sigma$.

Then the usual topological obstruction must vanish for enlarging the given operator to an elliptic one, namely the difference element in $K(S^*(\partial_2 X))$ associated with the Fredholm family on $S^*(\partial_2 X)$ must belong to $\pi_2^*(K(\partial_2 X)\subset\mathcal{K}(S^*(\partial_2 X))$(cf. [R 1] 3.2.1.)

4.3.6. Notes

The theory of 4.3. of pseudo-differential boundary value problems on manifolds with edges is a generalization and extension of the theory in the case of a smooth boundary (cf. Chapter 3). There are many papers devoted to the important special case of classical differential boundary value problems (cf. [M 7], [M 8], [M 9], [N 1], [G 7], [O 2], [R 12]), which continue the approach initiated by Kondrat'ev [K 5] (for a survey and a large bibliography cf. Kondrat'ev, Oleynik [K 7]). In the pseudo-differential framework algebras of operators are included in Melrose's calculus of totally characteristic ψDO's (cf. [M 5]). Compared with the more general classes of Melrose our calculus can give a finer information if all symbolic levels are employed. Working with smaller classes of negligible operators the parametrix of elliptic problems is described more precisely.

References

[A 1] S. Agmon, A. Douglis, L. Nirenberg
Estimates near the boundary for solutions of elliptic partial differential equations satisfying general boundary conditions I, II. Comm. Pure Appl. Math. 12 (1959) 623-727, 17 (1964)35-92

[A 2] M.S. Agranovič, M.I. Višik
Elliptic problems with parameter and parabolic problems of general type. Uspechi Mat.Nauk 19, 3 (1964) 53-161

[B 1] L. Boutet de Monvel
Boundary problems for pseudo-differential operators. Acta Math. 126 (1971) 11-51

[B 2] J. Bergh, J. Löfström
Interpolation spaces. Springer Verlag, Berlin-Heidelberg-New York 1976

[B 3] L. Bungart
Holomorphic functions with values in locally convex spaces and applications to integral formulas. Trans.Amer.Math. Soc. 111 (1964) 317-344

[B 4] Ju.A. Bryčkov, A.P. Prudnikov
Integral transformations of generalized functions. Nauka , Moscow 1977

[B 5] P. Bolley, J. Camus
Sur une classe d' opêrateurs elliptiques et dégéneres â plusieurs variables. Bull.Soc.Math. France, Mémoire 34 (1973) 55-140

[B 6] M. Baranowski
Totally characteristic pseudo-differential operators in Besov-Lizorkin-Triebel spaces. Ann. in Global Anal. and Geom. (to appear)

[C 1] H.O. Cordes, E.A. Herman
Pseudo-differential operators on a half-line. J. Math.Mech. 18 (1969) 893-908

[C 2] H.O. Cordes
Elliptic pseudo-differential operators - an abstract theory. Lecture Notes in Math. 756, Springer Verlag, Berlin-Heidelberg-New York 1979

[C 3] P. Collela, H.O. Cordes
The C* algebra of the elliptic boundary problem. Rocky Mountains J. 10, 1 (1980) 217-288

[C 4] M. Costabel
Singular integral operators on curves with corners. Int.Equ. and Operator Theory 3 (1980) 323-349

[C 5] M. Costabel
Boundary integral operators on curved polygons. Annali di Matematica pura et appl. 33 (1983) 305-326

[C 6] M. Costabel, E. Stephan, W.L. Wendland
On boundary integral equations of the first kind for the bi-Laplacian in a polygonal plane domain. Ann.Sc.Norm.Sup. Pisa, Ser. IV 10 (1983) 197-241

[D 1] E. Dubinsky
The structure of nuclear Fréchet spaces. Lecture Notes in Math. 720, Springer Verlag, Berlin-Heidelberg-New York 1979

[D 2] M. Dauge
Régularités et singularités des solutions de problémes aux limites elliptiques sur des domaines singuliers de type à coins. Thése, Univ.de Nantes 1986

[E 1] J. Elschner
Singular ordinary differential operators and pseudodifferential equations. Mat.Forsch. Bd. 22, Akademie-Verlag Berlin 1985

[E 2] G.I. Eskin
Boundary problems for elliptic pseudodifferential equations. Nauka,Moskva 1973

[E 3] G.I. Eskin
General mixed boundary problems for elliptic differential equations. CIME Course on Pseudodifferential Operators with Appl., Ligouri, Naples 1978

[E 4] G.I. Eskin
A transmission problem for equations of principal type in two independent variables. Trudy Mosk.Mat.Obšč. 21 (1970) 245-292

[F 1] E.B. Fabes, M. Jodeit, J.E. Lewis
Double layer potentials for domains with corners and edges. Indiana Univ. Math.J. 26 (1977) 95-114

[F 2] G. Fichera
Comportamento asintotico del campo elettrico e della densità elettrica in prossimità dei punti singolari della superficie conduttore. Uspechi Mat.Nauk 30, 3 (1975) 105-124

[F 3] Fung Kang
Generalized Mellin transforms. Acta Math. Sinica 7, 2 (1957) 242-267

[G 1] I.C. Gohberg, N.Ja. Krupnik
The algebra generated by the one dimensional singular integral operators with piecewise continuous coefficients. Funkcional. Anal. i.Prilozen. 4 3 (1970) 26-36

[G 2] I.C. Gohberg, N.Ja. Krupnik
Singular integral operators with piecewise continuous coefficients and their symbols. Izv. Akad.Nauk SSSR 35, 4 (1971) 940-964

[G 3] I.C. Gohberg, N.Ja. Krupnik
Einführung in die Theorie der eindimensionalen singulären Integraloperatoren. Birkhäuser Verlag Basel, Boston, Stuttgart 1979

[G 4] I.C. Gradštein, I.M. Ryžik
Tablicy integralov, summ, rjadov i proizvedenic. Gos. isd. Fiziko-Matem. literatury, Moskva 1963

[G 5] A. Grothendieck
Sur certains espaces de fonctions holomorphes I,II. J. für die reine und angew. Mathematik 192 (1953) 35-64, 77-95

[G 6] G. Grubb
Functional calculus of pseudo-differential boundary problems. Birkhäuser, Boston-Basel-Stuttgart 1986

[G 7] P. Grisvard
Elliptic problems in nonsmooth domains. Monographs and Studies in Math. 24 , Pitman Boston-London-Melbourne 1985

[H 1] L. Hörmander
Pseudo-differential operators and non-elliptic boundary problems. Ann. of Math. 83,1 (1966) 129-209

[H 2] L. Hörmander
An introduction to complex analysis in several variables. D. van Nostrand, Princeton, New Jersey, Toronto, New York, London 1966

[H 3] L. Hörmander
The analysis of linear partial differential operators I,II,III,IV. Springer Verlag, Berlin-Heidelberg-New York-Tokyo 1983-1985

[H 4] L. Hörmander
Pseudo-differential operators. Comm. Pure Appl. Math. 18, 3 (1965) 501-517

[H 5] L. Hörmander
Pseudo-differential operators and hypoelliptic equations. Amer. Math.Soc.Symp. on Singular Integral Operators (1966), 138-183

[J 1] P. Jeanquartier
Transformation de Mellin et developpements asymptotiques. L' Enseignement mathématique 25, 3-4 (1979) 285-308

[K 1] T. Kato
Perturbation theory for linear operators. Springer-Verlag Berlin-Heidelberg-New York 1966

[K 2] J.J. Kohn, L. Nirenberg
An algebra of pseudo-differential operators. Comm. Pure Appl. Math. 18, 1/2 (1965) 269-305

[K 3] A.I. Komeč
Elliptic boundary problems for pseudo-differential operators on manifolds with conical points. Mat. Sb. 86, 2 (1971) 268-298

[K 4] A.I. Komeč
Elliptic boundary problems on manifolds with piecewise smooth boundary. Mat.Sb. 92, 1 (1973) 89-134

[K 5] V.A. Kondrat'ev
Boundary problems for elliptic equations in domains with conical points. Trudy Mosk. Mat. Obšč. 16 (1967) 209-292

[K 6] V.A. Kondrat'ev
Singularities of solutions of the Dirichlet problem for elliptic equations of second order in a neighbourhood of a corner. Diff. Uravn. 13, 11 (1977) 2026-2032

[K 7] V.A. Kondrat'ev, O.A. Oleynik
Boundary problems for partial differential equations in non-smooth domains. Uspehi Mat.Nauk 38, 2 (1983) 3-76

[K 8] A.I. Komeč
Equations with homogeneous kernels and the Mellin transform of generalized functions. Teoretič. i matem. Fizika 27, 2 (1976) 149-162

[L 1] M.A. Lawrentjew, B.W.Schabat
Methoden der komplexen Funktionentheorie. Deutscher Verlag der Wissenschaften, Berlin 1967

[L 2] J.E. Lewis, C. Parenti
Pseudo-differential operators and Hardy kernels on $L^p(\mathbb{R}_+)$. Ann. Sc. Norm.Sup.Pisa, Ser. IV, 7 (1980) 481-503

[L 3] J.E. Lewis, C. Parenti
Pseudodifferential operators of Mellin type. Comm. Part.Diff. Equ. 8, 5 (1983) 477-544

[L 4] J.L. Lions, E. Magenes
Problémes aux limites non-homogénes et applications, I. Dunod, Paris 1968

[M 1] H. Mellin
Über die fundamentale Wichtigkeit des Satzes von Cauchy für die Theorien der Gamma- und der hypergeometrischen Funktionen. Acta Soc. Fennicae 21, 1 (1986) 1-115

[M 2] H. Mellin
Über den Zusammenhang zwischen den linearen Differential- und Differenzengleichungen. Acta Math. 25 (1902) 139-164

[M 3] R. Melrose
Transformation of boundary problems. Acta Math. 147 (1981) 149-236

[M 4] R. Melrose, G. Mendoza
Elliptic operators of totally characteristic type. Preprint MSRI 047-83 Berkeley 1983

[M 5] R. Melrose
Pseudo-differential operators on manifolds with corners. Manuscript MIT, Boston 1987

[M 6] P.A. Myškis
On the algebra generated by pseudodifferential operators with discontinuous symbols on a manifold. Uspehi Mat. Nauk 31 , 4 (1976) 269-270

[M 7] V.G. Maz'ja, B.A. Plamenevskij
On the coefficients in the asymptotics of solutions of elliptic boundary value problems near edges. Sov.Math. Dokl. 17, 4 (1976)

[M 8] V.G. Maz'ja, B.A. Plamenevskij
L_p estimates of solutions of elliptic boundary problems in domains with edges. Trudy Mosk.Mat.Obsc. 37 (1978) 49-93

[M 9] V.G. Maz'ja, N.F. Mosožov, B.A., B.A. Plamenevskij, L.Stupyalis
Elliptic boundary value problems. Amer. Math.Soc. Transl. Ser. 2 Vol. 123, 1984

[N 1] V.A. Nikiškin
Singularities of solutions of the Dirichlet problem for second order equations near a corner. Vestnik Mosk. Univ., Ser. Mat.34, 2 (1979) 51-62

[N 2] J. Nourrigat
Une classe d' opérateurs pseudodifférentiels a une variable. C.R.A.S. Series A 283 (1976) 457-460

[P 1] J. Peetre
Mixed problems for higher order elliptic equations in two variables, I, II. Ann. Sc. Norm. Sup. Pisa 15 (1961) 337-353, 17 (1963) 1-12

[P 2] A. Pietsch
Nukleare lokalkonvexe Räume. Schriftenreihe der Institute für Math. Reihe A, Heft 1, Akademie-Verlag,Berlin 1969

[P 3] B.A. Plamenevskii
On the boundedness of singular integrals in spaces with weight. Mat. Sb. 76 (1968) 573-592

[P 4] B.A. Plamenevskii
An algebra of pseudodifferential operators in spaces with weight. Mat.Sb. 106, 2 (1978) 297-320

[P 5] C. Pommerenke
Univalent functions. Vandenhoeck/ Ruprecht, Göttingen 1975

[P 6] A.P. Prudnik, Ju.A. Bryčkov, O.A. Maričev
Integrals and series. Nauka Moscow 1981

[P 7] B.A. Plamenevskii
Algebras of pseudo-differential operators. Moscow, Nauka 1986

[R 1] S. Rempel, B.-W. Schulze
Index theory of elliptic boundary problems. Akademie-Verlag Berlin 1982

[R 2] S. Rempel, B.-W. Schulze
Parametrices and boundary symbolic calculus for elliptic boundary problems without the transmission property. Math. Nachr. 105 (1982) 45-149

[R 3] S. Rempel, B.-W. Schulze
Complex powers for pseudo-differential boundary problems, I,II. Math. Nachr. 111 (1983) 41-109, 116 (1984) 269-314

[R 4] S. Rempel, B.-W. Schulze
Mixed boundary value problems for Lamé's system in three dimensions. Math. Nachr. 119 (1984) 265-290

[R 5] S. Rempel, B.-W. Schulze
A theory of pseudo-differential boundary value problems with discontinuous conditions, I,II. Ann. of Global Anal. and Geometry 2, 2(1984) 163-251, 2, 3(1984) 289-384

[R 6] S. Rempel, B.-W. Schulze
Complete Mellin and Green symbolic calculus in spaces with conormal asymptotics. Ann. of Global Anal. and Geom. 4, 2 (1986) 137-224

[R 7] S. Rempel, B.-W. Schulze
Branching of asymptotics for elliptic operators on manifolds with edges. Partial Diff. Equ., Banach Center Publ. Vol. 19 1984, pp. 231-236, PWN Polish Sc.Publ. Warsaw 1987

[R 8] S. Rempel, B.-W. Schulze
Complete Mellin symbols and the conormal asymptotics in boundary value problems. Proc. Journees "Equ. aux Dériv. Part" St. Jean de Monts 1984, Conf. N° V

[R 9] S. Rempel, B.-W. Schulze
Mellin symbolic calculus and asymptotics for boundary value problems. Seminar Analysis des Karl-Weierstraß-Instituts f. Math. 1984/85, Berlin 1985

[R 10] S. Rempel, B.-W. Schulze
Pseudo-differential and Mellin operators in spaces with conormal singularity (boundary symbols). Report Institute of Mathematics, Acad. Sci. 01/84 Berlin 1984

[R 11] S. Rempel
Gemischte Randwertprobleme und Kopplungsprobleme für elliptische Pseudodifferentialoperatoren. Doktor Dissertation Akad.d.Wiss. der DDR, Berlin 1983

[R 12] J. Roßmann
Gewichtete Sobolev-Slobodezkij-Räume und Anwendungen auf elliptische Randwertaufgaben in Gebieten mit Kanten. Doktor Dissertation Univ. Rostock 1988

[S 1] H.H. Schäfer
Topological vector spaces. MacMillan Company New York, London 1966

[S 2] P. Schapira
Théorie des hyperfonctions. Lecture Notes in Math. 126, Springer Verlag Berlin-Heidelberg-New York 1970

[S 3] M.A. Šubin
Pseudodifferential operators and spectral theory. Nauka, Moscow 1978

[S 4] B.-W. Schulze
Corner Mellin operators and reduction of orders with parameters. Ann. Sc. Norm. Sup. Pisa (to appear)

[S 5] B.-W. Schulze
Ellipticity and continuous conormal asymptotics on manifolds with conical singularities. Math. Nachr. 136 (1988) 7-57

[S 6] B.-W. Schulze
Regularity with continuous and branching asymptotics for elliptic operators on manifolds with edges. Int. Equ. and Operator Theory 11 (1988) 557-602

[S 7] R. Seeley
Complex powers of an elliptic operator. Proc. Symp.in Pure Math. 10 (1967) 288-307

[S 8] J. Sjöstrand
Parametrices for pseudodifferential operators with multiple characteristics. Arkiv för Math. 12 (1974), 85-130

[S 9] B.-W. Schulze
Pseudo-differential operators on manifolds with singularities. Akad.Verlagsges. Leipzig, North Holland (in preparation)

[T 1] E.C. Titchmarsh
The theory of functions.Second Ed. Oxford Univ. Press 1939

[T 2] F. Treves
Topological vector spaces, distributions and kernels. Academic Press New York, London 1967

[T 3] H. Triebel
Interpolation theory, function spaces, differential operators. Dt. Verlag d. Wiss. Berlin 1978

[T 4] F. Treves
Introduction to pseudo-differential and Fourier integral operators. Vols. 1,2, Plenum, New York 1985

[U 1] A. Unterberger
The calculus of pseudo-differential operators of Fuchs type. Comm. Part. Diff. Equ. 9, 12 (1984) 1179-1236

[V 1] M.I. Višik, G.I. Eskin
Boundary problems for general singular equations in a bounded domain. Dokl. Akad. Nauk SSSR 155, 1 (1964) 24-27

[V 2] M.I. Višik, G.I. Eskin
General boundary value problems with discontinuous boundary conditions. Dokl. Akad. Nauk SSSR 158 (1964) 25-28

[V 3] M.I. Višik, G.I. Eskin
Convolution equations in a bounded region. Uspehi Mat.Nauk 20, 3 (1965) 89-152

[V 4] M.I. Višik, G.I. Eskin
Convolution equations in bounded domains in spaces with weighted norms. Mat. Sb. 69, 1 (1966) 65-110

[V 5] M.I. Višik, G.I. Eskin
Sobolev-Slobodeckii spaces of variable order with weighted norms, and their applications to mixed boundary value problems. Sibirsk. Mat. Ž. 9 (1968) 973-997

[W 1] H. Widom
Singular integral equations in L_p. Trans.Amer.Math.Soc. 97 (1960) 131-159

[W 2] N. Wiener
Tauberian theorems. Ann. of Math. 33 (1932) 1-100

[Z 1] S. Zaremba
Sur un problême mixte relatif à l équation de Laplace. Bull. de l Academie des Sciences de Craconie, Class. math. et.nat. Ser. A (1910) 313-344

[Z 2] S. Zaremba
On a mixed problem for the Laplace equation. Uspehi Mat. Nauk 1, 3-4 (1946) 125-146

[Z 3] A.H. Zemanian
Distribution theory and transform analysis. Mc. Graw Hill, New York 1965

Index

Band 1

Inverse and Improperly Posed Problems in Differential Equations

Proceedings of the Conference on Mathematical and Numerical Methods, held in Halle/Saale (GDR) from May 29 to June 2, 1979

Edited by Gottfried Anger
296 Seiten, 1979 (vergriffen)

Band 2

Fundamentals of Computation Theory FCT '79

Proceedings of the Conference on Algebraic, Arithmetic and Categorial Methods in Computation Theory, held in Berlin/Wendisch-Rietz (GDR) September 17–21, 1979

Edited by Lothar Budach
576 Seiten, 1979

Band 3

Decidability and Generalized Quantifiers

By Andreas Baudisch, Detlef Seese, Hans-Peter Tuschik, and Martin Weese
XII, 235 Seiten, 1980

Band 4

Zur Elastizitäts- und Thermoelastizitätstheorie

Potentialtheoretische Methoden in Anwendung auf verschiedene Modelle

Von Lothar Jentsch und Johannes Maul
157 Seiten, 1980

Band 5

Systems Analysis and Simulation 1980

Proceedings of the International Symposium held at Berlin (GDR) September 1–5, 1980

Edited by Achim Sydow
496 Seiten, 1980 (vergriffen)

Band 6

Nichtlineare diskrete Optimierung

von Eberhard Girlich und Michail M. Kowaljow
218 Seiten, 1981

Band 7

Completeness Problems in the Structural Theory of Automata

By Jürgen Dassow
148 Seiten, 1981

Band 8

Darstellung von Lösungen linearer elliptischer Differentialgleichungen

Von Günther Wildenhain
92 Seiten, 1981

Band 9

Introduction to the Spectral Theory of Operators in Spaces with an Indefinite Metric

By I. S. Iohvidov, M. G. Krein, and H. Langer
120 Seiten, 1982 (vergriffen)

Band 10

Nonlinear Time-discrete Systems

A General Approach by Nonlinear Superposition

By Michael Gössel
115 Seiten, 1982 (vergriffen)

Band 11

Preprimal Algebras

By Klaus Denecke
162 Seiten, 1982

Band 12

Ergodic Theory and Related Topics

Proceedings of the Conference held in Vitte/Hiddensee (GDR) October 19–23, 1981

Edited by Horst Michel
232 Seiten, 1982

Band 13

Singularly Perturbed Differential Equations

By H. Goering, A. Felgenhauer, G. Lube, H.-G. Roos, and L. Tobiska
176 Seiten, 1983

Band 14

Leben wir in einer Volterra-Welt?

Ein ökologischer Zugang zur angewandten Systemanalyse

Von Manfred Peschel und Werner Mende
233 Seiten, 1983

Band 15

Markov Decision Problems with Countable State Spaces

Optimality Criteria – Algorithms – Clustering

By Hans Michael Dietz and Volker Nollau
174 Seiten, 1983

Band 16

Stochastische Verfahren zur Suche nach dem Minimum einer Funktion

Von Ryszard Zieliński und Peter Neumann
133 Seiten, 1983

Band 17

Invertibility and Asymptotics of Toeplitz Matrices

By Albrecht Böttcher and Bernd Silbermann
200 Seiten, 1983

Band 18

Structural Induction on Partial Algebras

Introduction to Theory and Application of Partial Algebras – Part II

By Horst Reichel
205 pages, 1984

Band 19

Algebraic Methods for Toeplitz-like Matrices and Operators

By Georg Heinig and Karla Rost
212 pages, 1984

Band 20

Frege Conference 1984

Proceedings of the International Conference held at Schwerin (GDR), September 10–14, 1984
Edited by Gerd Wechsung
408 pages, 1984

Band 21

Numerical Analysis of Selected Semilinear Differential Equations

Edited by Christian Großmann, Thomas Riedrich, and Hartmut Schönheinz
203 pages, 1984

Band 22

Singular Ordinary Differential Operators and Pseudodifferential Equations

By Johannes Elschner
200 pages, 1985

Band 23

Lotka-Volterra-Approach to Cooperation and Competition in Dynamic Systems

Proceedings of the 5th Meeting of UNESCO's Working Group on Systems Theory held on the Wartburg, Eisenach (GDR) at March 5–9, 1984
Edited by Werner Ebeling and Manfred Peschel
271 pages, 1985

Band 24

Convergence Structures 1984

Proceedings of the Conference on Convergence held in Bechyně (Czechoslovakia), September 24–28, 1984
Edited by Josef Novák, Werner Gähler, Horst Herrlich, and Jan Mikusiński
254 pages, 1985

Band 25

Parcella '84

Proceedings of the II. International Workshop on Parallel Processing by Cellular Automata and Arrays held in Berlin (GDR), September 25–27, 1984
Edited by Wolfgang Händler, Tamás Legendi, and Gottfried Wolf
239 pages, 1985 (vergriffen)

Band 26

Multiobjective and Stochastic Optimization Based on Parametric Optimization

By Jürgen Guddat, Francisco Guerra Vasquez, Klaus Tammer, and Klaus Wendler
175 pages, 1985

Band 27/28

Systems Analysis and Simulation 1985
I: Theory and Foundations
II: Applications

Proceedings of the International Symposium held in Berlin (GDR), August 26–31, 1985
Edited by Achim Sydow, Manfred Thoma, and Robert Vichnevetsky
per volume 495 pages, 1985

Band 29

Parcella '86

Proceedings of the III. International Workshop on Parallel Processing by Cellular Automata and Arrays held in Berlin (GDR), September 9–11, 1986
Edited by Tamás Legendi, Dennis Parkinson, Roland Vollmar, and Gottfried Wolf
279 pages, 1986

Band 30

Fuzzy Sets Applications, Methodological Approaches, and Results

Proceedings of the International Workshop on Fuzzy Sets Applications cosponsored by IIASA and convened in cooperation with IAEA held on the Wartburg, Eisenach (GDR), at March 3–8, 1985
Edited by Steffen Bocklisch, Sergei Orlovski, Manfred Peschel, and Yasushi Nishiwaki
233 pages, 1986

Band 31

Mathematical Methods of Specification and Synthesis of Software Systems '85

Proceedings of the International Spring School, held in Wendisch-Rietz (GDR), April 22–26, 1985
Edited by Wolfgang Bibel and Klaus P. Jantke
245 pages, 1986 (vergriffen)

Band 32

A Model Theoretic Oriented Approach to Partial Algebras

Introduction to Theory and Application of Partial Algebras – Part I
By Peter Burmeister
319 pages, 1986

Band 33

Finite Difference Method on Irregular Networks

A Generalized Approach to Elliptic Problems of Second Order
By Bernd Heinrich
206 pages, 1986

Band 34

Modern Algorithms for Large Sparse Eigenvalue Problems

By Arnd Meyer
125 pages, 1987

Band 35

Parametric Optimization and Related Topics

Edited by Jürgen Guddat, Hubertus Th. Jongen, Bernd Kummer, and František Nožička
407 pages, 1987

Band 36

Discretization in Differential Equations and Enclosures

Edited by Ernst Adams, Rainer Ansorge, Christian Großmann, and Hans-Görg Roos
250 pages, 1987

Band 37

Factorization of Measurable Matrix Functions

By Georgi Semenovich Litvinchuk and
Ilya Matveyevich Spitkovskii

372 pages, 1987

Band 38

Parallel Algorithms and Architectures

Proceedings of the International Workshop on Parallel Algorithms and Architectures held in Suhl (GDR), May 25–30, 1987

Edited by Andreas Albrecht, Hermann Jung, and Kurt Mehlhorn

205 pages, 1987

Band 39

Parametric Integer Optimization

By Bernd Bank and Reinhard Mandel

138 pages, 1988

Band 40

Computer Analysis of Images and Patterns 1987

Proceedings of the II. International Conference CAIP '87 on Automatic Image Processing, held in Wismar (GDR), September 2–4, 1987

Edited by Leonid P. Jaroslavskii,
Azriel Rosenfeld, and Wolfgang Wilhelmi

282 pages, 1987

Band 41

Multivariable Control

A Graph-theoretic Approach

By Kurt J. Reinschke

274 pages, 1988

Band 42

Equilibrium Distributions of Branching Processes

By Alfred Liemant, Klaus Matthes, and Anton Wakolbinger

240 pages, 1988

Band 43

Andreotti-Grauert Theory by Integral Formulas

By Gennadi Markovich Henkin and
Jürgen Leiterer

270 pages, 1988

Band 44

Continual Means and Boundary Value Problems in Function Spaces

By Efim Mikhailovich Polishchuk

164 pages, 1988

Band 45

Advances in Mathematical Optimization

Invited papers dedicated to Prof. Dr. Dr. h. c. F. Nožička on occasion of his 70th birthday

Edited by Jürgen Guddat et al.

332 pages, 1988

Band 46/47

Systems Analysis and Simulation 1988

Proceedings of the International Symposium held in Berlin (GDR),

September 12–16, 1988

I: Theory and Foundations
II: Applications

Edited by Achim Sydow, Spyros G. Tzafestas, and Robert Vichnevetsky

426/425 pages, 1988

Band 48

Parcella '88

Proceedings of the IV. International Workshop on Parallel Processing by Cellular Automata and Arrays, held in Berlin (GDR),

October 17–21, 1988

Edited by Gottfried Wolf, Tamás Legendi, and Udo Schendel

379 pages, 1988

Band 49

Algebraic and Logic Programming

Proceedings of an International Workshop held in Gaussig (GDR), November 14–18, 1988
Edited by Jan Grabowski, Pierre Lescanne, and Wolfgang Wechler

278 pages, 1988

Band 50

Asymptotics for Elliptic Mixed Boundary Problems

Pseudo-Differential and Mellin Operators in Spaces with Conormal Singularity

By Stephan Rempel and Bert-Wolfgang Schulze

418 pages, 1989

Band 51

Geobild '89

Proceedings of the 4th Workshop on Geometrical Problems of Image Processing held in Georgenthal (GDR), March 13–17, 1989

Edited by Albrecht Hübler, Werner Nagel,
Brian David Ripley, and Günter Werner

198 pages, 1989

Subject of the exposition is the theory of pseudo-differential boundary value problems in domains or on manifolds with edges, in particular the symbolic structure and the conormal asymptotics of solutions. The classical mixed problems such as the Zaremba problem are of this type, as well as crack problems (with smooth crack boundary), further the "usual" pseudo-differential boundary problems (with and without the transmission property). The calculus is based on the Mellin operator convention in terms of the distance to the singularities and classical achievements on pseudo-differential operators near conical singularities. The idea of symbolic hierarchies and associated ideal structures in the operator algebras leads to the tools for constructing parametrices of elliptic operators within the classes and the adequate analogues of the elliptic regularity, here with asymptotics.

Gegenstand der Darstellung ist die Theorie pseudo-differentieller Randwert-Probleme in Gebieten oder auf Mannigfaltigkeiten mit Kanten, insbesondere die Symbolstruktur und die Konormalasymptotik von Lösungen. Die klassischen gemischten Probleme wie das Zaremba-Problem sind von diesem Typ, ebenso wie Rißprobleme (mit glattem Rißrand), weiterhin die „gewöhnlichen" pseudo-differentiellen Randwert-Probleme (mit und ohne die Transmissionseigenschaft). Der Kalkül basiert auf der Mellin-Operator-Konvention in Termen des Abstands zu den Singularitäten und klassischen Errungenschaften über Pseudo-Differential-Operatoren nahe konischen Singularitäten. Die Idee der Symbolhierarchien und zugeordnete Idealstrukturen in den Operatoralgebren führt zu den Hilfsmitteln für die Konstruktion von Parametrices elliptischer Operatoren innerhalb der Klassen und zu den adäquaten Analogien der elliptischen Regularität, hier mit Asymptotik.

ISBN 3-05-500676-3
ISSN 0138-3019

Bei Fragen zur Produktsicherheit wenden Sie sich bitte an:
If you have any questions regarding product safety,
please contact:

Walter de Gruyter GmbH
Genthiner Straße 13
10785 Berlin
productsafety@degruyterbrill.com